HANDBOOK OF KNOT THEORY

Handbook of Knot Theory

Editors

WILLIAM MENASCO
Department of Mathematics
University at Buffalo
USA

MORWEN THISTLETHWAITE
Department of Mathematics
University of Tennessee
USA

2005

ELSEVIER

Amsterdam – Boston – Heidelberg – London – New York – Oxford – Paris
San Diego – San Francisco – Singapore – Sydney – Tokyo

ELSEVIER B.V.
Radarweg 29
P.O. Box 211, 1000 AE
Amsterdam, The Netherlands

ELSEVIER Inc.
525 B Street, Suite 1900
San Diego, CA 92101-4495
USA

ELSEVIER Ltd
The Boulevard, Langford Lane
Kidlington, Oxford OX5 1GB
UK

ELSEVIER Ltd
84 Theobalds Road
London WC1X 8RR
UK

First impression 2005

Library of Congress Cataloging in Publication Data
A catalog record is available from the Library of Congress.

British Library Cataloguing in Publication Data

Handbook of knot theory
1.Knot theory
I.Menasco, William W., 1954- II.Thistlethwaite, Morwen
514.2'242

ISBN: 0-444-51452-X

♾ The paper used in this publication meets the requirements of ANSI/NISO Z39.48-1992 (Permanence of Paper).
Printed in The Netherlands

Preface

Accessible to the layman, with immense raw geometric appeal, yet in close touch with some elevated disciplines, knot theory holds a special place in mathematics. From its humble beginnings as an empirical science in the mid-19th century, the subject has undergone an extraordinary evolution, greatly accelerated over the last 25 years. One of the first serious researchers in knot theory, the notable Scottish physicist Peter Guthrie Tait, began his celebrated tabulation of knots of small crossing-number with the hope of elucidating the Kelvin vortex theory of atoms, but soon became entranced with the subject in its own right. He observed, perceptively, that he could not ever prove that two knots were inequivalent; indeed, that had to wait until the advent of Poincaré's Fundamental Group in the early 1900s. Knot theorists now have a formidable arsenal of techniques for distinguishing knot and link types, but such considerations form only a narrow part of contemporary knot theory. Knots have found themselves involved with almost every major advance in low-dimensional topology, be it Papakyriakopoulos's proof of Dehn's Lemma in 1957 (whose original aim was to show that the group of a non-trivial knot could not be isomorphic to the integers), or Riley's discovery in the 1970s of a hyperbolic structure on the complement of the figure-eight knot and Thurston's subsequent profound work on geometric structures on 3-manifolds. Of equal importance are interactions with piecewise-linear 3-dimensional topology, with braids, with contact structures on manifolds, and with 4-dimensional topology, both classical and gauge-theoretical. It would be audacious to attempt even a summary of the full ramifications of knot theory, but mention must be made of Jones's astonishing discovery in 1984 of an entirely new breed of polynomial invariants, which strengthened knot theory's connection with the theory of braids and forged a completely new relationship with the mathematics of quantum field theory.

In this volume, we present a collection of survey articles by leading experts on a cross-section of present-day knot theory. For us, reading these articles and witnessing the vast range of knowledge contained therein was an inspiring and humbling experience.

We are grateful for this opportunity to thank the contributors for their sheer hard work and their willingness to share this knowledge. We would also like to thank the staff of Elsevier for their copious help with this project.

William Menasco
Morwen Thistlethwaite

List of Contributors

Prof. Adams, C., *Bronfman Science Center, Department of Mathematics, Williams College, Williamstown, MA 01267, USA, colin.adams@williams.edu* (Ch. 1)

Prof. Birman, J.S., *Department of Mathematics, Barnard College, Columbia University, 2990 Broadway, New York, NY 10027, USA, jb@math.columbia.edu* (Ch. 2)

Prof. Brendle, T.E., *Department of Mathematics, Louisiana State University, Baton Rouge, LA 70803-4918, USA, brendle@math.lsu.edu* (Ch. 2)

Prof. Etnyre, J.B., *University of Pennsylvania, Department of Mathematics, 209 South 33rd Street, Philadelphia, PA 19104-6395, USA, etnyre@math.upenn.edu* (Ch. 3)

Prof. Friedman, G., *Department of Mathematics, Yale University, 10 Hillhouse Ave/ P.O. Box 208283, New Haven, CT 06520-8283, USA, friedman@math.yale.edu* (Ch. 4)

Prof. Hoste, J., *Pitzer College, Department of Mathematics, 1050 N Mills Avenue, Claremont, CA 91711, USA, jhoste@pitzer.edu* (Ch. 5)

Prof. Kauffman, L.H., *Department of Mathematics, Statistics and Computer Science, University of Illinois, 851 South Morgan Street, Chicago, IL 60607-7045, USA, kauffmann@uic.edu* (Ch. 6)

Prof. Livingston, C., *Department of Mathematics, Indiana University, Bloomington, IN 47405, USA, livingst@indiana.edu (Ch. 7)*

Prof. Menasco, W., *Department of Mathematics, University at Buffalo, USA* (Editor)

Prof. Rudolph, L., *Department of Mathematics and Computer Science and Department of Psychology, Clark University, Worcester, MA 01610, USA, lrudolph@black.clarku.edu* (Ch. 8)

Prof. Scharlemann, M., *Department of Mathematics, South Hall, University of California, Santa Barbara, CA 93106, USA, mgscharl@math.ucsb.edu* (Ch. 9)

Prof. Thistlethwaite, M., *Department of Mathematics, University of Tennessee, USA* (Editor)

Prof. Weeks, J., *www.geometrygames.org/contact.html* (Ch. 10)

Contents

CHAPTER 1

Hyperbolic Knots

Colin Adams

Bronfman Science Center, Department of Mathematics, Williams College, Williamstown, MA 01267, USA
E-mail: colin.adams@williams.edu

Contents

HANDBOOK OF KNOT THEORY
Edited by William Menasco and Morwen Thistlethwaite

1. Introduction

In 1978, Thurston revolutionized low dimensional topology when he demonstrated that many 3-manifolds had hyperbolic metrics, or decomposed into pieces, many of which had hyperbolic metrics upon them. In view of the Mostow rigidity theorem [gM73], when the volume associated with the manifold is finite, these hyperbolic metrics are unique. Hence geometric invariants coming out of the hyperbolic structure can be utilized to potentially distinguish between manifolds.

One can either think of a hyperbolic 3-manifold as having a Riemannian metric of constant curvature -1, or equivalently, of there being a lift of the manifold to its universal cover, which is hyperbolic 3-space H^3, with the covering transformations acting as a discrete group of fixed point free isometries Γ. The manifold M is then homeomorphic to the quotient H^3/Γ.

A hyperbolic knot K in the 3-sphere S^3 is defined to be a knot such that $S^3 - K$ is a hyperbolic 3-manifold. Note that the complement is a finite volume but noncompact hyperbolic 3-manifold.

Given a hyperbolic knot, the SnapPea program, written by Jeffrey Weeks, can be utilized to determine the hyperbolic structure (see [jW03]). In particular, the program yields the volume of the manifold, the symmetry group, and a variety of invariants that are associated to the cusps of the manifold and that will be discussed in Section 5. Details of how the program determines the hyperbolic structure appear in this volume [jW04].

In addition, the program gives the option of determining whether or not two hyperbolic manifolds are isometric. Hence, since knots are known to be determined by their complements, (cf. [GL89]), we can use SnapPea to decide if two given knots are the same, assuming first of all that they are both hyperbolic and second of all that SnapPea successfully finds their hyperbolic structure. In practice, this is one of the fastest means for determining if two hyperbolic knots are identical. It was utilized in the tabulation of the prime knots of 16 and fewer crossings (cf. [HTW98]).

Many knot and link complements in the 3-sphere are in fact hyperbolic. Moreover, knot and link complements in other 3-manifolds are often hyperbolic as well. In this paper, we will discuss the prevalence of hyperbolic knot and link complements and the various invariants associated with the hyperbolic metric carried by the complement. We will focus on the geometric invariants. For an excellent overview of hyperbolic knots with emphasis on algebraic invariants, see [CR98]. That paper also contains a description of some of the applications, including the Smith Conjecture and the determination of symmetries of a knot.

We will not cover certain topics that are surveyed elsewhere. See [sB02] for details on Dehn surgery on hyperbolic knots in S^3. The paper [mS98] covers unknotting tunnels and their connections with the hyperbolic structure on a knot complement. Branched coverings and their relationship with hyperbolicity are covered in [CRV01] and [RZ01].

2. What knot and link complements are known to be hyperbolic?

The seminal work of Thurston in the 1970s and 1980s demonstrated that every knot in S^3 is either a torus knot, a satellite knot or a hyperbolic knot. (We include composite knots as satellite knots.) These three categories are mutually exclusive. Torus knots are well understood. Their fundamental groups have presentations of the form $\langle a, b : a^p = b^q \rangle$. They are exactly the knots such that their fundamental groups have nontrivial center. If the standardly embedded torus upon which a torus knot projects without crossings is cut open along the knot, the result is an essential annulus in the complement of the knot, which precludes a hyperbolic structure by fundamental group considerations.

Satellite knots have an incompressible non-boundary parallel torus in their complement, which also precludes a hyperbolic metric. This torus can be used to decompose the knot complement into simpler pieces, each of which may then be hyperbolic.

The remaining case is when the knot complement has neither an essential annulus nor a non-boundary parallel incompressible torus. Thurston's remarkable theorem demonstrates that such a knot must have a hyperbolic complement.

One might ask if it is likely that a randomly chosen knot is hyperbolic. If one uses a Gaussian distribution to select a cyclic sequence of n sticks glued end-to-end, forming a knot, then as n increases, the probability that the result is composite and hence non-hyperbolic, goes to one (cf. [DYS94]). If one restricts to prime knots, it is still the case that as n increases, the probability that the result is a satellite knot and hence non-hyperbolic, goes to one (cf. [dJ94]). So in some sense, hyperbolic knots are substantially less prevalent than non-hyperbolic knots.

But on the other hand, when these non-hyperbolic knot complements are cut open along essential annuli and tori, the resultant pieces have a high probability of being hyperbolic. And for small crossing number, the hyperbolic knots predominate. In fact, for the 2977 nontrivial prime knots through 12 crossings, the only non-hyperbolic knots are seven torus knots.

In addition, although many 3-manifolds are not hyperbolic, it was proved in [rM82] that every compact orientable 3-manifold contains a knot such that its complement is hyperbolic. In other words, every closed orientable 3-maifold is obtained by Dehn filling some hyperbolic 3-manifold. In this sense, hyperbolic knot complements are ubiquitous.

Although the decomposition of the set of knots into the three classes of torus, satellite and hyperbolic knots is fundamental, it does not necessarily allow us to easily determine whether a given knot is hyperbolic. It is often difficult to decide whether or not a given knot is a torus or satellite knot.

One approach is to input the knot into Jeff Week's SnapPea program, which attempts to find a hyperbolic metric on the complement. If the knot is hyperbolic and of reasonable crossing number, the program will almost always find the hyperbolic structure, thereby verifying that the knot or link is hyperbolic. However, if the program fails to find a hyperbolic structure, it could be that the knot is a torus knot or a satellite knot. Or it could be that the computations for determining a hyperbolic structure are too complex for the computer to handle with its limited memory.

There are certain categories of knots and links in S^3 that are known to be hyperbolic. It has been proved that their complements contain no essential spheres, annuli or tori. We list some of these categories below:

(1) Prime non-splittable alternating links that are not 2-braids are hyperbolic.

This was proved in [wM84]. This particular category of link is exceptionally easy to recognize as Menasco proved that an alternating link is splittable if and only if any and every alternating projection is disconnected. He also proved that an alternating link is composite if and only if any and every reduced alternating projection has a circle that crosses the link twice and that contains crossings to either side. A projection is *reduced* if there are no crossings as in Figure 1a.

By a *2-braid*, we mean two strands that twist around one another as in Figure 1b. If an alternating projection shows that the link is either splittable, composite or a 2-braid, then the complement of the link is not hyperbolic. Therefore, given an alternating projection of a link, we can immediately determine by examination whether or not it is hyperbolic. Note that the only reduced alternating projection of a 2-braid is the standard one, by the Flyping theorem of Menasco and Thistlethwaite (cf. [MT91] and [MT93]).

Two-bridge links are all known to be prime and alternating. Hence, assuming that we do not have a 2-braid (again, immediately obvious in an alternating 2-bridge representation), such a link is hyperbolic.

(2) The nontrivial prime non-torus almost alternating knots are hyperbolic (not true for links).

An almost alternating link is a non-alternating link with a projection such that one crossing change will make the projection alternating. This category of knot was introduced in [Aetal92], where it was proved that of the non-alternating knots of 11 or fewer crossings, all but at most three are almost alternating. Since then, one of those knots was shown to be almost alternating in [GHY01]. In [Aetal92], it was also proved that nontrivial prime non-torus almost alternating knots are hyperbolic. However, determining whether or not an almost alternating knot is prime, torus or even nontrivial remains difficult.

(3) Toroidally alternating knots that are prime and non-torus are hyperbolic.

A toroidally alternating link is a link such that it can be projected onto the surface of a standardly embedded torus so that the crossings alternate over and under as we travel around each component on the surface of the torus and any nontrivial closed curve on the torus intersects the projection. An example appears in Figure 2.

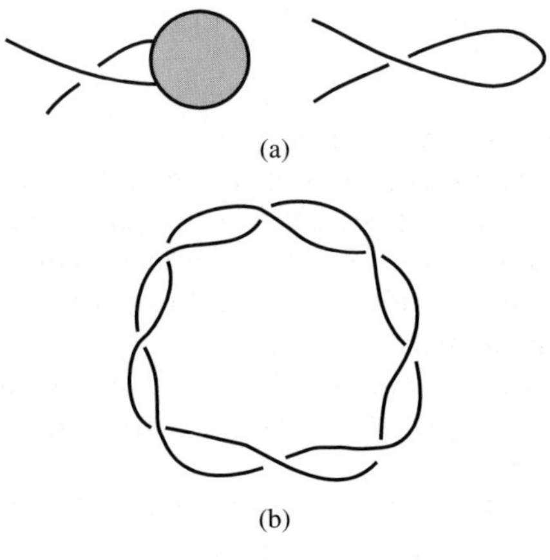

(a)

(b)

Fig. 1. Unreduced diagrams and a 2-braid.

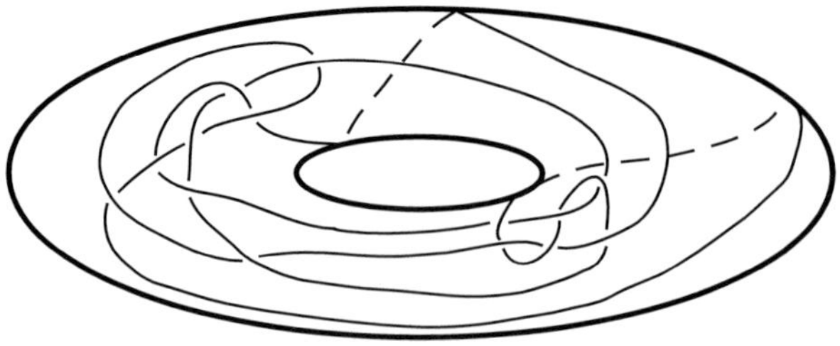

Fig. 2. A toroidally alternating knot.

Almost alternating knots are all toroidally alternating. In [cA94], it was proved that nontrivial prime non-torus toroidally alternating knots are hyperbolic. But again, showing that a toroidally alternating knot is not trivial, composite or torus remains a difficult question.

(4) Augmented alternating links are almost all hyperbolic.

Given a prime alternating link projection, one can add trivial components that bound disjoint disks, each of which is perpendicular to the projection plane and intersects the knot at exactly two points. If the resulting link is nonsplittable, and no two of these added components are parallel in the resulting link complement, we call the result an augmented alternating link. See Figure 3.

In [cA86], it was proved that if the initial link is not a 2-braid, then an augmented alternating link is hyperbolic. Note that in [mL03], Lackenby proved that the closure in the geometric topology of the set of all hyperbolic alternating links is the set of all hyperbolic alternating links and augmented alternating links.

In addition, one can cut open along a twice-punctured disk, twist any number of half-twists and reglue. If the number of half-twists is even, we simply obtain a new link whose complement is homeomorphic to the original. If the number of half-twists is odd, we obtain a new link complement. In [cA85], it was proved that if the original link complement is hyperbolic, so is the new link complement, and it will have the same volume.

So, given a minimal projection of any nontrivial nonsplit link complement other than the standard projection of the 2-braid or an obviously composite projection, one can add these trivial components around crossings, except where doing so yields

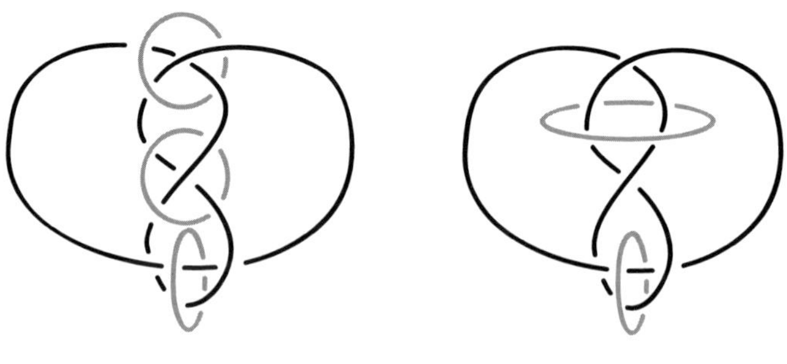

Fig. 3. Two augmented alternating links from the figure-eight knot.

parallel copies, to obtain a link, which is hyperbolic, since we can switch the crossings on the original link to make it alternating. We call such a link fully augmented. See[eS03] for more details.

(5) Arithmetic links.

Some of the first link complements known to be hyperbolic were arithmetic, meaning that their fundamental groups are commensurable (up to conjugacy in PSL(2,**C**)) with a Bianchi group PSL($2,O_d$), where d is a square-free positive integer, and O_d is the ring of integers in the imaginary quadratic field $\mathbb{Q}(\sqrt{-d})$. See for instance, [rR75,rR79,wT78,nW78,aH83,mB92,jS96,jS99,mB01]. Although it appears that arithmetic links are a small subset of the set of all links, Baker proved in [mB02] that every link in S^3 is a sub-link of an arithmetic link in S^3. In the case of knots, Reid proved that the only arithmetic knot is the figure-eight knot (cf. [aR191]).

(6) Montesinos links.

A Montesinos link (also called a star link) is obtained by connecting n rational tangles in a simple cyclic fashion. Such a knot or link is denoted $K(\frac{p_1}{q_1},\ldots,\frac{p_n}{q_n})$, where $\frac{p_i}{q_i}$ denotes the ith rational tangle. In [uO84], it was proved that a Montesinos link $K(\frac{p_1}{q_1},\ldots,\frac{p_n}{q_n})$ with $q_i \geq 2$, is hyperbolic if it is not a torus link and not equivalent to $K(1/2, 1/2, -1/2, -1/2)$, $K(2/3, -1/3, -1/3)$, $K(1/2, -1/4, -1/4)$, $K(1/2, -1/3, -1/6)$ or the mirror image of these links. In [BS80] the torus links that are Montesinos links are identified.

(7) Mutants of a hyperbolic link.

Given a knot or link in a projection and a circle in the projection plane that intersects the knot at four points and separates the knot into two tangles, one can perform the following operation. Cut the knot open at these four points and flip the interior tangle either around a vertical or horizontal axis, or rotate it 180 degrees about an axis perpendicular to the projection plane before reattaching it to the outside tangle at the four points. This operation is called *mutation* and the resulting knots are called *mutants* of the original. In [dR87], Ruberman demonstrated that a mutant of a hyperbolic knot or link is also hyperbolic and it has the same volume.

(8) Belted sums of hyperbolic links.

Given two links, each with a trivial component bounding a disk punctured twice by the link, one can cut them open along the disks in each, glue the resultant disks from the one complement to the disks from the other appropriately, to obtain a link complement as in Figure 4. In [cA85], it was proved that the belted sum of hyperbolic links is always hyperbolic with volume equal to the sum of the volumes of the original links.

(9) Hyperbolic tangles.

In [yW96], Wu determined all of the non-hyperbolic algebraic tangles, finding them to be a small subset of the set of all algebraic tangles. In particular, the hyperbolic algebraic tangles can be closed off to obtain algebraic knots and links that are hyperbolic. This gives a broad set of examples. Note that the tangles themselves can be thought of as complements of genus two handlebodies, and as such, are realized as hyperbolic sub-manifolds of S^3 with totally geodesic boundary.

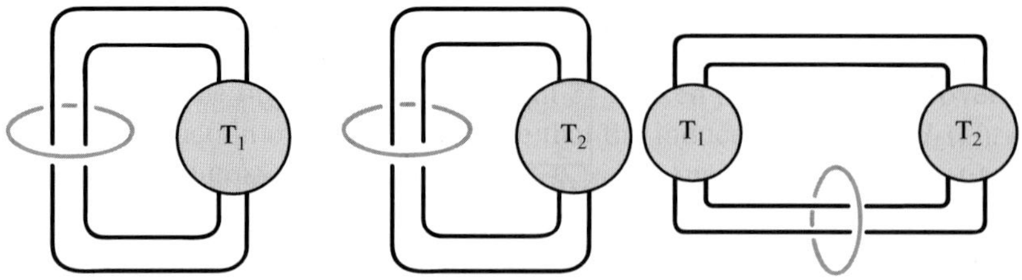

Fig. 4. The belted sum of two links.

In [kS99], it is shown that certain n-string alternating tangles are hyperbolic. These tangles can be utilized to show that prime semi-alternating links are hyperbolic. (See [LT88] for the definition of semi-alternating links.)

3. Volumes of knots

The most natural invariant associated to hyperbolic knots is the volume of the complement. Work of Thurston and Jørgensen shows that knot volumes are a well-ordered subset of $\mathbb{R}$. In [AHW91], lists of volumes of knots through ten crossings were calculated. Cao and Meyerhoff (cf. [CM01]) proved that the figure-eight knot has the smallest possible volume for a knot complement, the volume of which is $2.02988... = 2v_3$, where $v_3 = 1.01494...$ is the volume of an ideal regular tetrahedron. In fact, they proved that this is the smallest volume of any noncompact orientable hyperbolic 3-manifold, the volume being shared by the figure-eight knot complement and one other manifold known as the sibling of the figure-eight knot complement.

A hyperbolic n-component link is known to have volume at least nv_3 (cf. [cA88]), and a 2-component link is known to have volume at least $2.3952v_3$ (cf. [hY01]), although the expectation is that this bound is not sharp. The smallest known 2-component link is the Whitehead link, with volume 3.6638....

Results of Thurston and Jørgensen demonstrate that if one does (p,q)-Dehn filling on a hyperbolic knot or link complement, with $p^2 + q^2$ large enough, the resulting manifold will also be hyperbolic with volume less than the volume of the original manifold but approaching the volume of the original manifold as $p^2 + q^2$ approaches ∞ (see [wT78]).

In particular, we can take a component as in Figure 5, and do $(1,p)$-surgery to obtain a knot with arbitrarily many crossings but volume bounded by the original manifold.

On the other hand, there are knot complements with arbitrarily large volume. We will show this is true for the 2-bridge knots. Two-bridge knots are obtained from surgery on

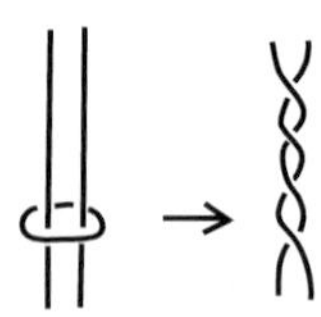

Fig. 5. Twisting by surgery.

the smaller components in the links depicted in Figure 6. Each of these links is a belted sum (see the previous section for the definition and properties) of Borromean rings.

The belted sum of n copies of the Borromean rings has volume equal to $n(7.3276\ldots)$. When we do high surgery on the augmenting components, we obtain 2-bridge knots with volume arbitrarily close to that of the belted sum. Hence, volumes of knot complements can be arbitrarily high.

How effective an invariant is volume for distinguishing between knots? In general, it is very good. The first example of two knots with the same volume does not occur until we consider knots of up to 12 crossings. The 5_2 knot and a 12-crossing knot do have the same volume.

However, the operation of mutation preserves volume. Hence, many knots do share their volumes with other knots. Moreover, the operation of twisting along a twice-punctured disk preserves volume as well, yielding a variety of link complements with the same volume.

Can we say anything about the volume of a particular knot or link complement simply by looking at its projection? In [cA83], it was proved that the volume of the complement of an n-crossing hyperbolic knot other than the figure-eight knot is bounded above by $(4n - 16)v_3$.

In [mL03], Lackenby defines a twist in an alternating projection to be a maximal chain of adjacent bigon regions (as in the second part of Figure 5), or to be a single crossing that is not adjacent to a bigon region. The twist number of a projection is the number of twists within it. He proves (with an improvement of his upper bound due to Ian Agol and Dylan Thurston) that a hyperbolic alternating knot or link in a prime alternating projection of twist number t satisfies

$$v_3\left(\frac{t-2}{2}\right) \leq \text{volume}(S^3 - K) < v_3(10t - 16)$$

By choosing a twist reduced projection, which always exists, the lower bound can be improved to $v_3(t - 2)$.

Is hyperbolic volume related to the more recently defined quantum invariants? In [rK97], Kashaev conjectured that the hyperbolic volume of a knot complement is determined by the asymptotic behavior of a link invariant that depends on the quantum dilogarithm and that was introduced by Kashaev in [rK95]. This is known as the Kashaev conjecture. It has been verified for a handful of knots.

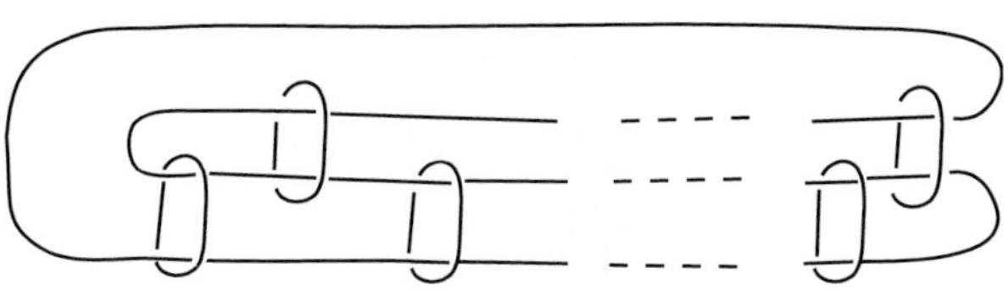

Fig. 6. Two-bridge knots come from surgery on these links.

4. Cusps

Given a hyperbolic knot in S^3, one can define a cusp C for the knot to be a neighborhood of the missing knot such that it lifts to a set of horoballs with disjoint interiors in the universal cover H^3. See Figure 7.

Topologically, a cusp is homeomorphic to $T^2 \times [0, 1)$ where $T^2 \times 0$ corresponds to the boundary of the neighborhood and the missing $T^2 \times 1$ corresponding to the knot itself. Choosing one of the covering horoballs to be the horoball centered at ∞ in the upper-half-space model of H^3, the subset of covering translations sending this horoball back to itself are all Euclidean translations, generated by two translations. A fundamental domain for their action is a parallelogram P in the horizontal boundary plane and the collection of horizontal parallelograms directly above this one. The parallelogram P projects to the torus boundary of the cusp. As we move up the z-axis in hyperbolic space, each horizontal parallelogram projects to a concentric torus around the missing knot, the set of them shrinking in toward the missing knot.

In $H^3 \cup \partial H^3$, the Euclidean translations fix a single point on the boundary, that being the point at ∞. We call an isometry that fixes a single point on the boundary a *parabolic* isometry. All such isometries will send the horoballs tangent to the boundary at that fixed point back to themselves. All other isometries in the group Γ of covering translations have two fixed points on the boundary. These are called *hyperbolic* isometries. They correspond

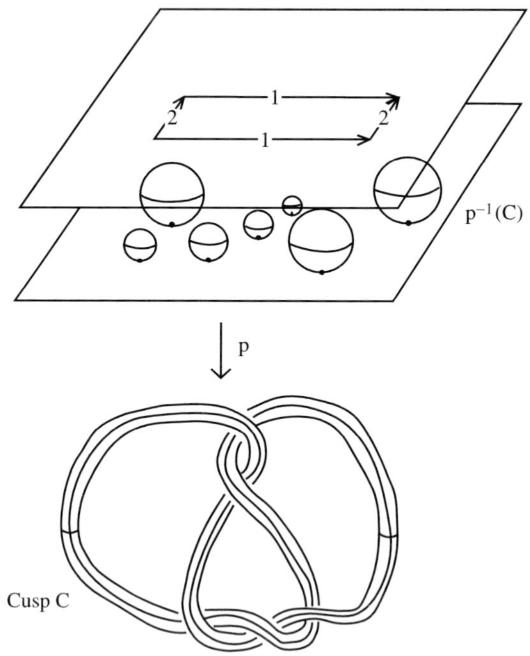

Fig. 7. The cusp of a knot.

to translations along the geodesic with the two fixed points as endpoints, together with a possible rotation about the geodesic.

The third type of orientation preserving isometry is a pure rotation about a geodesic. But as it is not fixed point free in H^3 it cannot appear as a covering translation in Γ. In the case of a knot, we have a single cusp in the complement. We can expand that cusp until it touches itself. Then, the set of horoballs that are the pre-image of C in H^3 will expand until two first touch. Since every horoball in $p^{-1}(C)$ is identified to every other one, the horizontal plane that is the boundary of the horoball centered at ∞ will touch other horoballs. We call this cusp a *maximal cusp*. In the case we have a link, we can expand the cusps until they touch each other or themselves, to obtain a maximal set of cusps. However, in different situations, it may be appropriate to do the expansions in different ways. For instance, we may insist that all the cusps have the same volume or the same length of shortest curve in their boundaries while we expand the set. Or we may choose to put the largest possible volume in the expanded set of cusps.

We define the *cusp density* of a hyperbolic knot or link complement to be the ratio of the largest possible volume in a maximal set of cusps divided by the total volume. Meyerhoff noted in [rM86] that this number is at most $0.853 = \frac{\sqrt{3}}{2v_3}$ where again, $v_3 = 1.01494...$ is the volume of an ideal regular tetrahedron. The cusp density of the figure-eight knot complement realizes this upper bound. In [Aetal02], it was proved that there are knots of arbitrarily small cusp density, although other methods for generating such knots were previously known, but had not appeared in the literature. Although it is known that the set of cusp densities for hyperbolic 3-manifolds are dense in the interval [0,.853...] (cf. [cA102]), it is not known whether the same holds true for cusp densities of hyperbolic link complements or perhaps, even for cusp densities of hyperbolic knot complements.

5. Meridians and other cusp invariants

The Thurston–Gromov 2π theorem implies that for any Dehn filling along a curve c in the cusp boundary of a hyperbolic 3-manifold, the resultant manifold will be negatively curved if c has length at least 2π. It is conjectured that in fact the resultant manifolds are hyperbolic with constant negative curvature, but this question remains open (see [BH96] for more details). If we define the length $|m|$ of a meridian for a hyperbolic knot K to be the length of its shortest representative in the maximal cusp boundary, then $|m| < 2\pi$, since Dehn filling along the meridian yields S^3, which is not negatively curved.

In [mL00] and [iA00], a variation on the 2π theorem, known as the 6-theorem, was proved. It shows that a manifold obtained by Dehn filling along a minimal length simple closed curve of length greater than 6 in the cusp boundary of a hyperbolic manifold is *hyperbolike*, which is to say that it is irreducible, boundary-irreducible, and has infinite fundamental group that is word hyperbolic. In other words, it has these attributes that one would expect of a negatively curved manifold. Agol also gave an example of a cusped manifold such that there was a curve of length 6 in the cusp boundary such that surgery on that curve yielded a non-hyperbolike manifold. So the 6 bound is sharp. In [Aetal03], examples were produced of knot complements with longitude of length exactly 6 such that surgery on the longitude yields non-hyperbolike manifolds. So the 6-theorem is sharp for knot complements in S^3.

Since Dehn filling along the meridian of a knot yields the 3-sphere, which has trivial fundamental group, the 6-Theorem implies every meridian of a knot must have length no greater than 6. Moreover, any nontrivial curve in the cusp boundary must have length at least 1. This is because a maximal cusp when lifted to the horoball centered at ∞, will have other horoballs tangent to it. The shortest parabolic isometry must shift the horoballs at least a distance 1 so that they do not overlap with one another.

So all meridian lengths fall in the interval [1,6]. In [cA202], it was proved that there is only one knot that realizes the lower bound. The length of the meridian in the figure-eight knot complement is exactly 1, and no other hyperbolic manifold has a nontrivial curve in its maximal cusp boundary this small. Moreover, in [cA03], it is proved that the next shortest meridian is that of the 5_2 knot, with a length of 1.150964..., and that there are no other knots of meridian length less than $\sqrt[4]{2} = 1.189207\ldots$.

What about the upper bound? In [iA00], Agol gives an example of a sequence of knots with meridian lengths approaching 4 from below. We show how to construct such a sequence of knots in Figure 8. As the number of times that the knot wraps around itself both vertically and horizontally increases, the meridian length approaches 4 from below. These are the largest known meridian lengths to date.

Utilizing ideas from [zH98], it was proved in [Aetal02] that the meridian length of a knot K is always at most $6 - \frac{7}{c}$, where c is the crossing number of K. The idea is to take the singular punctured surface obtained by coning the knot to a point below the projection plane. If it is not incompressible or boundary-incompressible, then we can do compressions and boundary-compressions that will only improve the ultimate bounds. Then we pleat the surface and it inherits a 2-dimensional hyperbolic metric from the 3-dimensional hyperbolic metric on the manifold. The length of the boundary curves of the surface intersected with the

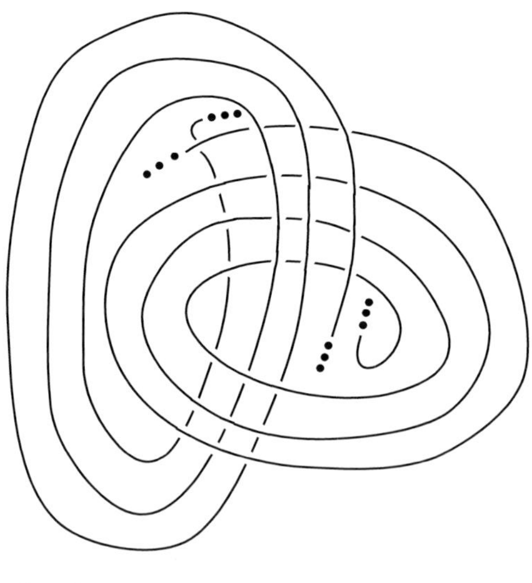

Fig. 8. These knots have meridian length approaching 4 from below.

cusp is equal to the area of the intersection of the cusp with the surface. This is bounded above by $\frac{3}{\pi}$ times the area of the surface, which is given by $2\pi|\chi(S)|$. From these inequalities, we can obtain bounds on meridian length and cusp volume.

One can obtain substantially better bounds for certain classes of knots. In [cA96], it was proved that the 2-bridge knots have meridian lengths in the interval [1,2), with sequences of meridians approaching 2.

In [Aetal02], upper bounds for alternating knots were determined. In particular, it was shown that the meridian of an alternating knot with crossing number c is bounded above by $3 - \frac{6}{c}$. The proof utilizes the checkerboard surfaces that come from any given projection of a knot. In the case of alternating knots, it was proved in [MT93] that these surfaces are incompressible and boundary-incompressible. Pleating these surfaces allows one to prove the result.

The expectation is that the actual upper bound for the meridian length of alternating knots is substantially smaller. For example, we say a property J is true for almost all hyperbolic 3-manifolds if for every fixed volume $V > 0, J$ is true for all but finitely many hyperbolic 3-manifolds with volume less than V. In [eS03], it is proved that almost all alternating hyperbolic knots have meridian length bounded above by 2. The expectation is that 2 is the correct bound for all alternating hyperbolic knots.

In addition, in [eS03], Schoenfeld proved that if we fully augment any reduced projection of a nontrivial knot other than a 2-braid, as in Section 2, the meridian of the original knot becomes 2 in the resultant link complement. In particular, this means that if we take any reduced projection of any non-2-braid knot and twist about each of its crossings to add bigons, the resultant knots will have meridian approaching 2.

Using different techniques, Purcell also shows that a fully augmented link will have meridian length 2 on the original component in [jP04]. Moreover, she used this to obtain a lower bound of $(0.8154)^2N$ on the ratio of the length of the longitude to the length of the meridian for any hyperbolic knot with at least N twist regions and a sufficient number of crossings in each twist region. This allowed her to further prove that if a hyperbolic knot has at least 85 twist regions and at least 73 full twists per region, all nontrivial Dehn fillings on the knot yield hyperbolic 3-manifolds. To know that the resulting manifold is hyperbolike, only 10 regions, each with at least 73 full twists are necessary.

In [zH98], an application of meridian length to the determination of crossing number of a knot is given. Specifically, it is shown that if C is a maximal cusp for a hyperbolic knot K, then the crossing number $\mathrm{cr}(K)$ satisfies

$$\mathrm{cr}(K) \geq \mathrm{area}(\partial C)/\big(|m|(2\pi - |m|)\big).$$

Moreover, if K' is a satellite of the hyperbolic knot K of degree p, then $cr(K') \geq p^2\mathrm{area}(\partial C)/(|m|(2\pi - |m|))$.

6. Geodesics and totally geodesic surfaces

Geodesics in hyperbolic knot complements may or may not be closed. If a geodesic is closed, it might intersect itself or not. However, every hyperbolic knot or link complement contains a simple closed geodesic [AHS99].

A *systole* is a shortest closed geodesic, and the *systole length* is the length of the shortest closed geodesic in a manifold. If we take two strands in a knot complement, and obtain a sequence of knots by twisting these two strands around one another as in Figure 5, then the length of the geodesic that wraps around the two strands will be approaching 0. Hence, this yields a sequence of knot complements with systole length shrinking to 0. So there is no lower bound for systole length in knot and link complements. On the other hand, there is an upper bound. Although the systole length can be arbitrarily large for hyperbolic 3-manifolds in general, it was proved in [AR02] that systole length for hyperbolic knots and links in S^3 is bounded above by 7.35534. For hyperbolic alternating knots, this was improved to 4.5 in [Aetal02]. It would be interesting to obtain better bounds on systole length for a variety of categories of knots and links.

Closed geodesics in knot complements can themselves be either unknotted or knotted as curves in S^3. See for instance, [sM01], where a variety of knotted geodesics in the figure-eight knot complement are displayed.

In [tS91] (see also [sK88]), it was proved that the complement of a simple closed geodesic or a set of disjoint simple closed geodesics in a hyperbolic manifold will itself yield a hyperbolic manifold. One can look at the link complements obtained from removing certain geodesics from a knot or link complement, in the hopes of decomposing the resultant manifolds into certain canonical pieces.

One might hope that the complement of a simple closed geodesic in a hyperbolic knot complement has minimal volume among all the complements of simple closed curves in the same free homotopy class. But counterexamples to this conjecture are given in the figure-eight knot complement in [sM01].

We turn now to surfaces in knot complements. A closed embedded incompressible surface in a hyperbolic knot complement can come in one of two varieties. The first possibility is that it is quasi-Fuchsian. This means that it lifts to the disjoint union of topological planes in H^3, each with a limit set that is a quasi-circle. In the case that the limit set is an actual circle, and the planes are geodesic, and we say that the surface is *Fuchsian* or *totally geodesic*.

The second possibility for a closed embedded incompressible surface is that there are simple closed curves on the surface that can be homotoped through the knot complement into a neighborhood of the missing knot. This means that the corresponding isometry, when we lift to hyperbolic space, is a parabolic isometry. We call such a curve an *accidental parabolic curve* and we call such a surface an *accidental surface*. In [IO200], it was proved that if an accidental parabolic curve exists for a given surface S, then it is isotopic to a unique curve on the boundary torus of the cusp. In particular, there is a well-defined accidental slope for each accidental surface. However, there can be more than one accidental slope for a given knot. It follows from [CGLS87] that accidental slopes must be meridional or integer.

In [AR93], explicit examples were given of closed quasi-Fuchsian surfaces in knot complements. These surfaces were shown not to be totally geodesic. In fact, in [MR92], it was conjectured that hyperbolic knot complements in S^3 do not contain any closed embedded totally geodesic surfaces. This has been proved for hyperbolic knots that are alternating knots, tunnel number one knots, 2-generator knots and knots of braid index three [MR92], almost alternating knots [Aetal92], toroidally alternating knots [cA94],

Montesinos knots [uO84], 3-bridge knots and double torus knots [IO100]) and knots of braid index three [LP85] and four [hM02]. Note that the conjecture does not hold for links. An explicit counterexample first appreared in [MR92]. In [cL04], examples are given of closed totally geodesic surfaces of arbitrary even genus at least 2 in the complements of 2-component links. Moreover, it is also shown there that if we do not insist on a metric of constant sectional curvature -1, but instead allow it an arbitrarily small amount of leeway to either side of -1, there do exist knots in S^3 with closed embedded totally geodesic surfaces of any genus $g \geq 3$.

There are closed totally geodesic surfaces immersed in hyperbolic knot complements. In [aR291], Reid showed that the figure-eight knot complement contains infinitely many such non-homotopic surfaces. Also, see [AR97] for two "dodecahedral" knots with immersed totally geodesic surfaces in their complements. However, at this time, only these three examples are known.

An incompressible boundary incompressible surface S with boundary properly embedded in a hyperbolic knot exterior can have one of three possible behaviors:

(1) S can be quasi-Fuchsian.

(2) S can be accidental.

(3) S can be a virtual fiber in a fibered knot, with limit set the entire boundary of H^3.

Specific examples of incompressible boundary-incompressible surfaces are afforded by minimal genus Seifert surfaces. In [sF98], it is proved that a minimal genus Seifert surface in a non-fibered hyperbolic knot complement must be quasi-Fuchsian. It can never be accidental.

Knots can have totally geodesic minimal genus Seifert surfaces. In [AS03], examples such as a (p,p,p) pretzel knot (Montesinos knot $K(1/p,\ 1/p,\ 1/p)$) are shown to have totally geodesic Seifert surfaces (see Figure 9). There are examples of hyperbolic knots with totally geodesic Seifert surfaces, where the Seifert surfaces are both free (the complement of a neighborhood of the Seifert surface in the knot complement is a handlebody) and non-free.

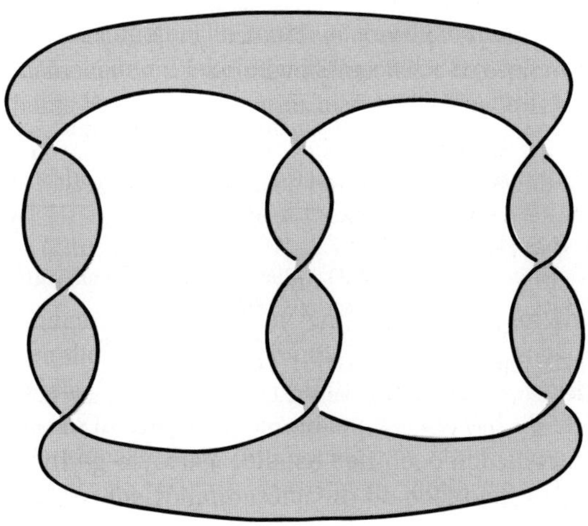

Fig. 9. A (3,3,3)-pretzel knot has a totally geodesic Seifert surface.

But the expectation is that knots with totally geodesic Seifert surfaces are the exception. For instance, in [AS03], it is proved that 2-bridge knots never have totally geodesic Seifert surfaces.

An interesting question is whether there are totally geodesic surfaces in knot complements other than such Seifert surfaces. This is one of many open questions that still remain in the theory of hyperbolic knots.

Acknowledgements

Thanks to Eric Schoenfeld for help with the figures.

References

[cA83] C. Adams, *Hyperbolic structures on knot and link complements*, Ph.D. Thesis, University of Wisconsin-Madison (1983).

[cA85] C. Adams, *Thrice-punctured spheres in hyperbolic 3-manifolds*, Trans. Amer. Math. Soc. **287** (2) (1985), 645–656.

[cA86] C. Adams, *Augmented alternating link complements are hyperbolic*, London Mathematical Society Lecture Notes Series, 112: Low Dimensional Topology and Kleinian Groups, Cambridge University Press, Cambridge (1986), 115–130.

[cA88] C. Adams, *Volumes of N-cusped hyperbolic 3-manifolds*, J. London Math. Soc. (2) **38** (3) (1988), 555–565.

[AHW91] C. Adams, M. Hildebrand and J. Weeks, *Hyperbolic invariants of knots and links*, Trans. Amer. Math. Soc. **326** (1) (1991), 1–56.

[Aetal92] C. Adams, J. Brock, J. Bugbee, T. Comar, K. Faigin, A. Huston, A. Joseph and D. Pesikoff, *Almost alternating links*, Topology Appl. **46** (2) (1992), 151–165.

[AR93] C. Adams and A. Reid, *Quasi-Fuchsian surfaces in hyperbolic knot complements*, J. Aust. Math. Soc. Ser. A **55** (1) (1993), 116–131.

[cA94] C. Adams, *Toroidally alternating knots and links*, Topology **33** (2) (1994), 353–369.

[cA96] C. Adams, *Hyperbolic 3-manifolds with two generators*, Comm. Anal. Geom. **4** (1–2) (1996), 181–206.

[AHS99] C. Adams, J. Hass and P. Scott, *Simple closed geodesics in hyperbolic 3-manifolds*, Bull. London Math. Soc. **31** (1) (1999), 81–86.

[cA102] C. Adams, *Cusp densities of hyperbolic 3-manifolds*, Proc. Edinburgh Math. Soc. **45** (2002), 277–284.

[cA202] C. Adams, *Waist size for cusps in hyperbolic 3-manifolds*, Topology **41** (2) (2002), 257–270.

[cA03] C. Adams, *Waist size for cusps in hyperbolic 3-manifolds II*, Preprint.

[AR02] C. Adams and A. Reid, *Systole length in hyperbolic 3-manifolds*, Math. Proc. Cambridge Philos. Soc. **128** (1) (2002), 103–110.

[Aetal02] C. Adams, A. Colestock, J. Fowler, W.D. Gillam and E. Katerman, *Cusp size bounds from singular surfaces in hyperbolic 3-manifolds*, Preprint (2002).

[AS03] C. Adams and E. Schoenfeld, *Totally geodesic Seifert surfaces in hyperbolic knot and link complements I*, Preprint, ArXiv: math.GT/0411162.

[Aetal03] C. Adams, H. Bennett, C. Davis, M. Jennings, J. Novak, N. Perry and E. Schoenfeld, *Totally geodesic surfaces in hyperbolic knot and link complements II*, Preprint, ArXiv: math.GT/0411358.

[iA00] I. Agol, *Bounds on exceptional dehn filling*, Geom. Topol. **4** (2000), 431–449.

[AR97] I. Aitchison and J.H. Rubinstein, *Geodesic surfaces in knot complements*, Exp. Math. **6** (2) (1997), 137–150.

[mB92] M. Baker, *Link complements and integer rings of class number greater than one*, Topology '90 (Columbus, OH, 1990), Ohio State Univ. Math. Res. Inst. Publ., 1, de Gruyter, Berlin (1992), 55–59.

[mB01] M. Baker, *Commensurability classes of arithmetic link complements*, J. Knot Theory Ramifications **10** (7) (2001), 943–957.

[mB02] M. Baker, *All links are sublinks of arithmetic links*, Pacific J. Math. **203** (2) (2002), 257–263.

[BH96] S. Bleiler and C. Hodgson, *Spherical space forms and Dehn filling*, Topology **35** (3) (1996), 809–833.

[BS80] M. Boileau and L. Siebenmann, *A planar classification of pretzel knots and Montesinos knots*, Orsay Preprint (1980).

[sB02] S. Boyer, *Dehn surgery on knots, Handbook of Geometric Topology*, North-Holland, Amsterdam (2002), 165–218.

[CR98] P.J. Callahan and A.W. Reid, *Hyperbolic structures on knot complements, Knot theory and its applications*, Chaos Solitons Fractals **9** (4–5) (1998), 705–738.

[CRV01] A. Cavicchioli, D. Repovš and A. Vesnin, *Recent results on topology of three-manifolds*, Atti Sem. Mat. Fis. Univ. Modena **49** (Suppl.) (2001), 31–71.

[CM01] C. Cao and G.R. Meyerhoff, *The orientable cusped hyperbolic 3-manifolds of minimum volume*, Invent. Math. **146** (3) (2001), 451–478.

[CGLS87] M. Culler, C.McA. Gordon, J. Luecke and P. Shalen, *Dehn surgery on knots*, Ann. Math. **125** (2) (1987), 237–300.

[DYS94] Y. Diao, N. Pippenger and D.W. Sumners, *On random knots. Random knotting and linking (Vancouver, BC, 1993)*, J. Knot Theory Ramifications **3** (3) (1994), 419–429.

[sF98] S. Fenley, *Quasi-Fuchsian Seifert surfaces*, Math. Z. **228** (2) (1998), 221–227.

[GHY01] H. Goda, M. Hirasawa and R. Yamamoto, *Almost alternating diagrams and fibered links in S^3*, Proc. London Math. Soc. (3) **83** (2) (2001), 472–492.

[GL89] C.McA. Gordon and J. Luecke, *Knots are determined by their complements*, J. Amer. Math. Soc. **2** (2) (1989), 371–415.

[aH83] A. Hatcher, *Hyperbolic structures of arithmetic type on some link complements*, J. London Math. Soc. (2) **27** (2) (1983), 345–355.

[zH98] Z.-X. He, *On the crossing number of high degree satellites of hyperbolic knots*, Math. Res. Lett. **5** (1–2) (1998), 235–245.

[HTW98] J. Hoste, M. Thistlethwaite and J. Weeks, *The first 1,701,936 knots*, Math. Intelligencer **20** (4) (1998), 33–48.

[IO100] K. Ichihara and M. Ozawa, *Hyperbolic knot complements without closed embedded totally geodesic surfaces*, J. Aust. Math. Soc. Ser. A **68** (3) (2000), 379–386.

[IO200] K. Ichihara and M. Ozawa, *Accidental surfaces in knot complements*, J. Knot Theory Ramifications **9** (6) (2000), 725–733.

[dJ94] D. Jungreis, *Gaussian random polygons are globally knotted*, J. Knot Theory Ramifications **3** (4) (1994), 455–464.

[rK95] R.M. Kashaev, *A link invariant from quantum dilogarithm*, Modern Phys. Lett. A **10** (19) (1995), 1409–1418.

[rK97] R.M. Kashaev, *The hyperbolic volume of knots from the quantum dilogarithm*, Lett. Math. Phys. **39** (3) (1997), 269–275.

[sK88] S. Kojima, *Isometry transformations of hyperbolic 3-manifolds*, Topology Appl. **29** (3) (1988), 297–307.

[mL00] M. Lackenby, *Word hyperbolic dehn surgery*, Invent. Math. **140** (2000), 243–282.

[mL03] M. Lackenby, *The volume of hyperbolic alternating knot complements*, With an appendix by Ian Agol and Dylan Thurston. Proc. London Math. Soc. (3) **88** (1) (2004), 204–224.

[cL04] C. Leininger, *Small curvature surfaces in hyperbolic 3-manifolds*, Preprint, ArXiv: math.GT/0409455.

[LT88] W.B.R. Lickorish and M.B. Thistlethwaite, *Some links with nontrivial polynomials and their crossing-numbers*, Comment. Math. Helv. **63** (4) (1988), 527–539.

[LP85] M.T. Lozano and J.H. Przytycki, *Incompressible surfaces in the exterior of a closed 3-braid I*, Math. Proc. Cambridge Philos. Soc. **98** (2) (1985), 275–299.

[hM02] H. Matsuda, *Complements of hyperbolic knots of braid index four contain no closed embedded totally geodesic surfaces*, Topology Appl. **119** (1) (2002), 1–15.

[wM84] W. Menasco, *Closed incompressible surfaces in alternating knot and link complements*, Topology **23** (1) (1984), 37–44.

[MT91] W. Menasco and M. Thistlethwaite, *The Tait flyping conjecture*, Bull. Amer. Math. Soc. (NS) **25** (2) (1991), 403–412.

[MR92] W. Menasco and A. Reid, *Totally geodesic surfaces in hyperbolic link complements, Topology '90 (Columbus, OH, 1990)*, Ohio State Univ. Math. Res. Inst. Publ., 1, de Gruyter, Berlin (1992), 215–226.

[MT93] W. Menasco and M. Thistlethwaite, *The classification of alternating links*, Ann. Math. **138** (1) (1993), 113–171.

[rM86] G.R. Meyerhoff, *Sphere-packing and volume in hyperbolic 3-space*, Comment. Math. Helv. **61** (2) (1986), 271–278.

[sM01] S. Miller, *Geodesic knots in the figure-eight knot complement*, Exp. Math. **10** (3) (2001), 419–436.

[gM73] G.D. Mostow, *Strong rigidity of locally symmetric spaces*, Ann. Math. Studies, 78, Princeton University Press, Princeton, NJ (1973).

[rM82] R. Myers, *Simple knots in compact orientable 3-manifolds*, Trans. AMS **273** (1982), 75–91.

[uO84] U. Oertel, *Closed incompressible surfaces in complements of star links*, Pacific J. Math. **111** (1) (1984), 209–230.

[jP04] J. Purcell, *Cusp shapes of hyperbolic link complements and Dehn filling*, Preprint, arXiv: math.GT/0410233v1.

[RZ01] M. Reni and B. Zimmermann, *Hyperbolic 3-manifolds and cyclic branched coverings of knots and links*, Atti Sem. Mat. Fis. Univ. Modena **49** (Suppl.) (2001), 135–153.

[aR191] A. Reid, *Arithmeticity of knot complements*, J. London Math. Soc. (2) **43** (1) (1991), 171–184.

[aR291] A. Reid, *Totally geodesic surfaces in hyperbolic 3-manifolds*, Proc. Edinburgh Math. Soc. (2) **34**(1) (1991), 77–88.

[rR75] R. Riley, *A quadratic parabolic group*, Math. Proc. Cambridge Philos. Soc. **77** (1975), 281–288.

[rR79] R. Riley, *An elliptical path from parabolic representations to hyperbolic structures, Topology of low-dimensional manifolds (Proc. Second Sussex Conf., Chelwood Gate, 1977)*, Lecture Notes in Math., 772, Springer, Berlin (1979), 99–133.

[dR87] D. Ruberman, *Mutation and volumes of knots in* S^3, Invent. Math. **90** (1) (1987), 189–215.

[tS91] T. Sakai, *Geodesic knots in a hyperbolic 3-manifold*, Kobe J. Math. **8** (1) (1991), 81–87.

[mS98] M. Sakuma, *The topology, geometry and algebra of unknotting tunnels, Knot theory and its applications*, Chaos Solitons Fractals **9** (4–5) (1998), 739–748.

[eS03] E. Schoenfeld, *Augmentations of knot and link complements*, Undergraduate Thesis, Williams College (2003).

[kS99] K. Shimokawa, *Hyperbolicity and ∂-irreducibility of alternating tangles*, Topology Appl. **96** (3) (1999), 217–239.

[jS96] J. Stephan, *Complementaires d'entrelacs dans* S^3 *et ordres maximaux des algbres de quaternions* $M_2(\mathbf{Q}[i\sqrt{d}])$, *C. R. Acad. Sci. Paris Sr. I Math. Link complements in* S^3 *and maximal orders of the quaternion algebras* $M_2(\mathbf{Q}[i\sqrt{d}])$ *322 (7) (1996), 685–688.*

[jS99] J. Stephan, *On arithmetic hyperbolic links*, J. Knot Theory Ramifications **8** (3) (1999), 373–389.

[wT78] W. Thurston, *The Geometry and Topology of 3-Manifolds*, Lecture Notes, Princeton University (1978).

[jW03] J. Weeks, *SnapPea, A computer program for creating and studying hyperbolic 3-manifolds*, available at http://www.geometrygames.org/SnapPea/.

[jW04] J. Weeks, *Computation of hyperbolic structures in knot theory*, This volume.

[nW78] N. Wielenberg, *The structure of certain subgroups of the Picard group*, Math. Proc. Cambridge Philos. Soc. **84** (3) (1978), 427–436.

[yW96] Y.-Q. Wu, *The classification of nonsimple algebraic tangles*, Math. Ann. **304** (3) (1996), 457–480.

[hY01] H. Yoshida, *Volumes of orientable 2-cusped hyperbolic 3-manifolds*, Kobe J. Math. **18** (2) (2001), 147–161.

CHAPTER 2

Braids: A Survey

Joan S. Birman*

Department of Mathematics, Barnard College, Columbia University,
2990 Broadway, New York, NY 10027, USA
E-mail: jb@math.columbia.edu

Tara E. Brendle†

Department of Mathematics, Louisiana State University, Baton Rouge, LA 70803-4918, USA
E-mail: brendle@math.lsu.edu

Contents

*The first author acknowledges partial support from the U.S. National Science Foundation under grant number 0405586.

†The second author was partially supported by a VIGRE postdoc under NSF grant number 9983660 to Cornell University.

HANDBOOK OF KNOT THEORY
Edited by William Menasco and Morwen Thistlethwaite

Abstract

This article is about Artin's braid group $\mathbf{B}_n$ and its role in knot theory. We set ourselves two goals: (i) to provide enough of the essential background so that our review would be accessible to graduate students, and (ii) to focus on those parts of the subject in which major progress was made, or interesting new proofs of known results were discovered, during the past 20 years. A central theme that we try to develop is to show ways in which structure first discovered in the braid groups generalizes to structure in Garside groups, Artin groups and surface mapping class groups. However, the literature is extensive, and for reasons of space our coverage necessarily omits many very interesting developments. Open problems are noted and so-labeled, as we encounter them. A guide to computer software is given together with an extensive bibliography.

1. Introduction

In a review article, one is obliged to begin with definitions. Braids can be defined by very simple pictures such as the ones in Figure 1. Our braids are illustrated as oriented from left to right, with the strands numbered $1, 2, \ldots, n$ from bottom to top. Whenever it is more convenient, we will also think of braids "vertically", i.e., oriented from top to bottom, with the strands numbered $1, 2, \ldots, n$ from left to right. Crossings are suggested as they are in a picture of a highway overpass on a map. The identity braid has a canonical representation in which two strands never cross. Multiplication of braids is by juxtaposition, concatenation, isotopy and rescaling.

Pictures like the ones in Figure 1 give an excellent intuitive feeling for the braid group, but one that quickly becomes complicated when one tries to pin down details. What is the ambient space? (It is a slice $\mathbb{R}^2 \times I$ of 3-space.) Are admissible isotopies constrained to $\mathbb{R}^2 \times I$? (Yes, we cannot allow isotopies in which the strands are allowed to loop over the initial points.) Does isotopy mean level-preserving isotopy? (No, it will not matter if we allow more general isotopies in $\mathbb{R}^2 \times I$, as long as strands do not pass through one another.) Are strands allowed to self-intersect? (No, to allow self-intersections would give the "homotopy braid group", a proper homomorphic image of the group that is our primary focus.) Can we replace the ambient space by the product of a more general surface and an interval, for example a sphere and an interval? (Yes, it will become obvious shortly how to modify the definition.)

We will bypass these questions and other related ones by giving several more sophisticated definitions. In Sections 1.1–1.3 we will define the braid group $\mathbf{B}_n$ and pure braid group $\mathbf{P}_n$ in three distinct ways. We will give a proof that two of them yield the same group. References to the literature establish the isomorphism in the remaining case. In Section 1.4 we will demonstrate the universality of "braiding" by describing four examples which show how braids have played a role in parts of mathematics which seem far away from knot theory.

1.1. $\mathbf{B}_n$ *and* $\mathbf{P}_n$ *via configuration spaces*

We define the topological concept of a braid and of a group of braids via the notion of a configuration space. This approach is nice because it gives, in a concise way, the appropriate equivalence relations and the group law, without any fuss.

The *configuration space* of n points on the complex plane $\mathbb{C}$ is

$$\mathcal{C}_{0,\hat{n}} = \mathcal{C}_{0,\hat{n}}(\mathbb{C}) = \{(z_1, \ldots, z_n) \in \mathbb{C} \times \cdots \times \mathbb{C} | z_i \neq z_j \text{ if } i \neq j\}.$$

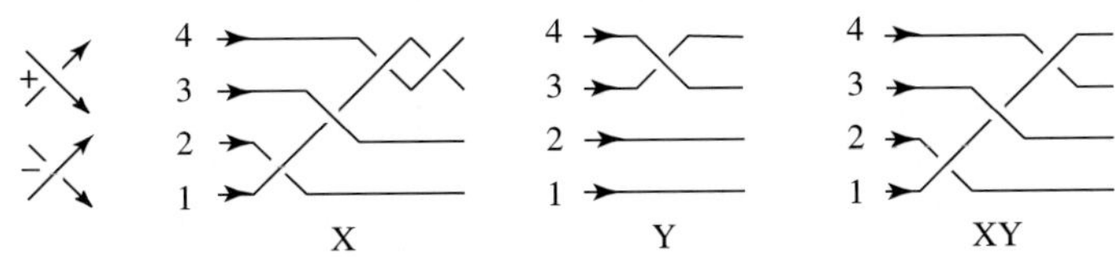

Fig. 1. Examples of 4-braids X, Y and their product XY.

A point on $\mathcal{C}_{0,\hat{n}}$ is denoted by a vector $\vec{z} = (z_1, \ldots, z_n)$. The symmetric group acts freely on $\mathcal{C}_{0,\hat{n}}$, permuting the coordinates in each $\vec{z} \in \mathcal{C}_{0,\hat{n}}$. The orbit space of the action is $\mathcal{C}_{0,n} = \mathcal{C}_{0,\hat{n}}/\Sigma_n$ and the orbit space projection is $\tau : \mathcal{C}_{0,\hat{n}} \to \mathcal{C}_{0,n}$. Choosing a fixed base point $\vec{p} = (p_1, \ldots, p_n)$, we define the *pure braid group* $\mathbf{P}_n$ on n strands and the *braid group* $\mathbf{B}_n$ on n strands to be the fundamental groups:

$$\mathbf{P}_n = \pi_1(\mathcal{C}_{0,\hat{n}}, \vec{p}), \qquad \mathbf{B}_n = \pi_1(\mathcal{C}_{0,n}, \tau(\vec{p})).$$

At first encounter $\pi_1(\mathcal{C}_{0,n}, \tau(\vec{p}))$ does not look as if it has much to do with braids as we illustrated them in Figure 1, but in fact there is a simple interpretation which reveals the intuitive picture. While the manifold $\mathcal{C}_{0,n}$ has dimension $2n$, the fact that the points $z_1, \ldots, z_n$ are pairwise distinct allows us to think of a point $\vec{z} \in \mathcal{C}_{0,n}$ as a set of n distinct points on $\mathbb{C}$. An element of $\pi_1(\mathcal{C}_{0,n}, \tau(\vec{p}))$ is then represented by a loop which lifts uniquely to a path $\vec{g} : I \to \mathcal{C}_{0,\hat{n}}$, where $\vec{g} = \langle g_1, \ldots, g_n \rangle$ consists of n coordinate functions $g_i : I \to \mathbb{C}$ satisfying $g_i(t) \neq g_j(t)$ if $i \neq j$, $t \in [0, 1]$, also $\vec{g}(0) = \vec{g}(1) = \vec{p}$, the base point. The graph of the n simultaneous functions $g_1, \ldots, g_n$ is a *geometric (pure) braid*. The appropriate equivalence relation on geometric braids is captured by simultaneous homotopy of the n simultaneous paths, rel their endpoints, in the configuration space.

The group $\mathbf{B}_n$ is the group which Artin set out to investigate in 1925 in his seminal paper [5] (however, he defined it in a less concise way); in the course of his investigations he was led almost immediately to study $\mathbf{P}_n$. Indeed, the two braid groups are related in a very simple way. Let τ_* be the homomorphism on fundamental groups which is induced by the orbit space projection $\tau : \mathcal{C}_{0,\hat{n}} \to \mathcal{C}_{0,n}$. Observe that the orbit space projection is a regular $n!$-sheeted covering space projection, with Σ_n as the group of covering translations. From this it follows that the group $\mathbf{P}_n$ is a subgroup of index $n!$ in $\mathbf{B}_n$, and there is a short exact sequence:

$$1 \to \mathbf{P}_n \xrightarrow{\tau_*} \mathbf{B}_n \to \Sigma_n \to 1. \tag{1.1}$$

1.2. $\mathbf{B}_n$ *and* $\mathbf{P}_n$ *via generators and relations*

We give two definitions of the group $\mathbf{B}_n$ by generators and relations. The *classical presentation* for $\mathbf{B}_n$ first appeared in [5]. We record it now, and will refer back to it many times later. It has generators $\sigma_1, \ldots, \sigma_{n-1}$ and defining relations:

$$\sigma_i\sigma_k = \sigma_k\sigma_i \quad \text{if } |i-k| \geq 2, \quad \sigma_i\sigma_{i+1}\sigma_i = \sigma_{i+1}\sigma_i\sigma_{i+1}. \tag{1.2}$$

The elementary braid σ_i is depicted in sketch (i) of Figure 2.

Many years after Artin did his fundamental work, Birman, Ko and Lee discovered a new presentation, which enlarged the set of generators to a more symmetric set. Let $\sigma_{s,t} = (\sigma_{t-1} \cdots \sigma_{s+1})\sigma_s(\sigma_{s+1}^{-1} \cdots \sigma_{t-1}^{-1})$, where $1 \leq s < t \leq n$. We define $\sigma_{s,t} = \sigma_{t,s}$, and adopt the convention that whenever it is convenient to do so we will write the smaller subscript first.

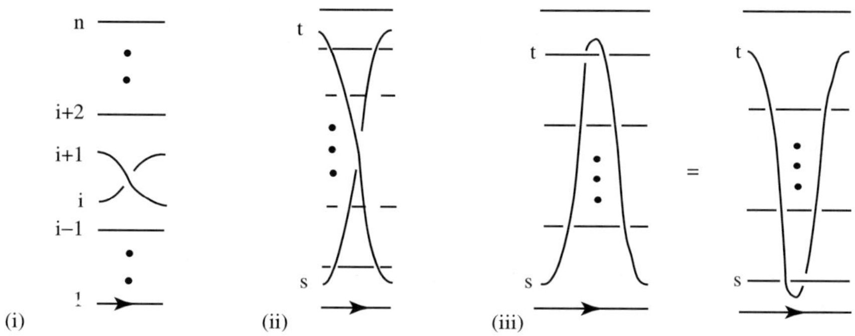

(i) (ii) (iii)

Fig. 2. (i) The elementary braid σ_i. (ii) The elementary braid $\sigma_{s,t}$. (iii) The pure braid $A_{s,t} = A_{s,t} = \sigma_{s,t}^2$.

The *new presentation* has generators $\{\sigma_{s,t},\ 1 \le s < t \le n\}$ and defining relations:

$$\begin{aligned} \sigma_{s,t}\sigma_{q,r} &= \sigma_{q,r}\sigma_{s,t} && \text{if } (t-r)(t-q)(s-r)(s-q) > 0, \\ \sigma_{s,t}\sigma_{r,s} &= \sigma_{r,t}\sigma_{s,t} = \sigma_{r,s}\sigma_{r,t} && \text{if } 1 \le r < s < t \le n. \end{aligned} \tag{1.3}$$

See sketch (ii) of Figure 2 for a picture of $\sigma_{s,t}$, and [19] for a proof that (1.2) and (1.3) define the same group. Both presentations will be needed in our work. Note that the new generators include the old ones as a proper subset, since $\sigma_i = \sigma_{i,i+1}$ for each $i = 1, 2, \ldots, n-1$.

By (1.1) the pure braid group $\mathbf{P}_n$ has index $n!$ in $\mathbf{B}_n$. Let $A_{s,t} = A_{t,s} = \sigma_{s,t}^2$. (See sketch (iii) of Figure 2.) The symmetry $A_{s,t} = A_{t,s}$ can be seen by tightening the Sth strand at the expense of loosening the tth strand. It is proved in [5] and also in [57] that $\mathbf{P}_n$ has a presentation with generators $A_{r,s}$, $1 \le r < s \le n$ and defining relations:

$$A_{r,s}^{-1}A_{i,j}A_{r,s} = \begin{cases} A_{i,j} & \text{if } 1 \le r < s < i < j \le n \\ & \text{or } 1 \le i < r < s < j \le n \\ A_{r,j}A_{i,j}A_{r,j}^{-1} & \text{if } 1 \le r < s = i < j \le n, \\ (A_{i,j}A_{s,j})A_{i,j}(A_{i,j}A_{s,j})^{-1} & \text{if } 1 \le r = i < s < j \le n, \\ (A_{r,j}, A_{s,j}A_{r,j}^{-1}A_{s,j}^{-1})A_{i,j}(A_{r,j}, A_{s,j}A_{r,j}^{-1}A_{s,j}^{-1})^{-1} & \text{if } 1 \le r < i < s < j \le n. \end{cases} \tag{1.4}$$

The relations in (1.4) come from the existence of a split short exact sequence, for every $k = 2, \ldots, n$:

$$\{1\} \to \mathbf{F}_{n-1} \to \mathbf{P}_n \xrightarrow{\pi_n^\star} \mathbf{P}_{n-1} \to \{1\}. \tag{1.5}$$

The map $\pi_n^\star$ is defined by pulling out the last braid strand, and the image of $\mathbf{P}_{n-1}$ under its inverse embeds $\mathbf{P}_{n-1}$ in $\mathbf{P}_n$, as the subgroup generated by pure braids on the first $n-1$ strands. The free subgroup F_{n-1} is generated by the braids $A_{1,n}, A_{2,n}, \ldots, A_{n-1,n}$. The pure braid group $\mathbf{P}_2$ is infinite cyclic and generated by $A_{1,2}$. Inducting on n, the structure of $\mathbf{P}_n$ via a sequence of semi-direct products is uncovered.

1.3. $\mathbf{B}_n$ *and* $\mathbf{P}_n$ *as mapping class groups*

Our announced goal in this review was to concentrate on areas where there have been new developments in recent years. While it has been known for a very long time that Artin's braid group is naturally isomorphic to the mapping class group of an n-times punctured disc, people have asked us many times for a simple proof of this fact. We do not know of a simple one in the literature, therefore we present one here. In this case "simple" does not mean intuitive and based upon first principles, rather it means using machinery which is normally available to a graduate student who has the tools learned in a first year graduate course in topology, and is preparing to begin research.

Let $S = S_{g,b,n}$ denote a 2-manifold of genus g with b boundary components and n punctures, and let $\mathrm{Diff}^+(S)$ denote the groups of all orientation preserving diffeomorphisms of S. Observe that we may assign the compact open topology to $\mathrm{Diff}^+(S)$, making it into a topological group. The *mapping class group* $\mathcal{M} = \mathcal{M}_{g,b,n}$ of S is $\pi_0(\mathrm{Diff}^+(S))$, that is, the quotient of $\mathrm{Diff}^+(S)$ modulo its subgroup of all diffeomorphisms of S which are isotopic to the identity rel ∂S. We allow diffeomorphisms in $\mathrm{Diff}^+(S)$ to permute the punctures, writing $\mathrm{Diff}^+(S_{g,b,\hat{n}})$ if they are to be fixed pointwise. Our interest in this article is mainly in the special case of $\mathcal{M}_{0,1,n}$.

THEOREM 1.1. *There are natural isomorphisms*:

$$\mathbf{B}_n \cong \mathcal{M}_{0,1,n} \text{ and } \mathbf{P}_n \cong \mathcal{M}_{0,1,\hat{n}}.$$

PROOF. We begin with an intuitive description of how to pass from diffeomorphisms to geometric braids and back again. Choose any $h \in \mathrm{Diff}^+(S_{0,1,n})$. While h is in general not isotopic to the identity, its image $i(h)$ in $\mathrm{Diff}^+(S_{0,1,0})$ under the inclusion map $i : \mathrm{Diff}^+(S_{0,1,n}) \to \mathrm{Diff}^+(S_{0,1,0})$ is, because $\mathrm{Diff}^+(S_{0,1,0}) = \{1\}$. Let h_t denote the isotopy. If the punctures in $S_{0,1,n}$ are at $(p_1, p_2, \ldots, p_n)$, then the n paths $(h_t(p_1), h_t(p_2), \ldots, h_t(p_n))$ defined by the traces of the points $(p_1, p_2, \ldots, p_n)$ under the isotopy sweep out a braid in $S_{0,1,0} \times [0, 1]$, and the equivalence class of this braid is the image of the mapping class $[h]$ in the braid group $\mathbf{B}_n$.

It is a little bit harder to understand the inverse isomorphism, from the braid group to the mapping class group. One chooses a geometric braid and imagines it as being located in a slice of 3-space, with the bottom endpoints of the n braid strands (which are oriented top to bottom) as being pinned to the distinguished points $p_1, \ldots, p_n$ on the punctured disc. If one is very careful the n braid strings can be laid down on the punctured disc so that they become n non-intersecting simple arcs, each of which begins and ends at a base point. One then constructs a homeomorphism of the punctured disc to itself in such a way that the trace of the isotopy to the identity is the given set of n non-intersecting simple arcs.

To prove the theorem, we begin by establishing the isomorphism between $\mathbf{P}_n$ and $\mathcal{M}_{0,1,\hat{n}}$. A good general reference for the underlying mathematics is Chapter 6 of the textbook [33]. As previously noted $\mathrm{Diff}^+(S_{0,1,0})$ is a topological group. Also, $\mathrm{Diff}^+(S_{0,1,\hat{n}})$ is a closed subgroup of $\mathrm{Diff}^+(S_{0,1,0})$. The *evaluation map* $\mathcal{E} : \mathrm{Diff}^+(S_{0,1,0}) \to \mathcal{C}_{0,\hat{n}}$ is defined by $\mathcal{E}(h) = (h(p_1), \ldots, h(p_n))$. It is clear that $\mathcal{E}$ is continuous with respect to the compact open topology on $\mathrm{Diff}^+(S_{0,1,\hat{n}})$ and the subspace topology for $\mathcal{C}_{0,\hat{n}} \subset \mathbb{C} \times \mathbb{C} \times \cdots \times \mathbb{C}$. The topological group $\mathrm{Diff}^+(S_{0,1,0})$ acts n-transitively on the disc in the sense that if $(p_1, \ldots, p_n)$ are n distinct points and $(w_1, \ldots, w_n)$ are n others then there is an $h \in \mathrm{Diff}^+(S_{0,1,0})$ such that $h(p_i) = w_i$, $i = 1, \ldots, n$. Observe that if $h \in \mathrm{Diff}^+(S_{0,1,\hat{n}})$ then $(h(p_1), \ldots, h(p_n)) = (p_1, \ldots, p_n)$ and if $h, h' \in \mathrm{Diff}^+(S_{0,1,0})$ with $\mathcal{E}(h) = \mathcal{E}(h')$, then h, h' are in the same left coset of $\mathrm{Diff}^+(S_{0,1,\hat{n}})$ in $\mathrm{Diff}^+(S_{0,1,0})$. In this situation it is shown in [109], part I, Sections 7.3 and 7.4, that the 3-tuple $(\mathcal{E}, \mathrm{Diff}^+(S_{0,1,0}), \mathcal{C}_{0,\hat{n}})$ is a fiber space, with total space $\mathrm{Diff}^+(S_{0,1,0})$ base space $\mathcal{C}_{0,\hat{n}}$, projection $\mathcal{E}$ and fiber $\mathrm{Diff}^+(S_{0,1,\hat{n}})$. (It is a good exercise for a graduate student to prove this directly by constructing the required local product structure in an explicit manner.) The long exact sequence of homotopy groups of a fibration then gives the following exact sequence of groups and homomorphisms, where we focus on the range that is of interest:

$$\cdots \to \pi_1(\mathrm{Diff}^+(S_{0,1,0})) \xrightarrow{\mathcal{E}_*} \pi_1(\mathcal{C}_{0,\hat{n}}) \xrightarrow{\partial_*} \pi_0(\mathrm{Diff}^+(S_{0,1,\hat{n}})) \xrightarrow{i_*} \pi_0(\mathrm{Diff}^+(S_{0,1,0})) \xrightarrow{\mathcal{E}_*} \cdots$$

The two end groups are trivial. The left middle group is $\mathbf{P}_n$ and the right middle group is $\mathcal{M}_{0,1,\hat{n}}$. The isomorphism of Theorem 1.1 is ∂_*. Tracing through the mathematics one finds that in fact its inverse is the map that we described right after we stated the theorem. The assertion about $\mathbf{P}_n$ is therefore true. The proof for $\mathbf{B}_n$ can then be completed by comparing the short exact sequence (1.1), which says that $\mathbf{B}_n$ is a finite extension of $\mathbf{P}_n$ with quotient the symmetric group with a related short exact sequence for the mapping class groups:

$$1 \to \mathcal{M}_{0,1,\hat{n}} \xrightarrow{j_*} \mathcal{M}_{0,1,n} \to \Sigma_n \to 1. \tag{1.6}$$

Thus, $\mathcal{M}_{0,1,n}$ is a finite extension of $\mathcal{M}_{0,1,\hat{n}}$ with quotient the symmetric group Σ_n. Comparing corresponding groups in the short exact sequences (1.1) and (1.6), we see that the first two and last two are isomorphic. The 5-Lemma then shows that the middle ones are too. This completes the proof of Theorem 1.1. □

1.4. *Some examples where braiding appears in mathematics, unexpectedly*

We discuss, briefly, a variety of examples, outside of knot theory, where "braiding" is an essential aspect of a mathematical or physical problem.

1.4.1. *Algebraic geometry.* Configuration spaces and the braid group appear in a natural way in algebraic geometry. Consider the complex polynomial

$$(X - z_1)(X - z_2)\ldots(X - z_n) = X^n + a_1X^{n-1} + \cdots + a_{n-1}X + a_n.$$

of degree n with n distinct complex roots $z_1, \ldots, z_n$. The coefficients $a_1, \ldots, a_n$ are the elementary symmetric polynomials in $\{z_1, \ldots, z_n\}$, and so we get a continuous map $\mathbb{C}^n \to \mathbb{C}^n$ which takes roots to coefficients. Two points have the same image if and only if they differ by a permutation, so we get the same identification as in the quotient map $\tau : \mathcal{C}_{0,\hat{n}} \to \mathcal{C}_{0,n}$, in quite a different way. Since we are requiring that our polynomial have n distinct roots, a point $\{a_1, \ldots, a_n\}$ is in the image of $\vec{z}$ under the root-to-coefficient map if and only if the polynomial $X^n + a_1X^{n-1} + \cdots + a_n$ has n distinct roots, i.e., if and only if its coefficients avoid the points where the discriminant

$$\Delta = \prod_{i<j} (z_i - z_j)^2,$$

expressed as a polynomial in $\{a_1, \ldots, a_n\}$, vanishes. Thus, $\mathcal{C}_{0,n}(\mathbb{C})$ can be interpreted as the complement in $\mathbb{C}^n$ of the algebraic hypersurface defined by the equation $\Delta = 0$, where Δ is rewritten as a polynomial in the coefficients $a_1, \ldots, a_n$. (For example, the polynomial $X^2 + a_1X + a_2$ has distinct roots precisely when $a_1^2 - 4a_2 = 0$). In this setting the base point $\tau(\vec{p})$ is regarded as the choice of a complex polynomial of degree n which has n distinct roots, and an element in the braid group is a choice of a continuous deformation of that polynomial along a path on which two roots never coincide. There is a substantial literature in this area, from which we mention only one paper, by Gorin and Lin [73]. We chose it because it contains a description of the commutator subgroup $\mathbf{B}_n'$ of the braid group, and many people have asked the first author for a reference on that over the years. While there may be other references they are unknown to us.

1.4.2. *Operator algebras.* Our next example, taken from the work of Jones [76,77], is interesting because it shows how "braiding" can appear in disguise, so that initially one misses the connection. We consider the theory, in operator algebras, of "type II_1 factors", ordered by inclusion. Let M denote a Von Neumann algebra, i.e., an algebra of bounded operators acting on a Hilbert space h. The algebra M is called a *factor* if its center consists only of scalar multiples of the identity. The factor is type II_1 if it admits a linear functional, called a trace, $\mathrm{tr} : M \to \mathbb{C}$, which satisfies the following three conditions: (i) $\mathrm{tr}(xy) = \mathrm{tr}(yx) \forall x, y \in M$, (ii) $\mathrm{tr}(1) = 1$, and $\mathrm{tr}(xx^*) > 0$, where x^* is the adjoint of x. In this situation it is known that the trace is unique, in the sense that it is the only linear functional satisfying the first two conditions. An old discovery of Murray and Von Neumann was that factors of type II_1 provide a type of "scale" by which one can measure the dimension of h. The notion of dimension which occurs here generalizes the familiar notion of integer-valued dimensions, because for appropriate M and h it can be any non-negative real number or ∞.

The starting point of Jones' work was the following question: if M_1 is a type II_1 factor and if $M_0 \subset M_1$ is a subfactor, is there any restriction on the real numbers which occur as the ratio $\lambda = \dim_{M_0}(h)/\dim_{M_1}(h)$? The question has the flavor of questions one studies in Galois theory. On the face of it, there was no reason to think that λ could not take on any value in $[1,\infty]$, so Jones' answer came as a complete surprise. He called λ the *index* $|M_1 : M_0|$ of M_0 in M_1, and proved a type of rigidity theorem about type II_1 factors and their subfactors:

THE JONES INDEX THEOREM. $\lambda \subset [4, \infty] \cup \{4 \cos 2\pi/p\}$, *where* $p \in \mathbb{Z}$, $p \geq 3$. *Moreover, each real number in the continuous part of the spectrum* $[4,\infty]$ *and in the discrete part* $\{\{4 \cos 2\pi/p\}, p \in \mathbb{Z}, p \geq 3\}$ *is realized.*

What does all this have to do with braids? To answer the question, we sketch the idea of the proof, which is to be found in [76]. Jones begins with the type II_1 factor M_1 and the subfactor M_0. There is also a tiny bit of additional structure: It turns out that in this setting there exists a map $e_1 : M_1 \longrightarrow M_0$, known as the *conditional expectation* of M_1 on M_0. The map e_1 is a projection, i.e., $e_1^2 = e_1$. His first step is to prove that the ratio λ is independent of the choice of the Hilbert space h. This allows him to choose an appropriate h so that the algebra M_2 generated by M_1 and e_1 makes sense. He then investigates M_2 and proves that it is another type II_1 factor, which contains M_1 as a subfactor, moreover $|M_2 : M_1| = |M_1 : M_0| = \lambda$. Having in hand another II_1 factor M_2 and its subfactor M_1, there is also a trace on M_2 (which by the uniqueness of the trace) coincides with the trace on M_1 when it is restricted to M_1, and another conditional expectation $e_2 : M_2 \longrightarrow M_1$. This allows Jones to iterate the construction, to build algebras $M_1, M_2, \ldots$ and from them a family of algebras $\{J_n,\ n = 1, 2, 3, \ldots\}$, where J_n is generated by $1,\ e_1, \ldots, e_{n-1}$.

Rewriting history a little bit in order to make the subsequent connection with braids a little more transparent, we now replace the projections e_k, which are not units, by a new set of generators which are units, defining: $g_k = te_k - (1 - e_k)$, where $(1 - t)(1 - t^{-1}) = 1/\lambda$. The g_k's generate J_n because the e_k's do, and we can solve for the e_k's in terms of the g_k's. So $J_n = J_n(t)$ is generated by $1,\ g_1, \ldots, g_{n-1}$ and we have a tower of algebras, $J_1(t) \subset J_2(t) \subset \cdots$, ordered by inclusion. The parameter t, which replaces the index λ, is the quantity now under investigation. It is woven into the construction of the tower. The algebra $J_n(t)$ has defining relations:

$$g_i g_k = g_k g_i \quad \text{if } |i - k| \geq 2,\ \ g_i g_{i+1} g_i = g_{i+1} g_i g_{i+1},\ \ g_i^2 = (t - 1) g_i + t,$$
$$1 + g_i + g_{i+1} + g_i g_{i+1} + g_{i+1} g_i + g_i g_{i+1} g_i = 0. \tag{1.7}$$

Of course there are braids lurking in the background. If we rename the g_i's, replacing g_i by σ_i, and declare the σ_i's to be generators of a group, then the first two relations are defining relations in the group algebra $\mathbb{C}\mathbf{B}_n$. The algebra $J_n(t)$ is thus a homomorphic image of the group algebra of the braid group. Loosely speaking, braids are encountered in Operator Algebras because they encode the way in which each type II_1 factor M_i acts on its subfactor M_{i-1}. Braiding is thus involved in defining the associated extensions. We shall see later, in Section 4.7 that a similar action, via group extensions, can be used to define representations of $\mathbf{B}_n$.

To see the connection with knots and links, recall that since M_n is type II_1 it supports a unique trace, and since J_n is a subalgebra it does too, by restriction. This trace is known as a *Markov trace*, i.e., it satisfies the important property:

$$\mathrm{tr}(wg_n) = f(t)\mathrm{tr}(w) \quad \text{if } w \in J_n, \tag{1.8}$$

where $f(t)$ is a fixed function of t. Thus, for each fixed value of f the trace is multiplied by a fixed scalar when one passes from one stage of the tower to the next, if one does so by multiplying an arbitrary element of J_n by the new generator g_n of J_{n+1}. The Jones trace is nothing more or less than the 1-variable Jones polynomial [78] associated to the knot or link which is obtained from the closed braid. We will have more to say about all this in Section 4.3.

1.4.3. *Homotopy groups of spheres.* As before, let $\mathbf{P}_{n+1}$ denote the pure braid group on $n+1$ strands. For each $i = 1, \ldots, n+1$ there is a natural homomorphism $p_i : \mathbf{P}_{n+1} \to \mathbf{P}_n$, defined by pulling out the ith strand. The group of *Brunnian* braids is

$$\mathrm{BR}_{n+1} = \cap_{i=1}^{i=n+1} \mathrm{kernel}(p_i),$$

i.e., a braid is in BR_{n+1} if and only if, on pulling out any strand, it becomes the identity braid on n strands. Brunnian braids have received some attention in knot theory.

Braids have played a role in homotopy theory for many years, most particularly in the work of F. Cohen and his students (see for example [9]), but during the past few years the connection was sharpened when it was discovered that there is an embedding of a free group F_n in P_{n+1} with the property that a well-defined quotient of $\mathrm{BR}_{n+1} \cap F_n$ (a little bit too complicated to describe here) is isomorphic to $\pi_{n+1}S^2$. It remains to be seen whether new knowledge about the unidentified higher homotopy groups of spheres can be obtained through the methods of Berrick et al. [9].

1.4.4. *Robotics.* Our fourth example is an application of configuration spaces to robotics. It shows the braid group popping up in an unexpected way (until you realize how natural it is). Robots, or AGVs (automatic guided vehicles), are required to travel across a factory floor that contains many obstacles, en route to a goal position (e.g., a loading dock or an assembly workstation). The problem is to design a control system which insures that the AGVs not collide with the obstacles, or with each other, and complete the task with efficiency with regard to various work functionals. Here is how configuration spaces appear: The underlying space in this simple example is the workspace floor X, from which a finite set $\mathcal{O}$ of obstacles are to be removed. The configuration space of n non-colliding AGVs is then precisely $\mathcal{C}_{0,n}(X - \mathcal{O})$. More generally, $X - \mathcal{O}$ is replaced by a finite graph Y, and the braid group $\mathbf{B}_n$ by the braid group $\pi_1(\mathcal{C}_{0,n}(Y))$ of the graph. There is a vast literature on this subject; we suggest [69] by Ghrist, as a starter.

1.4.5. *Public key cryptography.* In this example braids are important for rather different reasons than they were in our earlier examples. In our earlier examples the underlying phenomenon which was being investigated involved actual braiding, albeit sometimes in a concealed way. In the example that we now describe particular properties of the braid groups $\mathbf{B}_n$, $n = 1, 2, 3, \ldots$, rather than the actual interweaving of braid strands, are used in a clever way to construct a new method for encrypting data.

The problem which is the focus of "public key cryptography" will be familiar to everyone: the security of our online communications, for example our credit card purchases, our ATM transactions, our cell phone conversations and a host of other transactions that have become a part of everyday life in the 21st century. The basic problem is to encrypt or translate a secret message into a code that can be sent safely over a public system such as the internet, and decoded at the receiving end by the use of a secret piece of information known only to the sender and the recipient, the "key". The problem that must then be solved is to establish a private key that will be known only to the sender and the recipient, who will then be able to exchange information over an insecure channel. In recent years much work has been done on certain codes which are based upon the assumption that the word problem has polynomial growth as braid index n is increased, whereas the conjugacy problem does not. But in Section 5 we will review recent work on the word and conjugacy problems in the braid groups, and show that such an assumption seems problematic at best. See Section 5, and in particular the discussion in Section 5.5.

2. From knots to braids

In this chapter we will explore, for the benefit of readers who are new to the subject, the foundations of the close relationship between knots and braids. We will first describe the straightforward process of obtaining a knot or a link from a given braid by "closing" the braid. This leads us directly to formulate two fundamental questions about knots and braids. First, is it always possible to transform a given knot into a closed braid? This question will be answered in the affirmative in Theorem 2.1, first proved by Alexander in 1928 [2]. The correspondence between knots and braids is clearly not one-to-one (for example, conjugate braids yield equivalent knots), leading naturally to the second question: which closed braids represent the same knot type? That question is addressed in Theorem 2.8, first formulated by A. Markov [91], which gives "moves" relating any two closed braid representatives of a knot or link, while simultaneously preserving the closed braid structure.

Together, Theorems 2.1 and 2.8 form the cornerstone of any study of knots via closed braids, so we feel obliged to prove them. Among the many proofs that have been published of both over the years, we have chosen ones that we like but which do not seem to have appeared in any of the review articles that we know. The proof that we give of Alexander's Theorem is due to Shuji Yamada [117], with subsequent improvements by Vogel [116]. The algorithm is elementary enough to be accessible to a beginner, and has the advantage for experts of being suitable for programming. The proof that we present of Markov's theorem is due to Traczyk [111]. It is relatively brief, as it assumes Reidemeister's well-known theorem about the equivalence relation on any two diagrams of a knot, Theorem 2.7, building on methods introduced in the proof of Theorem 2.1.

2.1. *Closed braids*

For simplicity, let us begin with a planar diagram of a given geometric braid. To obtain a knot or link, one simply "closes up" the ends of the braid as in Figure 3. The pre-image in $\mathbb{R}^3$ of the "center point" shown in Figure 3 under the usual projection map is called the *axis* of the braid. (If one wishes to consider the knot in S^3, then we include the point at infinity so that the braid axis is an embedded S^1.) We then orient the resulting knot or link in such a way that the strands of the braid are all travelling counterclockwise about the braid axis.

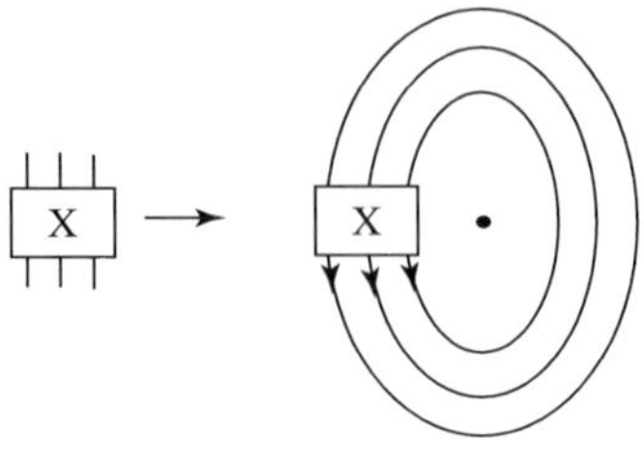

Fig. 3. The operation of closing a braid X to form a closed braid.

The knot or link type resulting from performing this operation on a braid X is known as the *closure* of X and will be denoted by $b(X)$. The same notation may also refer to the particular diagram as in Figure 3.

Equivalently, consider a knot $K \subset S^3$. Suppose there exists $A = h(S^1)$ where h is an embedding and Z is unknotted in S^3 and contained in the complement of K. Suppose further that we choose the point at infinity $\{\infty\}$ to be in A and, using standard cylindrical coordinates (ρ, θ, z) on $\mathbb{R}^3$, identify the resulting copy of $\mathbb{R} \cong A - \{\infty\}$ with the z-axis in $\mathbb{R}^3 \cong S^3 - \{\infty\}$. If we always have $\mathrm{d}\theta/\mathrm{d}t > 0$ as we travel about the knot K with an appropriate cylindrical parametrization, then we say that K is a *closed braid* with respect to the axis A. The closed braid diagram of Figure 3 is then obtained by projection parallel to the direction defined by A onto a plane that is orthogonal to A.

2.2. *Alexander's Theorem*

As we just observed, it is a simple matter to obtain a knot or link from a braid. The classical theorem of J. Alexander allows us to reverse this process, though not in a unique way:

THEOREM 2.1 (Alexander's Theorem [2]). *Every knot or link in S^3 can be represented as a closed braid.*

PROOF. Alexander's original proof was algorithmic, i.e., it gave an algorithm for transforming a knot or link into closed braid form. While it is straightforward, we do not know of any computer program based upon it. We shall give instead a rather different and newer algorithm originally due to Yamada [117], as later improved by Vogel [116]. We like it for two reasons: (1) It has a beautiful corollary (see Corollary 2.5) which reveals structure about knot diagrams that had not even been conjectured by any of the experts before 1987, even though there was abundant evidence of its truth; (2) It leads, very easily, to an efficient computer program for putting knots into braid form. In this regard we note that when Jones was writing the manuscript [77], which resulted in his award of the Fields medal, he computed closed braid representatives for the 249 knots of crossing number less than or equal to 10, constructing the first table known to us of closed braid representatives of knots. His list remains extremely useful to the workers in the area in 2004. Yamada's work was not yet known when he did that work, and there did not seem to him to be a good way to program the Alexander method for a computer, so he calculated them one at a time by hand. The amount of work that was involved can only be appreciated by the reader who is willing to try a few examples.

In order to prove Alexander's Theorem, we shall first present the Yamada–Vogel algorithm for transforming a knot into closed braid form in full, followed immediately by an illustrative example (Example 2.2). We shall then prove Alexander's Theorem by showing that it is always possible to perform the steps of the Yamada–Vogel algorithm for any given knot or link and that the algorithm always leads to a closed braid.

The Yamada–Vogel algorithm draws on Seifert's well-known algorithm for using a diagram of an oriented knot or link K to construct a Seifert surface for K (see [106] or [88], e.g., for a thorough treatment of Seifert's algorithm), and we will need some related terminology. Let C and C' be two oriented disjoint simple closed curves in S^2. Then C and C' cobound an annulus A. We say that C and C' are *coherent* (or *coherently oriented*) if C and C' represent the same element of $H_1(A)$. Otherwise we say that C and C' are *incoherent*. Following Traczyk [111], we define the *height* of a knot diagram D, denoted $h(D)$, to be the number of distinct pairs of incoherently oriented Seifert circles which arise from applying Seifert's algorithm to D. The height function gives us a useful characterization of a closed braid: *a diagram D represents a closed braid if and only if* $h(D) = 0$. (Recall that D lives in S^2.) □

THE YAMADA–VOGEL ALGORITHM.

(1) Let D be a diagram of an oriented knot K. Smooth all crossings of D as in Seifert's algorithm to obtain n Seifert circles $C_1, \ldots, C_n$. Record each original crossing with a signed arc: $(+)$ for a positive crossing (often called a *right-handed* crossing), $(-)$ for a negative (or *left-handed*) crossing (see Figure 2(iv)). The resulting diagram is the *Seifert picture* S corresponding to the diagram D. Note that any two circles joined by a signed arc in any Seifert picture are necessarily coherent. For an example that illustrates the construction of a Seifert diagram, see the passage from the bottom left to the top left sketches in Figure 4.

(2) If $h(D) = 0$, the knot K is already in closed braid form, and we are done. If $h(D) > 0$, we can find a *reducing arc* α, i.e., an arc joining an incoherent pair C_i, C_j such that α intersects S only at its endpoints. Reducing arcs are illustrated as heavy black arcs in the example in Figure 4. A component of $S^2 \setminus S$ which admits a reducing arc is called a *defect region*. Perform a *reducing move* along α, as shown in Figure 5, to obtain a new Seifert picture S' in which a pair of coherent Seifert circles, C_a and C_z, joined by two oppositely signed arcs, replaces the incoherent pair C_i, C_j. The corresponding move on the original diagram D is a Reidemeister move of type II in which we slide C_i over C_j in a small neighborhood of the arc α to obtain a new diagram D' with two new crossings. Note that if we instead slide C_i under C_j, we obtain the same two new Seifert circles but the signs of the two new signed arcs are now switched.

(3) Continue performing reducing moves on incoherent pairs until a diagram with height zero is obtained.

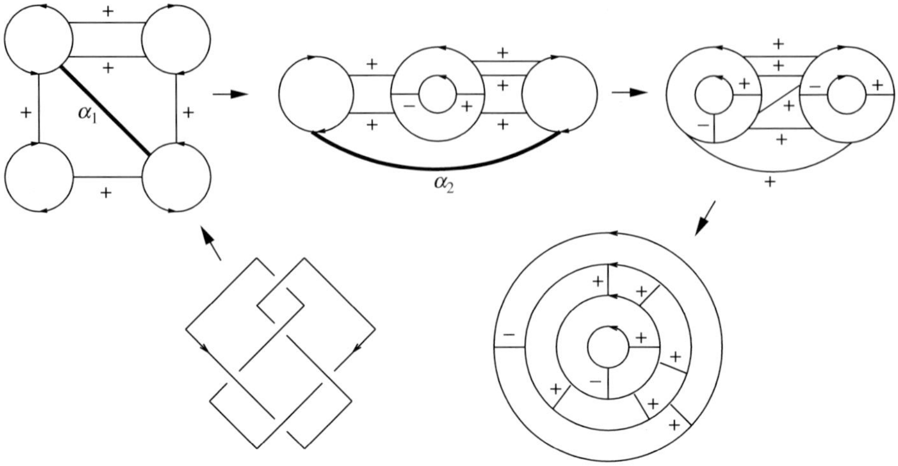

Fig. 4. The Yamada–Vogel algorithm performed on the knot 5_2.

EXAMPLE 2.2. We apply the Yamada–Vogel algorithm to the diagram of the knot 5_2, pictured in Figure 4. (The reader can check that this is the first example in the knot tables of a knot diagram with height greater than zero.) The first stage shows the Seifert picture associated to the original diagram, consisting of four Seifert circles, with five signed arcs (all positive) recording the original crossings. We see that the original knot diagram has height 2. The figure shows a choice of reducing arc, α_1, joining one of the two pairs of incoherent circles.

In the third sketch of Figure 4, we see the new Seifert picture resulting from performing the reducing move along α_1. Note that we have introduced two new crossings of opposite sign. We also see a new reducing arc, α_2, joining the only remaining pair of incoherent circles. We see in the fourth sketch the Seifert picture with height zero resulting from the second reducing move performed along α_2. At this point, we are done, but in the final sketch we see a different planar projection of the same Seifert picture which allows us easily to read off a braid word associated to the knot: beginning with the positive signed arc in the "twelve o'clock" position and reading counterclockwise, we see that the knot 5_2 is equivalent to $b(X)$, where $X = \sigma_2\sigma_1^{-1}\sigma_2\sigma_3^{-1}\sigma_2\sigma_1\sigma_2\sigma_3\sigma_2$.

We may learn several things from this simple example. Applying the braid relations to the word defined by X, we see that $X = \sigma_2\sigma_1^{-1}\sigma_2\sigma_3^{-1}\sigma_2\sigma_1\sigma_2\sigma_3\sigma_2 = \sigma_2\sigma_1^{-1}\sigma_2\sigma_3^{-1}\sigma_1\sigma_2\sigma_1\sigma_3\sigma_2 = \sigma_2\sigma_1^{-1}\sigma_2\sigma_1\sigma_3^{-1}\sigma_2\sigma_3\sigma_1\sigma_2 = \sigma_2\sigma_1^{-1}\sigma_2\sigma_1\sigma_2\sigma_3\sigma_2^{-1}\sigma_1\sigma_2$. Since this braid only involves σ_3 once, we may delete a trivial loop to get the 8-crossing 3-braid

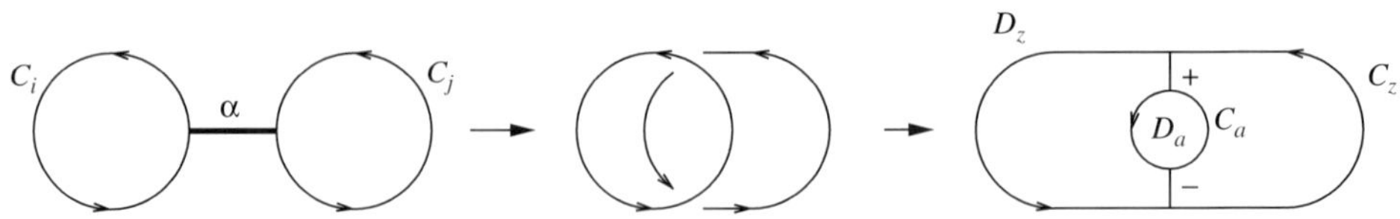

Fig. 5. The local picture of a reducing move.

$\sigma_2\sigma_1^{-1}\sigma_2\sigma_1\sigma_2\sigma_2^{-1}\sigma_1\sigma_2$, so the algorithm did not give us minimum braid index. The algorithm also does not give shortest words because our 8-crossing braid may be shortened to the 6-crossing 3-braid $\sigma_2\sigma_1^{-1}\sigma_2\sigma_1^2\sigma_2$. In fact, 6 is minimal, because the crossing number of a 3-braid knot must be even, and this knot has no diagram with fewer than 5 crossings, so 6 is minimal. So we may deduce one more fact: when we use the Yamada–Vogel algorithm to change a knot which is not in closed braid form to one which is, the crossing number goes up.

To prove Alexander's Theorem, we first need to show that a reducing move strictly decreases the height of a diagram. This lemma is sometimes stated as "obvious" in the literature, but the question arises frequently enough to warrant a short but thorough argument.

LEMMA 2.3. *Suppose a reducing move is performed which transforms a diagram D to a diagram D'. Then $h(D') = h(D) - 1$.*

PROOF. Let $C_1, \ldots, C_n$ be the Seifert circles in the Seifert picture S corresponding to the diagram D. Let C_i, C_j be an incoherent pair. The union $C_i \cup C_j$ separates the 2-sphere into three components: an annulus A cobounded by C_i and C_j, and two disks, $\mathcal{D}_i$ and $\mathcal{D}_j$ bounded by C_i and C_j, respectively. Suppose that A admits a reducing arc α. A reducing move along α preserves circles C_p, $p \neq i,j$ and replaces C_i and C_j with two new circles, one of which necessarily bounds a disk $\mathcal{D}_a$ containing no other Seifert circles in the new Seifert picture S'. We denote this circle by C_a (see Figure 5). The other new circle, denoted C_z, bounds a disk $\mathcal{D}_z$ containing all Seifert circles originally contained in the annulus A in S.

To simplify the bookkeeping, we shall write $(C_r, C_s) = 1$ if the pair C_r, C_s is coherent, or else $(C_r, C_s) = -1$ if C_r, C_s are incoherent. Obviously, if $\{p,q\} \cap \{i,j\} = \emptyset$, then (C_p, C_q) is unchanged by the reducing move, so we need only consider the effect of the reducing move on (C_p, C_x), where $x = i$ or $x = j$ and $p \neq i,j$. Now if C_p is contained in the annulus A in S (and hence in $\mathcal{D}_z$ in S'), then clearly $(C_p, C_z) = (C_p, C_a) = (C_p, C_i) = (C_p, C_j)$. Also, if $C_p \subset \mathcal{D}_i$ in S, then $(C_p, C_z) = (C_p, C_i)$ and $(C_p, C_a) = (C_p, C_j)$. Similarly, if $C_p \subset \mathcal{D}_j$ in S, then $(C_p, C_z) = (C_p, C_j)$ and $(C_p, C_a) = (C_p, C_i)$. Therefore, the number of distinct incoherent pairs C_r, C_s in S with $\{r,s\} \neq \{i,j\}$ is equal to the total number of distinct incoherent pairs in S' of the form C_r, C_s with $\{r,s\} \neq \{a,z\}$. By construction, however, we have replaced $(C_i, C_j) = -1$ with $(C_z, C_a) = 1$. Thus, $h(D') = h(D) - 1$. □

The previous lemma tells us that the Yamada–Vogel algorithm will always lead to a diagram of height zero, i.e., a closed braid, as long as it is always possible to perform Step 3. Therefore, the following lemma, whose proof was suggested to us by Bryant Adams, will conclude the proof of Alexander's Theorem.

LEMMA 2.4 [117]. *Let D be a knot or link diagram. If $h(D) > 0$, then the Seifert picture S associated to D contains a defect region.*

PROOF. Each component $\mathcal{R}$ of $S^2 \setminus S$ is a surface of genus 0 with $k \geq 1$ boundary components. Each boundary component of $\mathcal{R}$ is a union of some number of signed arcs

(possibly zero) and subarcs of Seifert circles. We call the collection of Seifert circles in S which form part or all of a boundary component of $\mathcal{R}$ the *exposed circles* of $\mathcal{R}$.

Let us now examine the possible ways in which $\mathcal{R}$ could fail to be a defect region. Certainly, if $\mathcal{R}$ has one exposed circle, then it is not a defect region. If $\mathcal{R}$ has two exposed circles, then $\mathcal{R}$ is either a disk or an annulus. In this case, if $\mathcal{R}$ is a disk, then its two exposed circles are joined by at least one signed arc; hence the two circles are coherent and $\mathcal{R}$ is not a defect region. If $\mathcal{R}$ is an annulus, either its two exposed circles are incoherent, in which case it is a defect region, or else they are coherent and $\mathcal{R}$ is not a defect region. If $\mathcal{R}$ has three or more exposed circles, then there is necessarily one incoherent pair among them, and $\mathcal{R}$ is a defect region.

Suppose that no component of $S^2 \setminus S$ is a defect region and hence that each component is of one of the three types of non-defect regions described above. It is clear that we must have at least one region of the second type, since otherwise $h(D)$ is clearly zero. We can think of such a region as lying between two nested, coherent circles joined by at least one signed arc.

Let us now start with such a region, and try to build a diagram with no defect regions. We cannot add any circles in the annulus cobounded by the two nested circles, since this necessarily gives rise to at least one component with three or more exposed circles. In fact, our only option is to add coherent circles which nest with the original two circles (and as many signed arcs between adjacent pairs as we like). However, such a diagram has height zero. Therefore, $h(D) > 0$ implies that a defect region exists. This completes the proof of Lemma 2.3, and so also of Theorem 2.1. □

The *braid index* of a knot or link K is the minimum number n such that there exists a braid $X \in \mathbf{B}_n$ whose closure $b(X)$ represents K. (We note that it is also common to refer to the *index* of a braid or a closed braid, meaning simply the number of its strands or the number of times it travels around its axis, respectively.) It is clear that the minimum number of Seifert circles in any diagram of a knot or link K is bounded above by the braid index of K. It is equally clear from the Yamada–Vogel algorithm that the reverse inequality holds. Thus, we obtain the following corollary, which is due to Yamada [117]. It seems remarkable that it was not noticed long before 1987.

COROLLARY 2.5 [117]. *The minimum number of Seifert circles in any diagram of a knot or link K is equal to the braid index of K.*

It also follows that we have a measure of the complexity of the process of transforming a knot into closed braid form as follows.

COROLLARY 2.6 ([111,116]). *Let N denote the length of any sequence of reducing moves required to transform a diagram D into closed braid form. Then we have*

$$N = h(D) \leq \frac{(n-1)(n-2)}{2} \tag{2.1}$$

where n is the number of Seifert circles associated to D.

Open Problem 1. It is an open problem to determine, among all regular diagrams for a given knot or link, the minimum number of Seifert circles that are needed. By Corollary 2.5 this is the same as the minimum braid index, among all closed braid representatives of a given knot or link. We know of only one general result relating to this problem, namely the Morton–Franks–Williams inequality of [98] and [65]. It will be discussed briefly in Section 4.3. The literature also contains an assorted collection of ad hoc techniques for determining the braid index of individual knots. For example, see the methods used in [26] to prove that the 6-braid template in Figure 20 actually has braid index 6, which rests on the fact that non-trivial braid-preserving flypes always have braid index at least 3.

2.3. *Markov's Theorem*

To introduce the main goal of this section, we begin by recalling for the reader Reidemeister's theorem, which dates from the earliest days of knot theory. It was assumed and used (as a folk theorem) long before anybody wrote down a formal statement and proof.

Theorem 2.7 (Reidemeister's Theorem). *Let D, D' be any two (in general not closed braid) diagrams of the same knot or link K. Then there exists a sequence of diagrams $D = D_1 \to D_2 \to \cdots \to D_k = D'$ such that any D_{i+1} in the sequence is obtained from D_i by one of the three Reidemeister moves, depicted in Figure* 6.

Proof. We refer the reader to [38] for a complete proof. □

Alexander's Theorem, proved in the last section, guarantees us that closed braid representatives of a knot exist, but as previously noted, they are certainly not unique. Markov's Theorem, first stated in [91] with a sketch of a proof, gives us a certain amount of control over different closed braid representatives of the same knot. It asserts that any two are related by a finite sequence of elementary moves and serves as the analogue for closed braids of the Reidemeister Theorem for knots.

One of the moves of the Markov Theorem is *braid isotopy*. From the point of view of a topologist, braid isotopy means isotopy of the closed braid, through braids, in the complement of the braid axis. Morton has proved that if two braids have closures that are braid isotopic, then they are conjugate in $\mathbf{B}_n$ [95]. The other two moves that we need are mutually inverse, and are illustrated in Figure 7 as a move on certain $(w + 2)$-braids. We call them *destabilization* and *stabilization*, where the former decreases braid index by one and the latter increases it by one. The *weight* w that is attached to one of the braid strands in

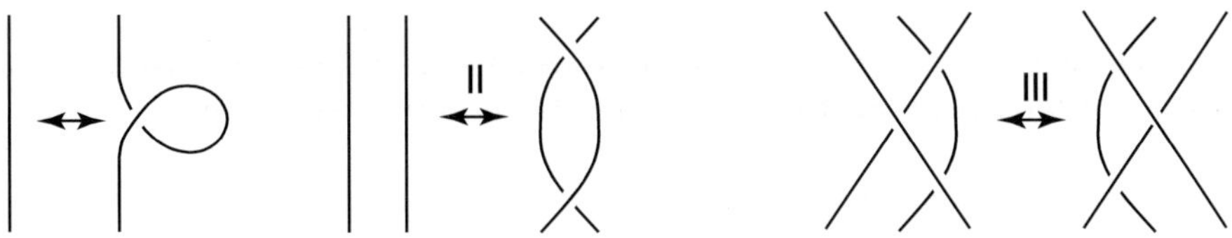

Fig. 6. The three Reidemeister moves. The 1, 2 or 3 strands in the left sketch of each have arbitrary orientations, also we give only one of the possible choices for the signs of the crossings, for each move.

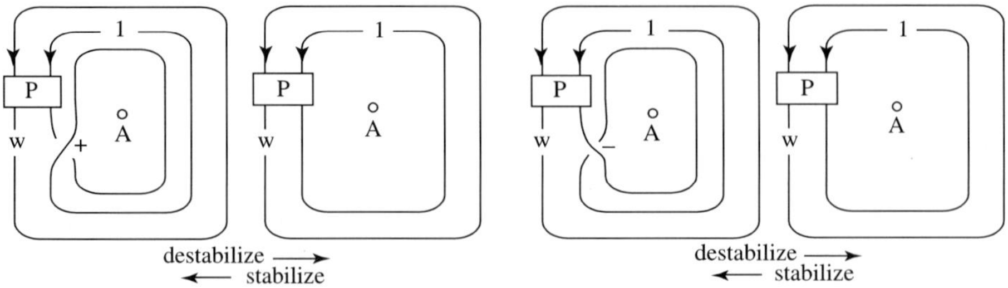

Fig. 7. The destabilization and stabilization moves.

Figure 7 denotes that many "parallel" strands, where parallel means in the framing defined by the given projection. The braid inside the box which is labeled P is an arbitrary $(w+1)$-braid. Later, it will be necessary to distinguish between positive and negative destabilizations, so we illustrate both now.

THEOREM 2.8 (Markov's Theorem). *Let X, X' be closed braid representatives of the same oriented link type K in oriented 3-space. Then there exists a sequence of closed braid representatives of K:*

$$X = X_1 \rightarrow X_2 \rightarrow \cdots \rightarrow X_r = X'$$

taking such that each X_{i+1} is obtained from X_i by either (i) *braid isotopy, or* (ii) *a single stabilization or destabilization.*

We call the moves of Theorem 2.8 *Markov moves*, and say that closed braids that are related by a sequence of Markov moves are *Markov equivalent*.

Forty years after Markov's theorem was announced, the first detailed proof was published in [15]. At least five essentially different proofs exist today. See for example [97], in which Morton gives his beautiful threading construction for knots and braids which also yields an alternate proof of Alexander's Theorem. Here we shall present a proof due to Traczyk [111]. It begins with Reidemeister's theorem, and uses the circle of ideas that were described in the previous section, and so it is particularly appropriate for us.

PROOF. We are given closed braids X, X' which represent the same oriented knot type K. Without loss of generality we may assume that X and X' are defined by closed braid diagrams Y, Y' of height $h(Y) = h(Y') = 0$. By Theorem 2.7 we know there is a sequence of knot diagrams $Y = Y_1 \rightarrow Y_2 \rightarrow \cdots \rightarrow Y_k = Y'$, where in general $h(Y_i) \geq 0$ for $i = 2, \ldots, k-1$, such that any two diagrams in the sequence are related by a single Reidemeister move of type I, II or III. The first step in Traczyk's proof is to reduce the proof to sequences of knot diagrams which are related by Yamada–Vogel reducing moves. □

LEMMA 2.9. *It suffices to prove Theorem 2.8 for closed braid diagrams Y, Y' which are related by sequences $Y = Y_1 \rightarrow Y_2 \rightarrow \cdots \rightarrow Y_q = Y'$ with the properties* (i) $h(Y) = h(Y') = 0$, (ii) $h(Y_i) > 0$ *for $i = 2, \ldots, q-1$, and* (iii) Y_{i+1} *is obtained from Y_i by a single Yamada–Vogel reducing move or the inverse of a reducing move.*

PROOF. We may always assume that the diagrams $Y_2, \ldots, Y_{q-1}$ have height > 0, for if not we simply replace the given sequences by the subsequences joining any two intermediate diagrams of height zero.

We say that a Reidemeister move is *braid-like* if the strands that are involved in it are locally oriented in a coherent fashion, as they would be if the diagram is a closed braid. In particular, any Reidemeister move of type I is braid like. To begin the proof of the lemma, we establish a somewhat weaker result: we claim that we can get from Y to Y' via a finite sequence of the following four types of moves and their inverses:

- a braid-like Reidemeister move of type I, denoted type I^{b},
- a braid-like Reidemeister move of type II, denoted type II^{b},
- a braid-like Reidemeister move of type III, denoted type $\mathrm{III}^{\mathrm{b}}$,
- a Yamada–Vogel reducing move, denoted type $\mathcal{Y}$.

To prove the claim, it is enough to show that non-braid-like Reidemeister moves, which we denote by the symbols I^{nb}, $\mathrm{II}^{\mathrm{nb}}$ and $\mathrm{III}^{\mathrm{nb}}$, can be achieved via a finite sequence of moves of type I^{b}, II^{b}, $\mathrm{III}^{\mathrm{b}}$ and $\mathcal{Y}$. To prove this, we examine the cases I^{nb}, $\mathrm{III}^{\mathrm{nb}}$ and $\mathrm{II}^{\mathrm{nb}}$ in that order:

(1) As previously noted, any type I Reidemeister move is of type I^{b}.

(2) A type $\mathrm{III}^{\mathrm{nb}}$ Reidemeister move involves three arcs of the knot or braid. There are many different cases, depending on the local orientations and the signs of the three crossings, but they are all similar. One of the possible cases is given by the first and last sketches of Figure 8, where an arc passing under a crossing formed by the other two arcs is locally oriented opposite to the other two strands. The replacement sequence that is given in Figure 8 shows that our type $\mathrm{III}^{\mathrm{nb}}$ Reidemeister move can be achieved by a sequence consisting of a type $\mathrm{II}^{\mathrm{nb}}$ move, an isotopy, a type $\mathrm{III}^{\mathrm{b}}$ move and finally another type $\mathrm{II}^{\mathrm{nb}}$ move. We leave the other type $\mathrm{III}^{\mathrm{nb}}$ cases to the reader, and we have reduced to the case of moves of type $\mathrm{II}^{\mathrm{nb}}$.

(3) A move of type $\mathrm{II}^{\mathrm{nb}}$ may be regarded as a move of type $\mathcal{Y}^{\pm}$ if the arcs that are involved belong to distinct Seifert circles, so we only need to handle the case where they are subarcs of the same Seifert circle. This is done in Figure 9, where it is shown that the move can be replaced by two moves of type I^{b} (which create two new Seifert circles) followed by a move of type $\mathcal{Y}$ and another of type $\mathcal{Y}^{-1}$. This proves the claim.

We are thus reduced to the case in which each diagram in the sequence taking Y to Y' is either type I^{b}, II^{b}, $\mathrm{III}^{\mathrm{b}}$ or $\mathcal{Y}^{\pm}$. To complete the proof of Lemma 2.9, let t be a braid-like Reidemeister move to be performed on diagram Y_i. Suppose that $h_i = h(Y_i) > 0$. Then we

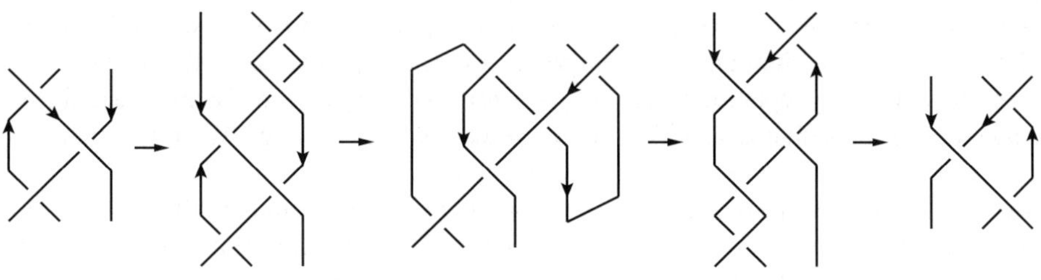

Fig. 8. Replacing moves of type $\mathrm{III}^{\mathrm{nb}}$.

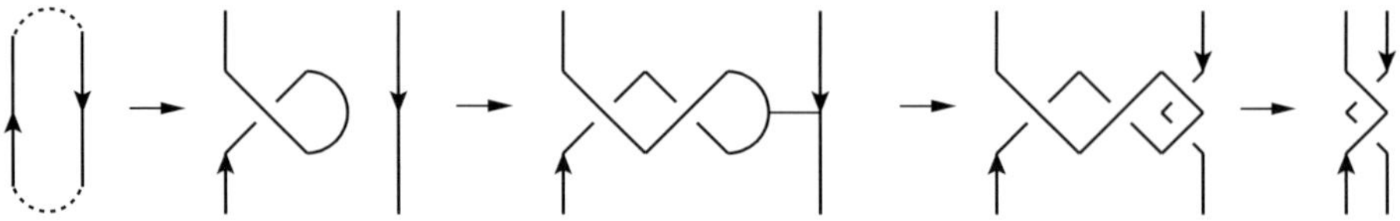

Fig. 9. Replacing moves of type $\mathrm{II}^{\mathrm{nb}}$.

can find a sequence of reducing moves $r_1, \dots, r_{h_1}$ such that $r_{h_i} \circ \cdots \circ r_1(Y_i)$ is a braid and such that the associated reducing arcs $\alpha_1, \dots, \alpha_{h_i}$ are each disjoint from the region in which t is to be performed. Thus, each reducing move r_j commutes with t, and we can replace t with its "conjugate" $r_1^{-1} \circ \cdots \circ r_{h_i}^{-1} \circ t \circ r_{h_i} \circ \cdots \circ r_1$, so that t is now performed at height 0, i.e., on a braid.

If t is of type II^{b} or $\mathrm{III}^{\mathrm{b}}$, then we are done, since a braid-like move of type II or type III performed on a braid is a braid isotopy. If t is of type I^{b}, it is a stabilization (up to isotopy) only if it is performed on the braid strand "nearest" to the braid axis. However, it is not hard to see how to realize a type I^{b} move on an arbitrary strand in terms of Markov moves: simply push the strand under the others via type II^{b} moves, and perform the required stabilization in a neighborhood of the braid axis. Note that to pass the resulting "kink" back under a neighboring strand in a braid requires first a type $\mathrm{III}^{\mathrm{b}}$ move followed by a type $\mathcal{Y}^{-1}$ move (the "kink" is always its own Seifert circles, so two distinct circles are necessarily involved). Thus, we can return the strand with the "kink" in it back to its original position by repeated applications of this two-step process. Since we have just seen that any type II^{b} or type $\mathrm{III}^{\mathrm{b}}$ move can be realized by a finite sequence of type $\mathcal{Y}^{\pm}$ moves and braid isotopies, this means that a type I^{b} move can also be replaced by a finite sequence of type $\mathcal{Y}^{\pm}$ moves and braid isotopies. We can handle inverse moves of type I^{b} in a similar fashion.

We have thus replaced our original sequence relating Y to Y' by a new one which is in general much longer, but which consists entirely of Markov moves (performed, by definition, on braid diagrams, i.e., on diagrams of height zero) and moves of type $\mathcal{Y}^{\pm}$. To be precise, our original sequence from Y to Y' may be replaced by a sequence of the form $Y = Y_0, \dots, Y_{a_1}, \dots, Y_{a_2}, \dots, Y_{a_n} = Y'$ where $h(Y_{a_i}) = 0$ for all i and in each subsequence $Y_{a_i}, \dots, Y_{a_{i+1}}$, either (1) each diagram in the subsequence has height zero and adjacent diagrams are related by a single Markov move, or (2) all the intermediate diagrams have strictly positive height and adjacent diagrams are related by a single move of type $\mathcal{Y}$ or $\mathcal{Y}^{-1}$. Therefore, in order to prove Markov's theorem, it suffices to consider only sequences of the second type, and the proof of Lemma 2.9 is complete. □

REMARK 2.10. The astute reader will have noticed the following: we have eliminated Reidemeister moves completely (they will not appear in the arguments that follow), nevertheless they played an important role already. We started with a sequence relating the given braids X and X' that consisted entirely of Reidemeister moves. We replaced it with a sequence of reducing moves and braid-like Reidemeister moves. The latter are in general not applied to diagrams of height zero, but we changed them to apply to diagrams of height zero. That is the moment when Traczyk's braid-like Reidemeister moves were changed to Markov moves. The modified sequence from X to X' has changed to a series of subsequences, each of which starts with a closed braid and ends with a closed braid, after which the ending braid is modified (by Markov moves) to a new closed braid, which is the initial closed braid in the next subsequence. As will be seen, Markov moves will not be used explicitly again until the proof of Lemma 2.15, where they are used (without the help of Reidemeister's Theorem) to relate very special closed braid diagrams.

Consider now a sequence of diagrams $Y = Y_1, \ldots, Y_n = Y'$ satisfying the criteria of Lemma 2.9. We note that, as in the above proof, we shall not in general distinguish between a diagram and its associated Seifert picture. Thus, we shall make reference to "a Seifert circle in the diagram Y_i", for example, meaning a Seifert circle in the Seifert picture associated to Y_i. In fact, we can think of each circle in a Seifert picture as forming part of the associated diagram, except in a small neighborhood of signed arcs, which correspond to crossings. Since reducing arcs avoid signed arcs, there is no ambiguity when referring to "a reducing arc in a diagram".

We now wish to consider the graph of the height function on our sequence. The graph will begin and end at height zero; each "step" in between will either take us up 1 or down 1 since we have reduced to the case where all moves are reducing moves (or their inverses). We will examine local maxima in the height function. Let $Y(r)$, $\hat{Y}$, $Y(s)$ be three consecutive diagrams in our sequence such that the height function has a local maximum at $\hat{Y}$. In other words, we have two reducing moves r, s with corresponding arcs α_r, α_s in $\hat{Y}$ such that reducing $\hat{Y}$ along α_r (resp. α_s) results in the diagram $Y(r)$ (resp. $Y(s)$), and it makes sense discuss $\alpha_r \cup \alpha_s$. We will call such a triple $\{Y(r), \hat{Y}, Y(s)\}$ a *peak* in the height function of our sequence. We define the *height of the peak* to be $h(\hat{Y})$ and define the *height of the sequence* to be the maximum value attained by the height function on the sequence, in other words, the maximum over the height of all the peaks in the sequence. In order to prove Theorem 2.8, we are going to induct on the height of the sequence.

LEMMA 2.11. *We may assume that the reducing arcs involved in any peak in the height function of our sequence are disjoint. Further, the adjustments in our sequence of reducing moves which are required preserve the height of the sequence.*

PROOF. Let $\{Y(r), \hat{Y}, Y(s)\}$ be a peak in the height function with associated reducing arcs α_r and α_s. We may always assume that the arcs intersect transversally and minimally. Suppose that $|\alpha_r \cap \alpha_s| = n \geq 2$. By smoothing out one or more of the points of intersection, we find a new reducing arc $\alpha_{r'}$ with the same endpoints as α_r such that $\alpha_r \cap \alpha_{r'} = \emptyset$ and $|\alpha_{r'} \cap \alpha_s| < n$. We can then replace the given peak $\{Y(r), \hat{Y}, Y(s)\}$ with two consecutive peaks $\{Y(r), \hat{Y}, Y_{r'}\}$ and $\{Y_{r'}, \hat{Y}, Y(s)\}$. We call this procedure *inserting the reducing operation* r' at $\hat{Y}$, and it essentially amounts to replacing one peak with two peaks of the same height. In this way, we continue on until the intersection numbers of all adjacent pairs is at most 1.

Now suppose the arcs α_r, α_s associated to a given peak $\{Y(r), \hat{Y}, Y(s)\}$ have intersection number 1. If there exists a reducing arc α_t such that $\alpha_t \cap \alpha_r = \alpha_t \cap \alpha_s = \emptyset$, then we can insert the reducing operation t at $\hat{Y}$ to produce two peaks, each with a disjoint pair of associated reducing arcs. Suppose that the defect region which supports α_r and α_s contains no third reducing arc which is disjoint from both α_r and α_s. There is only one possible arrangement for such a defect region, shown in Figure 10.

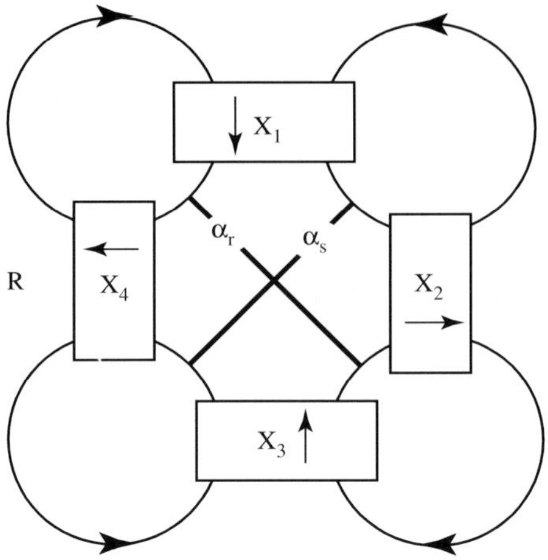

Fig. 10. The case where two reducing arcs at a peak intersect once.

With this arrangement, the reducing arcs α_r and α_s must act on four distinct Seifert circles, possibly joined by signed arcs. We will now examine the region labeled R which lies "outside" of the circles involved in our defect region and the signed arcs which join them. If R contains a Seifert circle, then it is easy to check that there must be a region somewhere in the diagram with three exposed circle. Such a region, as observed in the previous section, is necessarily a defect region which would contain a reducing arc disjoint from both α_r and α_s. If, on the other hand, R contains no Seifert circles, then it contains no signed arcs either, since all possible signed arcs between the four exposed circles of region R already appear in Figure 10. Thus, we can join either pair of diagonally opposed circles by a reducing arc in R.

We conclude that if a peak $\{Y(r), \hat{Y}, Y(s)\}$ has associated arcs of intersection 1, we can always find a third reducing arc α_t so that we can insert the reducing operation t at $\hat{Y}$, thereby replacing the original peak with two peaks, each having a disjoint pair of reducing arcs. Since the operation of inserting a reducing operation at a peak preserves the height of the sequence, this finishes the proof of the lemma. □

Thanks to the previous lemma, we can assume from now on that each peak in the graph of our height function corresponds to a disjoint pair of reducing arcs. Before we can state the next lemma, which concerns the peaks in the height function, we need to introduce a few some concepts. Note that when the reducing arcs involved in a peak $\{Y(r), \hat{Y}, Y(s)\}$ are disjoint, the reducing moves commute, that is, they can be performed in either order, starting with the diagram $\hat{Y}$ and resulting in the same diagram Y'. Further, as long as the reducing arcs α_r, α_s act on three or four distinct Seifert circles, we can perform the two reducing moves in either order with the same result. Observe that, since the reducing arcs α_r, α_s are disjoint in the diagram $\hat{Y}$, it makes sense to talk about the arc α_s (resp. α_r) in the context of the diagram $Y(r)$ (resp. $Y(s)$) obtained by reducing $\hat{Y}$ along α_r (resp. α_s). In this case we say we have a "commuting pair" of reducing moves associated to the peak. If a peak $\{Y(r), \hat{Y}, Y(s)\}$ has a commuting pair, then we may replace it by a "valley", that is, a subsequence $\{Y(r), \hat{Y}, Y(s)\}$ where $h(Y') = h(Y) - 2$ and $Y' = Y(s \circ r) = Y(r \circ s)$ is the result of reducing $Y(r)$ along α_s, or equivalently reducing $Y(s)$ along α_r. Thus, we can eliminate any peak corresponding to a commuting pair (such a peak necessarily has height at least 2).

In the case where two reducing arcs at a peak act on the same two circles, then after one move is performed, the second Reidemeister move will no longer be a reducing move; we call this a "non-commuting pair" of reducing moves. Let $\{Y(r), \hat{Y}, Y(s)\}$ be a peak corresponding to a non-commuting pair of reducing arcs, and let C_1, C_2 be the two Seifert circles involved. Suppose there is a reducing arc α_t such that $\alpha_t \cap \alpha_r = \alpha_t \cap \alpha_s = \emptyset$ and such that t involves a circle other than C_1 or C_2, then we can insert t at $\hat{Y}$ to replace our peak $\{Y(r), \hat{Y}, Y(s)\}$ with two new peaks with commuting pairs of reducing moves: $\{Y(r), \hat{Y}, Y(t)\}$ and $\{Y(t), \hat{Y}, Y(s)\}$. As above, we now replace each peak with a "valley": $\{Y(r), Y', Y(t)\}$ and $\{Y(t), Y'', Y(s)\}$, respectively, where $Y' = Y(t \circ r) = Y(r \circ t)$ is the diagram resulting from reducing $Y(r)$ by t (or equivalently, from reducing $Y(t)$ by r, and Y'' is the diagram resulting from reducing $Y(s)$ by t (or equivalently, from reducing $Y(t)$ by s). Again, this implies that the height of the original non-commuting peak $\{Y(r), \hat{Y}, Y(s)\}$ was at least 2. Thus, we can replace such a peak with peaks of strictly smaller height and repeat this process until all peaks either have height 1 or do not admit such a reducing arc α_t as above; we call a peak of the latter type *irreducible*.

We can now state the next lemma:

LEMMA 2.12. *We may assume that any peak in the height function of our sequence either has height* 1 *or is irreducible.*

PROOF. The proof is clear. In the discussion that preceded the statement of the Lemma we defined a peak to be irreducible in such a way that it subsumed all possibilities which did not allow us to reduce to height 1. □

We note that each condition of Lemma 2.12 necessarily implies the non-commuting condition.

LEMMA 2.13. *We may assume that no peaks in the height function of our sequence have height* 1.

PROOF. Let $\{Y(r), \hat{Y}, Y(s)\}$ be a peak of height 1. We recall that height 1 implies that the two reducing arcs are non-commutative and hence involve precisely two circles. It is an easy exercise to show that these two circles must live either on the "inside" or on the "outside" of a band of circles, and that α_r and α_s are in fact equivalent as reducing moves. Therefore, the diagram $Y(r)$ is equivalent to $Y(s)$ and we can simply eliminate this peak from our sequence. □

It remains to deal with irreducible peaks in the height function of our sequence. Fortunately, it turns out that they only occur in a very particular way. To describe the particular way, define a *weighted Seifert circle* in a manner which is similar to the weights that we attach to closed braid diagrams, e.g., as in Figure 7. That is, a Seifert circle with weight w attached means a collection of w coherently oriented, nested, parallel Seifert circles. We use the term *band* for a Seifert circle with an attached weight.

LEMMA 2.14. *If* $\{Y(r), \hat{Y}, Y(s)\}$ *is an irreducible peak in the height function of our sequence, then the diagram* $\hat{Y}$ *contains at most four bands, arranged as in Figure* 11.

PROOF. Let $\{Y(r), \hat{Y}, Y(s)\}$ be an irreducible peak, and let C_1, C_2 be the two circles involved in the reducing arcs α_r and α_s. See Figure 11(i). For $i = 1, 2$, let D_i denote the disk bounded by C_i in S^2 which does not contain the reducing arcs α_r and α_s. Then $S^2 \backslash (D_1 \cup D_2 \cup \alpha_r \cup \alpha_s)$ has two components. By assumption, neither component can contain a defect region (or else we could find a reducing arc α_t as above, contradicting the irreducibility of our peak). Thus, if either component contains any Seifert circles, the circles must form a band, oriented oppositely to C_1 and C_2. We allow the possibility that the weight of these bands is zero. The same reasoning shows that D_i cannot contain a defect region for $i = 1, 2$ and hence that C_i must be the outer circle of a band. It is of course possible that some braiding takes place between adjacent coherent circles, as indicated in Figure 11(i). The braids joining the various bands are labeled X_i. Thus, we have a diagram of the given form, in which each circle pictured represents a band. The lemma is proved. □

The following lemma will allow us to replace irreducible peaks in the height function of our sequence with peaks of strictly smaller height.

LEMMA 2.15. *Let* $\{Y(r), \hat{Y}, Y(s)\}$ *be an irreducible peak of height* $n + 1$ *in the height function of our sequence. Then there exist sequences of diagrams* $Y(r) = Y_1^r, \ldots, Y_n^r = Y(p \circ r)$ *and* $Y(s) = Y_1^s, \ldots, Y_n^s = Y(u \circ s)$ *such that* Y_{i+1}^r *(resp.* Y_{i+1}^s*) is obtained from* Y_i^r *(resp.* Y_i^s*) by a reducing move and such that* $h(Y(p \circ r)) = h(Y(u \circ s)) = 0$ *and* $Y(p \circ r)$ *and* $Y(u \circ s))$ *are Markov equivalent.*

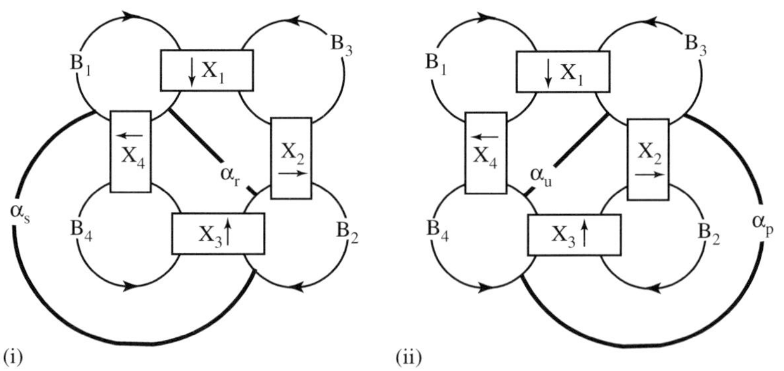

Fig. 11. (i) The diagram corresponding to the irreducible peak $\{Y(r), \hat{Y}, Y(s)\}$. (ii) Two additional reducing arcs used to eliminate irreducible peaks. Each pair α_r, α_s and α_p, α_u is a possible non-commuting pair associated to the peak. The blocks X_1, X_2, X_3, X_4 indicate (possibly empty) collections of signed arcs.

Before discussing the proof of this lemma, we show how to use it to prove the Markov theorem. Using Lemmas 2.9–2.14, we have reduced the proof of Markov's Theorem to the situation of two closed braid diagrams X and X' related by a sequence of diagrams related by reducing moves (and their inverses) such that the height of each intermediate diagram is strictly positive and such that any peak in the height function of the sequence is irreducible (of height at least 2). Let $\{Y(r), \hat{Y}, Y(s)\}$ be an irreducible peak of height n in the height function of our sequence. By Lemma 2.15, we can replace this subsequence with a subsequence of strictly smaller height (possibly including a subsequence entirely at height 0 related by Markov moves, which we can remove from consideration as before). In doing so, we create new peaks whose height is *strictly lower* than the height of the peak being replaced. If we perform this operation at every irreducible peak, then we obtain a new sequence relating our closed braid diagrams Y, Y'.

The new peaks may or may not be irreducible; in fact, their corresponding arcs may not even be disjoint, but now we are back in the same situation as we were before Lemma 2.11 except that we are starting with a sequence of lower height. Thus, by induction on the height of the sequence (with base case provided by Lemma 2.13), we can replace any sequence as described in Lemma 2.9 with a sequence consisting entirely of diagrams of height zero. This completes the proof of Theorem 2.8, modulo the proof of Lemma 2.15.

SKETCH OF THE PROOF OF LEMMA 2.15. By Lemma 2.14, the diagram $\hat{Y}$ has at most four bands which we label B_1, B_2, B_3, and B_4, joined by four (possibly trivial) braids X_1, X_2, X_3, X_4, as in Figure 11(i) and (ii). Let w_i denote the weight of the band B_i. Note that, e.g., the braid X_1 has $w_1 + w_3$ strands, and similarly for the other X_i. If w_i or w_j is greater than 1, then a reducing arc joining B_i, B_j is understood to indicate a sequence of $w_i w_j$ reducing moves. There are several ways to construct such a sequence; we adopt the convention that we choose reducing moves in such a way that the strands of one band all slide under or else all slide over the strands of the other band.

Referring again to Figure 11, we first perform the reducing move r along the arc α_r and then reduce again via the arc α_p involving B_3 and B_4, giving us some number of reducing moves depending on the weights of the bands involved. The resulting diagram $Y(p \circ r)$ is the closure of the first braid shown in Figure 12, up to a choice of arcs passing over or under in the various Reidemeister moves of type II. This gives us the first sequence of the lemma.

To get the second sequence of the lemma, we begin instead with the reducing move s along the arc α_s and then reduce again via the arc α_u. The resulting diagram $Y(u \circ s)$ is the closure of the second braid shown in Figure 12.

We have found our two sequences of reducing moves, and we have reduced the proof of Markov's Theorem to just one specific calculation, namely, showing that the closed braids $Y(p \circ r)$ and $Y(u \circ s)$ corresponding to the two braids of Figure 12 are M-equivalent. The two braids can be related by a sequence consisting of several braid isotopies as well as two stabilizations and two destabilizations; the reader is referred to [111] for the details of this calculation. This concludes the proof of Lemma 2.15, and so also of Theorem 2.8. □

REMARK 2.16. Our choice of "over" or "under " in the reducing moves p and u leads to possible ambiguity, but the various diagrams which would result from different choices are all related by "exchange moves", which are defined and discussed at the beginning of the

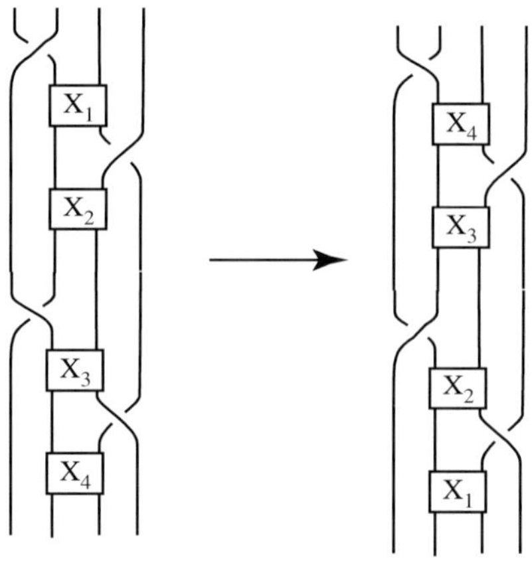

Fig. 12. Two braids whose respective closures are the result of performing the two pairs of reducing moves r, p and s, u indicated in Figure 11.

next section. We will show in the next section that an exchange move replaces a sequence of four Markov moves: a braid isotopy, a stabilization, a second braid isotopy and a destabilization.

The essential groundwork has been laid regarding the connection between braids and knots. From this point on, all knot diagrams will be assumed to be in closed braid form, i.e., in the form that was indicated earlier, in Figure 3 In the sections that follow we will examine many consequences.

3. Braid foliations

We begin our study of new results which relate to the study of knots via closed braids by presenting some results that use the theory of braid foliations of a Seifert surface bounded by a knot which is represented as a closed braid. We will develop three applications of braid foliations: The first is Theorem 3.1, in Section 3.1. We give an essentially complete proof of the "Markov Theorem Without Stabilization" (MTWS) in the special case of the unknot, based upon the presentation in [17]. In Section 3.2 we state the MTWS, Theorem 3.5, in the general case. A full proof of that theorem can be found in [25]. In Section 3.3, we give an application of the MTWS to contact topology.

3.1. *The Markov Theorem Without Stabilization (special case: the unknot)*

After describing the basic ideas about braid foliations, we will apply them to the study of a classical problem in topology, the unknot recognition problem. Alexander's Theorem (Theorem 2.1) tells us that every link K may be represented as a closed n-braid, for some n. Markov's Theorem (Theorem 2.8) tells us how any two closed braid representatives of the same knot or link are related. Looking for a way to simplify a given closed braid

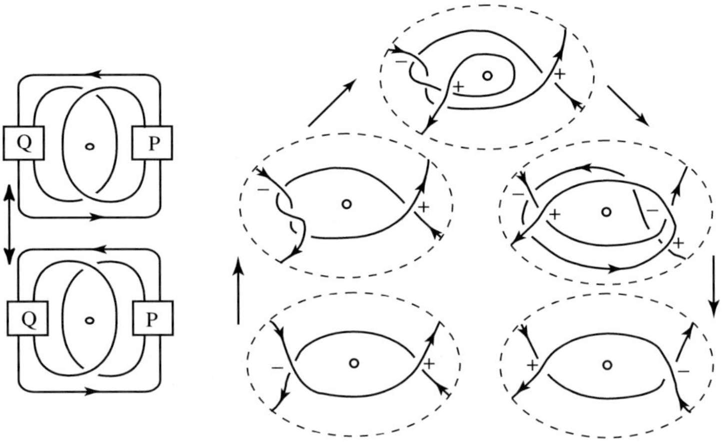

Fig. 13. The left top and bottom sketches define the exchange move. The right sequence of five sketches shows how it replaces a sequence of Markov moves which include braid isotopy, a single stabilization, additional braid isotopy and a single destabilization.

representative of a knot or link systematically, the first author and Menasco were lead to the study of the unknot as a key example. There is an obvious choice of a simplest representative, namely a 1-braid representative. The MTWS for the unknot, which is stated below as Theorem 3.1, asserts that, in the special case of the unknot, the stabilization move of Markov's Theorem can be eliminated, at the expense of adding the exchange move. Therefore, we begin with a discussion of the *exchange move*, which is defined in Figure 13, and the reason why it is so important.

A natural question to ask is what is accomplished by stabilization, braid isotopy and destabilization? Figure 14 shows that exchange moves are the obstruction to sliding a trivial loop around a braid [29]. However, there is more at issue than just sliding trivial loops around a braid. As is illustrated in Figure 15 a sequence of exchange moves, together with braid isotopy, can create infinitely many closed braid representatives of a single knot type, all of the same braid index. This bypasses a key question: do exchange moves actually change conjugacy class? The answer is "yes", and in fact the phenomenon first appears in the study of closed 4-braid representatives of the unknot. In Section 5 we show that there is a definitive test for proving it.

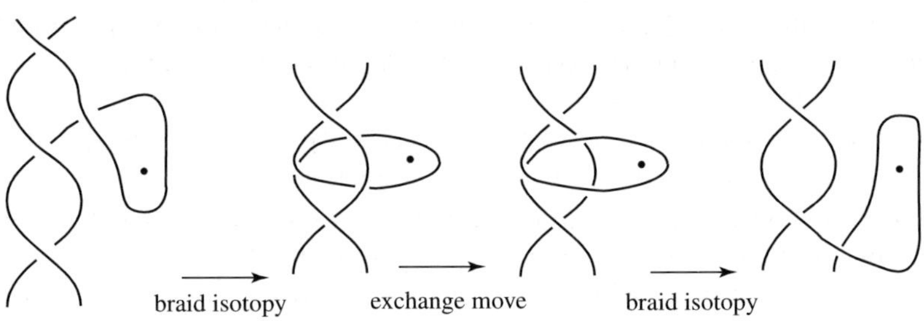

Fig. 14. Exchange moves are the obstruction to sliding a trivial loop around a closed braid.

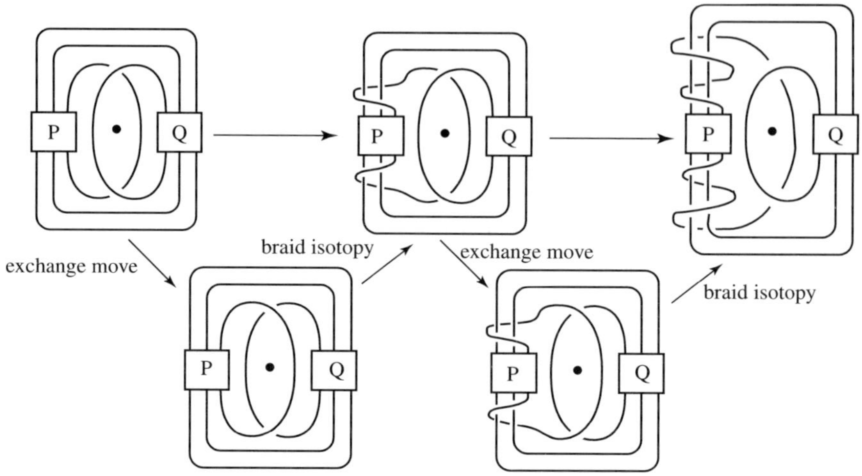

Fig. 15. A sequence of exchange moves and braid isotopies with non-trivial consequences.

Keeping Figures 13–15 in mind, we are ready to state the MTWS for the unknot:

THEOREM 3.1 [24]. *Every closed braid representative K of the unknot $\mathcal{U}$ may be reduced to the standard 1-braid representative U_1, by a finite sequence of braid isotopies, destabilizations and exchange moves. Moreover there is a complexity function associated to closed braid representative in the sequence, such that each destabilization and exchange move is strictly complexity-reducing.*

The first proof of Theorem 3.1 was the one in [24]. A somewhat different and slicker proof can be found in [17], but it requires more machinery than was necessary for present purposes. We follow the proof in [24]. However, parts of our presentation and most of our figures were essentially lifted (with the permission of both authors) from [17], a review article on braid foliation techniques. Our initial goal is to set up the machinery needed for the proof.

Recalling the definition of a closed braid from Section 2.1, we adopt some additional structure and say that K is in closed n-braid form if there is an unknot A in $S^3 \setminus K$, and a choice of fibration H of the solid torus $S^3 \setminus A$ by meridian disks, such that K intersects each fiber of H transversely. Sometimes it is convenient to replace S^3 by $\mathbb{R}^3$ and to think of the fibration H as being by half-planes $\{H_\theta;\ \theta \in [0, 2\pi]\}$ of constant polar angle θ through the z-axis. Note that K intersects each fiber H_θ in the same number of points, that number being the *index* n of the closed braid K. We may always assume that K and A can be oriented so that K travels around A in the positive direction, using the right-hand rule. Now let K be a closed braid representative of the unknot, and let $\mathcal{D}$ denote a disk spanned by K, oriented so that the positive normal bundle to each component has the orientation induced by that on $K = \partial\mathcal{D}$. In this section, we will describe a set of ideas which shows that there is a very simple method that changes the pair $(K, \mathcal{D})$ to a planar circle that bounds a planar disc, via closed braids, moreover there is an associated complexity function that is strictly reducing. The ideas that we describe come from [24], however our main reference will be to the review article [17].

The braid axis A and the fibers of H will serve as a coordinate system in 3-space in which to study $\mathcal{D}$. A singular foliation of $\mathcal{D}$ is induced by its intersection with fibers of H. A *singular leaf* in the foliation is one which contains a point of tangency with a fiber of H. All other leaves are *non-singular.*

It follows from standard general position arguments that the disk $\mathcal{D}$ can be chosen to be "nice" with respect to our fibration. More precisely, we can assume the following:

(i) The intersections of A and $\mathcal{D}$ are finite in number and transverse.

(ii) There is a neighborhood N_A of A in $R^3 \setminus K$ such that each component of $\mathcal{D} \cap N_A$ is a disk, and each disk is radially foliated by its arcs of intersection with fibers of H. There is also a neighborhood N_K of K in $\mathbb{R}^3$ such that $N_K \cap \mathcal{D}$ is foliated by arcs of intersection with fibers of H which are transverse to K.

(iii) All but finitely many fibers H_θ of H meet $\mathcal{D}$ transversely, and those which do not (the *singular fibers*) are each tangent to $\mathcal{D}$ at exactly one point in the interior of both $\mathcal{D}$ and H_θ. Moreover, each point of tangency is a saddle point (with respect to the parameter θ). Finally, each singular fiber contains exactly one singularity of the foliation, each of which is a saddle point.

It can also be assumed (see [17] for details) that

(iv) Each non-singular leaf is either an *a-arc*, which has one endpoint on A and one on $K = \partial\mathcal{D}$ or a *b-arc*, which has both endpoints on A.

(v) Each b-arc in a fiber H_θ separates that fiber into two components. Call the b-arc *essential* if each of these components is pierced at least once by K, and *inessential* otherwise. Then we also have that all b-arcs in the foliation of $\mathcal{D}$ may be assumed to be essential.

Note that $\mathcal{D}$ cannot be foliated entirely by b-arcs, since such a surface is necessarily a 2-sphere. Thus, if the foliation of $\mathcal{D}$ contains no singularities, $\mathcal{D}$ is foliated entirely by a-arcs. Otherwise, let U be the union of all the singular leaves in the foliation of $\mathcal{D}$. Moving forward through the fibration, we see that any singular leaf in the foliation is formed by non-singular leaves moving together to touch at a saddle singularity. The three types of singular leaves which can occur are labeled aa, ab or bb, corresponding to the non-singular leaves associated to them. Now each singular leaf λ in U has a foliated neighborhood N_λ in $\mathcal{D}$ such that $N_\lambda \cap U = \lambda$. According to whether λ has type aa, ab or bb, N_λ is one of the foliated open 2-cells shown in Figure 16(i), with the arrows indicating the direction of increasing θ.

The complement of U in $\mathcal{D}$ is a union $B_1 \cup B_2 \cup \cdots \cup B_k$, where each B_i is foliated entirely by a-arcs or entirely by b-arcs. Choose one non-singular leaf in each B_i and declare it to be a *boundary arc* of *type a* or *type b* according to whether it is an a-arc or b-arc, respectively. Then the union of all boundary arcs determines a *tiling* of $\mathcal{D}$, that is, a decomposition into regions called *tiles*, each of which is a foliated neighborhood of one singular leaf. Each tile has type aa, ab and bb, according to the type of its unique singularity. Note that a tiling of $\mathcal{D}$ is a foliated cell-decomposition. We further define the *sign of a singular point s* in the foliation of $\mathcal{D}$ to be positive if the positive normal to $\mathcal{D}$ points in the direction of increasing θ in the fibration, negative otherwise. The *sign of a tile* is then defined to be the sign of its singularity.

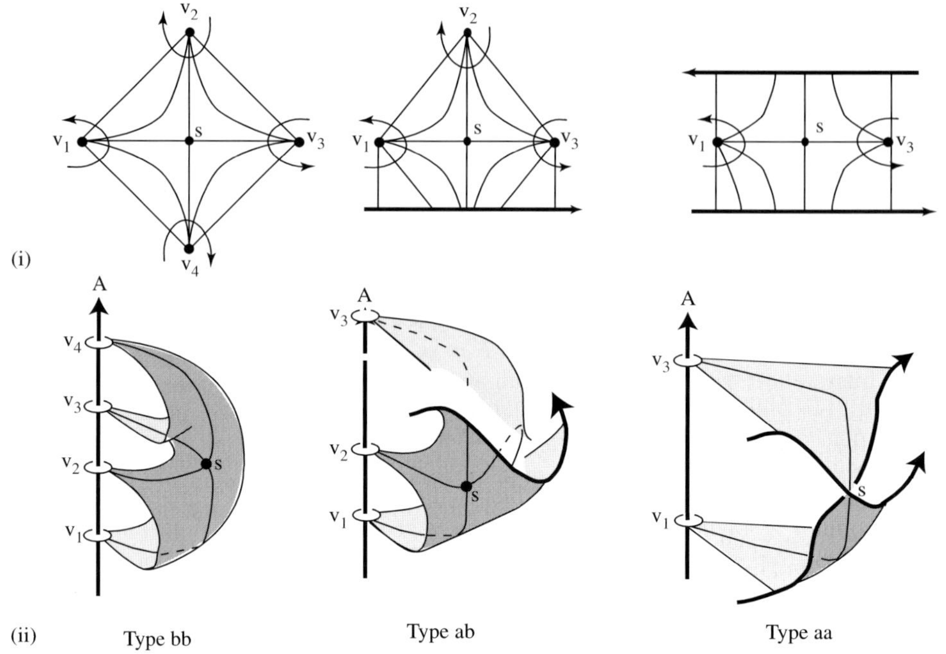

Fig. 16. (i) The three types of tiles in the decomposition of $\mathcal{D}$; (ii)The canonical embedding of each type of tile.

The axis A intersects the surface in a finite number of points, called *vertices* of the tiles. Each vertex v is an endpoint of finitely many boundary arcs in the surface decomposition. Let the *type* of v be the cyclic sequence $(x_1, \ldots, x_r)$, where each x_i is either a or b, and the sequence lists the types of boundary arcs meeting at v in the cyclic order in which they occur in the fibration. Three examples are given in Figure 17. The *valence of a vertex v* is the number of distinct tiles intersecting at v. The *sign of a vertex v* is the cyclic array of signs of the tiles meeting at v. See Figure 17.

The *parity* of v is said to be positive (resp. negative) if the outward-drawn normal to the surface has the same (resp. opposite) orientation as the braid axis at the vertex. Thus, when we view the positive side of the surface, the sense of increasing θ around a vertex will be

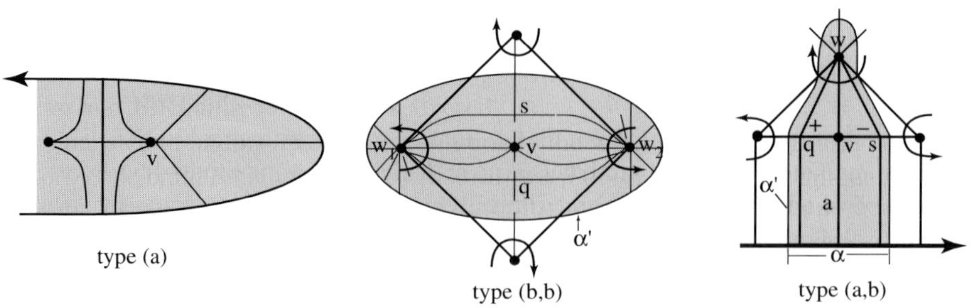

Fig. 17. A vertex v of type (a), (a, b) and (b, b).

counterclockwise (resp. clockwise) when the vertex is positive (resp. negative), as illustrated in Figure 16(i).

Finally, we note that (again, see [17] for a proof) that, up to a choice of their sign, tiles of type *aa*, *ab*, *bb*, each have a canonical embedding in 3-space, which is determined up to an isotopy of 3-space which preserves the axis A and each fiber of H setwise. The canonical embedding for each tile is shown in Figure 16(ii).

The decompositions of the disk $\mathcal{D}$ which we have just described are not unique. We shall now describe three ways in which they can be changed. In each of the three cases the possibility of making the change is indicated by examining the combinatorics of the tiling. The change is realized by an isotopy of the disk which is supported in a neighborhood N of a specified small number of tiles of type *aa*, *ab* or *bb*, leaving the decomposition of $\mathcal{D}$ unchanged outside N.

For a spanning disk $\mathcal{D}$ and a tiling $\mathcal{T}$, we denote the *complexity* of the tiling by $c(\mathcal{D},\mathcal{T})$, where we define $c(\mathcal{D},\mathcal{T}) = (n, S)$, where n is the braid index of the boundary and S is the number of singularities in the tiling. Setting V_+ and V_- equal to the number of positive and negative vertices, one has $n = V_+ - V_-$, so n is also determined by the tiling. Tiled discs $(\mathcal{D}, \mathcal{T})$ can then be ordered, using lexicographical ordering of the associated pair $c(\mathcal{D},\mathcal{T})$. Each of our three moves will replace the given disk and tiling $\mathcal{D}$, $\mathcal{T}$ with some $\mathcal{D}'$, $\mathcal{T}'$ with the following effects on complexity:

- Change in foliation: $(n', S') = (n, S)$.
- Destabilization: $(n', S') = (n - 1, S - 1)$.
- Exchange moves (two types): $(n', S') = (n, S - 2)$.

We focus here only on the combinatorics of tilings which admit one of the above moves, as well as the effect of each move on the embedding of $\mathcal{D}$ and the new tiling of $\mathcal{D}$ which results. For proof and further details, see [17].

- *Changes in foliation.* The choice of a foliation of $\mathcal{D}$ is not unique, and our first move involves ways in which the surface decomposition can be changed by an isotopy of $\mathcal{D}$ or, equivalently, by an isotopy of the fibers of H keeping $\mathcal{D}$ fixed. This particular change was introduced in [23] for 2-spheres and was modified in [24] for certain spanning surfaces. In what follows, we say that two tiles are *adjacent* if they have a common boundary arc in the given tiling.

  A tiling admits a change in foliation whenever there are two tiles T_1, T_2 of the same sign adjacent at a b-arc. Roughly speaking, a change in foliation is a local isotopy of the surface which pushes two saddle points past each other. Locally, the disk $\mathcal{D}$ is embedded as in the left sketch in Figure 18(i). A *change in foliation* is defined as the passage from the embedding left to the right embedding. Figure 18(ii) the effect of this move on the local foliation, while Figure 18(iii) illustrates the effect of a change in foliation on the tiling, in the case of two adjacent (ab)-tiles.
- *Destabilization via a type (a) vertex.* A vertex v of valence 1 in a tiling of $\mathcal{D}$ occurs when two of the edges in a single tile T are identified in $\mathcal{D}$. Such a vertex must have type (a), and therefore T is an *aa*-tile (see Figure 17). Since there is a canonical embedding for an *aa*-tile in 3-space (Figure 16), identifying two edges of an *aa*-tile with endpoints on a common vertex yields the canonical embedding for T shown in Figure 19(i). Notice that there is a radially foliated disk D in T cut off from $\mathcal{D}$ by the arc of the singular leaf of T with both endpoints on K. Hence there must be a trivial loop in the braid representation

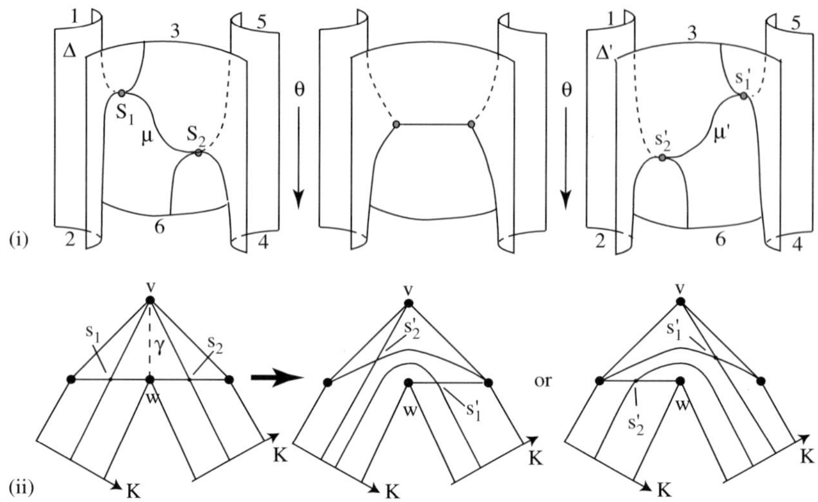

Fig. 18. Sketch (i) shows an isotopy of $\mathcal{D}$ that induces a change in foliation. Sketch (ii) shows the effect of the change on two *ab*-tiles.

of K, and we can modify the braid and the disk $\mathcal{D}$ in the manner illustrated. We call this modification *destabilization* via a type (a) vertex. The effect on the tiling of $\mathcal{D}$ is that the *aa*-tile T and its type (a) vertex v are deleted, while the tiling outside T is unaltered. The braid index is decreased by one. See the right sketch in Figure 19(i).

- *Exchange moves*. The destabilization move that we just described can be accomplished if a vertex of valence 1 exists in our tiling. Exchange moves are based upon the existence of a vertex of valence 2. An example of the type of move we would like to achieve on the braid itself was shown in sketch (ii) of Figure 19(ii), and also in the second row from the top in Figure 20. There are two different types of tiling patterns, and hence two different types of local embeddings of $\mathcal{D}$, which support exchange moves:

  Exchange move, type (bb): We now consider a vertex v with sign $(\pm, \mp)$ and type (b, b). In this situation, two *bb*-tiles are adjacent along consecutive b-arcs.

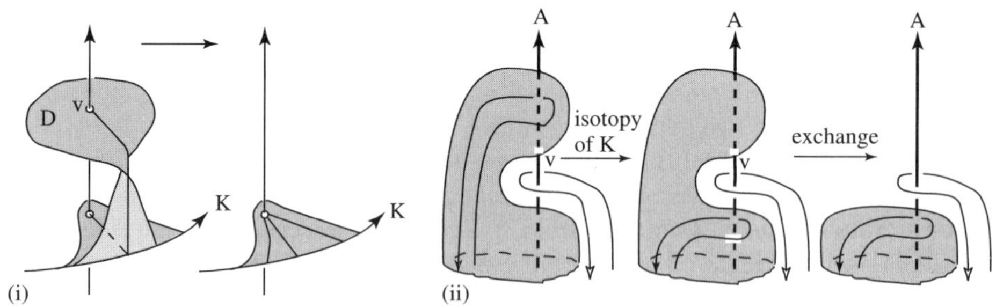

Fig. 19. The effect of (i) a destabilization, and (ii) an exchange move of type (*bb*), on the embedded (and foliated) foliated disc $\mathcal{D}$ and on the braid K that is its boundary.

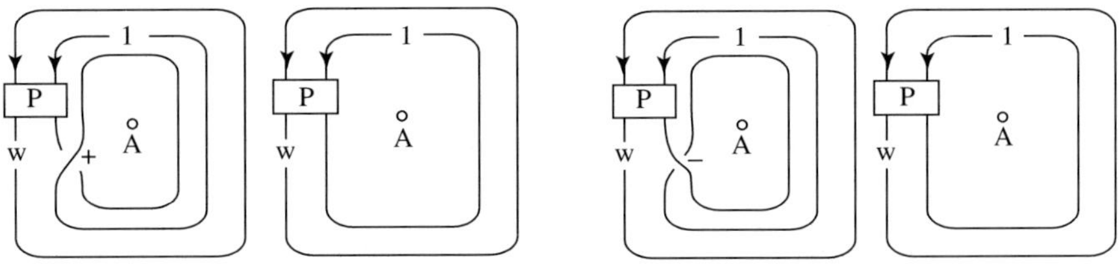

The two destabilization templates

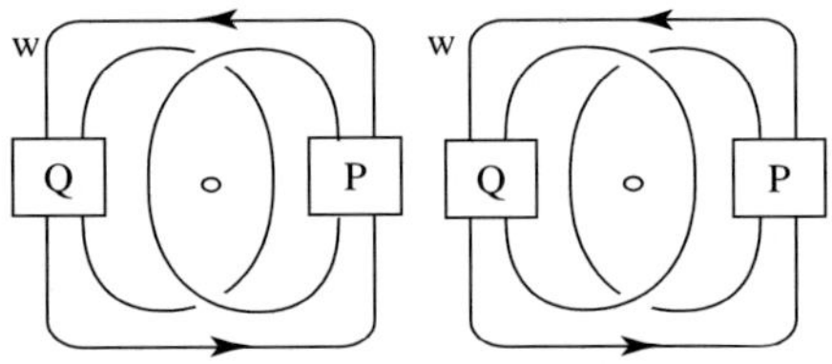

The exchange move template

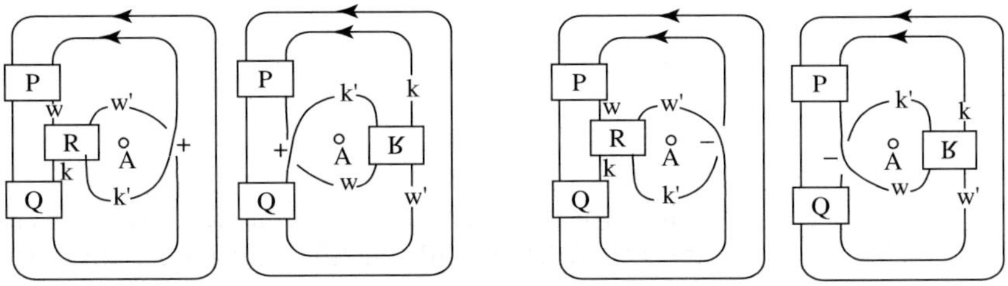

The two flype templates

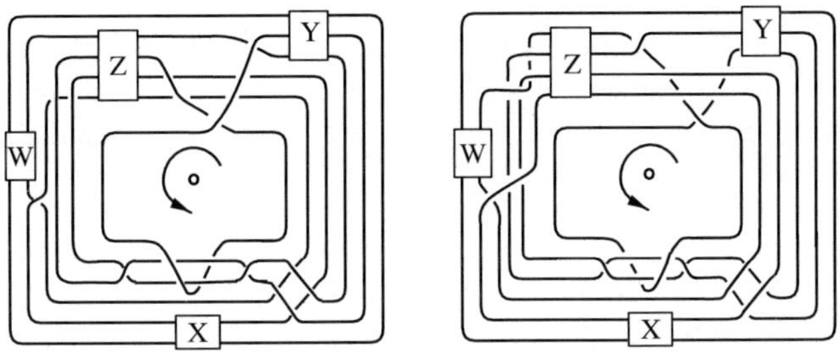

A 6-braid template

Fig. 20. Examples of templates.

The canonical embeddings of the tiles in this case are shown in Figure 16, and the overall modifications to $\mathcal{D}$ are suggested in Figure 19(ii). The end result is a new surface $\mathcal{D}'$ whose decomposition has at least two fewer vertices (in particular, the vertex v is deleted) and at least two fewer regions than the decomposition of $\mathcal{D}$.

Exchange move, type (ab): An exchange move of type (ab) is possible whenever the tiling of $\mathcal{D}$ has a vertex v of valence 2, type (a, b), and sign ($\pm$, $\mp$). Such a vertex occurs only when two ab-tiles are adjacent along corresponding a and b edges, as in Figure 17. The isotopy of $\partial\mathcal{D} = K$ is achieved by pushing a subarc α of K across a disk which is contained in the union of two (ab) tiles T_1, T_2 to a new arc α'. The effect on the tiling is that the ab-tiles T_1 and T_2 are deleted and any adjacent ab-tiles (resp. bb-tiles) become aa-tiles (resp. ab-tiles).

In order to prove the MTWS for the unknot, we must locate the places on the tiling of $\mathcal{D}$ where the three moves just described can be made. We begin with two lemmas. The first of the two is implicit in [8], but we refer the reader to [17] for an explicit proof.

LEMMA 3.2 [8]. *Let v be a vertex of type$(b, b, \ldots, b)$. Then the set of all singularities s which lie on a singular leaf ending at v contains both positive and negative singular points.*

LEMMA 3.3 [24]. *Suppose that the disk $\mathcal{D}$ is nontrivially tiled and has no vertices of valence 1. Then the tiling of $\mathcal{D}$ contains a vertex of type $(ab), (bb), or (bbb)$.*

PROOF. The proof is essentially an Euler characteristic argument. A tiling $\mathcal{T}$ of the disk $\mathcal{D}$ corresponds to a cell decomposition of the 2-sphere in the following way. Let V be the number of vertices in $\mathcal{T}$, let E be the number of boundary arcs in $\mathcal{T}$, which we will think of as *edges*, and let F be the number of tiles in $\mathcal{T}$. Now let Σ be the 2-sphere obtained by collapsing $\partial\mathcal{D} = K$ to a point. This gives a cell decomposition of Σ with $V + 1$ 0-cells (vertices), E 1-cells (edges), and F 2-cells (faces). We know that the Euler characteristic $\chi(\Sigma) = 2$, and therefore we have that $V - E + F = 1$. We also know that every face in our cell decomposition of Σ has four edges, and that every edge has two adjacent faces, and so $E = 2F$. We therefore have the following equation.

$$2V - E = 2. \tag{3.1}$$

Now let $V(\alpha, \beta)$ denote the number of vertices in $\mathcal{T}$ with α adjacent a-arcs and β adjacent b-arcs. If v is the valence of such a vertex, then $v = \alpha + \beta$, and so $V(\alpha, v - \alpha)$ denotes the number of vertices in $\mathcal{T}$ with valence v and α adjacent a-arcs. Since by hypothesis $v \neq 1$, we can now count the number of vertices in $\mathcal{T}$ using the following summation:

$$V = \sum_{v=2}^{\infty} \sum_{\alpha=0}^{v} V(\alpha, v - \alpha).$$

Now write $E = E_a + E_b$, where E_a denotes the number of edges in $\mathcal{T}$ of type a, and similarly for E_b. Since an a-edge is incident at one vertex in $\mathcal{T}$, and a b-edge is incident at two vertices in $\mathcal{T}$, we can write:

$$E_a = \sum_{v=2}^{\infty} \sum_{\alpha=0}^{v} \alpha V(\alpha, v - \alpha) \tag{3.2}$$

$$E_b = \frac{1}{2} \sum_{v=2}^{\infty} \sum_{\alpha=0}^{v} (v - \alpha) V(\alpha, v - \alpha). \tag{3.3}$$

Substituting into (3.1), we now have

$$\sum_{v=2}^{\infty}\sum_{\alpha=0}^{v}(4-v-\alpha)V(\alpha,v-\alpha)=4.$$

Observe now that if $v \geq 4$, the coefficient $(4-v-\alpha) \leq 0$. Writing out the terms corresponding to $v=2$ and $v=3$ gives, respectively, $2V(0,2)+V(1,1)$ and $V(0,3)-V(2,1)-2V(3,0)$. Each $V(\alpha,\beta) \geq 0$ by definition. Hence we can move terms around in the above equation so that each term is non-negative, as follows:

$$\begin{aligned}&2V(0,2)+V(1,1)+V(0,3)\\&\quad=4+V(2,1)+2V(3,0)+\sum_{v=4}^{\infty}\sum_{\alpha=0}^{v}(4-v-\alpha)V(\alpha,v-\alpha).\end{aligned}$$

The right-hand side of the equation is clearly ≥ 4, and therefore at least one term on the left-hand side is non-zero. Since $V(0,2)$, $V(1,1)$, and $V(0,3)$ record the number of type (bb), type (ab), and type (bbb) vertices, respectively, the lemma is proved. □

We are now ready to prove Theorem 3.1, the MTWS in the case of the unknot.

Proof of Theorem 2.8. We start with an arbitrary closed braid representative K of the unknot. Our closed braid K is the boundary of a disk $\mathcal{D}$ which admits a tiling $\mathcal{T}$. Note that if the complexity of the tiling $c(\mathcal{D},\mathcal{T})$ is equal to (1,0), then our disk is radially foliated by a-arcs and K is the standard embedding of the unknot. Therefore, we assume $c(\mathcal{D},\mathcal{T}) > (1,0)$. We will use induction on $c(\mathcal{D},\mathcal{T})$ to show that after a finite sequence of our three moves, $\mathcal{D}$ is radially foliated by a-arcs.

We first observe that if there exists a vertex of type (a) in $\mathcal{T}$, then we can destabilize along this vertex, thereby reducing $c(\mathcal{D},\mathcal{T})$. Hence we can assume that all vertices in $\mathcal{T}$ have valence at least 2. By Lemma 3.3, $\mathcal{D}$ must have a vertex of type (ab), (bb), or (bbb). If there is a type (bbb) vertex, then at least two of the adjacent tiles must have the same sign. In this case we can do a change in foliation, replacing our (bbb) vertex with a (bb) vertex. Therefore, we may assume that $\mathcal{D}$ contains a vertex of type (ab) or type (bb).

If $\mathcal{D}$ has a type (ab) vertex with sign $(+,+)$ or $(-,-)$, then we can apply a change in foliation which replaces this vertex with a type (a) vertex. We now destabilize along this type (a) vertex in order to reduce the complexity. If $\mathcal{D}$ has a type (ab) vertex with sign $(+,-)$, then we can do an exchange move of type (ab), thereby reducing complexity. If $\mathcal{D}$ has a type (bb) vertex, then it is an interior vertex, and by Lemma 3.2, it has sign $(+,-)$. Then we can do an exchange move of type (bb), which reduces complexity.

Thus, if $c(\mathcal{D},\mathcal{T}) > (1,0)$, we can always reduce until $c(\mathcal{D},\mathcal{T}) = (1,0)$, and the theorem is proved. □

Example 3.4 (Morton's irreducible 4-braid). We now give an example illustrating the necessity of exchange moves. In [96], Morton gave the following example of an *irreducible* braid:

$$X=\sigma_3^{-2}\sigma_2\sigma_3^{-1}\sigma_2\sigma_1^{3}\sigma_2^{-1}\sigma_1\sigma_2^{-1},$$

i.e., a braid whose closure represents the unknot but cannot be isotoped to the unknot in the complement of the braid axis. Now, it is clear that if a braid in $\mathbf{B}_n$ admits a factorization of the form $w_1\sigma_{n-1}w_2\sigma_{n-1}^{-1}$, where w_1 and w_2 are words in $\sigma_1, \ldots, \sigma_{n-2}$, then its closure admits an exchange move. As observed in [24], the conjugate braid $(\sigma_3\sigma_2\sigma_1)X(\sigma_1^{-1}\sigma_2^{-1}\sigma_3^{-1})$ is isotopic to the braid

$$X' = (\sigma_2^{-2}\sigma_1\sigma_2^{-1})\sigma_3(\sigma_2^3\sigma_1^{-1}\sigma_2)\sigma_3^{-1}.$$

Thus X' admits an exchange move to obtain the new braid word $(\sigma_2^{-2}\sigma_1\sigma_2^{-1})\sigma_3^{-1}(\sigma_2^3\sigma_1^{-1}\sigma_2)\sigma_3$, which can be cyclically rewritten as

$$X'' = (\sigma_2\sigma_3\sigma_2^{-2})\sigma_1(\sigma_2^{-1}\sigma_3^{-1}\sigma_2^3)\sigma_1^{-1}.$$

If we now interchange the axis of our braid with the point at infinity, we can perform another exchange move to obtain

$$X''' = (\sigma_2\sigma_3\sigma_2^{-2})\sigma_1^{-1}(\sigma_2^{-1}\sigma_3^{-1}\sigma_2^3)\sigma_1$$

which can in fact be isotoped to the unknot in the complement of its braid axis [24].

Wright has given a foliated disk corresponding to this braid in [115]. The reader is cautioned that Wright uses a different notation convention for braid words than that used in this paper (see p. 98 of [115]). We also refer the reader to [27] for further examples.

We note that Theorem 3.1 has been used as the basis for an algorithm for recognizing the unknot. See [18]. This algorithm has been put on a computer [27], however it needs more work before it can become a practical tool for recognizing the unknot.

OPEN PROBLEM 2. At this writing the development of a sound and practical algorithm for unknot recognition remains one of the major open problems in low dimensional topology.

With regard to Problem 2, we note that Theorem 3.1 proves the existence of a monotonic simplification process which begins with an arbitrary closed n-braid representative of the unknot and ends with a 1-braid representative, but the complexity function $c(\mathcal{D}, \mathcal{T})$ which gives instructions for the process is "hidden" in the tiling of the surface $\mathcal{D}$. We need it in order to know when complexity-reducing destabilizations and exchange moves are possible. The reader who is interested in this problem might wish to consult [18], where a somewhat different approach suggests itself. Instead of working with the tiled surface, one may work with the "extended boundary word" of [18]. The latter is a closed braid which is obtained from an arbitrary closed braid representative of the given knot by threading in additional 1-braids, and it seems likely that it will give an alternative monotonic reduction process. Thus, we suggest:

OPEN PROBLEM 3. Investigate the monotonic reduction process of Theorem 3.1 from the point of view of the extended boundary word of [18].

3.2. *The Markov Theorem Without Stabilization, general case*

In this section we state the generalized version of Theorem 3.1 which was established in [25] for arbitrary closed braid representatives of arbitrary knots and links. The moves which are needed are of course much more complicated than in the case of the unknot. They are described in terms of "block-strand diagrams" and "templates". The concept of a block-strand diagram is fairly easy to understand from the eight examples in Figure 20. The first important feature of a block-strand diagram is that after an assignment of a braided tangle to each block, it becomes a closed braid that represents a specific knot or link. Thus, each block-strand diagram determines infinitely many closed braid representatives of (presumably) infinitely many knots and links. A *template* is a pair of block-strand diagrams, both of which represent the same knot or link, when we make the same braiding assignments to corresponding blocks. Note that there are two destabilization templates (they differ in the sign of the "trivial loop" that is removed). There are also two flype templates, which differ in the sign of the single crossing which is outside all the braid boxes. Templates always have an associated braid index, namely the braid index of any knot or link of minimum braid index that they carry. Destabilization templates occur for every braid index ≥ 2. Since the exchange move can be realized by braid isotopy when the braid index is ≤ 3, it does not play a role until braid index ≥ 4. The flype templates occur for braid index ≥ 3. The method by which the 6-braid template in Figure 20 was constructed is described in the manuscript [25].

The two block strand diagrams in a template are always related by a sequence of Markov moves, however the sequence may be quite complicated, and so the isotopy that takes the left diagram to the right diagram is in general not obvious. This fact is illustrated by the 6-braid template of Figure 20. (See [25] for the isotopy, as an explicit sequence of Markov moves.)

With this brief introduction, we are able to state the MTWS in the general case:

THEOREM 3.5 [25]. *Let $\mathcal{B}$ be the collection of all braid isotopy classes of closed braid representatives of oriented knot and link types in oriented 3-space. Among these, consider the subcollection $\mathcal{B}(K)$ of representatives of a fixed link type K. Among these, let $\mathcal{B}_{\min}(K)$ be the subcollection of representatives whose braid index is equal to the braid index of K. Choose any $X_+ \in \mathcal{B}(K)$ and any $X_- \in \mathcal{B}_{\min}(K)$. Then there is a complexity function which is associated to X_+, X_- and for each braid index m a finite set $\mathcal{T}(m)$ of templates is introduced, each template determining a move which is non-increasing on braid index, such that the following hold: First, there are initial sequences which modify $X_- \to X'_-$ and $X_+ \to X'_+$:*

$$X_- = X_-^1 \to \cdots \to X_-^p = X'_-, \quad X_+ = X_+^1 \to \cdots \to X_+^q = X'_+.$$

Each passage $X_-^j \to X_-^{j+1}$ is strictly complexity reducing and is realized by an exchange move, so that $b(X_-^{j+1}) = b(X_-^j)$. These moves "unwind" X_-, if it is wound up as in the top right sketch in Figure 15. Each passage $X_+^j \to X_+^{j+1}$ is strictly complexity-reducing and is realized by either an exchange move or a destabilization, so that $b(X_+^{j+1}) \leq b(X_+^j)$. Replacing X_+ with X'_+ and X_- with X'_-, there is an additional sequence which modifies X'_+, *keeping X'_- fixed:*

$$X'_+ = X^q \longrightarrow \cdots \longrightarrow X^r = X'_-.$$

Each passage $X^j \longrightarrow X^{j+1}$ in this sequence is also strictly complexity-reducing. It is realized by one of the moves defined by a template $\mathcal{T}$ in the finite set $\mathcal{T}(m)$, where $m = b(X_+)$. The inequality $b(X^{j+1}) \leq b(X^j)$ holds for each $j = q, \ldots, r-1$ and so also for each $j = 1, \ldots, r-1$.

The proof of Theorem 3.5 uses the braid foliation techniques that were used in the proof of Theorem 3.1, but in a more complicated setting. Instead of looking at a foliated embedded disc which is bounded by a given unknotted closed braid, we are given two closed braids, X_+ and X_-, and an isotopy that takes X_+ to X_-. The trace of the isotopy sweeps out an annulus, but in general it is not embedded. The proof begins by showing that the given isotopy can be split into two parts, over which we have some control. An intermediate link X_0 which represents the same link type K as X_+ and X_- is constructed, such that the trace of the isotopy from X_+ to X_0 is an embedded annulus $\mathcal{A}_+$. Also the trace of the isotopy from X_0 to X_- is a second embedded annulus $\mathcal{A}_-$. The union of these two embedded annuli $\mathcal{TA} = \mathcal{A}_+ \cup \mathcal{A}_-$ is an immersed annulus, but its self-intersection set is controlled, and is a finite number of clasp arcs. The main tool in the proof of Theorem 3.5 is the study of the braid foliation of the immersed annulus $\mathcal{TA}$. In Section 3.3, we will see how Theorem 3.5 was used to settle a long-standing problem about contact structures.

3.3. *Braids and contact structures*

In this section we describe how Theorem 3.5 was used in [26] to settle a problem about contact structures on $\mathbb{R}^3$ and S^3.

Let **A** be the z-axis in $\mathbb{R}^3$, with standard cylindrical coordinates (ρ, θ, z) and let **H** be the collection of all half-planes H_θ through **A**. The pair (**A**, **H**) defines the *standard braid* structure on $\mathbb{R}^3$. Using the same cylindrical coordinates, let α be the 1-form $\alpha = \rho^2 \, d\theta + dz$. The kernel ξ of the 1-form α defines a *contact structure* on $\mathbb{R}^3$. We can visualize ξ by imagining that there is a 2-plane (spanned by $\partial/\partial x$ and $\partial/\partial \rho$) attached to every point in $\mathbb{R}^3$. Figure 21 shows both the braid structure and the polar contact structure, for comparison. The family of 2-planes that define ξ twist (to the left) as one moves along the x-axis from 0 to ∞. The family is invariant under rotation of 3-space about the z-axis and under translation of 3-space along rays parallel to the z-axis. Its salient feature is that it is totally non-integrable, that is there is no surface in $\mathbb{R}^3$ which is everywhere tangent to the 2-planes of (ξ) in any neighborhood of any point in $\mathbb{R}^3$. (Of course this makes it hard to visualize.) The twisting is generic in the sense that, if p is a point in a contact 3-manifold M^3, then in every neighborhood of p in M^3 the contact structure is locally like the one we depicted in Figure 21.

Let K be a knot (for simplicity we restrict to knots here, but everything works equally well for links) which is parametrized by cylindrical coordinates $(\rho(t), \theta(t), z(t))$, where $t \in [0, 2\pi]$. Then, as defined in Section 2.1, K is a closed braid if $\rho(t) > 0$ and $d\theta/dt > 0$

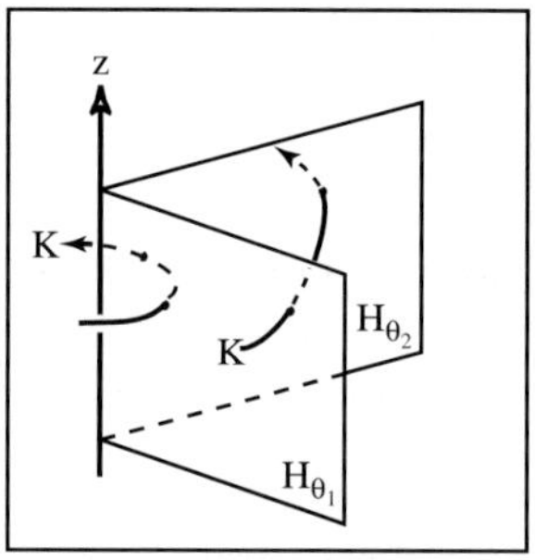

standard braid structure

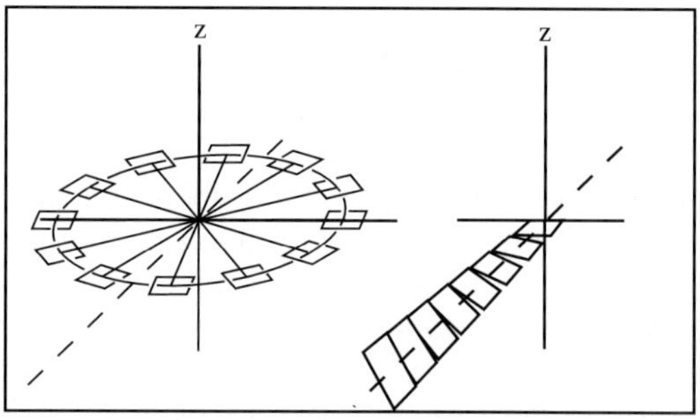

standard tight polar contact structure

Fig. 21. The standard braid structure and tight contact structure on $\mathbb{R}^3$.

for all t. On the other hand, K is a *Legendrian* (resp. *transversal*) knot if it is everywhere (resp. nowhere) tangent to the 2-planes of ξ. In the Legendrian case this means that on K we have $\mathrm{d}\theta/\mathrm{d}t = (-1/\rho^2)(\mathrm{d}z/\mathrm{d}t)$. In the transversal case we require that $\mathrm{d}\theta/\mathrm{d}t > (-1/\rho^2)$ $(\mathrm{d}z/\mathrm{d}t)$ at every point of $K(t)$. We are interested here primarily in the transversal case.

The *total twist* of the contact structure is the number of multiples of $\pi/2$ as one traverses the positive real axis from the origin to ∞. The case when the total twist angle is π is known as the *standard* (polar) contact structure. We call it ξ_π. While the braid structure is very different from the standard polar contact structure near **A**, for large values of ρ the 2-planes in ξ_π are very close to the half-planes H_θ of the braid structure. In fact, assume that X is a closed braid which represents a knot K, and that X bounds a Seifert surface of minimum genus which supports a foliation as in Section 3.1. Assume further that the closed braid X intersects every 2-plane in the contact structure transversally. Then, in the complement of a tubular neighborhood of the braid axis, from the point of view of a topologist, the braid foliation and the foliation induced by the contact structure will be "the same".

Recall that Theorem 2.1 (Alexander's Theorem) was first proved in 1925. Sixty years later, Bennequin adapted Alexander's original proof (which is different from the proof given in this article) to the setting of transversal knots in [8], where he showed that every transversal knot is isotopic, through transversal knots, to a closed braid. In 2002 Orevkov and Shevchishin extended Bennequin's ideas and proved a version of Theorem 2.8 (Markov's Theorem) which holds in the transversal setting:

THEOREM 3.6 [101]. *Let TX_+, TX_- be closed braid representatives of the same oriented link type K in oriented 3-space. Then there exists a sequence of closed braid representatives of $\mathcal{TK}$:*

$$TX_+ = TX_1 \to TX_2 \to \cdots \to TX_r = TX_1$$

such that, up to braid isotopy, each TX_{i+1} is obtained from TX_i by a single positive stabilization or destabilization.

Is the Transverse Markov Theorem really different from the Markov Theorem? Are there transversal knots which are isotopic as topological knots but are not transversally isotopic? To answer this question we take a small detour and review the contributions of Bennequin [8].

Why did topologists begin to think about contact structures, and analysts begin to think about knots? While we might wish that analysts suddenly became overwhelmed with the beauty of knots, there was a more specific and focused reason. At the time that Bennequin did his foundational work in [8] it was not known whether a 3-manifold could support more than one isotopy class of contact structures. Bennequin answered this question in the affirmative, in the case of contact structures on $\mathbb{R}^3$ or S^3 which were known to be homotopic to the standard one. His tool for answering it was highly original, and it had to do with braids and knots.

Let TK be a transversal knot. By the above, we can assume without loss of generality that it is a closed braid. Let $\mathcal{TK}$ be its transversal knot type, i.e., its knot type under transversal isotopies, and let $[\mathcal{TK}]_{\text{top}}$ be its topological knot type. Choose a representative TX of $\mathcal{TK}$, which (by Bennequin's transversal version of Alexander's theorem) is always possible. Choose a Seifert surface F of minimal genus, with $TX = \partial F$. Bennequin studied the foliation of F which is induced by the intersections of F with the plane field determined by ξ_π. Let $n(TX)$ be the braid index and let $e(TX)$ be the algebraic crossing number of a projection of TX onto the plane $z = 0$. Both can be determined from the foliation. Bennequin found an invariant of $\mathcal{TK}$, given by the formula $\beta(\mathcal{TK}) = e(TX) - n(TX)$. Of course if he had known Theorem 3.6 the proof that $\beta(\mathcal{TK})$ is an invariant of $\mathcal{TK}$ would have been trivial, but he did not have that tool. He then showed a little bit more: he showed that $\beta(\mathcal{TK})$ is bounded above by the negative of the Euler characteristic of F in ξ_π. He then showed that this bound fails in one of the contact structures $\xi_{>\pi}$. In this way he proved that the contact structures $\xi_{>\pi}$ cannot be isotopic to ξ_π.

To knot theorists, Bennequin's proof should seem intuitively natural, because the invariant $\beta(\mathcal{TK})$ is a self-linking number of a representative $TX \in \mathcal{TK}$ (the sense of push-off being determined by ξ), and the more twisting there is the higher this number can be. For an explanation of the self-linking, and lots more about Legendrian and transversal knots we refer the reader to John Etnyre's excellent review article [55]. The basic idea is that TX bounds a Seifert surface, and this Seifert surface is foliated by the plane field associated to ξ. Call this foliation the *characteristic foliation.* Near the boundary, the characteristic foliation is transverse to the boundary. The Bennequin invariant is the linking number of TX, TX', where TX' is a copy of TX, obtained by pushing TX off itself onto F, using the direction determined by the characteristic foliation of F.

Bennequin's paper was truly important. Shortly after it was written Eliashberg [51] showed that the phenomenon of an infinite sequence of contact structures related to a single one of minimal twist angle occurred generically in every 3-manifold, and introduced the term "tight" and "overtwisted" to distinguish the two cases. Here too, there is a reason that will seem natural to topologists. In 2003 Giroux proved [70] that every contact structure on every closed, orientable 3-manifold M^3 can be obtained in the following way: represent M^3 as a branched covering space of S^3, branched over a knot or link, and lift the standard and overtwisted contact structures on S^3 to M^3.

Returning to knot theory, the invariant $\beta(\mathcal{TK})$ allows us to answer a fundamental question: is the equivalence relation on knots that is defined by transversal isotopy really different from the equivalence relation defined by topological isotopy? The Bennequin invariant will be used to answer this question in the affirmative:

THEOREM 3.7 [8]. *There are infinitely many distinct transversal knot types associated to each topological knot type.*

PROOF. Choose a transversal knot type $\mathcal{TK}$ and a closed braid representative TX_0. Stabilizing the closed braid TX_0 once negatively (recall the definition of positive/negative stabilizations given in Section 2.3, Figure 7), we obtain the transverse closed braid TX_1, with $e(TX_1) = e(TX_0) - 1$ and $n(TX_1) = n(TX_0) + 1$, so that $\beta(TX_1) = \beta(TX_0) - 2$. Iterating, we obtain transverse closed braids $TX_2, TX_3, \ldots$, defining transverse knot types $\mathcal{TK}_1, \mathcal{TK}_2, \mathcal{TK}_3, \ldots$, and no two have the same Bennequin invariant. Since stabilization does not change the topological knot type, the assertion follows. □

At this writing, it is an open problem to find computable invariants of $\mathcal{TK}$ which are not determined by $[\mathcal{TK}]_{top}$ and $\beta(\mathcal{TK})$. A hint that the problem might turn out to be quite subtle was in the paper by Fuchs and Tabachnikov [63], who proved that while ragbags filled with polynomial and finite type invariants of transversal knot types $\mathcal{TK}$ exist, based upon the work of Arnold in [4], they are all determined by $[\mathcal{TK}]_{top}$ and $\beta(\mathcal{TK})$. Thus, the seemingly new invariants that many people had discovered by using Arnold's ideas were just a fancy way of encoding $[\mathcal{TK}]_{top}$ and $\beta(\mathcal{TK})$.

This leads naturally to a question: Are there computable invariants of transversal knots which are *not* determined by $[\mathcal{TK}]_{top}$ and $\beta(\mathcal{TK})$? A similar question arises in the setting of Legendrian knots. Each Legendrian knot $\mathcal{LK}$ determines a topological knot type $[\mathcal{LK}]_{top}$, and just as in the transverse case it is an invariant of the Legendrian knot type. There are also two numerical invariants of $\mathcal{LK}$: the Thurston–Bennequin invariant $tb(\mathcal{LK})$ (a self-linking number) and the Maslov index $M(\mathcal{LK})$ (a rotation number). So until a few years ago the same question existed in the Legendrian setting, but the Legendrian case has recently been settled by Yuri Chekanov:

THEOREM 3.8 [40]. *There exist distinct Legendrian knot types which have the same topological knot type $[\mathcal{LK}]_{top}$, and also the same Thurston–Bennequin invariant $tb(\mathcal{LK})$ and Maslov index $M(\mathcal{LK})$.*

The analogous result for transversal knots proved to be quite difficult, so to begin to understand whether something could be done via braid theory, the first author and Nancy Wrinkle asked an easier question, which they answered in part in [29]: are there transversal knot types which *are determined* by their topological knot type and Bennequin number? This question leads to a definition: a transversal knot type $\mathcal{TK}$ is *transversally simple* if it is determined by $[\mathcal{TK}]_{top}$ and $\beta(\mathcal{TK})$. So the question is: are there transversally simple knots? The manuscript [29] gives a purely topological (in fact braid-theoretic) criterion which enables one to answer the question affirmatively, adding one more piece of evidence that topology and analysis walk hand in hand. To explain it, recall the destabilization and exchange moves, and the two flypes, depicted in Figure 20. A topological knot or link type K is said to be *exchange reducible* if an arbitrary closed braid representative X of K can be changed to an arbitrary representative of minimum braid index by braid isotopy, positive and negative destabilizations and exchange moves. We have:

THEOREM 3.9 [29]. *If a knot type K is exchange reducible, then any transversal knot type $\mathcal{TK}$ which has $[\mathcal{TK}]_{\text{top}} = K$ is transversally simple.*

This theorem was used to give a new proof of a theorem of Eliashberg [52], which asserts that the unlink is transversally simple, and also (with the help of [92]) to prove the then-new result that most iterated torus knots are transversally simple.

The rest of this section will be directed at explaining the main result of [26]:

THEOREM 3.10 [26]. *There exist transversal knot types which are not transversally simple. Explicitly, the transverse closed 3-braids $TX_+ = \sigma_1^5\sigma_2^4\sigma_1^6\sigma_2^{-1}$ and $TX_- = \sigma_1^5\sigma_2^{-1}\sigma_1^6\sigma_2^4$ determine transverse knot types $\mathcal{TK}_+$, $\mathcal{TK}_-$ with $[\mathcal{TK}_+]_{\text{top}} = [\mathcal{TK}_-]_{\text{top}}$ and $\beta(\mathcal{TK}_+) = \beta(\mathcal{TK}_-)$, but $\mathcal{TK}_+ \neq \mathcal{TK}_-$.*

SKETCH OF THE PROOF OF THEOREM 3.10. See [26] for all details. The examples in Theorem 3.10 were obtained by choosing all the weights in the negative flype template to be 1, and assigning explicit 2-braids to the blocks P, Q, R of the negative flype template of Figure 20. If the weights are all chosen to be 1, the blocks P, Q, R are 2-braids and (except in very special cases) the flype templates have braid index 3. First one must show that the examples satisfy the conditions of the theorem. The topological knot types defined by the closed 3-braids $\sigma_1^5\sigma_2^4\sigma_1^6\sigma_2^{-1}$ and $\sigma_1^5\sigma_2^{-1}\sigma_1^6\sigma_2^4$ coincide because they are carried by the block-strand diagrams for the negative flype template of Figure 20. The Bennequin invariant can be computed as the exponent sum of the braid word (14 in both cases) minus the braid index (three in both cases). So the examples have the required properties.

The hard part is the establishment of a special version of Theorem 3.5 which is applicable to the situation. Its special features are as follows:

(1) Both X_+ and X_- have braid index 3.
(2) Since it is well known that exchange moves can be replaced by braid isotopy for 3-braids, the first two sequences in Theorem 3.5 are vacuous, i.e., $X_\pm = X'_\pm$.
(3) Because of the special assumption just noted, the templates that are needed, in the topological setting, can be enumerated explicitly: they are the positive and negative destabilization and the positive and negative flype templates. No others are needed.

It is proved in [26] that if X_- and X_+ are transversal closed braids TX_+ and TX_-, then the isotopy that takes TX_+ to TX_- may be assumed to be transversal. So, suppose that a transversal isotopy exists from the transverse closed braid TX_+ to the transverse closed braid TX_-. Then there is a 3-braid template that carries the braids $\sigma_1^5\sigma_2^4\sigma_1^6\sigma_2^{-1}$ and $\sigma_1^5\sigma_2^{-1}\sigma_1^6\sigma_2^4$. This is the first key fact that we use from Theorem 3.5. Instead of having to consider *all* possible transversal isotopies from TX_+ to TX_-, we only need to consider those that relate the left and right block-strand diagrams in one of the four 3-braid templates. So the braids in question are carried either by one of the two destabilization templates or by one of the two flype templates. If it was one of the destabilization templates, then the knots in question could be represented by 2 or 1-braids, i.e., they would be type (2,n) torus knots or the unknot, however an easy argument shows that the knots in Theorem 3.10 are neither type (2,n) torus knots or the unknot. The positive flype templates are ruled out in different way: topologically, our closed braids admit a negative flype, so if they are also carried by the positive flype template they admit flypes of both signs. However, the manuscript [81] gives conditions under which a closed 3-braid admits flypes of both signs, and the examples were chosen explicitly to rule out that possibility.

We are reduced to isotopies that are supported by the negative flype template. It is straightforward to show that the obvious isotopy is not transversal, but maybe there is some *other* isotopy which is transversal. Here a key fact about the definition of a template is used (and this is a second very strong aspect of the MTWS). If such a transversal isotopy exists, then it exists for every knot or link defined by a fixed choice of braiding assignments to the blocks. Choose the braiding assignments σ_1^3, σ_2^4, σ_1^{-5} to the blocks P, R, Q. This braiding assignment gives a 2-component link $L_1 \sqcup L_2$ which has two distinct isotopy classes of closed 3-braid representatives. If L_1 is the component associated to the left strand entering the block P, then $\beta(L_1) = -1$ and $\beta(L_2) = -3$ before the flype, but after the flype the representative will be $\sigma_1^3\sigma_2^{-1}\sigma_1^{-5}\sigma_2^4$, with $\beta(L_1) = -3$ and $\beta(L_2) = -1$. However, by Eliashberg's isotopy extension theorem (Proposition 2.1.2 of [52]) a transversal isotopy of a knot/link extends to an ambient transversal isotopy of the 3-sphere. Any transversal isotopy of $L_1 \sqcup L_2$ must preserve the β-invariants of the components. It follows that no such transversal isotopy exists, a contradiction of our assumption that TX_+ and TX_- are transversally isotopic. □

Other examples of a similar nature were discovered by Etnyre and Honda [56] after the proof of Theorem 3.10 was posted on the arXiv. Their methods are very different from the proof that we just described (being based on contact theory techniques rather than topological techniques), but are equally indirect. They do not produce explicit examples, rather they present a bag of pairs of transverse knots and prove that at least one pair in the bag exists with the properties given by Theorem 3.10. Therefore we pose, as an important open problem:

OPEN PROBLEM 4. Find new computable invariants of transversal knot types. Here "new" means an invariant which is not determined by the topological knot type $\mathcal{TX}$ and the Bennequin invariant $\beta(\mathcal{TX})$. "Computable" means that it should be computable from either a closed braid diagram or some other representation of the transversal knot. Braid groups seem to be a natural setting for investigating this problem.

REMARK 3.11. The connections between closed braids and contact structures does not end with transversal knots. There are fundamental relationships between open book structures on 3-manifolds and contact structures on those manifolds, both untwisted and twisted. See [55] for an introduction to this interesting new area, and see [70] for a review of the mathematics and a discussion of many open problems waiting to be investigated.

4. Representations of the braid groups

Before the discovery of Hecke algbra representations of the braid group (discussed in this section) very little was known about finite dimensional but infinite representations of $\mathbf{B}_n$, except for the ubiquitous Burau representation. That matter changed dramatically in 1987 with the publication of [78]. Suddenly, we had more knot invariants and with them more braid group representations than anyone could deal with, and the issue became one of organizing them. However, we shall not attempt to give a comprehensive overview of the rich theory of representations of braid groups in this section. Instead, we focus here on the representations of $\mathbf{B}_n$ which have played the greatest roles in the development of that theory: the Burau representation, the Hecke algebra representations, and, more recently, the Lawrence–Krammer representation.

4.1. *A brief look at representations of Σ_n*

The fact that the representation theory of $\mathbf{B}_n$ is rooted in the representation theory of the symmetric group Σ_n is a consequence of the surjection $\mathbf{B}_n \rightarrow \Sigma_n$, given by mapping the elementary braid σ_i to the transposition $s_i = (i, i+1)$. While the kernel of this homomorphism is very big (it is the entire pure braid group $\mathbf{P}_n$), it nevertheless turns out that a great deal can be learned about representations of $\mathbf{B}_n$ by studying the collection of irreducible representations of Σ_n, and attempting to lift them to representations of $\mathbf{B}_n$ by "deforming" them, and hoping that you get something new. For the record, we now note that the group algebra $\mathbb{C}\Sigma_n$ has generators $1, s_1, \ldots, s_{n-1}$ and defining relations:

$$s_i s_k = s_k s_i \quad \text{if } |i-k| \geq 2, \quad s_i s_{i+1} s_i = s_{i+1} s_i s_{i+1}, \quad s_i^2 = 1, \tag{4.1}$$

where $1 \leq i \neq k \leq n-1$. As a vector space, $\mathbb{C}\Sigma_n$ is spanned by $n!$ reduced words [30] in the transpositions s_i:

$$\{(s_{i_1} s_{i_1-1} \cdots s_{i_1-k_1})(s_{i_2} s_{i_2-1} \cdots s_{i_2-k_2}) \cdots (s_{i_r} x_{i_r-1} \cdots s_{i_r-k_r})\} \tag{4.2}$$

where $1 \leq i_1 < i_2 < \cdots < i_r \leq n-1$ and $i_j - k_j \geq 1$.

Irreducible representations of Σ_n are parametrized by *Young diagrams*, which in turn are parametrized by partitions of n. (See [64] for one of the many good discussions of this subject in the literature.) A Young diagram consists of stacked rows of boxes, aligned on the left, with the number of boxes in each row strictly non-increasing as you go from top to bottom. The number of boxes in a row corresponds to a term in a given partition of n. For example, the Young diagram shown in Figure 22 corresponds to the partition $8 = 4 + 2 + 1 + 1$.

As it turns out, the irreducible representations of the symmetric group are in 1–1 correspondence with Young diagrams, or equivalently with partitions of n. There is a row–column symmetry, and we adopt the convention that the Young diagram which has one row of n boxes, which comes from the partition $n = n$, corresponds to the trivial representation. Then the Young diagram which has n rows consisting of 1 box per row,

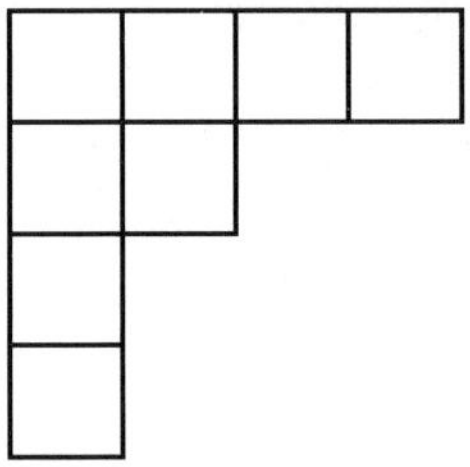

Fig. 22. The Young diagram corresponding to the partition $4 + 2 + 1 + 1$.

which comes from the partition $n = 1 + 1 + \cdots + 1$, corresponds to the parity representation, mapping each element of Σ_n to its sign. The first interesting representation of a braid group occurs when $n = 3$, corresponding to the partition $3 = 2 + 1$. More generally, we consider the partition $n = (n - 1) + 1$, or the Young diagram with two rows, the top row having $(n - 1)$ boxes and the bottom having a single box. This Young diagram corresponds to the *standard representation* $\Sigma_n \to \mathrm{GL}_n(\mathbb{C})$, given by the usual action of Σ_n on $\mathbb{C}^n$ by permuting basis vectors. Thus, if s_i denotes the transposition $(i \ i+1) \in \Sigma_n$, the standard representation sends s_i to the following:

$$I_{i-1} \oplus \begin{pmatrix} 0 & 1 \\ 1 & 0 \end{pmatrix} \oplus I_{n-i-1}$$

where I_k denotes the $k \times k$ identity matrix.

The standard representation is reducible; it is easy to see that it fixes a 1-dimensional subspace, namely, the span of the sum of the basis vectors. The complementary $(n - 1)$-dimensional subspace is the set of all points $(z_1, \ldots, z_n) \in \mathbb{C}^n$ such that $z_1 + \cdots + z_n = 0$. This subspace corresponds to the irreducible representation given by the Young diagram in question. One can use the *hook length formula* to calculate directly that the dimension of the representation corresponding to this Young diagram is indeed $n - 1$ (we refer the reader again to [64] for an explanation of this formula).

4.2. *The Burau representation and polynomial invariants of knots*

Burau first introduced his representation of the braid group in 1936 [37]. Much later, it was realized that it could be thought of as a deformation of the standard representation of Σ_n corresponding to the partition $n = (n - 1) + 1$. For many years it was the focus of the representation theory of braid groups. We define the *Burau representation* $\rho : \mathbf{B}_n \to \mathrm{GL}_n(\mathbb{Z}[t, t^{-1}])$ as follows:

$$\sigma_i \mapsto I_{i-1} \oplus \begin{pmatrix} 1 - t & t \\ 1 & 0 \end{pmatrix} \oplus I_{n-i-1}.$$

Note that substituting $t = 1$ gives back the representation (of $\mathbf{B}_n$ factoring through Σ_n), and this is why we say that it is a deformation of the standard representation of Σ_n. Like the representation of Σ_n, the Burau representation splits into a 1-dimensional representation and an $(n-1)$-dimensional irreducible representation known as the *reduced Burau representation* which we denote by $\bar{\rho}: \mathbf{B}_n \to \mathrm{GL}_n(\mathbb{Z}[t,t^{-1}])$ as follows:

$$\sigma_i \mapsto I_{i-2} \oplus \begin{pmatrix} 1 & -t & 0 \\ 0 & -t & 0 \\ 0 & -1 & 1 \end{pmatrix} \oplus I_{n-i-2}$$

where the $-t$ in the middle of the 3×3 matrix is always in the (i,i)th spot.

It has been known for a long time that the Burau representation is faithful for $n \leq 3$ (see [15], for example), and for many years the representation was held to be a reasonable candidate for a faithful linear representation of $\mathbf{B}_n$ for all n. However, in 1991, Moody showed that for $n \geq 9$, the Burau representation is *not* faithful [94]. Long and Paton [90] later improved Moody's result to $n \geq 6$. Bigelow further improved this to $n \geq 5$ [12]. At the time of this writing, the case $n = 4$ remains open.

Despite such results, the Burau representation continues to play an important role in the study of representations of $\mathbf{B}_n$. It was known classically that the Alexander polynomial $\Delta_K(t)$ of a knot or link K can be calculated directly from the image under the reduced Burau representation of a braid X such that $b(X)$ represents K, as follows (see [15], e.g., for a proof based upon Theorem 2.8, the Markov Theorem):

$$\Delta_{b(X)}(t) = \frac{\det(\bar{\rho}(X) - I_{n-1})}{1 + t + \cdots + t^{n-1}}. \tag{4.3}$$

Thus, the Alexander polynomial of the closed braid associated to the open braid X, i.e., $\Delta_{b(X)}(t)$, is a rescaling of the characteristic polynomial of the image of X in the reduced representation. In what follows, we shall see how a property of the Burau representation motivated the definition of a variant on the two-variable HOMFLY polynomial [66], a knot invariant of which both the Alexander polynomial and the Jones polynomial are specializations. Further, in Section 4.4 we shall give a topological interpretation of the Burau representation which naturally leads to the definition of a faithful linear representation of $\mathbf{B}_n$ known as the Lawrence–Krammer representation, which will be our focus in Section 4.5.

4.3. *Hecke algebras representations of braid groups and polynomial invariants of knots*

A simple calculation, together with the Cayley–Hamilton theorem, shows that the image of each of our braid group generators under the Burau representation, $\rho(\sigma_i)$, satisfies the characteristic equation $x^2 = (1-t)x + t$ and thus has two distinct eigenvalues. This prompted Jones to study all representations $\rho: \mathbf{B}_n \to \mathrm{GL}_n(\mathbb{C})$ which have at most two distinct eigenvalues [78]. Let $x_i = \rho(\sigma_i)$. Then for all i, x_i must satisfy a quadratic equation of the form $x_i^2 + ax_i + b = 0$. By rescaling, we may assume that one of the eigenvalues is 1 and eliminate one of the variables, e.g., $a = -(1+b)$. Note that by rewriting our quadratic equation and making the substitution $b = -t$ we regain the characteristic equation from the Burau representation. However, the convention in the literature seems to be to rescale our representation by (-1) so that the equation takes the form $x_i^2 = (t-1)x_i + t$. With this motivation, we define the *Hecke algebra* $H_n(t)$ to be the algebra with generators $1, x_1, \ldots, x_{n-1}$ and defining relations as follows:

$$x_i x_k = x_k x_i \quad \text{if } |i-k| \geq 2, \; x_i x_{i+1} x_i = x_{i+1} x_i x_{i+1}, \; x_i^2 = (t-1)x_i + t, \tag{4.4}$$

where $1 \leq i \neq k \leq n-1$. Comparing the relations in (4.1) and (4.4), we see that $H_n(1) \cong \mathbb{C}\Sigma_n$, the group algebra of the symmetric group. Hence we can think of $H_n(t)$ as a "deformation" of $\mathbb{C}\Sigma_n$.

The connection between $H_n(t)$ and $\mathbb{C}\Sigma_n$ is made even more transparent by noting that as a vector space, $H_n(t)$ is spanned by $n!$ lifts of a system of reduced words in the transpositions $s_i \in \Sigma_n$. For example, we can take as a spanning set

$$\{(x_{i_1} x_{i_1 - 1} \cdots x_{i_1 - k_1})(x_{i_2} x_{i_2 - 1} \cdots x_{i_2 - k_2}) \cdots (x_{i_r} x_{i_r - 1} \cdots x_{i_r - k_r})\} \tag{4.5}$$

where $1 \leq i_1 < i_2 < \cdots < i_r \leq n-1$ and $i_j - k_j \geq 1$ [30,79].

REMARK 4.1. In Section 1.4.2 we defined an algebra $J_n(t)$ with generators $1, g_1, \ldots, g_{n-1}$ and defining relations (1.7), the *Jones algebra*. As it turns out, its irreducible summands are in 1–1 correspondence with the irreducible representations of the Hecke algebra that are parametrized by Young diagrams with exactly two rows. The Hecke algebra, as defined above, has generators maps $1, x_1, \ldots, x_{n-1}$ with defining relations (4.4), so that the map $\xi : H_n(t) \to J_n(t)$ that is defined by $\xi(x_i) = g_i$ is a homorphism of algebras.

Our main purpose in this section is to outline Jones' development in [78] of a two-variable polynomial knot invariant arising from representations of the Hecke algebras $H_n(t)$. This polynomial is essentially the well-known HOMFLY polynomial, and includes the Jones polynomial as a specialization. We have just seen that the Jones algebra is the quotient of the Hecke algebra $H_n(t)$ by one extra relation (as an aside, we note that this extra relation is satisfied by the image of the transpositions generators $s_i = (i, i+1)$ in Σ_n under all representations arising from Young diagrams with at most two rows.) To pick up a second variable, we introduce an additional parameter by allowing a 1-parameter family of traces on Hecke algebras. We now pursue this point of view, and we also refer the reader to [74] for another exposition which follows the same point of view.

We begin by defining a function $f : \mathbf{B}_n \to H_n(t)$ by $f(\sigma_i) = x_i$. The function f is well defined on reduced words in the generators σ_i and commutes with the natural inclusions $\mathbf{B}_{n-1} \subset \mathbf{B}_n$ and $H_{n-1}(t) \subset H_n(t)$, although in general f fails to be a homomorphism. We can then apply the following result due to Adrian Ocneanu which appeared in [66] and was proved inductively in [78] using the n-element basis given above.

THEOREM 4.2 [66,78]. *For each $z \in \mathbb{C}^*$ (and each $t \in \mathbb{C}^*$), there exists a unique trace function* $\mathrm{tr} : \cup_{n=1}^{\infty} H_n(t) \to \mathbb{C}$ *such that*

(1) $\mathrm{tr}(1) = 1$
(2) $\mathrm{tr}(ab) = \mathrm{tr}(ba)$

(3) tr *is* $\mathbb{C}$*-linear*
(4) $\mathrm{tr}(ux_{n-1}v) = z\,\mathrm{tr}(uv)$ *for all* $u, v \in H_{n-1}(t)$.

Theorem 4.2 gives us a one-parameter family of trace functions on a one-parameter family of algebras. In fact, using the properties of the trace function given in theorem it is possible to compute $\mathrm{tr}(f(X))$ for all $X \in \mathbf{B}_n$. (We note the fact that for any $w \in H_n(t)$ such that $w \notin H_{n-1}(t)$, there is a unique reduced word $w = x_{i_1} \cdots x_{i_r}$ in which x_{n-1} appears exactly once [78].) In practice, the third relation of $H_n(t)$ is quite useful for computing the trace function tr, both in its original form and in the following:

$$x_i^{-1} = t^{-1}x_i + (t^{-1} - 1).$$

EXAMPLE 4.3. Let $X_1 = \sigma_1^3 \in \mathbf{B}_2$, and let $X_2 = \sigma_1\sigma_2^{-1}\sigma_1\sigma_2^{-1} \in \mathbf{B}_3$. Note that $\hat{X}_1$ is the right-handed trefoil knot and that $\hat{X}_2$ is the Figure 8 knot. We invite the reader to check that

$$\mathrm{tr}(f(X_1)) = (t^2 - t + 1)z + t(t - 1)$$

and that

$$\mathrm{tr}(f(X_2)) = (3 - t^{-1} - t)t^{-1}z^2 + (3 - t^{-1} - t)(t^{-1} - 1)z - (2 - t^{-1} - t).$$

We also note that the second property of the trace function given in Theorem 4.2 implies that $\mathrm{tr} \circ f$ is invariant on conjugacy classes in $\mathbf{B}_n$. It remains to tweak the function a bit in order to obtain from a given braid a two-variable polynomial which is also invariant under stabilization and destabilization moves as defined in Section 2.3; such a polynomial will be Markov invariant and hence an invariant of the knot type of the closed braid. Algebraically, stabilization and destabilization each take the form $X \to X\sigma_n^{\pm 1}$, the only difference being appropriate conditions on the braid X. We would like to rescale our representation f in such a way that both versions of stabilization (resp. destabilization) have the same effect on the trace function. Suppose there exists a complex number k such that $\mathrm{tr}(kx_i) = \mathrm{tr}((kx_i)^{-1})$. Then we can find a "formula" for k as follows:

$$k^2\mathrm{tr}(x_i) = \mathrm{tr}(x_i^{-1})$$

$$k^2 z = \mathrm{tr}(t^{-1}x_i + t^{-1} - 1)$$

$$k^2 = \frac{t^{-1}z + t^{-1} - 1}{z}$$

$$k^2 = \frac{1 + z - t}{tz}.$$

Solving this for z, we obtain

$$z = -\frac{1 - t}{1 - k^2 t}.$$

We set $\kappa = k^2$, and define $f_\kappa : \mathbf{B}_n \to H_n(t)$ by $f_\kappa(\sigma_i) = \sqrt{\kappa}\sigma_i$. Now we have

$$\mathrm{tr}(f_\kappa(\sigma_n)) = \sqrt{\kappa}z = -\sqrt{\kappa}\frac{1-t}{1-\kappa t}.$$

We would like to define a map $\mathbf{B}_n \to \mathbb{Z}[t^{\pm 1}, \kappa^{\pm 1}]$ which is Markov invariant. At the moment, we have that

$$\mathrm{tr}(f(w\sqrt{\kappa}\sigma_n)) = -\sqrt{\kappa}\frac{1-t}{1-\kappa t}\mathrm{tr}(f(w)) = \mathrm{tr}\left(f\left(w\frac{1}{\sqrt{\kappa}}\sigma_n^{-1}\right)\right)$$

for any $w \in \mathbf{B}_n$. Now we simply define

$$\begin{aligned} F(X) = F_X(t,\kappa) &= \left(-\frac{1}{\sqrt{\kappa}}\frac{1-\kappa t}{1-t}\right)^{n-1}\mathrm{tr}(f_\kappa(X)) \\ &= \left(-\frac{1}{\sqrt{\kappa}}\frac{1-\kappa t}{1-t}\right)^{n-1}(\sqrt{\kappa})^E\mathrm{tr}(f(X)) \end{aligned}$$

for $X \in \mathbf{B}_n$, where E is the exponent sum of X as a word in $\sigma_1, \ldots, \sigma_{n-1}$. It is clear that $F(X)$ depends only on the knot type of $b(X)$. We now reparametrize one last time, setting

$$l = \sqrt{\kappa}\sqrt{t}, \quad m = \sqrt{t} - \frac{1}{\sqrt{t}}.$$

With this substitution, we obtain a Laurent polynomial in two variables l and m, which we denote $P_{b(X)}(l,m) = P_K(l,m)$, where K is the (oriented) knot or link type of $b(X)$. Furthermore, $P_K(l,m)$ satisfies the skein relation

$$mP_{K_0} = l^{-1}P_{K_+} - lP_{K_-} \tag{4.6}$$

where K_0, K_+, and K_- are oriented knots with identical diagrams except in a neighborhood of one crossing, where they have a diagram as given in Figure 23 (Proposition 6.2 of [78]). Thus, by beginning with $P_\mathrm{U} = 1$, where U denotes the unknot, it is possible to calculate P_K for any knot or link K using only the skein relation, which is often simpler than using the trace function.

Fig. 23. Crossings for K_+, K_-, and K_0, respectively, in the skein relation given in (4.6).

REMARK 4.4. We note that the polynomial $P_K(l, m)$ obtained in this way is essentially the same as the two-variable polynomial known as the HOMFLY polynomial [66] which is usually reparametrized as $P_K(il^{-1}, im)$.

EXAMPLE 4.5. Let X_1, X_2 be the braids defined in Example 4.3 whose closures are $K_1 =$ the right-handed trefoil knot and $K_2 =$ the Figure 8 knot, respectively. We leave it as an exercise for the reader to check that

$$F_{X_1}(t, \kappa) = \kappa(1 + t^2 - \kappa t^2) = \kappa t(t + t^{-1} - \kappa t) = \kappa t(2 - \kappa t + t + t^{-1} - 2)$$
$$= \kappa t\left(2 - \kappa t + \left(\sqrt{t} - \frac{1}{\sqrt{t}}\right)^2\right)$$

and hence we have

$$P_{K_1}(l, m) = 2l^2 - l^4 + l^2 m^2.$$

Similarly, the reader can check that

$$F_{X_2}(t, \kappa) = \frac{1 - \kappa(1 - t + t^2) + \kappa^2 t^2}{t\kappa}$$

$$P_{K_2}(l, m) = l^{-2} - m^2 - 1 + l^2.$$

For explicit calculations of F_{X_2} and P_{K_2} using the trace function, see p. 350 of [78].

REMARK 4.6. There is also a 1-variable knot polynomial, the *Jones polynomial*, associated to the algebra $J_n(t)$ generated by $1, g_1, \ldots, g_{n-1}$ with defining relations (1.7). In the situation of the Jones algebra the trace is unique, whereas in the situation of the Hecke algebra, as we presented it here, there is a 1-parameter family of traces. The 1-variable Jones polynomial was discovered before the 2-variable HOMFLY polynomial.

This two-variable knot polynomial has been much studied and reviewed in the literature. For the sake of completeness we list here a few of its noteworthy properties and applications.

(1) *Connect sums*: $P_{K_1 \# K_2} = P_{K_1} P_{K_2}$
(2) *Disjoint unions*: $P_{K_1 \coprod K_2} = ((l^{-1} - l)/m) P_{K_1} P_{K_2}$.
(3) *Orientation*: $P_{\bar{K}} = P_K$, where $\bar{K}$ denotes the link obtained by reversing the orientation of every component of the link K.
(4) *Chirality*: $P_{\tilde{K}}(l, m) = P_K(l^{-1}, -m)$, where $\tilde{K}$ denotes the mirror image of the link K.
(5) *Alexander polynomial*: Note that $F_X(1, \kappa)$ is not defined. It comes as something of surprise, then, that the specialization $l = 1$, $m = \sqrt{t} - (1/\sqrt{t})$ gives the Alexander polynomial $\Delta_K(t)$. Jones shows how to avoid the singularity by

exploiting an alternate method of calculating the trace function using weighted sums of traces (see [114] and [79] as well as [66] and [78]). A by-product of this alternate method is another derivation of (4.3) showing how to calculate $\Delta_K(t)$ from the Burau representation.

(6) *Jones polynomial*: The famous Jones polynomial can be obtained from the two-variable polynomial by setting

$$V_K(t) = P\Big(t, \sqrt{t} - \frac{1}{\sqrt{t}}\Big).$$

Note that we are abusing notation by reusing the variable t here and above in $\Delta_K(t)$.

(7) *A lower bound for braid index*: At the end of Section 2.2 we remarked that it is an open problem to determine the braid index of a knot algorithmically. However, the HOMFLY polynomial does give a remarkably useful lower bound, via a famous inequality which is known as the Morton–Franks–Williams inequality. It was proved simultaneously and independently by Hugh Morton in [98] and by John Franks and Robert Williams in [65]. While it proved to be sharp on all but five of the knots in the standard tables of knots having at most 10 crossings, there are also infinitely many knots on which it fails to be sharp. The first author came to a new appreciation of the importance of this problem when she was faced with the problem of determining, precisely, the braid index of certain 6-braid knots on which it failed. Consulting many people, it became apparent that there was essentially no other useful result on this basic question.

4.4. *A topological interpretation of the Burau representation*

We have seen that the Burau representation is deeply connected with the topology of closed braids. We shall now give a fully topological definition of the Burau representation. We outline here ideas given in [12]; for full details see that or Turaev's excellent survey article [112], among others.

Let D_n be an n-punctured disk, which we will think of as a disk D with n distinguished points $q_1, \ldots, q_n$. Choose a point $d_0 \in \partial D_n$ to serve as the basepoint. Then $\pi_1(D_n, d_0)$ is free on n generators which can be represented by loops x_i based at d_0 travelling counterclockwise about the puncture q_i for $i = 1, \ldots, n$. We define a surjective map $\epsilon : \pi_1(D_n, d_0) \rightarrow \mathbb{Z}$ as follows. Let $\gamma = x_{i_1}^{n_1} \cdots x_{i_r}^{n_r} \in \pi_1(D_n, d_0)$. Then we define the exponent sum $\epsilon(\gamma) = \sum_{i=1}^{r} n_i$. The integer $\epsilon(\gamma)$ can be interpreted as the *total algebraic winding number* of γ about the punctures $\{q_i\}$, i.e., the sum over all $i = 1, \ldots, n$ of the winding number of γ about q_i. Now there is a regular covering space $\tilde{D}_n$ of D_n corresponding to the kernel of the map ϵ. Since $\epsilon : \pi_1(D_n, d_0) \rightarrow \mathbb{Z}$ is surjective, the group of covering transformations $\mathrm{Aut}(\tilde{D}_n) \cong \mathbb{Z}$. Let τ be a generator of $\mathbb{Z}$, and let $\Lambda = Z[t, t^{-1}]$. Then $H_1(\tilde{D}_n)$ inherits a Λ-module structure from the action of the covering transformations: we simply set $t\gamma = \tau_*(\gamma)$, where $\gamma \in H_1(\tilde{D}_n)$ and τ_* denotes the induced action on homology. We note that as a Λ-module, $H_1(\tilde{D}_n)$ is free of rank $n - 1$.

In what follows it will be convenient to think of $\mathbf{B}_n$ as in Section 1.3, i.e., as the mapping class group of D_n where ∂D_n is fixed pointwise, while the punctures may be permuted.

We shall abuse terminology by not distinguishing between a mapping class and a diffeomorphism which represents it. Therefore, we think of any $X \in \mathbf{B}_n$ as a map $X : D_n \to D_n$. Then X lifts uniquely to a map $\tilde{X} : \tilde{D}_n \to \tilde{D}_n$ which fixes the fiber over the basepoint d_0 pointwise. Furthermore, $\tilde{X}$ induces a Λ-module automorphism $\tilde{X}_*$ of $H_1(\tilde{D}_n)$. Since $H_1(\tilde{D}_n)$ is a free Λ-module of rank $n - 1$, we can now define a map $\mathbf{B}_n \to \mathrm{GL}_{n-1}(\Lambda)$ by $X \mapsto \tilde{X}_*$. This map turns out to be equivalent to the reduced Burau representation defined previously (see [59] for a classification of linear representations of the braid group of degree at most $n - 1$).

The main idea of Stephen Bigelow's proof of the non-faithfulness of the Burau representation in the case $n = 5$ is contained in the following theorem.

THEOREM 4.7 [12]. *For $n \geq 3$, the Burau representation $\bar{\rho} : \mathbf{B}_n \to \mathrm{GL}_{n-1}(\Lambda)$ is not faithful if and only if there exist arcs α, β embedded in D_n satisfying*:

(1) $\partial\alpha = \{q_1, q_2\}$ *and* $\partial\beta = \{d_0, q_3\}$ *or* $\{q_3, q_4\}$.
(2) α *intersects* β *non-trivially* (*more precisely, there exists no isotopy rel endpoints which carries* α *off* β).
(3) *For some choice of lifts* $\tilde{\alpha}, \tilde{\beta}$, *we have* $\sum_{k\in\mathbb{Z}} (t^k\tilde{\alpha}, \tilde{\beta})t^k = 0$, *where* (x, y) *denotes the algebraic intersection number of two* (*oriented*) *arcs in* $\tilde{D}_n$.

We note that the case $\partial\beta = \{d_0, q_3\}$ follows from Theorem 1.5 of [90]. Bigelow has produced an explicit example of arcs α and β satisfying the criteria of Theorem 4.7 in the case $n = 5$ (see p. 402 of [12]). It follows that the Burau representation of $\mathbf{B}_n$ is not faithful for $n \geq 5$. It has been known for many years that it *is* faithful for $n = 3$.

OPEN PROBLEM 5. Is the Burau representation of $\mathbf{B}_4$ faithful?

In the next section, we will explore another representation of $\mathbf{B}_n$, with a topological interpretation analogous to that of the Burau representation.

4.5. *The Lawrence–Krammer representation*

In 1990, Ruth Lawrence introduced a family of representations of $\mathbf{B}_n$ corresponding to Young diagrams with two rows which arise out of a topological construction of representations of the Hecke algebras $H_n(t)$ [87]. Later, Krammer gave an entirely algebraic definition of one of these representations:

$$\lambda : \mathbf{B}_n \to \mathrm{GL}_r(A)$$

where $A = \mathbb{Z}[t^{\pm 1}, q^{\pm 1}]$ and showed that it was faithful for $n = 4$ [84]. The representation λ has become known as the Lawrence–Krammer representation. Shortly after Krammer's result appeared, Bigelow was able to use topological methods to show that λ is faithful for all n [13]. Krammer later gave an algebraic proof of the same result [85]. Therefore, we now have:

THEOREM 4.8 [13,85]. *The map λ is a faithful representation of $\mathbf{B}_n$, and hence $\mathbf{B}_n$ is a linear group for all n.*

We observe that Krammer's result in [84] also implies that $\mathrm{Aut}(F_2)$, the group of automorphisms of a free group of rank 2, is linear.

We shall not present here a full proof of Theorem 4.8; our aim is to define the representation λ, to present the key ingredients in Bigelow's proof: "forks", "noodles", and the pairing between them, and to describe Bigelow's characterization of the kernel of λ in terms of these ingredients.

As mentioned previously, Bigelow's topological definition of λ is somewhat analogous to the topological definition of the Burau representation given in the previous section. This time, we begin the construction with a certain configuration space of a punctured disk rather than the punctured disk itself. In the notation of Section 1.1, we let $C = \mathcal{C}_{0,2}(D_n)$, where D_n is the n-punctured disk described in the previous section. Elements of C will be denoted $\{z_1, z_2\} = \{z_2, z_1\}$, where $z_1 \neq z_2 \in D_n$. Now we choose two distinct points $d_1, d_2 \in \partial D_n$, and let $c_0 = \{d_1, d_2\}$ be the basepoint of C. We are now in a position to define a map $\phi : \pi_1(C, c_0) \to \mathbb{Z} \times \mathbb{Z}$ as follows. Let $[\gamma] \in \pi_1(C, c_0)$. Then the loop $\gamma(s)$ in C can be expressed as $\gamma(s) = \{\gamma_1(s), \gamma_2(s)\}$, where γ_i is an arc in D_n for $i = 1, 2$.

We define two integers a, b as follows.

$$a = \frac{1}{2\pi i} \sum_{j=1}^{n} \int_{\gamma_1} \frac{\mathrm{d}z}{z - q_j} + \int_{\gamma_2} \frac{\mathrm{d}z}{z - q_j}, \quad b = \frac{1}{\pi i} \int_{\gamma_1 - \gamma_2} \frac{\mathrm{d}z}{z}.$$

Since γ is a loop in C, we either have that both γ_1 and γ_2 are both loops, i.e., $\gamma_1(0) = \gamma_1(1)$ and $\gamma_2(0) = \gamma_2(1)$, or else, as Bigelow puts it, the γ_i are arcs which "switch places", i.e., $\gamma_1(1) = \gamma_2(0)$ and $\gamma_2(1) = \gamma_1(0)$. In the first case, Bigelow observes that we can interpret a as the sum of the total algebraic winding numbers of γ_1 and γ_2 about the punctures q_i and b as twice the winding number of γ_1 and γ_2 about each other. In the second case, the product $\gamma_1\gamma_2$ is a loop, and so we still have a nice interpretation of a as the total algebraic winding number of $\gamma_1\gamma_2$ about the punctures q_i. Bigelow also points out that in this case the condition $(\gamma_1 - \gamma_2)(1) = -(\gamma_1 - \gamma_2)(0)$ implies that b is an odd integer. It is worth noting that Turaev [112] interprets the integer b in a different way (in both cases). The composition of the map $\pi_1(C) \to S^1$ defined by

$$\gamma \mapsto \frac{\gamma_1(s) - \gamma_2(s)}{|\gamma_1(s) - \gamma_2(s)|}$$

with the usual projection from S^1 onto $\mathbb{R}P^1$ sends our loop γ to a loop in $\mathbb{R}P^1$. Let $\bar{\gamma}$ denote the homology class of this loop in $H_1(\mathbb{R}P^1)$. Up to a choice of generator u for $H_1(\mathbb{R}P^1) \cong \mathbb{Z}$, we have $\bar{\gamma} = u^b$. We define the map ϕ by setting $\phi(\gamma) = q^a t^b \in \mathbb{Z} \times \mathbb{Z}$, thought of as the free abelian group with basis $\{q, t\}$.

Now we proceed more or less as in the case of the Burau representation. Let $\tilde{C}$ be the regular cover of C which corresponds to the kernel of ϕ in $\pi_1(C, c_0)$. Let A denote the ring $\mathbb{Z}[t^{\pm 1}, q^{\pm 1}]$. The homology groups of $\tilde{C}$ naturally inherit an A-module structure via the action of $\mathrm{Aut}(\tilde{C})$.

It is clear that any homeomorphism $X : D_n \to D_n$ induces a homeomorphism $X' : C \to C$ defined by $X'(\{z_1, z_2\}) = \{X(z_1), X(z_2)\}$. Thus X' necessarily fixes the basepoint c_0.

If X'_* denotes the induced action on $\pi_1(C)$, then it is an easy exercise to check that X'_* preserves the values of the integers a and b defined above. Then X' lifts uniquely to a map $\tilde{X} : \tilde{C} \to \tilde{C}$ such that the fiber over the basepoint c_0 is fixed pointwise. Moreover, since $\tilde{X}$ commutes with the action of $\mathrm{Aut}(\tilde{C})$, we have that the induced action on $H_2(\tilde{C})$, denoted $\tilde{X}_*$, is an A-module automorphism. (Note that $\tilde{C}$ is a 4-manifold.) Now $H_2(\tilde{C})$ is a free A-module of rank

$$r = \binom{n}{2}.$$

(Theorem 4.1 of [13]) and thus we can think of $\tilde{X}_* \in A$. We therefore define the *Lawrence–Krammer representation* as follows:

$$\lambda : \mathbf{B}_n \to \mathrm{GL}_r(A)$$

$$X \mapsto \tilde{X}_*$$

If we choose q and t to be algebraically independent in $\mathbb{C}$, we obtain a faithful representation $\lambda : \mathbf{B}_n \to \mathrm{GL}_r(\mathbb{C})$.

Recall that $d_1, d_2 = c_0$ is the basepoint we have chosen for the configuration space C. A *fork* is a tree F embedded in the disk D such that

(1) F has four vertices: d_1, z, q_i and q_j,
(2) the three edges of F share z as a common vertex,
(3) $F \cap \partial D_n = d_1$, and
(4) $F \cap \{q_1, \ldots, q_n\} = \{q_i, q_j\}$

We note that Krammer's algebraic definition of the representation λ depends on the induced action on an A-module generated forks.

The edge of F which contains the vertex d_1 is called the *handle* of F. The union of the other two edges is called the *tine edge* of F, denoted $T(F)$, and may contain punctures q_k, $k \neq i,j$. For any fork F we define a *parallel copy* F' to be a copy of the tree F embedded in D, with vertices $\{d_2$, z', q_i and q_j such that $d_1 \neq d_2 \in \partial D_n$, $z \neq z'$, and F' is isotopic to F rel $\{q_i, q_j\}$. Given a fork F, we also define a *noodle* to be an arc N properly embedded in D_n with $\partial N = \{d_1, d_2\}$ and oriented so that its initial point is d_1 and its terminal point is d_2.

We next construct surfaces in $\tilde{C}$ associated to forks and noodles with which we will define a certain pairing on forks and noodles. For each fork F, we choose a parallel copy F' and first define a surface $S(F)$ as the set of all points of the form $\{x, y\} \in C$ with $x \in T(F) \backslash \{q_1, \ldots, q_n\}$ and $y \in T(F') \backslash \{q_1, \ldots, q_n\}$. Let ζ (resp. ζ') denote the handle of F (resp. F'), oriented from d_1 to z (resp. d_2 to z'). Define $\tilde{\zeta}$ to be the unique lift of $\{\zeta(s), \zeta'(s)\} \in C$ with initial point $\tilde{c}_0$, a fixed point in the fiber over c_0. Now let $\tilde{S}(F) \subset \tilde{C}$ be the unique lift of $S(F)$ which contains the terminal point of $\tilde{\zeta}$. Similarly, we can associate to each noodle N a surface $\tilde{S}(N) \subset \tilde{C}$. Define $S(N) \subset C$ as the set of all points of the form $\{x, y\} \in C$ such that $x, y \in N$, $x \neq y$, and let $\tilde{S}(N)$ be the unique lift of $S(N)$ which contains $\tilde{c}_0$.

Given a noodle N and a fork F, we define their *pairing*, denoted $\langle N, F\rangle$, to be an element of the ring A as follows. We can assume that (up to isotopy rel endpoints) $T(F)$ and N intersect transversely in a finite number of points $\{z_1, ..., z_r\}$ and similarly that $T(F')$ intersects N transversely at $\{z'_1, ..., z'_r\}$. Then each pair z_i, z'_i cobounds an arc in N which lies between $T(F)$ and $T(F')$. Now for each $i, j = 1, ..., r$, there is a unique monomial $m_{i,j} = q^{a_{i,j}}t^{b_{i,j}}$ such that $m_{i,j}\tilde{S}(N)$ intersects the surface $\tilde{S}(F)$ at a point in the fiber over $z_i, z'_j \in C$. Letting $\epsilon_{i,j}$ denote the sign of that intersection, we define

$$\langle N, F\rangle = \sum_{i=1}^{r}\sum_{j=1}^{r}\epsilon_{i,j}q^{a_{i,j}}t^{b_{i,j}}.$$

Bigelow supplies a method for explicit calculation of the pairing for a given noodle and fork and shows that it is well defined on isotopy classes of forks rel q_i, q_j.

The proof that λ is faithful relies on two important lemmas (the "Basic Lemma" and "Key Lemma" of [13]), which we present here. As usual, we shall abuse notation and not distinguish between a mapping class and specific representatives.

LEMMA 4.9 [13]. *If $\lambda(X) = 1$, then $\langle N, F\rangle = \langle N, X(F)\rangle$ for every noodle N and every fork F.*

LEMMA 4.10 [13]. *For a noodle N and a fork F, the pairing $\langle N, F\rangle = 0$ if and only if $T(F)$ can be isotoped off N relative to q_i, q_j.*

Thus, it is the pairing on noodles and forks which gives the essential characterization of the kernel of λ.

We end this section by noting two facts about the Lawrence–Krammer representation. First, Budney has shown that λ is unitary[1] for an appropriate choice of q and t (still algebraically independent) [36]. In other words, we have

$$\lambda : \mathbf{B}_n \to U_r(\mathbb{C})$$

where $U_r = \{X \in \mathrm{GL}_r(\mathbb{C}) | \bar{X}^T = X^{-1}\}$. Thus, the conjugacy class of $\lambda(X) \in U_r(\mathbb{C})$ is determined by its eigenvalues, and one could hope for an efficient solution to the conjugacy problem for $\mathbf{B}_n$ by passing to the whole of $U_r(\mathbb{C})$. However, Budney has given examples of non-conjugate braids $X_1, X_2 \in \mathbf{B}_n$ such that $\lambda(X_1)$ and $\lambda(X_2)$ are conjugate in $U_r(\mathbb{C})$ (see Section 4 of [36]). We will return to these examples in a bit more detail in Section 5.7.

Second, Matthew Zinno has shown a connection between the Lawrence–Krammer representation and the Birman–Murakami–Wenzl (BMW) algebra. The BMW algebra is related to Kauffman's knot polynomial and can be thought of as a deformation of the Brauer algebra in the same way that the Hecke algebra can be thought of as a deformation of the group algebra $\mathbb{C}\Sigma_n$ (see [28,100]). Braid groups map homomorphically into the BMW algebra, giving rise to irreducible representations of $\mathbf{B}_n$. Zinno has identified a summand of the BMW algebra which corresponds exactly to the Lawrence–Krammer representation.

[1] Squier had previously shown that the Burau representation is unitary [107].

THEOREM 4.11 [118]. *The Lawrence–Krammer representation of* $\mathbf{B}_n$ *is equivalent to the* $(n-2)\times 1$ *irreducible representation of the BMW algebra.*

It follows immediately from Theorem 4.11 that the Lawrence–Krammer representation λ is irreducible and also that the regular representation of the BMW algebra is faithful.

4.6. *Representations of other mapping class groups*

The first theorem that we proved in this review article was Theorem 1.1, which asserted that the braid group $\mathbf{B}_{n-1}$ has a natural interpretation as the mapping class group $\mathcal{M}_{0,1,n-1}$ of the n-times punctured disc. One wonders, then, whether the linearity of $\mathbf{B}_n$ extends to a more general statement about the linearity of other mapping class groups $\mathcal{M}_{g,b,n}$?

The manuscripts of Korkmaz [83] and of Bigelow and Budney [14] exploit the very special connection between the braid group $\mathbf{B}_{n-1} = \mathcal{M}_{0,1,n-1}$ and the mapping class group $\mathcal{M}_{0,0,n}$, and between the mapping class groups $\mathcal{M}_{0,0,2g+2}$ and $\mathcal{M}_{g,0,0}$, to produce faithful finite dimensional representations of $\mathcal{M}_{0,0,n}$ for every n, also of $\mathcal{M}_{2,0,0}$, and finally of a particular subgroup of $\mathcal{M}_{g,0,0}$ for every $g \geq 3$. In the case of $\mathcal{M}_{2,0,0}$, the basic fact used by both Korkmaz and Bigelow–Budney is a theorem proved by Birman and Hilden in 1973 (see [15]), which asserts that the mapping class group $\mathcal{M}_{2,0,0}$ is a $\mathbb{Z}_2$ central extension of $\mathcal{M}_{0,0,6}$. This theorem is special to genus 2, which is the only group among the mapping class groups $\mathcal{M}_{g,0,0}$ which has a center. For $g \geq 3$ the so-called "hyperelliptic involution" which generates the center of $\mathcal{M}_{2,0,0}$ generalizes to an involution whose centralizer is a subgroup of infinite index in $\mathcal{M}_{g,0,0}$. The same circle of ideas yield faithful matrix representations of those subgroups.

We now explain how Bigelow and Budney used the Lawrence–Krammer representation of the braid groups to obtain faithful matrix representations of $\mathcal{M}_{0,0,n}$. If one considers any surface $S_{g,b,n}$ and caps one of the boundary components by a disc, one obtains a geometrically induced *disc-filling* homomorphism $d_\star : \mathcal{M}_{g,b,n-1} \to \mathcal{M}_{g,b-1,n}$. Its kernel is the Dehn twist about the distinguished boundary component (see Section 2.8 of [75]); a distinguished point in the interior of the new disc becomes the new fixed point. Applying these ideas to the braid group, one then sees that there is a natural homomorphism $d_\star : \mathbf{B}_{n-1} \to \mathcal{M}_{0,0,n}$. It turns out that image $(d_\star)$ is not the full group $\mathcal{M}_{0,1,n}$, but the stabilizer of the new fixed point, and kernel $(d_\star)$ is the infinite cyclic subgroup of $\mathbf{B}_{n-1}$ that is generated by the braid $C = (\sigma_{n-1}\sigma_{n-2}\cdots\sigma_2\sigma_1)^n$, a full twist of all of the n braid strands. Since the image $\lambda(C)$ in the Lawrence–Krammer representation is a scalar matrix (the diagonal entries are $q^{2(n-1)}t^2$ in the representation as it is defined in [13]), one obtains a faithful representation of image $(d_\star)$ by rescaling the Lawrence–Krammer matrices, setting $t^{-2} = q^{2(n-1)}$. The group image $(d_\star)$ is of finite index in $\mathcal{M}_{0,0,n}$, which therefore is a linear group. The dimension of the explicit representation of $\mathcal{M}_{0,0,n}$ constructed in [14] is

$$n\binom{n-1}{2}^2.$$

It leads to a related representation of dimension 64 of $\mathcal{M}_{2,0,0}$. In this regard we note that, while Korkmaz uses the identical geometry, he uses less care with regard to dimension, and his representation of $\mathcal{M}_{2,0,0}$ has dimension $2^{10}3^5 5^3$, which is very much bigger than 64.

OPEN PROBLEM 6. Is there a faithful finite dimensional matrix representation of the mapping class group $\mathcal{M}_{g,b,n}$ for any values of the triplet (g, b, n) other than $(0, 1, n)$, $(0, 0, n)$, (2,0,0), (1,0,0) and (1,1,0)? A folklore conjecture is that most of the mapping class groups are, in fact linear. See [34] for evidence in this regard. New ideas seem to be needed to construct candidates.

In view of the very large dimension of the representation of $\mathcal{M}_{2,0,0}$ which we just discussed, we note that there is an interesting 5-dimensional representation of the same group which occurs as one of the summands in the Hecke algebra representation of $\mathbf{B}_n$, namely the one belonging to the Young diagram with two rows and three columns. It is discussed in [78]. At this writing its kernel does not seem to be known, although it is known to be infinite.

4.7. *Additional representations of* $\mathbf{B}_n$

We end our discussion of representations of the braid groups by describing a construction which yields infinitely many finite dimensional representations of $\mathbf{B}_n$, each one over a ring $\mathbb{C}[t_1, t_1^{-1}, t_2, t_2^{-1}, \ldots, t_k, t_k^{-1}]$, where $t_1, \ldots, t_k$ are parameter, for some $k \geq 1$. The construction includes all the summands in the Temperley–Lieb algebra and the Lawrence–Krammer representation too, and in addition infinitely many presumably new faithful representations of $\mathbf{B}_n$. It was first described in [21], generalizing ideas in [87]. It is due to Moody, with details first worked out by Long in [89].

We will be interested in the braid group $\mathbf{B}_n$ on n-strands, but to describe our construction it will be convenient to regard $\mathbf{B}_n$ as a subgroup of $\mathbf{B}_{n+1}$. Number the strands in the latter group as $0, 1, \ldots, n$. Let $\mathbf{B}_{1,n} \subset \mathbf{B}_{n+1}$ be the subgroup of braids in $\mathbf{B}_{n+1}$ whose associated permutation fixes the letter 0. Its relationship to $\mathbf{B}_n$ is given by the group extension

$$1 \to \mathbf{F}_n \to \mathbf{B}_{1,n} \to \mathbf{B}_n \to 1, \tag{4.7}$$

where the homomorphism $\mathbf{B}_{1,n} \to \mathbf{B}_n$ is defined by pulling out the zeroth braid strand. There is a cross section which is defined by mapping $\mathbf{B}_n$ to the subgroup of braids on strands $1, \ldots, n$ in $\mathbf{B}_{1,n}$. Therefore, we may identify $\mathbf{B}_{1,n}$ with $\mathbf{F}_n \times \mathbf{B}_n$. The semi-direct product structure arises when we regard $\mathbf{B}_n$ as a subgroup of the automorphism group of a free group. The action of $\mathbf{B}_n$ on F_n is well known and is given in [5], also [57], and also in [15]. Thinking of F_n as the fundamental group $\pi_1(S_{0,1,n})$ of the n-times punctured disc, the action of the elementary braid σ_i is given explicitly by

$$\sigma_i \mathbf{x}_j \sigma_i^{-1} = \begin{cases} \mathbf{x}_{i+1} & \text{if } j = i, \\ \mathbf{x}_{i+1}^{-1} \mathbf{x}_i \mathbf{x}_{i+1} & \text{if } j = i + 1, \\ \mathbf{x}_j & \text{otherwise.} \end{cases} \tag{4.8}$$

Since $\mathbf{B}_n$ is a subgroup of $\mathbf{B}_{n+1}$ we see in this way that the groups $\mathbf{B}_n$, $\mathbf{F}_n$ and also $\mathbf{F}_n \times \mathbf{B}_n$ are all subgroups of $\mathbf{B}_{n+1}$.

In order to describe the idea behind the construction we recall the notion of homology or cohomology of a space with coefficients in a flat vector bundle. Suppose that X is a manifold and that we are given a representation $\rho : \pi_1(X) \to \mathrm{GL}(V)$. This enables us to define a flat vector bundle E_ρ: Let $\tilde{X}$ be the universal covering of X. The group $\pi_1(X)$ acts on $\tilde{X} \times V$ by $g\cdot(\tilde{x}, \mathbf{v}) = (g\cdot\tilde{x}, \rho(g)\cdot\mathbf{v})$. Then E_ρ is the quotient of $\tilde{X} \times V$ by this action. We now form the cohomology groups of 1-forms with coefficients in E_ρ, denoting these by $H^1(X;\rho)$ or $H^1_c(X;\rho)$ for compactly supported cochains. In order to get an action of the braid groups, we use (4.8). The action gives a canonical way of forming a split extension $F_n \times \mathbf{B}_n$. It turns out that in order to get an action on the twisted cohomology group what is required is exactly a representation of this split extension. Since $\mathbf{B}_{1,n}$ is a subgroup of $\mathbf{B}_{n+1}$, any representation of the latter will of course do the job, and that is why we think of $\mathbf{B}_n$ as a subgroup of $\mathbf{B}_{n+1}$.

THEOREM 4.12 [24]. *Given a representation $\rho : F_n \times \mathbf{B}_n \to \mathrm{GL}(V)$ we may construct another representation $\rho_t^+ : \mathbf{B}_n \to H^1_c(S_{0,1,n};\rho)$ where t is a new parameter. In particular, given any representation $\rho : B_{n+1} \to \mathrm{GL}(V)$, we may construct a representation $\rho_t^+ : \mathbf{B}_n \to H^1_c(S_{0,1,n};\rho)$.*

This works in exactly the way one might expect. The representation restricted to the free factor gives rise to the local system on the punctured disc and thus the twisted cohomology group and the compatibility condition provided by the split extension structure gives the braid group action.

A comment is in order concerning Theorem 4.12. Although the theorem is stated abstractly, (4.8) gives a concrete recipe which enables one to write down the description of ρ_t^+ given ρ. Moreover, as we have noted, any time that we have a representation of $\mathbf{B}_{n+1}$ we also have one of the semi-direct product, which we identify with the subgroup $\mathbf{B}_{1,n}$.

The theorem shows that given a k parameter representation of the braid group, the construction yields a $k+1$ parameter representation, apparently in a non-trivial way. For example, if one starts with the (zero parameter) trivial representation of $F_n \times \mathbf{B}_n$, the theorem produces the Burau representation, and starting with the Burau representation, the construction produces the Lawrence–Krammer representation [89]. However, the role of this extra parameter is not purely to add extra complication, it also adds extra structure. For there is a natural notion of what it should mean for a representation of a braid group to be unitary (see [107], for example) and the results of Deligne–Mostow and Kohno imply:

THEOREM 4.13 [89]. *In the above notation, if ρ is unitary, then for generic values of s, so is ρ_s^+.*

OPEN PROBLEM 7. This problem is somewhat vague. It begins with a suggestion that Long's construction be studied in greater detail, and goes on to ask whether (a wild guess) all finite dimensional unitary matrix representations of $\mathbf{B}_n$ arise in a manner which is related to the construction of Theorem 4.12?

5. The word and conjugacy problems in the braid groups

Let G be any finitely generated group. Fix a set of generators for G. A *word* ω is a word in the given set of generators and their inverses. The element of G that it represents will be denoted $[\omega]$, and its conjugacy class will be denoted $\{\omega\}$. We consider two problems. The *word problem* begins with $\omega, \omega' \in G$, and asks for an algorithm that will decide whether $[\omega] = [\omega']$? The *conjugacy problem* asks for an algorithm to decide whether $\{\omega\} = \{\omega'\}$, i.e., whether there exists α such that $\{\omega'\} = \{\alpha^{-1}\omega\alpha\}$? A sharper version asks for a procedure for finding α, if it exists. In both cases we are interested in the complexity of the algorithm, and ask whether it is polynomial in either braid index n or word length $|\omega|$ or both? The word problem and the conjugacy are two of the three classical "decision problems" first posed by Dehn [45]. In the case we consider here of $G = \mathbf{B}_n$, the set of generators we choose will be those of either the classical or the new presentation. Artin gave the first solution to the word problem for the braid group in 1925 [5], and a number of fundamentally different solutions to the word problem in $\mathbf{B}_n$ exist today, with at least two of them being polynomial in both n and $|\omega|$. We focus here on an approach due to Garside [67] and improved on by a number of others, and briefly describe other methods at the end of this section.

On the other hand, we know of only one definitive solution to the conjugacy problem, namely the combinatorial solution that was discovered by Garside. He found a finite set of conjugates of an arbitrary element $[\omega] \in \mathbf{B}_n$, the "Summit Set" $\mathbf{S}_\omega$ of ω, with the properties that if $\{\omega\} = \{\beta\}$, then $\mathbf{S}_\omega = \mathbf{S}_\beta$, whereas if $\{\omega\} \neq \{\beta\}$, then $\mathbf{S}_\omega \cap \mathbf{S}_\beta = \emptyset$. If $\mathbf{S}_\omega = \mathbf{S}_\beta$, his methods also find α such that $\beta = \alpha\omega\alpha^{-1}$. While his algorithm has been improved in major ways over the years, at this writing it is exponential in both n and $|\omega|$. The principle difficulties may be explained in the following way:

($\star$) The entire set $\mathbf{S}_\omega$ and a single element in $\mathbf{S}_\beta$ must be computed to decide whether ω and β are or are not conjugate. While the calculation of a single element can now be done rapidly, in the general case $\mathbf{S}_\omega$ has unpredictable size. The combinatorics that determine the size of $\mathbf{S}_\omega$ are particularly subtle and difficult to understand, and at this writing only partial progress has been made, in spite of much effort by experts.

The improvements that have been made over the years have included the replacement of $\mathbf{S}_\omega$ by a proper subset S_ω, the "Super Summit Set", via the work of ElRifai and Morton in [53]. However, while S_ω is very much smaller than $\mathbf{S}_\omega$, the principle difficulty ($\star$) is unchanged. Very recently Gebhardt found a still smaller subset, the "Ultra Summit Set U_ω", to replace S_ω [68]. The difficulty is unfortunately the same as it was for the summit set and the super summit set, but can be made more specific: the thing that one needs to understand is the number of "closed cycling orbits" in U_ω and the length of each orbit, and how distinct orbits are related. Lee [86] has given examples to show that the number of orbits and their size can be arbitrarily large, and that it is in no way clear how distinct orbits are related, except in special cases. More work remains to be done.

It was shown in Theorem 1.1, proved earlier in this article, that there is a faithful action of the braid group $\mathbf{B}_n$ on the n-times punctured disc. Investigating this action (in the more general setting of the action of the mapping class group of a 2-manifold on the 2-manifold),

Thurston proposed in the early 1980s a very different approach to the conjugacy problem which is based upon the dynamical properties of that action. These ideas were investigated in Mosher's PhD thesis [99]. The curious fact is that while the dynamical picture quickly yields several very interesting class invariants not easily seen in the combinatorial picture, namely a graph (known as a "train track") which is embedded in the surface and is invariant under the action of properly chosen representatives of $\{\omega\}$, and a real number λ, however they are not enough to determine conjugacy. Generic elements in the mapping class group of a surface have an invariant train track, but it is not unique. It is known that there are finitely many possible train tracks associated to a given element, but their enumeration remains an unsolved problem, and that is what gets in the way of a nice solution to the conjugacy problem. In fact, when one begins to understand the details, the entire picture suggests difficulties very much like those in ($\star$) again.

We have just one more remark of a general nature, before we proceed to review all these matters in more detail. One of the reasons that the Garside approach to the conjugacy problem is interesting is because the techniques that were developed for the braid groups revealed unexpected structures in $\mathbf{B}_n$ that generalize to related unexpected structures in Artin groups (see Section 5.4), and in the less well-known class of "Garside groups" (see Section 5.2). There has been major activity in recent years regarding these structures, and there are also many open problems. The same can also be said for the Thurston approach, as it too generalizes from an action of $\mathbf{B}_n$ on the n-times punctured disc to actions of mapping class groups on curves on surfaces, discussed in Section 5.6. All of these matters, and other related ones, will be discussed below.

5.1. *The Garside approach, as improved over the years*

We assume, initially, that the group $\mathbf{B}_n$ is defined by the classical presentation (1.2), with generators $\sigma_1, \ldots, \sigma_{n-1}$. In this subsection we describe the solution to the word and conjugacy problem which was discovered by Garside in 1968 (see [67]) and subsequently sharpened and expanded, in many ways, through the contributions of others, in particular Thurston [54], ElRifai and Morton [53], Birman et al. [19,20], Franco and Gonzales-Meneses [61] and most recently Gebhardt [68]. Other relevant papers are [71] and [72].

In (G1)–(G6) we describe a constructive method for finding a normal form for words. In (G7)–(G10) we describe how to find a unique finite set of words in normal form that characterizes $\{\omega\}$.

(G1) Elements of $\mathbf{B}_n$ which can be represented by braid words which only involve positive powers of the σ_i are called *positive braids*. A key fact which was observed by Garside [67] is that the presentation (1.2) not only defines the group $\mathbf{B}_n$, it also defines a monoid $\mathbf{B}_n^+$. He then went on to prove that this monoid embeds in the obvious natural way in $\mathbf{B}_n$, in the strong sense that two positive words in $\mathbf{B}_n$ define the same element of $\mathbf{B}_n$ if and only if they also define the same element in the monoid $\mathbf{B}_n^+$. The monoid of positive braids is particularly useful in studying $\mathbf{B}_n$ because the defining relations all preserve word length, so that the number of candidates for a positive word which represents a positive braid is finite. However, all this is of little use unless we can show that the monoid of positive braids is more than just a small and very special subset of $\mathbf{B}_n$. In (G3) we show that this is indeed the case.

(**G2**) A special role is played by the *Garside braid* Δ in $\mathbf{B}_n$. It is a positive half-twist of all the braid strands, and is defined by the braid word

$$\Delta_n = (\sigma_{n-1}\sigma_{n-2}\cdots\sigma_1)(\sigma_{n-1}\sigma_{n-2}\cdots\sigma_2)\cdots(\sigma_{n-1}\sigma_{n-2})(\sigma_{n-1}).$$

It is illustrated in sketch (i) of Figure 24, for $n = 5$. The square of Δ_n (a full twist of all the braid strands) generates the infinite cyclic center of $\mathbf{B}_n$. The inner automorphism $\tau : \mathbf{B}_n \to \mathbf{B}_n$ which is defined by $\tau(X) = \Delta^{-1}X\Delta$ is the symmetry which sends each σ_i to σ_{n-i}. Of course Δ itself is invariant under this symmetry. Later we will show that almost all of the combinatorics which we are describing hold equally well for the new presentation (1.3). The role of Δ in the classical presentation is replaced by that of the braid $\delta = \sigma_{n-1,n}\cdots\sigma_{2,3}\sigma_{1,2}$ which is depicted in Figure 24(ii) (see Section 5.3).

(**G3**) The Garside braid Δ is very rich in elementary braid transformations. In particular, for each $i = 1,\dots,n-1$ there exist (non-unique) positive braids L_i, and R_i such that $\Delta = L_i\sigma_i = \sigma_i R_i$. This implies that $\sigma_i^{-1} = \Delta^{-1}L_i = R_i\Delta^{-1}$ for each $i = 1,\dots,n-1$, so that an arbitrary word $\omega = \sigma_{\mu_1}^{\epsilon_1}\cdots\sigma_{\mu_r}^{\epsilon_r}$, $\epsilon_q = \pm 1$, can be converted to a word which uses only positive braid generators, at the expense of inserting arbitrary powers of Δ^{-1}. Since, by (G2) we have $\beta\Delta^{\epsilon} = \Delta^{\epsilon}\tau(\beta)$ for every braid β, all powers of Δ^{-1} can be moved to the left (or right). It follows that $[\omega]$ is also defined by a word of the form $\Delta^r T$ where T is positive. This representation is non-unique.

(**G4**) To begin to find unique aspects, Garside observes that there is a maximum value, say i, of r with the property that the braid defined by ω is also represented by a word $\Delta^i Z$ where Z is positive and i is maximal for all braids of this form. Note that this means that Z cannot be written in the form $Z_1\Delta Z_2$ for any positive words Z_1, Z_2, because if it could, then it would be possible, by (G2), to rewrite $\Delta^i Z$ as $\Delta^{i+1}\tau(Z_1)Z_2$, with $\tau(Z_1)Z_2 \in \mathbf{B}_n^+$, contradicting the maximality of i. The integer i is known as the *infimum* of $[\omega]$, and written $\inf(\omega)$. It is an invariant of $[\omega]$. Garside's complete invariant of $[\omega]$ is $\Delta^i Z_0$, where among all positive words $Z_0, Z_1, \dots, Z_q$ that define the same element as Z, one chooses the unique word Z_0 whose subscript array (as a positive word in $\sigma_1,\dots,\sigma_{n-1}$) is

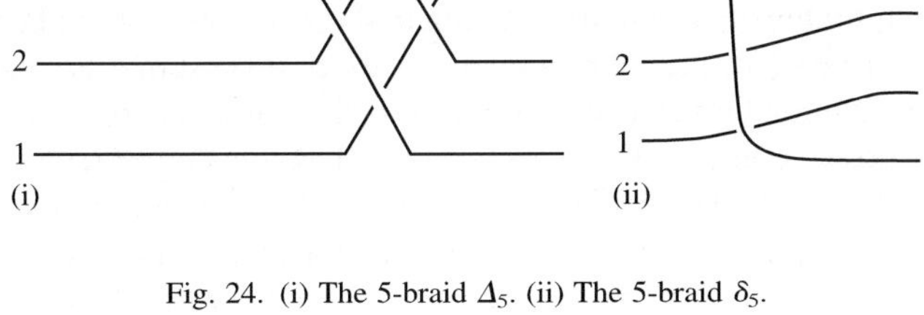

Fig. 24. (i) The 5-braid Δ_5. (ii) The 5-braid δ_5.

lexicographically minimal. This is Garside's solution to the word problem in $\mathbf{B}_n$. It is exponential in both n and $|\omega|$ because there exist braids which admit very many of the elementary braid transformations in (1.2). To make this very explicit, we note that by [108] the number of positive words of letter length $(n)(n-1)/2$ in the standard braid generators that represents Δ_n is the same as the number of standard Young tableaux of shape $(n, n-1, \ldots, 1)$. It is given by the hook length formula as $((n)(n-1)/2)!/(1^{n-1}3^{n-1}5^{n-2}\cdots(2n-3))$. Calculating, we learn that Δ_5 can be represented by 768 positive words, all of letter length 10, in the σ_i's, giving a good idea of the problem. Moreover, this formula shows that the number of words increases exponentially with n. Of course the elements that are of interest to us, that is the positive braids defined by words like Z above, do not contain Δ, however there is not reason why they cannot contain, for example, arbitrarily high powers of large subwords of Δ.

(G5) Let l and r be positive words such that there is a factorization of Δ in the form $\Delta = lr$. We call l a *left divisor* of Δ and r a *right divisor* of Δ. Let $\mathcal{P}$ (respectively, $\mathcal{P}'$) be the collection of all left (resp. right) divisors of Δ. Note that we already used the fact that each of the generators σ_i belongs to both $\mathcal{P}$ and $\mathcal{P}'$. As it turns out, the sets $\mathcal{P}$ and $\mathcal{P}'$ coincide. The set $\mathcal{P}$ will be seen to play an essential role in Garside's solution to the conjugacy problem, but before we discuss that we describe how to use it to get a very fast solution to the word problem. The solution to the word problem was first discovered by Adyan [1] in 1984, but Adyan's work was not well known in the West. It was not referenced by either Thurston (see Chapter 5 of [54]) or ElRifai and Morton [53], who rediscovered Adyan's work 8–10 years later, and added to it. Nevertheless, we use [53] and [54] as our main source rather than [1], because the point of view in those papers leads us more naturally to recent generalizations.

(G6) A key observation is that there are $n!$ braids in $\mathcal{P}$, and that these braids are in 1–1 correspondence with the permutations of their end-points, under the correspondence defined by sending each σ_i to the transposition $(i, i+1)$. For this reason $\mathcal{P}$ is known as the set of *permutation braids*. This, combined with the fact that the crossings are always positive, makes it possible to reconstruct any permutation braid from its permutation, and so to obtain a unique element, even though its representation as a word is highly non-unique. Permutation braids have a key property: after the braid is tightened, any two strands cross at most once, positively. To understand the importance of this fact, we note that any positive 5-braid in which two strands cross at most once, whose associated permutation is (1,2,3,4,5) → (5,4,3,2,1) must be Δ_5, a criterion which is an enormous improvement over searching through the 768 distinct positive words that the hook length formula showed us represent Δ_5, giving a good idea of the simplification that Adyan, Thurston and ElRifai and Morton discovered.

The final step in finding a very rapid solution to the word problem in $\mathbf{B}_n$ is a unique way to factorize the braid P in the partial normal form $\Delta^i P$ of (G4) above as a unique product of finitely many permutation braids. If the strands in P cross at most once, then $P \in \mathcal{P}$ and we may choose any positive braid word, say l_1, which has the same permutation as P as its representative. If not, set $P = l_1 l_1^\star$ where l_1 is a positive braid of maximal length in which two strands cross at most once and $l^\star$ is the rest of P. If no two strands in $l_1^\star$ cross more than once, set $l_2 = l_1^\star$, and stop. If not, repeat, setting $l_1^\star = l_2 l_2^\star$, where l_2 has maximal length among all positive braids whose strands cross at most once, and so forth to obtain $P = l_1 l_2 \cdots l_s$ where each $l_i \in \mathcal{P}$ and each l_i has maximal length for all factorizations of $l_{i-1}^{-1} \cdots l_1^{-1} L$ as $l_i l_i^\star$ with $l_i \in \mathcal{P}$ and $l_i^\star$ positive. This representation is unique, up to the choices of the words which represent $l_1, \ldots, l_s$. As we remarked earlier, each l_i is determined uniquely as an element in $\mathbf{B}_n^+$ by the permutation of its strands.

The factorization of an arbitrary braid word as $\Delta^i P = \Delta^i l_1 l_2 \cdots l_s$ is the *left greedy* normal form, where the term "left greedy" suggests the fact that each l_i is a maximal permutation subbraid relative to the positive braid to its right. The left greedy normal form solves the word problem. The integer i was already defined to be $\inf(\omega)$. The integer $i + s$ is known as the *supremum* of ω, written $\sup(\omega)$. The integer s is the *canonical length* $L(\omega)$.

REMARK 5.1. The Adyan–Thurston–ElRifai–Morton solution to the word problem is shown in [19] to be $\mathcal{O}(|\omega|^2 n \log n)$, where $|\omega|$ is the word length of the initial representative of the braid ω, as a word in the standard generators of $\mathbf{B}_n$.

EXAMPLE 5.2. We illustrate how to find the left greedy normal form for a positive braid via an example. Assume that we are given the positive braid

$$P = \sigma_1 \sigma_3 \sigma_2^2 \sigma_1 \sigma_3^2 \sigma_2 \sigma_3 \sigma_2.$$

The first step is to factor P as a product of permutation braids by working along the braid word from left to right and inserting the next partition whenever two strands are about to cross a second time since the beginning of the current partition. The reader may wish to draw pictures to go with this example. This gives the factorization:

$$(\sigma_1 \sigma_3 \sigma_2)(\sigma_2 \sigma_1 \sigma_3)(\sigma_3 \sigma_2 \sigma_3)(\sigma_2).$$

Next, start at the right end of the braid word, and ask whether the elementary braid relations can be used to move crossings from one partition to the partition on its immediate left, without forcing an adjacent pair of strands in a factor to cross. This process is repeated three times. The first time one applies the braid relations to factor 3, and then to adjacent letters in factors 2 and 3, to increase the length of factor 2 at the expense of decreasing the length of factor 3:

$$(\sigma_1 \sigma_3 \sigma_2)(\sigma_2 \sigma_1 \sigma_3 \sigma_2)(\sigma_3 \sigma_2)(\sigma_2).$$

In fact one can push one more crossing from factor 3 to factor 2:

$$(\sigma_1 \sigma_3 \sigma_2)(\sigma_2 \sigma_1 \sigma_2 \sigma_3 \sigma_2)(\sigma_2)(\sigma_2).$$

Now it is possible to move a crossing from factor 2 to factor 1, giving:

$$(\sigma_1 \sigma_3 \sigma_2 \sigma_1)(\sigma_2 \sigma_1 \sigma_3 \sigma_2)(\sigma_2)(\sigma_2).$$

No further changes are possible, we have achieved left greedy normal form. If it had happened that P contained Δ_4, our factorization would reveal it because $\Delta_4 \in \mathcal{P}$. Each canonical factor represents a braid in $\mathcal{P} \subset \mathbf{B}_4^+$ and so has length strictly less than $|\Delta_4| = 6$. The words which represent the canonical factors are non-unique. Their associated permutations determine the canonical factors uniquely.

(**G7**) We pass to the conjugacy problem, which builds on the solution just given to the word problem. From now on a word ω will always be assumed to be in left greedy normal form. Notice that the abelianizing map $e : \mathbf{B}_n \to \mathbb{Z}$ has infinite cyclic image, and that $e(\omega)$ is the exponent sum of a representing word in the σ_i. Clearly $e(\omega)$ is an invariant of both word and conjugacy class. It follows that there are only finitely many braids ω' in the conjugacy class $\{\omega\}$ which have left greedy normal form with $\inf(\omega') > \inf(\omega)$, because $e(\Delta) = (n)(n-1)/2$ and any increase in $\inf(\omega)$ must come at the expense of a corresponding decrease in word length of the positive word that remains after all powers of Δ have been pushed to the left. Let $\mathrm{Inf}(\omega)$ be the maximum value of $\inf(\omega')$ for all braids ω' in the conjugacy class $\{\omega\}$. Assume that a representative ω_1 of $\{\omega\}$ is given and that it has left greedy normal form $\Delta^I L_1 L_2 \cdots L_S$, where $I = \mathrm{Inf}(\omega)$ and S is minimal for all braids in normal form which are conjugate to ω. Let $\mathrm{Sup}(\omega)$ be the integer $I + S$. Then $\inf(\omega)$ and $\mathrm{Sup}(\omega)$ are class invariants of ω. The ElRifai–Morton *super summit set* S_ω is the collection of all elements in left greedy normal form which realize $\mathrm{Inf}(\omega)$ and $\mathrm{Sup}(\omega)$. It is a major improvement over Garside's *summit set* $\mathbf{S}(\omega)$, which is the larger set of all elements which realize $\mathrm{Inf}(\omega)$ but not $\mathrm{Sup}(\omega)$, with left greedy form replaced by the subscript ordering described in (G4).

One might wonder how the normal forms which we just described for words and conjugacy classes relate to length functions on $\mathbf{B}_n$. In this regard, Charney has introduced a concept of "geodesic length" in [39]. As above, let $\mathcal{P}$ be the set of all permutation braids. Charney defines the *geodesic length* of a braid ω to be the smallest integer $K = K(\omega)$ such that there is a word $L_1^{\epsilon_1} L_2^{\epsilon_2} \cdots L_K^{\epsilon_K}$, where each $\epsilon_i = \pm 1$, which represents the conjugacy class $\{\omega\}$, with each $L_i \in \mathcal{P}$ and each L_i^{-1} the inverse of a word in $\mathcal{P}$. If $\Delta^u L_1 L_2 \cdots L_s$ is an arbitrary element in the super summit set of ω, it is not difficult to show that the geodesic length of $\{\omega\}$ is the maximum of the three integers $(s+u, -u, s)$. We remark that Charney's geodesic length is defined in [39] for all Artin groups of finite type. This and other ways in which the Garside machinery generalizes to other classes of groups, including all Artin groups of finite type, will be discussed (much too briefly) in Section 5.2.

Open Problem 8. In the manuscript [85] Krammer gives several proofs of the faithfulness of the Lawrence–Krammer representation of $\mathbf{B}_n$. One of his very interesting proofs shows that if $[\omega] \in \mathbf{B}_n$, then it is a very simple matter to read Charney's geodesic length from the matrix representation of ω. Since the unique element of geodesic length zero is the identity, it follows that the representation is faithful. Our suggestion for future work is to investigate the Garside solution to the conjugacy problem and its improvements (to be described below) via the Lawrence–Krammer representation of $\mathbf{B}_n$. We mention this because we feel that this aspect of Krammer's work has received very little attention.

(**G8**) Here is a fast constructive procedure for finding Inf(ω) and Sup(ω), due to ElRifai and Morton [53] and to Birman et al. [20]. Starting with any element in $\{\omega\}$, define the cycling of $\omega = \Delta^i l_1 l_2 \cdots l_k$ to be the braid $c(\omega) = \tau^{-i}(l_1)\omega\tau(l_1)$ and the *decycling* of ω to be the braid $d(\omega) = l_k\omega(l_k)^{-1}$. Note that $c(\omega)$ and $d(\omega) \in \{\omega\}$. Putting $c(\omega)$ and $d(\omega)$ into left greedy normal form, one obtains braids which have at least i powers of Δ, and possibly more because it can happen that after cycling the braid $\Delta^i l_1 l_2 \cdots l_k$ will change to one whose left greedy normal form is $\Delta^{i'} l'_1 l_2 \cdots l'_k s'$ with $i' \geq i$. This would, of course, reduce s if it increases i. ElRifai and Morton [53] proved that if inf(ω) is not maximal for the conjugacy class, then it can be increased by repeated cycling. Also, if sup(ω) is not minimal, then it can be decreased by decycling.

They could not say, however, how many times one might have to cycle or decycle before being sure that no further improvement was possible. The solution to that problem was found by Birman et al. [20]. It was shown that if inf(ω) is not maximal for the conjugacy class if it will be increased after fewer than $(n)(n-l)/2$ cyclings, and similarly if sup(ω) is not minimal it will be decreased after fewer than $(n-l)(n)/2$ decyclings. One then has a tool for increasing inf(ω) and decreasing sup(ω), and also a test which tells, definitively, when no further increase or decrease is possible.

REMARK 5.3. The fact that everything we do to compute Inf(ω) also applies to the computation of Sup(ω) is not surprising because ElRifai and Morton showed that $\text{Sup}(\omega) = -\text{Inf}(\omega^{-1})$.

(**G9**) We now come to Gebhardt's very new work. Following the steps given in (G8), one will have on hand the summit set $S(\omega)$, i.e., the set of all braids in the conjugacy class $\{\omega\}$ which have left greedy normal form $\Delta^I L_1 L_2 \cdots L_S$, where $I = \text{Inf}(\omega)$ is maximal for the class and $I + S = \text{Sup}(\omega)$ is minimal for the class. This set is still very big. Gebhardt's improvement is to show that it suffices to consider only the subset of braids which in a closed orbit under cycling. This finite set of words is called the *ultra summit set* U_ω. It has been proved by Gebhardt [68] that $\{\omega\} = \{\omega'\}$ if and only if $U_\omega = U_{\omega'}$. Note that Gebhardt does not need to use decycling, he proves that it suffices to consider the closed orbits under cycling.

(**G10**) To compute U_ω, Gebhardt also shows that if $\omega_i, \omega_j \in U_\omega$, then there is a finite chain $\omega_i = \omega_{i,1} \to \omega_{i,2} \to \cdots \to \omega_{i,q} = \omega_j$ of braids, with each $\omega_{i,j}$ in U_ω such that each $\omega_{i,t}$ is obtained from $\omega_{i,t-1}$ by conjugating by a single element in $\mathcal{P}$. Thus, the following steps suffice to compute U_ω, after one knows a single element $\rho \in U_\omega$: One first computes the conjugates of ρ by the $n!$ elements in $\mathcal{P}$. One then puts each into left greedy normal form, and discards any braid that either does not (a) realize Inf and Sup, or (b) realizes Inf and Sup but is not in a closed orbit under cycling, or (c) realizes Inf and Sup and is in a closed orbit under cycling but is not a new element in U_ω. Ultimately, the list of elements so obtained closes to give U_ω.

Note that in doing this computation, one not only learns, for each $\omega_i, \omega_j \in U_\omega$, that $\{\omega_j\} = \{\omega_j\}$, but one also computes an explicit element α such that $\omega_j = \alpha^{-1}\omega_i\alpha$.

An inefficient part of this computation is the constant need to access the $n! - 2$ nontrivial elements in $\mathcal{P}$. A more efficient process is known (see (G11)), but to describe it, we need new notions, which will be introduced after we discuss a wider class of groups, known as Garside groups.

5.2. *Generalizations: from* $\mathbf{B}_n$ *to Garside groups*

The first person to realize that the structure described in (G1)–(G10) is not restricted to braids was Garside himself [67], but his generalizations were limited to examples. Soon after his paper was published, the ideas were shown to go through, appropriately modified, in all *finite type* Artin groups, i.e., Artin groups whose associated Coxeter group is finite, by Breiskorn and Saito [35] and by Deligne [48]. See [93] for a survey (and reworking) of the results first proved in [35,48]. (See Section 5.4 for definitions of Artin and Coxeter groups in general.) They used explicit properties of finite reflection groups in their proof, but Paris and Dehornoy were thinking more generally and defined a broader class which they called "Garside groups". The class included all finite type Artin groups. Over the last few years, several tentative definitions of the term "Garside group", referring to various classes of groups that generalize the braid group in this way, were proposed and appear in the literature (see [47,105], for example) before the one that we give below was agreed upon by many, although the search for the most general class of such groups continues and the definition of "Garside groups" is likely to continue to be in flux for some time. As we proceed through the definitions of "Garside monoids", "Garside structures", and "Garside groups", the reader can look to the classical presentation of $\mathbf{B}_n$ for examples.

Given a finitely generated monoid G^+ with identity e, we can define partial orders on its elements. Let $a, b \in G^+$. We say that $a < b$ if a is a left divisor of b, i.e., there exists $c \in M, c \neq e$, with $ac = b$. Also $b > a$ if there exists c such that $b = ca$. Caution: the two orderings are really different, i.e., $a < b$ does not imply that $b > a$. An interesting example in the positive braid monoid $\mathbf{B}_n^+$ is the partial order induced on the elements in the sets $\mathcal{P}$ of left divisors of Δ_n, which gives it the structure of a lattice. The reader who wishes to get a feeling for the ordering might wish to construct the lattice in the cases Δ_3 and Δ_4. The lattice for Δ_n has $n!$ elements.

Now given $a, b \in G^+$ we can define in a natural way the (left) greatest common divisor of a and b, if it exists, written $d = a \wedge b$, as follows: $d < a$, $d < b$ and if, for any x, it happens that $x < a$ and $x < b$ then $x < d$. Similarly, we define the (left) least common multiple of a, b, denoted $m = a \vee b$, if $a < m$, $b < m$ and if for any x it happens that $a < x$ and $b < x$ then $m < x$.

It turns out that in the case of the braid monoid $\mathbf{B}_n^+$ the left partial ordering extends to a right-invariant ordering on the full braid group $\mathbf{B}_n$, a matter which we will discuss in Section 5.7.

We continue with our description of the features of the monoid $\mathbf{B}_n^+ \subset \mathbf{B}_n$ which will lead us to define more general monoids G^+ and their associated groups G. An element $x \in G^+$ is an *atom* if $x \neq e$ and if x has no proper left or right divisors. For example, the generators $\sigma_1, \ldots, \sigma_{n-1}$ are the atoms in the monoid G^+. Note that by (G3), the atoms in $\mathbf{B}_n^+$ are left and right divisors of Δ, also they generate $\mathbf{B}_n^+$.

While we have not had occasion to introduce a key property of the Garside braid Δ before this, we do so now: As noted earlier, its sets $\mathcal{P}$ and $\mathcal{P}'$ of left and right divisors of Δ coincide. This property was used in the proofs of the facts that we described in (G1)–(G10), and is part of a long story about symmetries in the braid group.

A monoid G^+ is a Garside monoid if

(1) G^+ is generated by its atoms,

(2) For every $x \in G^+$ there exists an integer $l(x) > 0$ such that x cannot be written as a product of more than $l(x)$ atoms,

(3) G^+ is left and right cancellative, and every pair of elements in G^+ admits a left and also a right least common multiple and greatest common divisor.

(4) There is an element $\Delta \in G^+$, the *Garside element*, whose left divisors and right divisors coincide, also they form a lattice, also each generates G^+.

All of the data just listed is called a *Garside structure*. It follows (via the work of Ore, described in Volume 1 of the book [41]) that every Garside monoid embeds in its group of fractions, which is defined to be a *Garside group*. It is not difficult to prove that the Garside element in a Garside monoid is the least common multiple of its set of atoms. An interesting property is that a Garside monoid admits a presentation $\langle S|R\rangle$, where for every pair of generators x, y there is a relation of the form $x\cdots = y\cdots$ that prescribes how to complete x and y on the right in order to obtain equal elements.

Examples abound. While we shall see that the braid groups are torsion free, there are examples of Garside groups which have torsion. Whereas the relations in the braid monoid all preserve word length (which results in various finiteness aspects of the algorithms that we described) there are examples of Garside monoids in which this is not the case. As noted above, every Artin group of finite type is a Garside group. Torus knot groups are Garside groups, as are fundamental groups of complements of complex lines through the origin [47]. See [105] for additional examples.

Our reason for introducing Garside groups is that the solutions to the word problem and conjugacy problem in $\mathbf{B}_n$ described in Section 5.1, suitably modified, generalize to the class of Garside groups [47,105], as do the simplifications of (G10) which we now describe. These improvements were first discovered by Franco and Gonzales-Meneses [61], and later improved by Gebhardt [68].

We return to the braid group, with the partial ordering of $\mathbf{B}_n$ on hand:

(G11) We already learned that if we begin with an arbitrary braid ω, then after a bounded number of cyclings and decyclings we will obtain a braid in $\{\omega\}$ which realizes $\mathrm{Inf}(\omega)$ and $\mathrm{Sup}(\omega)$. Continuing to cycle, we will arrive at an element, say ρ, in U_ω. The remaining task is the computation of the full set U_ω, and the method described in (G10) is inefficient. The difficulty is that, starting with $\rho \in U_\omega$ one is forced to compute its conjugates by *all* the $n!$ elements in $\mathcal{P}$, even though many of those will turn out to either not realize $\mathrm{Inf}(\omega)$ and/or $\mathrm{Sup}(\omega)$, and so will be discarded, whereas others will turn out to be duplicates of ones already computed. Moreover, this inefficient step is done repeatedly.

The good news is that, following ideas first introduced by Franco and Gonzalez-Meneses [61], Gebgardt proves in Theorem 1.17 of [68], that if $\rho \in U_\omega$, and if $a, b \in \mathcal{P}$, with $a^{-1}\rho a$ and $b^{-1}\rho b \in U_\omega$, then $c^{-1}\rho c \in U_\omega$, where $c = a \wedge b$. See [61], and then [68], for a systematic procedure that allows one to use this fact to find all the orbits in U_ω efficiently. In this regard we remark that the work of Gebhardt is very new. The major open problem that remains is to improve it to an algorithm which will be polynomial in the word length of ω:

Open Problem 9. The bad news is that, like the Garside's summit set and ElRifai and Morton's super summit set, Gebhardt's ultra summit set also has the key difficulty ($\star$). More work remains to be done.

5.3. *The new presentation and multiple Garside structures*

As it turns out, essentially all of the structure that we just described also exists with respect to the second presentation of $\mathbf{B}_n$ which was given in Section 1 of this paper. This is the main result of [23]. The fundamental braid Δ_n is replaced by $\delta_n = \sigma_{n-1,n}\sigma_{n-2,n-1}\cdots\sigma_{2,3}\sigma_{1,2}$. It is proved in [20] that the associated monoid $\mathbf{B}_n^+$ embeds in $\mathbf{B}_n$, and that all the results described in (G1)–(G11) above have curious (and very surprising) variations which hold in the new situation.

For example, the elements in the set of left divisors of δ_n are in 1–1 correspondence with a set of permutations, namely permutations which are products of non-interlacing descending cycles. The set of left divisors of δ_n turns out to have order equal to the nth Catalan number, $(2n!)/(n!(n+1)!)$, whereas the left divisors of Δ have order $n!$. The fact that the order of the set of permutation braids is smaller in the new presentation than in the classical presentation had led to the hope by the authors of [20], when they first discovered the new presentation, that it would result in a polynomial algorithm for the conjugacy problem, but the Catalan numbers grow exponentially with index, and once again ($\star$) proved to be a fundamental obstacle. Thus, while the new presentation is extremely interesting in its own right, and does lead to faster word and conjugacy algorithms, the improvement in that regard does not address the fundamental underlying difficulties.

Open Problem 10. Curiously, it appears very likely that the classical and new presentations of $\mathbf{B}_n$ are the only positive presentations of $\mathbf{B}_n$ in which the Garside structure exists, although that has not been proved at this time, and is an interesting open problem. For partial results in this direction, see [82].

There is a different aspect of the dual presentations which we mention now, which involves a small detour. Before we can explain it, recall that one of our earliest definitions of the braid group $\mathbf{P}_n$, given in Section 1.1 was as the fundamental group $\pi_1(\mathcal{C}_{0,\hat{n}}, \vec{p})$, of the space formed from $\mathbb{C}^n$ by deleting the hyperplanes along which two or more complex coordinates coincide. Of course this gives a natural cell decomposition for $\mathcal{C}_{0,\hat{n}}$ as a union of (open) cells of real dimension $2n$. The braid group $\mathbf{B}_n$, as we defined it in Section 1.1, is the fundamental group of the quotient $\mathcal{C}_{0,n} = \mathcal{C}_{0,\hat{n}}/\Sigma_n$, where the symmetric group Σ_n acts on $\mathcal{C}_{0,\hat{n}}$ by permuting coordinates. In the interesting manuscript [60], Fox and Neuwirth used this natural cell decomposition of $\mathcal{C}_{0,\hat{n}}$ to find a presentation for $\mathbf{B}_n$, arriving at the classical presentation (1.2) for the braid group $\mathbf{B}_n$. See Section C of Chapter 10 of [38] for a succinct presentation of the results in [60]. Fox and Neuwirth also prove that $C_{0,\hat{n}}$ is aspherical, and use this to give the first proof that $\mathbf{B}_n$ is torsion free. See Section 6.4 for a different and very easily understood proof of that same fact.

Around the same time that the first author, together with Ko and Lee, wrote the paper [19], which introduced the dual presentation of $\mathbf{B}_n$, Thomas Brady was thinking about other complexes which, like the one just described might serve as a $K(\pi, 1)$ for the group $\mathbf{B}_n$. See [31], which describes the construction of a finite CW-complex K_n of dimension $n - 1$ which is homotopy equivalent to $C_{0,n}$ and which is defined combinatorially, using the partial ordering on Σ_n, as described in Section 5.2 in the discussion of the lattice of simple words in $\mathbf{B}_n$. Brady then used his complex to determine a presentation for $\mathbf{B}_n$, arriving at the dual presentation (1.3).

As it turned out, the new presentation was important for other reasons too. The braid groups inherit a second Garside structure from the new presentation (1.3). The fact that the same was true for many other Garside groups played a major role in the discovery of the appropriate definitions. The Garside structure on the braid group arising from the new presentation ends up revealing even more structure: it is dual to that coming from the standard presentation, in the sense of an action on a complex and a dual complex. In fact, all finite type Artin groups of finite type also have dual Garside structures, as proved by Bessis [10] (this reference also contains the details of this dual structure). See also [32]. We also refer the reader to [105] for explicit presentations for the dual monoids associated to the Artin groups of finite type, and also for an interesting table that gives the number of simple elements defined by the left divisors of the Garside element in the classical and dual monoids for the finite type Artin groups.

OPEN PROBLEM 11. This one is a very big set of problems. It is not known whether all Garside groups have dual presentations, in fact, it is also not known how many distinct Garside structures a given Garside group may have.

5.4. *Artin monoids and their groups*

There is another class of groups which is intimately related to the braid group and its associated monoid, but it is much less well understood than the Garside groups. Let $S = \{u, \dots, \dots, t, \dots, v\}$ be a finite set. A *Coxeter graph* Γ over S is a graph whose vertices are in 1–1 correspondence with the elements of S. There are no edges joining a vertex s to itself. There may or may not be an edge joining a vertex s to a vertex $t \neq s$. Each pair of vertices $(s, t) = (t, s)$ is labeled by a non-negative integer $m(s, t)$. There are two types of labels: The label $m(s, t)$ is 2 if Γ does not have an edge that joins s and t, and it is $\in \{3, 4, \dots, \infty\}$ if there is an edge joining s and t.

The *Artin group* $A(\Gamma)$ associated to Γ has generators $\{\sigma_s,\ s \in S\}$. There is a relation for each label $m(s, t) < \infty$, namely $\sigma_s\sigma_t\sigma_s\sigma_t\cdots = \sigma_t\sigma_s\sigma_t\sigma_s\cdots$, where there are $m(s, t)$ terms on each side, $2 \leq m(s, t) < \infty$. The *Coxeter group* $C(\Gamma)$ associated to the Artin group $A(\Gamma)$ is obtained by adding the relations $\sigma_s^2 = 1$ for every $s \in S$. As previously mentioned, we say that the group $A(\Gamma)$ has finite type if its associated Coxeter group is finite. The braid group $\mathbf{B}_n$ is an example of an Artin group of finite type; its associated Coxeter group being the symmetric group Σ_n. Note that by definition the defining relations in the presentations that we just described for Artin groups all preserve word length.

One of the many interesting properties of Artin groups is that for every Artin group there is an associated monoid A^+, and just as every Garside monoid embeds in its group, it was proved by Paris in [102] that every Artin monoid injects in its group. The injectivity property holds in the following strong sense too: If two positive words P, P' represent the same element of A, then they represent the same element of A^+. Paris's proof is completely different from Garside's proof of injectivity in the case of the braid group, which is the basis for the known proofs of the same fact for Garside groups. On the other hand, the other properties that are needed to obtain a Garside structure may or may not hold, for example it is definitely not true that every element in an Artin group can be written in the form NP, where N is negative and P is positive.

REMARK 5.4. An interesting special case of Artin groups are the *right-angled* Artin groups. They are Artin groups in which the defining relations are all of the form $\sigma_s\sigma_t = \sigma_t\sigma_s$. For example, the right-angled Artin group associated to the braid group is defined by the presentation:

$$\langle \sigma_1, \ldots, \sigma_{n-1} | \sigma_i\sigma_k = \sigma_k\sigma_i \text{ if } |i-k| \geq 2\rangle. \tag{5.1}$$

OPEN PROBLEM 12. In Section 4.5 we explained the fairly recent proof that the braid groups are linear. This leads one to ask, immediately, whether the same is true for its natural generalizations, e.g., Garside groups and Artin groups. It turns out that, like the braid groups, all right-angled Artin groups have faithful matrix representations [44,50]. Of course any Artin group that injects as a subgroup of a related braid group is also linear, e.g., Artin groups of type $\mathbf{B}_n$. It was proved recently by Cohen and Wales [42] and simultaneously by Digne [50] that all finite type Artin groups are linear. Going beyond that, the matter seems to be wide open and interesting.

REMARK 5.5. For reasons of space, we have not included any significant discussion of the vast literature on Artin groups and associated complexes on which they act. It is a pity to omit it, because it is a major area, and much of it had its origins in work on $\mathbf{B}_n$.

5.5. *Braid groups and public key cryptography*

The problem which is the focus of "public key cryptography" was mentioned, very briefly, in Section 1.4.5. The basic issue is how to send information, securely, over an insecure channel. The solution is always to use some sort of code whose main features are known to the sender and recipient, but which cannot be deduced by a viewer who lacks knowledge of the shared keys. To the best of our knowledge, all solutions to this problem rest on the same underlying idea: they make use of problems which have a precise answer, which is known to both the sender and recipient, but one which is deemed to be so difficult to compute that it is, in effect, unavailable to a viewer, even though the viewer has all the necessary data to deduce it. The earliest such schemes were based upon the difficulty of factorizing large integers into a product of primes. The individuals who wish to exchange data over a public system are denoted A and B. In a vastly oversimplified version, both A and B have agreed, privately, on the choice of a prime number p. The sender chooses another prime q and transmits the product $n = pq$. A viewer may learn n, but because of the difficulty of factorizing n into primes cannot deduce p and q. The recipient, who knows both p and n, has no problem computing q.

A more recent approach is due to Diffie and Hellman [49]. As before, both A and B have agreed, privately, on the choice of a prime number p and a generator g of the finite cyclic group $\mathbb{Z}/p\mathbb{Z}$. A chooses a number a at random and computes $g^a(\text{mod } p)$, which she sends to B on the public channel. As for B he chooses a number b at random, computes $g^b(\text{mod } p)$ and sends in to A on the public channel. Since A now knows both g^b and a she can compute $(g^a)^b = g^{ab}(\text{mod } p)$. Similarly, since B now knows both g^a and b, he can compute $(g^b)^a = g^{ba}(\text{mod } p) = g^{ab}(\text{mod } p)$. So both know $g^{ab}(\text{mod } p)$. As for the viewer, he knows both $g^a(\text{mod } p)$ and $g^b(\text{mod } p)$, however, because of the known difficulty of computing discrete logs, the crucial information $g^{ab}(\text{mod } p)$ is in effect unavailable to the viewer.

A rather different set of ideas was proposed in [3,80], and this is where the braid group comes in. The security of a system that is based upon braid groups relies upon the assumption that the word and conjugacy problems in the braid group have both been solved, but the conjugacy problem is computationally intractable whereas the word problem is not. However, as we have seen, that matter seems to be wide open at this moment. We have described the underlying mathematics behind the ElRifai–Morton solution to the word problem, and the best of the current solutions to the word and conjugacy problem, namely that of Gebhardt. The solution to the word problem is used in the conjugacy problem. The reason that the conjugacy problem is not polynomial is that we do not understand enough about the structure of the summit set, the super summit set and the ultra summit set. Our strong belief is that these matters will be settled. The assumption that our current lack of understanding of aspects of the mathematics of braids means that they cannot be understood seems unwarranted.

5.6. *The Nielsen–Thurston approach to the conjugacy problem in* $\mathbf{B}_n$

In this subsection we consider the conjugacy problem in $\mathbf{B}_n$ from a new point of view. We return again to the interpretation given in Theorem 1.1 of Section 1.3 of the braid group $\mathbf{B}_n$ as the mapping class group $\mathcal{M}_{0,1,n}$ of the punctured disc $S_{0,1,n} = D^2_n$. In this context, then, we emphasize that the term *braid* refers to a mapping class, that is, an isotopy class of diffeomorphisms.

The Nielsen–Thurston classification of mapping classes of a surface is probably the single most important advance in this theory in the last century, and we review it here. For simplicity, we shall focus for now on the mapping class group of a closed surface $\mathcal{M}_g = \mathcal{M}_{g,0,0}$ in the notation of Section 1.3. We note that the groups $\mathcal{M}_g$ have trivial center for all $g \geq 3$.

We make the tentative definition that an element of $\mathcal{M}_g$ is "reducible" if it preserves up to isotopy a family of disjoint non-trivial curves on the punctured disc. Such a family of curves is known as a *reduction system*. (Throughout this discussion, we use "curve" to refer to the isotopy class of a curve. By "disjoint curves", we mean two distinct isotopy classes of curves with respective representatives which are disjoint.) It was proved in [22] that for a reducible map ϕ, there exists a essential reduction system, denoted ERS(ϕ). A curve $c \in$ ERS(ϕ) if and only if

(1) There exists an integer k such that $\phi^k(c) = c$.

(2) If a curve x has non-zero geometric intersection with c, then $\phi^m(x) \neq x$ for all integers m. In particular, any curve x which has non-zero geometric intersection with a curve $c \in$ ERS(ϕ) is not in ERS(ϕ).

It is proved in [22] that the curves in ERS(ϕ) are contained in every reducing system for ϕ, and are a minimal reduction system for ϕ. Also, the system of curves ERS(ϕ) is unique. Keeping all this in mind, we now define an element $\phi \in \mathcal{M}_g$ to be *reducible* if it fixes the curves in an essential reduction system, i.e., in ERS(ϕ).

A mapping class $\phi \in \mathcal{M}_g$ is *periodic* if $\phi^k = 1$ in $\mathcal{M}_g$.

We note that a reducible mapping class always contains a representative diffeomorphism which fixes a given reduction system setwise. Likewise, Nielsen showed that a periodic mapping class always contains a representative which is periodic as a diffeomorphism. Finally, a mapping class which is neither reducible nor periodic is *pseudo-Anosov* (abbreviated as PA).

Braid groups, as the mapping class group of a punctured disc, admit a similar classification. The group $\mathbf{B}_n$ is torsion free, but by analogy with the above case of mapping class groups of closed surfaces of genus at least two, which are centerless, we say that an element $\omega \in \mathbf{B}_n$ is *periodic* if for some integer k, ω^k is isotopic to a full Dehn twist on the boundary of the disc D_n^2. In terms of generators and relations this is equivalent to saying that the braid is a root of Δ^2, where Δ is the Garside braid. (Recall from Section 5.1 that the square of Δ_n generates the center of $\mathbf{B}_n$.)

The definitions of reducible braids and pseudo-Anosov (PA) braids require no alteration other than replacing the surface S_g with the punctured disc.

REMARK 5.6. It is clear for $\mathbf{B}_n$ and $\mathcal{M}_g$ alike, we also have a classification of elements as reducible, periodic and pseudo-Anosov up to conjugacy.

Each of these three possibilities reveals new structure, so we consider them one at a time. A major reference for us is the paper [7], which gives an algorithm for finding a system of reducing curves if they exist, and for recognizing periodic braids. This paper did not receive much attention at the time that it was written because it had the misfortune to be written simultaneously and independently with the ground breaking and much more general papers of Bestvina and Handel [11]. However, as is often the case, one learns very different things by examining a particular case of a phenomenon in detail, and by proving a broad generalization of the same phenomenon, and that is what happened here.

Periodic braids. The first author learned from [7] that the classification of periodic braids had been solved via the work of Kerekjarto (1919) and Eilenberg (1935), who proved that up to conjugacy a periodic braid is a power of either δ or α, where $\delta = \sigma_1\sigma_2\cdots\sigma_{n-1}$ and $\alpha = \sigma_1\sigma_2\cdots\sigma_{n-1}\sigma_1$. One verifies, using the elementary braid relations in the classical presentation, that $\delta^n = \alpha^{n-1} = \Delta^2$. To visualize the assertion for δ geometrically, place the punctures at polar angles $2\pi k/n$ around a circle of radius $r < 1$, and think of δ as a rotation of D_n^2 of angle $2\pi/n$. To visualize α, do the same, only now place one of the punctures at the origin and arrange the remaining $n-1$ punctures symmetrically at polar angles $2\pi k/(n-1)$.

Reducible braids. Let $\omega = \Delta^l L_1 L_2\cdots L_s$ be a reducible mapping class in left greedy normal form, as described in Section 5.1. Our initial model for D_n^2 will be the unit disc, with the n punctures arranged along the real axis, and placed symmetrically so that they divide the interval $[-1, 1] \subset \mathbb{R}$ into $n + 1$ equal line segments. It is shown in [7] that one can choose a braid ω' conjugate to ω and in the super summit set of ω which fixes the simplest possible family of closed curves, namely a family C of geometric ellipses whose centers are on a horizontal "axis" through the punctures, and which are chosen in such a way that the axis bisects the discs that the curves in $\mathrm{CRS}(\omega')$ bound.

OPEN PROBLEM 13. Is it true that, if a reducible braid is in its ultra summit set, then its invariant multicurves are can always be chosen to be geometric ellipses?

The remaining braiding may then be thought of as going on inside tubes, also the tubes may braid with other tubes. With this model it should be intuitively clear that the choice of such a representative for the conjugacy class of a braid which permutes the tubes in the required fashion may require very different choices (again up to conjugacy) for the braids which are inside the tubes, and that the sensible way to approach the problem is to cut the initial disc open along the reducing curves and focus on the periodic or PA maps inside the tubes. This is, of course, a very different approach from the one that was considered in Section 5.1, where no such considerations entered the picture.

Pseudo-Anosov braids. In the PA case (which is the generic case), there is additional structure, and now our description becomes very incomplete. In this case [110] there exist two projective measured foliations $\mathcal{F}^u$ and $\mathcal{F}^s$, which are preserved by an appropriate representative w of $\{\omega\}$. Moreover the action of w on $\mathcal{F}^u$ (the unstable foliation) scales its measure by a real number $\lambda > 1$, whereas the action on $\mathcal{F}^s$ (the stable foliation) scales its measure by $1/\lambda$. These two foliations and the scaling factor λ are uniquely determined by the conjugacy class of $\{\omega\}$, however the triplet $(\mathcal{F}^u, \mathcal{F}^s, \lambda)$ does not determine $\{\omega\}$. To explain the missing pieces, we replace the invariant foliations by an invariant "train track", with weights associated to the various branches.

In the case of pseudo-Anosov mapping classes acting on a once-punctured surface, a method for enumerating the train tracks is in Mosher's unpublished PhD thesis [99]. But Mosher's work was incomplete and remained unpublished for many years, even as many of the ideas in it were developed and even expanded, leading to some confusions in the literature about exactly what is known and what remains open. At this writing a complete solution to the conjugacy problem for braid groups or more generally surface mapping class groups, based upon the Nielsen–Thurston machinery, does not exist in the literature. Mosher has a partially completed monograph in preparation, *Train track expansions of measured foliations*, which promises to give such a solution, but for more general surfaces, i.e., for $S_{g,b,n}$ very little seems to be known. We therefore pose it as an open problem:

OPEN PROBLEM 14. Investigate the conjugacy problem in the mapping class groups $\mathcal{M}_{g,b,n}$ with the goals of (i) pinning down precisely what is known for various triplets (g, b, n); (ii) describing all cases in which there is a complete solution; and (iii) describing what remains to be done in the simplest cases, that is $\mathcal{M}_{g,0,0}$ and $\mathcal{M}_{g,0,1}$.

Essentially nothing is known, at this writing, about the interface between the dynamic and combinatorial solutions to the conjugacy problem in the braid groups, and still less in the general case of the more general mapping class groups $\mathcal{M}_{g,b,n}$, to which the entire Thurston machinery applies. We know of nothing which even hints at related dynamic structures in Artin groups or Garside groups.

OPEN PROBLEM 15. Does the Garside approach to the conjugacy problem in $\mathcal{M}_{0,1,n}$ generalize to a related approach to the problem in $\mathcal{M}_{g,1,0}$ for any $g > 0$?

OPEN PROBLEM 16. Does the Nielsen–Thurston approach to the conjugacy problem in $\mathcal{M}_{0,1,n}$ generalize to a related approach to the problem in any other Artin group?

5.7. *Other solutions to the word problem*

We review, very briefly, other ways that the word and conjugacy problems have been solved in the braid groups. We restrict ourselves to results which revealed aspects of the structure of $\mathbf{B}_n$ that has had major implications for our understanding of braid groups, even when the implications for the word and/or conjugacy problems fall short of that criterion.

(1) *Artin's solution to the word problem.* The earliest solution to the word problem was discovered in 1925 by Artin, in his first paper on braids [5]. It was based upon his analysis of the structure of the pure braid group, which we described in Section 1.2, and in particular in the defining relations (1.4) and the sequence given just below it:

$$\{1\} \to \mathbf{F}_{n-1} \to \mathbf{P}_n \stackrel{\pi_n^\star}{\to} \mathbf{P}_{n-1} \to \{1\}.$$

The resulting normal form, described in [5] as *combing* a braid, is well known. We refer the reader to Artin's original paper for a very beautiful example. In spite of much effort over the years nobody has managed, to this day, to use related techniques to solve the conjugacy problem in $\mathbf{B}_n$, except in very special cases.

(2) *The Lawrence–Krammer representation.* See Section 4. When a group admits a faithful matrix representation, there exists a fast way to solve the word problem.

It is interesting to note that Krammer's first proof of linearity, in the case of $\mathbf{B}_4$, used the solution to the word problem which came from the presentation (1.3) in Section 1 of this paper. See [84]. Also, when he proved linearity in the general case, in [85], he gives two proofs. The second proof, Theorem 6.1 of [85], shows that the Lawrence–Krammer matrices detect the infimum and supremum of a braid. It follows as an immediate corollary (noted in [85]) that the representation is faithful, because any braid which has non-zero infimum and supremum cannot be the identity braid.

As regards the conjugacy problem, there is a difficulty. The image of $\mathbf{B}_n$ under the isomorphism given by the Lawrence–Krammer representation yields a group $\mathcal{B}_n$ which is a subgroup of the general linear group $\mathrm{GL}_m(\mathbb{Z}[q^{\pm 1}, t^{\pm 1}])$, where $m = (n) \times (n-1)/2$. To the best of our knowledge it is unknown at this time how to describe $\mathcal{B}_n$ in any way that allows one to test membership in $\mathcal{B}_n$. Lacking such a test, any class invariants which one finds in this way will be limited in usefulness, because if $\omega \in \mathbf{B}_n$, there will be no way to distinguish between class invariants which arise from conjugation by elements in $\mathcal{B}_n$ from those which arise from conjugation by more general elements of $\mathrm{GL}_m(\mathbb{Z}[q^{\pm 1}, t^{\pm 1}])$ from invariants thus one cannot hope for a complete solution.

Open Problem 17. Investigate the eigenvalues and the trace of the Lawrence–Krammer matrices.

(3) *The Dehornoy ordering*. A group or a monoid has a *right-invariant* (resp. *left-invariant*) ordering if there exists a strict linear ordering of its elements, denoted $<$, with the property that if $f, g, h \in G$, then $f < g$ implies $fh < gh$ (resp. $hf < hg$). To the best of our knowledge nobody had considered the question of whether $\mathbf{B}_n$ had this property before 1982, when Patrick Dehornoy announced his discovery that the groups $\mathbf{B}_n$ admit such an ordering, and that it can be chosen to be either right invariant or left invariant, but not both. In the 5-author paper [58] the Dehornoy ordering was shown to have the following topological meaning. We restrict to the right-invariant case, the two results being essentially identical. We regard $\mathbf{B}_n$ as $\mathcal{M}_{0,1,n}$, where the surface $S_{0,1,n}$ is the unit disc in the complex plane and the punctures lie on the real axis in the interval $(-1,1)$. The punctures divide $[-1,1] \subset \mathbb{R}$ into $n+1$ segments which we label $\xi_1, \xi_2, \ldots, \xi_{n+1}$ in order, where ξ_1 joins $\{-1\}$ to the first puncture. Choose $[\omega_1], [\omega_2] \in \mathbf{B}_n$, and representatives ω_1, ω_2 with the property that $\omega_1([-1,1])$ and $\omega([-1,1])$ intersect minimally. Note that $\omega_k([-1,1])$ divides $S_{0,1,n}$ into two halves, and it makes sense to talk about the upper and lower half of $\omega_2([-1,1])$ because $\{-1\}$ and $\{1\}$ are on $\partial S_{0,1,n}$ and $\omega_k|\partial S_{0,1,n}$ is the identity map. Then $[\omega_2] > [\omega_1]$ if $\omega_1(\xi_i) = \omega_2(\xi_i)$ for $i = 1, \ldots, j-1$ and an initial segment of $\omega_2(\xi_j)$ lies in the upper component of $S_{0,1,n} - \omega_1([-1,1])$.

It was proved in [113] that the resulting left-invariant ordering of $\mathbf{B}_n$ extends the ElRifai–Morton left partial ordering that we described in Section 5.2. It is proved in [46] that there are infinitely many other left orderings.

The subject blossomed after the paper [58] appeared. In fact so much work resulted that there is now a 4-author monograph on the subject [46], written by some of those who were the main contributors, containing an excellent review of what has been learned during the 10 years since the initial discovery. We do not wish to repeat what is readily available elsewhere, especially because we are not experts, so we point the reader to Chapter 3 of [46], where it is shown that the Dehornoy ordering leads to a solution to the word problem. Unfortunately, the solution so-obtained is exponential in $|\omega|$, whereas the solution described in (G1)–(G6) is quadratic in $|\omega|$.

(4) $\mathbf{B}_n$ *as a subgroup of* $\mathrm{Aut}(\mathbf{F}_n)$. It is well known that $\mathbf{B}_n$ has a faithful representation as a subgroup of $\mathrm{Aut}(\mathbf{F}_n)$, for example see [15] for a proof. This of course gives a solution to the word problem, but (like the Dehornoy solution) it is exponential in $|\omega|$. On the other hand, it is extremely interesting that the Nielsen–Thurston machinery, described in Section 5.6, generalizes to the entire group $\mathrm{Aut}(\mathbf{F}_n)$, giving yet another instance where the braid group is at the intersection of two rather different parts of mathematics, and could be said to have pointed the way to structure in the second based upon known structure in the first.

In this section we have described several solutions to the word problem, and noted that one of them (the modified Garside approach of (G1)–(G6)) is $\mathcal{O}(|\omega|^2(n)(\log(n))$, where $|\omega|$ is the letter length of two arbitrary representatives of elements of $\mathbf{B}_n$, using the classical presentation (1.2) for $\mathbf{B}_n$. The word problem seems to be one short step away from the *non-minimal braid* problem: given a word ω in the generators $\sigma_1, \ldots, \sigma_{n-1}$ and their inverses, determine whether there is a shorter word ω' in the same generators which represents the same element of $\mathbf{B}_n$? For, it is clear that a decision process exists: list all the words that are shorter than the given one, thin the list by eliminating as many candidates as possible by simple criteria such as preserving length mod 2, an obvious invariant, and then test, one by one, whether the survivors represent the same element of $\mathbf{B}_n$ as the given word ω? It therefore seemed totally surprising to us that in 1991 M.S. Patterson and A.A. Razborov proved:

THEOREM 5.7 [104]. *The non-minimal braid problem is NP complete.*

Thus, if one could find an algorithm to decide whether a given word ω is non-minimal, and if the algorithm was polynomial in $|\omega|$, one would have proved that $P = NP$! To the best of our knowledge, essentially nothing has been done on this problem. In this regard we suggest two research problems:

OPEN PROBLEM 18. The proof that is given in [104] is very specific to the classical presentation (1.2) for $\mathbf{B}_n$. Can it be adapted to the new presentation? To other presentations?

OPEN PROBLEM 19. Investigate the shortest word problem in the braid group $\mathbf{B}_n$, using the classical presentation.

In Section 3.1 we discussed and proved the MTWS in the special case of the unknot. See Theorem 3.1, which asserts that there is a complexity measure on closed braid representatives of the unknot, and using it a sequence of strictly complexity-reducing destabilizations and exchange moves that reduce a closed braid diagram for the unknot to a round planar circle. In [18], Theorem 3.1 was used to develop an algorithm for unknot recognition. In [27] the first steps were taken to develop a computer program to realize the resulting algorithm. The algorithm is far from being practical in even simple cases, however we are now in a position to describe three open problems, all closely related to (G1)–(G11), which would lead to an effective solution to the unknot recognition problem:

OPEN PROBLEM 20. Develop an algorithm that will detect when the conjugacy class of a closed braid admits a destabilization.

An n-braid W admits an exchange move if it is conjugate to a braid of the form $U\sigma_{n-1}V\sigma_{n-1}^{-1}$, where U and V depend only on $\sigma_1, \ldots, \sigma_{n-2}$. Thus, up to conjugation, W is a product of two reducible braids, one positively reducible and the other negatively reducible. By Theorem 3.1, it may be necessary to modify the given conjugacy class by exchange moves in order to jump from the given class to a new class (if it exists) which admits a destabilization. Thus, there is an unknown complexity measure on conjugacy classes, with the ones which admit destabilizations being especially nice. This leads us to our second open problem.

OPEN PROBLEM 21. Develop an algorithm that will detect when the conjugacy class of a closed braid admits an exchange move, and when a collection of classes are exchange equivalent.

This would still not be a complete tool, because exchange moves can be either be complexity reducing or complexity increasing, and unless we can tell the difference (that is the content of Open Problem 3) we are left with the option of trying a sequence of exchange moves of unpredictable length in our search for destabilizations.

OPEN PROBLEM 22. The complexity measure that was introduced in Section 3.1 is "hidden" in the braid foliation of an incompressible surface whose boundary is the given knot. This is a highly non-trivial matter, because if we had the foliated surface in hand, then we would be able to compute its Euler characteristic and would know whether we had the unknot. On the other hand, the braid foliation determines the embedding of its boundary (see Theorem 4.1 of [17]), so there is no essential obstacle to "seeing" the complexity measure in the given braid. That is the essence of our third open problem, which asks that we recognize how to translate the complexity measure that was used in the proof of Theorem 3.1 into a complexity measure on closed braids which will be able to distinguish exchange moves that reduce complexity from those which do not.

6. A potpourri of miscellaneous results

This section is for leftovers – topics on which there have been interesting new discoveries which did not seem to fit well anywhere else.

6.1. *Centralizers of braids and roots of braids*

The *mixed braid groups* are defined in [72] to be the braids which preserve a given partition of the puncture points on the disc. In [72] Gonzales-Meneses and Wiest gave full descriptions of the centralizer of a braid in terms of semi-direct and direct products of mixed braid groups, and found sharp bounds on the number of generators of the centralizer of a braid.

In a related paper [71] Gonzales-Meneses proved that, up to conjugacy, braids have unique roots. That is, if $\omega \in \mathbf{B}_n$ and if $\alpha, \beta \in \mathbf{B}_n$ have the property $\alpha^k = \beta^k = \omega$, then $\{\alpha\} = \{\beta\}$.

6.2. *Singular braids, the singular braid monoid, and the desingularization map*

The singular braid monoid $\mathbf{SB}_n$ is a monoid extension of the braid group $\mathbf{B}_n$. It was introduced in [6] and, simultaneously and independently, in [16]. Its definition was suggested by the mathematics which revolved about Vassiliev invariants of knots and links. To define it, we need to describe a presentation for $\mathbf{SB}_n$, taken from [16]. There are three types of generators, which we call σ_i, σ_i^{-1} and τ_i, $1 \leq i \leq n-1$. Here, the $\sigma_i^{\pm 1}$ are to be thought of as the classical positive and negative elementary braids and the τ_i are to be thought of as *elementary singular braids*. The braid τ_i is obtained from σ_i (see Figure 24(ii)) by identifying strands i and $i+1$ as they cross. Defining relations in $\mathbf{SB}_n$ are:

$$\sigma_i\sigma_i^{-1} = \sigma_i^{-1}\sigma_i = 1, \qquad \sigma_i\tau_i = \tau_i\sigma_i,$$

$$\sigma_i\sigma_j = \sigma_j\sigma_i, \quad \sigma_i\tau_j = \tau_j\sigma_i, \quad \tau_i\tau_j = \tau_j\tau_i \qquad \text{if } |i-j| \geq 2$$

$$\sigma_i\sigma_j\sigma_i = \sigma_j\sigma_i\sigma_j, \quad \sigma_i\sigma_j\tau_i = \tau_j\sigma_i\sigma_j \qquad \text{if } |i-j| = 1.$$

The *desingularization map* is a homomorphism from $\mathbf{SB}_n$ to the group ring $\mathbb{Z}\mathbf{B}_n$ of the braid group, defined by $\psi(\sigma_i^{\pm 1}) = \sigma_i^{\pm 1}$, $\psi(\tau_i) = \sigma_i - \sigma_i^{-1}$. Birman used this map to develop the relationship between Vassiliev invariants and quantum groups. It was conjectured in [16] that the map ψ is an embedding. After several proofs of special cases of the conjecture, it was settled by Luis Paris in the affirmative in 2003 [103].

During the 10-year interval after the introduction of the singular braid monoid, and before the proof of the embedding theorem, it came as quite a surprise when it was discovered that there was a new group on the scene – the *singular braid group* of [62]. To this day we are unsure of its significance, although its existence is unquestioned! It is most easily defined via generators and relations, starting with the presentation that we just gave for the singular braid monoid, and then adding one new generator $\bar{\tau}$ (to suggest that it behaves the way that the inverse of τ ought to behave). Defining relations for $\mathbf{GB}_n$ are all the relations in $\mathbf{SB}_n$, plus ones satisfied by the new generator. The latter are "monoid relations" which are the same as those in the singular braid monoid, but substituting $\bar{\tau}$ for τ, and two additional relations, namely $\tau\bar{\tau} = \bar{\tau}\tau = 1$. Pictorially, one has two types of singular crossings, and they annihilate one another. The main result in [62] is that $\mathbf{SB}_n$ embeds in $\mathbf{GB}_n$.

6.3. *The Tits conjecture*

The Tits conjecture is very easy to state, in its simplest form. Consider the classical presentation of the braid group $\mathbf{B}_n$, i.e., the presentation (1.2). The Tits conjecture, in the special case of the braid group, is that the subgroup G of $\mathbf{B}_n$ generated by the elements $T_i = \sigma_i^2$ has the presentation $\langle T_1, \dots, T_{n-1} | T_iT_j = T_jT_i \text{ if } |i-j| \geq 2\rangle$. A generalized version of the conjecture (the generalization relates to the arbitrary choices of the powers) was proved by Crisp and Paris in 2001, settling a question which had plagued the experts for many years:

THEOREM 6.1 [43]. *Let S be a finite set, let Γ be a Coxeter graph over S, and let $A(\Gamma)$ be the associated Artin group, as defined in Section 5.4, with generating set $\{\sigma_s, s \in S\}$. Associate further to each $s \in S$ an integer $m_s \geq 2$. Consider the subgroup G of $A(\Gamma)$ generated by the elements $\{T_{s,m_s} = \sigma_s^{m_s}\}$. Then defining relations among the generators of the subgroup G are "the obvious ones", namely that T_{s,m_s} and T_{t,m_t} commute in G if and only if they commute in $A(\Gamma)$. No other relations are needed.*

The proof is interesting, because it introduces a technique which is very closely related to the themes that we have explored in this article. The basic idea is that, for a key subclass of Artin groups there is a representation f of $A(\Gamma)$ into the mapping class group $\mathcal{M}(S)$ of a connected surface S with boundary that is associated to the graph Γ, which induces an action of $A(\Gamma)$ on a monoid determined by Γ. Since the group $H(\Gamma)$ which is presented in the theorem has an obvious homomorphism onto $A(\Gamma)$, and the proof shows that the restriction of the action to $H(\Gamma)$ gives the desired isomorphism.

6.4. *Braid groups are torsion free: a new proof*

As we stated in Section 1, it was necessary to make choices in the writing of this review, and our decision was to be guided by the principle of focusing on new results or new proofs of known results. During the years since [15] was written, many people have written to the first author with questions about braids, and a regular question has been "Isn't there a *simple* proof that the braid groups have no elements of finite order?" It is therefore fitting that we end this review with exactly that – a beautiful simple proof, based upon the discovery, due to Dehornoy, that the braid groups admit a left-invariant ordering:

THEOREM 6.2. *The groups $\mathbf{B}_n$, $n = 1, 2, 3, \dots$ are all torsion free.*

PROOF. See Section 5.7 for a discussion of the Dehornoy left-invariant ordering of the braid groups. Choose any element $g \in \mathbf{B}_n$, $g \neq 1$. Replacing g if necessary with g^{-1}

we may assume that $1 < g$. Since the ordering is left invariant we then have that $g < g^2$ and $g^{-1} < 1$. Iterating, $\cdots < g^{-3} < g^{-2} < g^{-1} < 1 < g < g^2 < g^3 < \cdots$ □

Acknowledgements

We thank Tahl Nowik, who suggested, during a course that the first author gave on mapping class groups, that techniques she had used for other purposes could be adapted to give the proof we presented here of Theorem 1.1. We also thank Robert Bell, David Bessis, Nathan Broaddus, John Cannon, Ruth Charney, Fred Cohen, Patrick Dehornoy, Roger Fenn, Elizabeth Finkelstein, Daan Krammer, Lee Mosher, Luis Paris, Richard Stanley, Morwen Thistlethwaite and Bert Wiest for their help in chasing down facts and references, and helping us fill gaps in our knowledge. In addition, we are grateful to all the students who attended the course "Knots and Braids" in the Fall of 2003 at Cornell University, given while the second author was a VIGRE postdoc, who were gracious guinea pigs for large parts of this article. In particular, we are indebted to Heather Armstrong, whose careful attention and diligence significantly improved the manuscript, and Bryant Adams, who suggested the proof of Lemma 2.4.

Appendix. Computer programs

In 2004, it is almost as important to know about computer tools as it is to have a guide to the literature, so we supplement our bibliography with a guide to the computer tools that we know about and which have been useful to us and colleagues.

Changing knots and links to closed braids: Vogel's proof of his method for changing arbitrary knot diagrams to closed braid diagrams is ideal for computer programming. We refer the reader to the URL http://www.layer8.co.uk/maths/braids/, for a program, due to Andrew Bartholomew and Roger Fenn, which does this and much more.

Garside's algorithm for the word and conjugacy problems: Many people have programmed Garside's algorithm for the word and conjugacy problem, but the one we know best is the program of Juan Gonzalez-Meneses, which can be downloaded from http:www.personal.us.es/meneses. The very robust version that the reader will find there computes Garside's normal forms for braids, and the ultra-summit set of a braid, a complete invariant of conjugacy.

Nielsen–Thurston classification of mapping classes in $\mathcal{M}_{0,n+1}$: We know of two very useful computer programs, all based upon the Bestvina–Handel algorithm [11]. Both assume that the boundary of the n-times punctured disc has been capped with a disc, so that they compute in the mapping class group of $\mathbf{B}_n$/modulo its center, rather than in $\mathbf{B}_n$. Equivalently, they work with the mapping class group of an $(n+1)$-times punctured sphere, where admissible maps fix the distinguished point.

The first, due to W. Menasco and J. Ringland, can be downloaded from http://orange.math.buffalo.edu/software.html, by following the link to "BH2.1 An Implementation of the Bestvina–Handel Algorithm". The second, due to T. Hall, can be downloaded from http://www.liv.ac.uk/Maths/pure and following the links to the research group on Dynamical Systems, and then to the home page of Toby Hall. Both determine whether an input map is pseudo-Anosov, reducible or finite order and both find an invariant train track and a train track map (which makes it possible to calculate essential dynamics in the class by Markov partition techniques). In the reducible case Hall's program provides a set of reducing curves. Recently, Hall updated his program to adapt it to MacIntosh OS-X computers, versions 10.2 and above.

Other software: We mention M. Thistlethwaite's Knotscape, because the program has a reputation for being very versatile and user-friendly. It is available for download from http://www.math.utk.edu/morwen. It accepts as input knots that are defined as closed braids (and also knots defined by the Dowker code or mouse-drawn diagrams), and locates it in the tables if it has at most 16 crossings. The program computes numerous invariants, including the Alexander, Jones, Homfly and Kauffman polynomials, and hyperbolic invariants (assuming that the knot is hyperbolic). The hyperbolic routines in Knotscape were taken with permission from Jeff Weeks's program SnapPea; the procedures for calculating polynomials were supplied by Bruce Ewing and Ken Millett, and the procedure for producing a knot picture from Dowker code is part of Ken Stephenson's Circlepack program.

References

[1] S.I. Adyan, *Fragments of the word Δ in a braid group*, Mat. Zametki **36** (1) (1984), 25–34, English Translation Math. Notes, **36** (1–2) (1984), 505–510.

[2] J. Alexander, *A lemma on a system of knotted curves*, Proc. Nat. Acad. Sci. USA **9** (1923), 93–95.

[3] I. Anshel, M. Anshel and D. Goldfeld, *An algebraic method for public key crypography*, Math Res. Lett. **6** (3–4) (1999), 287–291.

[4] I. Anshel, M. Anshel and D. Goldfeld, *Topological invariants of plane curves and caustics*, AMS University Lecture Ser. **5** (1994).

[5] E. Artin, *Theorie der Zopfe*, Abh. Math. Sem. Hamburg **4** (1925), 47–72.

[6] J. Baez, *Link invariants of finite type and perturbation theory*, Lett. Math. Phys. **26** (1992), 43–51.

[7] D. Benardete, M. Guitierrez and Z. Nitecki, *Braids and the Nielsen-Thurston classification*, J. Knot Theory Ramificat. **4** (1995), 549–618.

[8] D. Bennequin, *Entrelacements et équations de Pfaff*, Asterisque (1983) 87-161. English version: Russian Math. Surveys **107–108** (1989), 87–161, English version: Russian Math. Surveys **44** (3) (1989), 1–65.

[9] A. Berrick, F. Cohen, Y. Wong, J. Wu, *Configurations, braids and homotopy theory*, preprint January 2003.

[10] D. Bessis, *The dual braid monoid*, Ann. Sci. Ecole Normal Sup. (4) **36** (5) (2003), 647–683.

[11] M. Bestvina and M. Handel, *Train tracks for surface automorphisms*, Topology **34** (1995), 109–140.

[12] S. Bigelow, *The Burau representation of the braid group B_n is not faithful for $n = 5$*, Geom. Topol. **3** (1999), 397–404.

[13] S. Bigelow, *Braid groups are linear*, J. Am. Math. Soc. **14** (2) (2001), 471–486.

[14] S. Bigelow and R. Budney, *The mapping class group of a genus 2 surface is linear*, Algebraic Geom. Topol. **1** (2001), 699–708.

[15] S. Bigelow and R. Budney, *Braids, links and mapping class groups*, Ann. Math. Studies **82** (1974).

[16] S. Bigelow and R. Budney, *New points of view in knot theory*, Bull. AMS **28** (1993), 253–287.

[17] J. Birman and E. Finkelstein, *Studying surfaces via closed braids*, J. Knot Theory Its Ramificat. **7** (3) (1998), 267–334.

[18] J. Birman and M. Hirsch, *A new algorithm for recognizing the unknot*, Geom. Topol. **2** (1998), 175–220.

[19] J. Birman, K.Y. Ko and S.J. Lee, *A new approach to the word and conjugacy problems in the braid groups*, Adv. Math. **139** (2) (1998), 322–353.

[20] J. Birman, K.Y. Ko and S.J. Lee, *The infimum, supremum and geodesic length of a braid conjugacy class*, Adv. Math. **164** (1) (2001), 41–56.

[21] J. Birman, D. Long and J. Moody, *Finite-dimensional representations of Artin's braid group*, in *"The Mathematical Legacy of Wilhelm Magnus"*, Contemporary Math, **169** (1994), 123–132.

[22] J. Birman, A. Lubotzky and J. McCarthy, *Abelian and solvable subgroups of the mapping class groups*, Duke Math. J. **50** (1983), 1107–1120.

[23] J. Birman, A. Lubotzky and J. McCarthy, *Studying links via closed braids IV: Split links and composite links*, Inventiones mathematicae **102** (1990), 115–139.

[24] J. Birman, A. Lubotzky and J. McCarthy, *Studying links via closed braids V: The unlink*, Trans. AMS **329** (2) (1992), 585–606.

[25] J. Birman, A. Lubotzky and J. McCarthy, *Stabilization in the braid groups I: MTWS*, preprint, submitted for publication, revised July 2003, arXiv math. GT/0310279.

[26] J. Birman, A. Lubotzky and J. McCarthy, *Stabilization in the braid groups II: On transversal simplicity of transversal knots*, preprint, submitted for publication, revised October 2003, arXiv math. GT/0310280.

[27] J. Birman, M. Rampichini, P. Boldi and S. Vigna, *Towards an implementation of the B-H algorithm for recognizing the unknot*, J. Knot Theory Its Ramificat. **11** (4) (2002), 601–645.

[28] J. Birman and H. Wenzl, *Braids, link polynomials, and a new algebra*, Trans. AMS **313** (1) (1989), 249–273.

[29] J. Birman and N. Wrinkle, *On transversally simple knots*, J. Diff. Geom. **55** (2000), 325–354.

[30] N. Bourbaki, *Groups Et Algebres De Lie*, Hermann, Paris (1968), Chapitres 4–6.

[31] T. Brady, *A partial order on the Symmetric Group and new $K(\pi, 1)'s$ for the braid groups*, Adv. Math. **161** (2001), 20–40.

[32] T. Brady and C. Watt, *$K(\pi, 1)'s$ for Artin groups of finite type, Geom.* Dedicata **94** (2002), 225–250.

[33] G. Bredon, *Topology and Geometry*, Graduate Texts in Mathematics, Vol. 139, Springer, Berlin (1991).

[34] T. Brendle and H. Hamidi-Tehrani, *On the linearity problem for mapping class groups*, Algebr. Geom. Topol. **1** (2001), 445–468.

[35] E. Brieskorn and K. Saito, *Artin-Gruppen und Coxeter-Gruppen*, Invent. Math. **17** (1972), 245–272.

[36] R. Budney, *On the image of the Lawrence-Krammer representation*, arXiv:math. GT/0202246.

[37] W. Burau, *Über Zopfgruppen und gleischsinning verdrillte Verkettungen*, Abh. Math. Sem. Ham. II (1936), 171–178.

[38] G. Burde and H. Zieschang, *Knots*, de Gruyter, Berlin (1985).

[39] R. Charney, *Geodesic automation and growth functions for Artin groups of finite type*, Math. Ann. **301** (1995), 307–324.

[40] Y. Chekanov, *Differential algebra of Legendrian knots*, Invent. Math. **105** (3) (2002), 441–483.

[41] A. Clifford and G. Preston, *The Algebraic Theory of Semigroups*, AMS Surveys 7, Vol. 1, (1961).

[42] A. Cohen and D. Wales, *Linearity of Artin groups of finite type*, Israel J. Math. **131** (2002), 101–123.

[43] J. Crisp and L. Paris, *The solution to a conjecture of Tits on the subgroup generated by the squares of the generators of an Artin group*, Invent. Math. **145** (2001), 19–36.

[44] M. Davis and T. Januszkiewicz, *Right-angled Artin groups are commensurable with right-angled Coxeter groups*, J. Pure Appl. Algebra **153** (3) (2000), 229–235.

[45] M. Dehn, *Über unendliche discontinuerliche Gruppen*, Math. Ann. **69** (1911), 116–144.

[46] M. Dehn, I. Dynnikov, D. Rolfsen and B. Wiest, *Why are braids orderable?*, Panoramas et Synthéses **14** (2002), Sociètè Mathèmatique de France.

[47] M. Dehn and L. Paris, *Gaussian groups and Garside groups, two generalizations of Artin groups*, Proc. London Math. Soc. **79** (3) (1999), 569–604.

[48] P. Deligne, *Les immeubles des groupes de tresses generalises*, Invent. Math. **17** (1972), 273–302.

[49] W. Diffie and M. Hellman, *New directions in cryptography*, IEEE Trans. Inform. Theory **IT-22** (6) (1977), 74–84.

[50] F. Digne, *On the linearity of Artin braid groups*, J. Algebra **268** (1) (2003), 39–57.

[51] Y. Eliashberg, *Classification of overtwisted contact structures on 3-manifolds*, Invent. Math. **98** (1989), 623–637.
[52] Y. Eliashberg, *Legendrian and transversal knots in tight contact 3-manifolds*, Topological methods in Modern Mathematics, Proceedings of a Symposium at SUNY Stonybrook in 1991, Publish or Perish, Houston, TX (1993).
[53] E. ElRifai and H. Morton, *Algorithms for positive braids*, Quart. J. Math. Oxford Ser. (2) **45** (180) (1994), 479–497.
[54] D. Epstein, J. Cannon, F. Holt, S. Levy, M. Patterson and W. Thurston, *Word Processing in Groups*, Jones and Bartlett, Boston, MA (1992), Chapter 9.
[55] J. Etnyre, *Legendrian and Transversal knots*, arXiv math. SG/0306256.
[56] J. Etnyre and K. Honda, *Cabling and transverse simplicity*, arXiv:math. SG(GT)/0306330.
[57] E. Fadell and J. Van Buskirk, *The braid groups of E^2 and S^2*, Duke Math J. **29** (1962), 243–257.
[58] R. Fenn, M. Greene, D. Rolfsen, C. Rourke and B. Wiest, *Ordering the braid groups*, Pac. J. Math. **191** (1) (1999), 49–74.
[59] E. Formanek, *Braid group representations of low degree*, Proc. London Math Soc. (3) **73** (2) (1996), 279–322.
[60] R. Fox and L. Neuwirth, *The braid groups*, Math. Scand. **10** (1962), 119–126.
[61] N. Franco and J. Gonzalez-Meneses, *The conjugacy problem for braid groups and Garside groups*, J. Algebra **266** (1) (2003), 112–132.
[62] R. Fenn, E. Keyman and C. Rourke, *The singular braid monoid embeds in a group*, J. Knot Theory Ramificat. **7** (7) (1998), 881–892.
[63] D. Fuchs and S. Tabachinikov, *Invariants of Legendrian and transverse knots in the standard contact space*, Topology **36** (5) (1997), 1025–1053.
[64] W. Fulton and J. Harris, *Representation Theory*, Springer, Berlin (1991).
[65] J. Franks and B. Williams, *Braids and the Jones-Conway polynomial*, Trans. AMS **303** (1) (1987), 97–108.
[66] P. Freyd, D. Yetter, J. Hoste, W. Lickorish, K. Millett and A. Ocneanu, *A new polynomial invariant of knots and links*, Bull. Am. Math. Soc. **12** (1985), 183–312.
[67] F. Garside, *The braid group and other groups*, Quart. J. Math Oxford **20** (1969), 235–254.
[68] V. Gebhardt, *A new approach to the conjugacy problem in Garside groups*, preprint arXiv:math. GT/0306199 v2, 21 October 2003.
[69] R. Ghrist, Configuration spaces and braid groups on graphs in Robotics, in *Knots, Braids and Mapping Class Groups*, AMS/IP Studies in Advanced Mathematics, American Mathematical Society and International Press (2001), 29–40.
[70] E. Giroux, *Géométrie de Contact:de la Dimension Trois vers les Dimensions Supé rieures*, Proc. ICM (Beijing 2002) **II** (2002), 405–414.
[71] J. Gonzalez-Meneses, *The nth root of a braid is unique up to conjugacy*, Algebr. Geom. Topol. **3** (2003), 1103–1118.
[72] J. Gonzalez-Meneses, and B. Wiest, *On the structure of the centralizer of a braid*, preprint, arXiv math. GT/0305156.
[73] E. Gorin and V. Lin, *Algebraic equations with continuous coefficients and some problems in the algebraic theory of braids*, Mat. Sbornik **7** (4) (1969).
[74] P. de la Harpe, M. Kervaire and C. Weber, *On the Jones polynomial*, L'Enseignement Mathematique **32** (1986), 271–335.
[75] N. Ivanov, *Mapping Class Groups, Handbook of Geometric Topology*, R. Daverman and R. Sher, eds, Elsevier, Amsterdam (2002), 523–633.
[76] V. Jones, *Index for subfactors*, Invent. Math. **72** (1983), 1–25.
[77] V. Jones, *Braid groups, Hecke algebras and subfactors*, Geom. Methods Operator Algebras, Pitman Res. Notes Math. **123** (1986), 242–273.
[78] V. Jones, *Hecke algebra representations of braid groups and link polynomials*, Ann. Math. **126** (1987), 335–388.
[79] D. Kazdan and G. Lusztig, *Representations of Coxeter groups and Hecke algebras*, Invent. Math. **53** (1979), 165–184.

[80] K.H. Ko, S.J. Lee, J.H. Cheon, J.W. Han, J. Kang and C. Park, *New Public Key Cryptosystems Using Braid Groups*, Lecture Notes in Computer Science, Vol. 1880, Springer, Berlin (2000).

[81] K.Y. Ko and S.J. Lee, *Flypes of closed 3-braids in the standard contact space*, J. Kor. Math. Soc. **36** (1999), 51–71.

[82] K.Y. Ko and J.W. Han, *Positive presentations of the braid groups and the embedding problem*, Math. Zeit. **240** (1) (2002), 211–232.

[83] M. Korkmaz, *On the linearity of certain mapping class groups*, Turkish J. Math. **24** (4) (2000), 367–371.

[84] D. Krammer, *The braid group B_4 is linear*, Invent. Math. **142** (3) (2000), 451–486.

[85] D. Krammer, *Braid groups are linear*, Ann. of Math. (2) **155** (1) (2002), 131–156.

[86] S.J. Lee, data presented at a talk at the First East Asian Conference on Knots and Related Topics, Seoul, Korea, February, 2004 16–19.

[87] R. Lawrence, *Homological representations of the Hecke algebra*, Comm. Math. Phys. **135** (1) (1990), 141–191.

[88] W.B.R. Lickorish, *An Introduction to Knot Theory*, Graduate Texts in Mathematics, Vol. 175, Springer, Berlin (1997).

[89] D. Long, *Constructing representations of braid groups*, Comm. Math. Anal. Geom. **2** (1994), 217–238.

[90] D. Long and M. Paton, *The Burau representation of the braid group B_n is not faithful for $n \geq 6$*, Topology **32** (1993), 439–447.

[91] A. Markov, *Über die freie Äquivqlenz geschlossener Zöpfe*, Recueil Math., Moscou (1935), 1.

[92] W. Menasco, *On iterated torus knots and transversal knots*, Geom. Top. **5** (2001), 651–682.

[93] J. Michel, *A note on words in braid monoids*, J. Algebra **215** (1) (1999), 366–377.

[94] J. Moody, *The Burau representation of the braid group B_n is not faithful for large n*, Bull. Am. Math. Soc. **25** (1991), 379–384.

[95] H. Morton, *Infinitely many fibered knots having the same Alexander polynomial*, Topology **17** (1978), 101–104.

[96] H. Morton, *An irreducible 4-string braid with unknotted closure*, Math. Proc. Camb. Phil. Soc. **93** (1983), 259–261.

[97] H. Morton, *Threading knot diagrams*, Math. Proc. Camb. Phil. Soc. **99** (1986), 247–260.

[98] H. Morton, *Seifert circles and knot polynomials*, Math. Proc. Camb. Phil. Soc. **99** (1986), 107–109.

[99] L. Mosher, *Pseudo-Anosov's on punctured surfaces*, PhD thesis, Princeton University (1983).

[100] J. Murakami, *The Kauffman polynomial of links and representation theory*, Osaka J. Math. **24** (4) (1987), 745–758.

[101] S. Orevkov and V. Shevchisin, *Markov Theorem for Transversal Links*, preprint February 2003, math.GT/0112207.

[102] L. Paris, *Artin monoids embed in their groups*, Cooment. Math. Helv. **77** (3) (2002), 609–637.

[103] L. Paris, *The proof of Birman's conjecture on singular braid monoids*, arXiv:math. GR/0306422.

[104] M. Patterson and A. Razborov, *The set of minimal braids is Co-NP-complete*, J. Algorithms **12** (1991), 393–408.

[105] M. Picantin, *The conjugacy problem in small Gaussian groups*, Comm. Algebra **29** (3) (2001), 1021–1039.

[106] D. Rolfsen, *Knots and Links*, American Mathematical Society (2003).

[107] C. Squier, *The Burau representation is unitary*, Proc. Am. Math. Soc. **90** (2) (1984), 199–202.

[108] R. Stanley, *On the number of reduced decompositions of elements of Coxeter groups*, Eur. J. Comb. **5** (1984), 359–372.

[109] N. Steenrod, *The topology of Fibre Bundles*, Princeton University Press, Princeton, NJ (1974).

[110] W. Thurston, *On the topology and geometry of diffeomorphisms of surfaces*, Bull. Am. Math. Soc. **19** (1988), 109–140.

[111] P. Traczyk, *A new proof of Markov's braid theorem*, Knot Theory, Vol. 42, Banach Center Publications, Warsaw (1998).

[112] V. Turaev, *Faithful linear representations of the braid groups*, arXiv math. GT/0006202.

[113] B. Wiest, *Dehornoy's ordering of the braid groups extends the subword ordering*, Pacific J. Math. **191** (1999), 183–188.

[114] H. Wenzl, *Representations of Hecke algebras and subfactors*, PhD thesis, University of Pennsylvania (1985).

[115] G. Wright, *A foliated disk whose boundary is Morton's irreducible 4-braid*, Math. Proc. Camb. Phil. Soc. **128** (2000), 95–101.

[116] P. Vogel, *Representation of links by braids: A new algorithm*, Comment. Math. Helvetici **65** (1) (1990), 104–113.

[117] S. Yamada, *The minimal number of Seifert circles equals the braid index of a link*, Invent. Math. **89** (2) (1987), 347–356.

[118] M. Zinno, *On Krammer's Representation of the Braid Group*, Math. Ann. **321** (1) (2001), 192–211.

CHAPTER 3

Legendrian and Transversal Knots

John B. Etnyre

University of Pennsylvania, Department of Mathematics, 209 South 33rd Street, Philadelphia, PA 19104-6395, USA
E-mail: etnyre@math.upenn.edu

Contents

HANDBOOK OF KNOT THEORY
Edited by William Menasco and Morwen Thistlethwaite

1. Introduction

Contact structures on manifolds and Legendrian and transversal knots in them are very natural objects, born over a century ago, in the work of Huygens, Hamilton and Jacobi on geometric optics and work of Lie on partial differential equations. They touch on diverse areas of mathematics and physics, and have deep connections with topology and dynamics in low dimensions. The study of Legendrian knots is now a rich and beautiful theory with many applications. This survey is an introduction to, and overview of the current state of knot theory in contact geometry. For a discussion of the driving questions in the field see Ref. [37] and for a more historical discussion of contact geometry see Ref. [46].

This chapter will concentrate on Legendrian and transversal knots in dimension three where their theory is most fully developed and where they are most intimately tied to topology. Moreover, in this dimension one may use a predominately topological and combinatorial approach to their study. In Section 6 we will make our only excursion into the study of higher dimensional Legendrian knots.

Throughout this survey we assume the reader is familiar with basic topology at the level of [71]. We have tried to keep the contact geometry prerequisites to a minimum, but it would certainly be helpful to have had some prior exposure to the basics as can be found in [33,47]. Some of the proofs in Section 5 rely on convex surface theory which can be found in [33], but references to convex surfaces can be largely ignored without serious loss of continuity.

2. Definitions and examples

A *contact structure* on an oriented 3-manifold M is a completely non-integrable plane field ξ in the tangent bundle of M. That ξ is a plane field simply means that at each point $x \in M$, ξ_x is a 2 dimensional subspace of T_xM. Near each point one may always describe a plane field as the kernel of a (locally defined) 1-form, i.e., there is a 1-form α so that $\xi_x = \ker(\alpha_x)$. The plane filed ξ is completely non-integrable if any 1-form α defining ξ satisfies $\alpha \wedge d\alpha \neq 0$. We will require that $\alpha \wedge d\alpha$ defines the given orientation on M. This orientation compatibility has become a fairly standard part of the definition of contact structure, but in the past such a contact structure was called a positive contact structure. The condition $\alpha \wedge d\alpha \neq 0$ implies that ξ is not everywhere tangent to any surface. Intuitively one can see this in Figure 1. There one sees the plane fields twisting. This twisting prevents the planes from being everywhere tangent to a surface. To make this "not everywhere tangent" condition precise one should consult the Frobenius Theorem [17].

2.1. *The standard contact structure on* $\mathbb{R}^3$

The simplest example of a contact structures is on $\mathbb{R}^3$ (with Cartesian coordinates (x, y, z)) and is given by

$$\xi_{\text{std}} = \text{span}\left\{\frac{\partial}{\partial y}, \frac{\partial}{\partial x} + y\frac{\partial}{\partial z}\right\}.$$

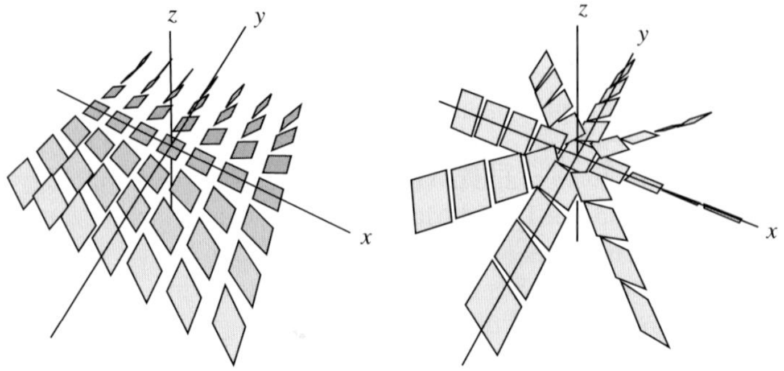

Fig. 1. The contact structure ξ_{std} (left) and ξ_{sym} (right) on $\mathbb{R}^3$. (Figures courtesy of S. Schönenberger.)

Clearly ξ_{std} is the kernel of the 1-from $\alpha = \mathrm{d}z - y\mathrm{d}x$ and thus one may easily verify that it is a contact structure. The plane field is indicated in Figure 1. We observe a few things about this plane field that will be useful later. Note that the planes along the xz-plane are all horizontal (i.e., parallel to the xy-plane). Moving out along the y axis the planes start horizontal and "twist" around the y axis in a left handed manner. Traversing the y axis from the origin "to infinity" the planes will make a 90° twist. Somehow it is this twisting that makes the plane field contact.

There are many other contact structures on $\mathbb{R}^3$ and all 3-manifolds have many contact structures, but Darboux's theorem [1,59] says that all of them locally look like this one. By this we mean that any point in a manifold has a neighborhood that is diffeomorphic to a neighborhood of the origin in $\mathbb{R}^3$ by a diffeomorphism that takes the contact structure on the manifold to the one described above. Throughout much of this article we will restrict attention to the contact structure above. This is largely to simply the discussion. In a few places it will be essential that we are in $(\mathbb{R}^3, \xi_{\mathrm{std}})$ and we will make that clear then.

2.2. *Other contact structures*

We begin with a symmetric version of ξ_{std} on $\mathbb{R}^3$.

EXAMPLE 2.1. Let $\alpha = \mathrm{d}z + x\mathrm{d}y - y\mathrm{d}x$ or in cylindrical coordinates $\alpha = \mathrm{d}z + r^2\mathrm{d}\theta$. The contact structure $\xi_{\mathrm{sym}} = \ker \alpha$ is as shown on the right hand side of Figure 1. One can find a diffeomorphism of $\mathbb{R}^3$ taking ξ_{std} to ξ_{sym}. Such a diffeomorphism is called a *contactomorphism*. Thus in some sense they are the same contact structure, but for various purposes one is sometimes easier to work with than the other.

EXAMPLE 2.2. In $\mathbb{R}^3$ with cylindrical coordinates consider $\alpha = \cos r\mathrm{d}z + r \sin r\mathrm{d}\theta$. The contact structure $\xi_{\mathrm{ot}} = \ker \alpha$ is as shown in Figure 2. Note the planes are similar to

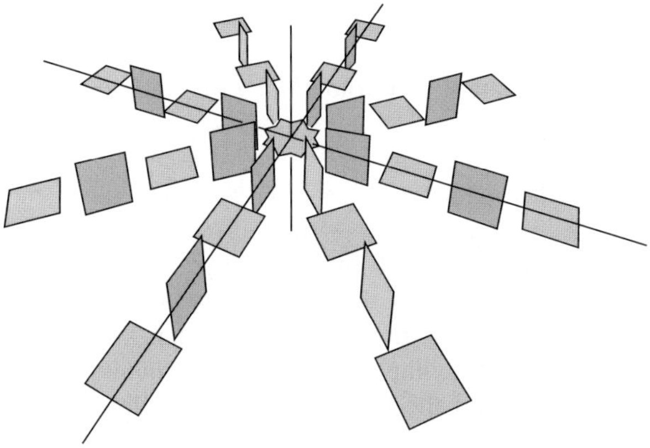

Fig. 2. The contact structure ξ_{ot} on $\mathbb{R}^3$. (Figures courtesy of S. Schönenberger.)

the ones in ξ_{sym} but as you move out along rays perpendicular to the z-axis the planes twist around many times (infinitely often) whereas for ξ_{sym} they twist only 90°.

Notice in this last example the disk $D = \{(r, \theta, z) | z = 0, r \leq \pi\}$ is tangent to the contact planes along the boundary. Such a disk is called an *overtwisted disk*. A contact structure is called *overtwisted* if there is an embedded overtwisted disk, otherwise it is called *tight*. From the discussion above we see that the plane field must twist to be a contact structure, but ξ_{ot} twists much more that is necessary, hence the name "overtwisted". Clearly ξ_{ot} is overtwisted and below we will see that ξ_{std} and ξ_{sym} are tight. Moreover, we will also see that tight contact structures tell us interesting things about knots related to them. For more discussion of tight and overtwisted contact structures see [23,33]. Throughout this survey we will be mainly concerned with tight contact structures with a real focus on $(\mathbb{R}^3, \xi_{std})$.

We end this section with a contact structure on a closed manifold.

EXAMPLE 2.3. Let S^3 be the unit three sphere in $\mathbb{R}^4$. Let

$$\alpha = i^* \frac{1}{2}(x_1 \mathrm{d}y_2 - y_1 \mathrm{d}x_1 + x_2 \mathrm{d}y_2 - y_2 \mathrm{d}x_2)$$

where $i : S^3 \to R^4$ is the inclusion map. One may readily check that α is a contact form so $\xi = \ker \alpha$ is a contact structure on S^3. If one removes a point form S^3 the contact structure ξ is contactomorphic to ξ_{std} on $\mathbb{R}^3$. For a more thorough discussion of this and similar examples see [33].

For other examples of contact structures on closed manifolds see [1,33,51,59].

2.3. *Legendrian knots*

A *Legendrian knot* L in a contact manifold (M^3, ξ) is an embedded S^1 that is always tangent to ξ:

$$T_xL \in \xi_x, \quad x \in L.$$

For the rest of this section we restrict attention to $(\mathbb{R}^3, \xi_{std})$. This is for two reasons. First, according to Darboux's theorem all contact structures look locally like $(\mathbb{R}^3, \xi_{std})$ and thus we are studying the "local Legendrian knot theory" in any contact manifold. Secondly, when studying this contact structure we can use various projections to help us understand the Legendrian knots, allowing us to "get a feel for Legendrian knots".

Below it will be convenient to have a parameterization of L. So throughout this section

$$\phi : S^1 \to \mathbb{R}^3 : \theta \mapsto (x(\theta), y(\theta), z(\theta))$$

will be a parameterization of L. Moreover, we will assume that ϕ is at least a C^1 immersion. Now the fact that L is tangent to ξ can be easily expressed by

$$\phi'(\theta) \in \xi_{\phi(\theta)}$$

or since $\xi = \ker(dz - y dx)$

$$z'(\theta) - y(\theta)x'(\theta) = 0. \tag{2.1}$$

There are two ways to picture Legendrian knots in $(\mathbb{R}^3, \xi_{std})$, that is via the *front projection* and the *Lagrangian projection*. We begin with the front projection. Let

$$\Pi : \mathbb{R}^3 \to \mathbb{R}^2 : (x, y, z) \to (x, z).$$

The image, $\Pi(L)$, of L under the map Π is called the front projection of L. If ϕ above parameterizes L then

$$\phi_\Pi : S^1 \to \mathbb{R}^2 : \theta \mapsto (x(\theta), z(\theta))$$

parameterizes $\Pi(L)$. While ϕ was an immersion (in fact an embedding), ϕ_Π will certainly not be one. To see this note that (2.1) implies that $z'(\theta) = y(\theta)x'(\theta)$ thus anytime $x'(\theta)$ vanishes so must $z'(\theta)$. So if ψ_Π is to be an immersion $x'(\theta)$ must never vanish. But this implies that $\Pi(L)$ has no vertical tangencies and of course any immersion of S^1 into $\mathbb{R}^2$ must have vertical tangencies. (Here and below "vertical" means spanned by $\partial/\partial z$.) This brings us to our first important fact about front projections.

FF 1. Front projections $\Pi(L)$ have no vertical tangencies.

So how does ψ_Π fail to be an immersion? Note the discussion above implies that $z'(\theta)$ always vanishes to at least the order of $x'(\theta)$ Thus we can always recover the y coordinate of ϕ from ϕ_Π by rewriting (2.1) as

$$y(\theta) = \frac{z'(\theta)}{x'(\theta)}, \tag{2.2}$$

if $x'(\theta)$ is non-zero. If x' is zero at some θ_0 then we have

$$y(\theta_0) = \lim_{\theta \to \theta_0} \frac{z'(\theta)}{x'(\theta)}. \tag{2.3}$$

There are Legendrian knots for which $x'(\theta) = 0$ on open intervals, see top of Figure 3, but this is not a stable phenomena. We can easily make the cusp point in the projection "sharper", see bottom of Figure 3. This leads to a Legendrian isotopic arc where $x'(\theta) = 0$ only at one point. From this it is easy to convince oneself that for a generic C^1 smooth Legendrian embedding in $\mathbb{R}^3$, $x'(\theta)$ can only vanish at isolated points. Moreover, at these isolated points there is a well-defined tangent line in the front projection. Thus we may assume that

FF 2. Front projections may be parameterized by a map that is an immersion except at a finite number of points, at which there is still a well-defined tangent line. Such points are called *generalized cusps*.

Actually this condition only guarantees the y coordinate defined by (2.2) is a C^0 function. We need to add conditions on the second derivatives of x and z to get the y coordinate to be C^1. We will not concern ourselves here with this. It is interesting to note that if we demand that all the coordinates be C^∞ then generically our cusps must be

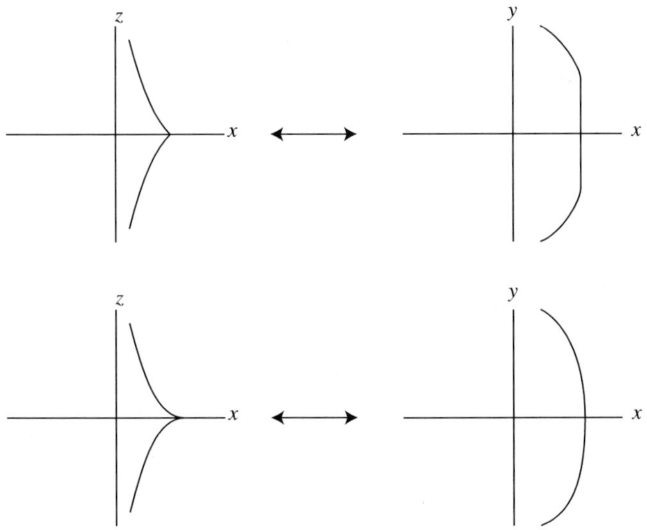

Fig. 3. Top row shows two projections of a Legendrian arc with $x'(\theta) = 0$ on a open interval. Bottom row shows a near by Legendrian arc with $x'(\theta) = 0$ only at one point.

"semi-cubic parabolas". By this we mean that after a change of coordinates $z(\theta) = 3\theta^3$ and $x(\theta) = 2\theta^2$.

Our above discussion implies that *FF 1* and *FF 2* characterize front projections. In particular, any map $f : S^1 \to xz\text{-plane} : \theta \mapsto (x_f(\theta), z_f(\theta))$ that satisfies *FF 1* and *FF 2* represents a Legendrian knot, since under these conditions we can always define $y(\theta)$ by (2.2). Then the image of the map $\phi(\theta) = (x_f(\theta), y(\theta), z_f(\theta))$ will be a Legendrian knot.

Interpreting this in terms of knot diagrams one sees that given any knot diagram

(1) that has no vertical tangencies;
(2) the only non-smooth points are generalized cusps; and
(3) at each crossing the slope of the overcrossing is smaller (that is, more negative) than the undercrossing;

represents the front projection of a Legendrian knot. To understand the condition on the overcrossing recall that in order to have the standard orientation on $\mathbb{R}^3$ the positive y axis goes into the page.

EXAMPLE 2.4. In Figure 4 we show front diagrams for Legendrian knots realizing the unknot, right and left trefoil knots and the figure eight knot.

Using these observations one can easily prove.

THEOREM 2.5. *Given any topological knot K there is a Legendrian knot C^0 close to it. In particular, there are Legendrian knots representing any topological knot type.*

PROOF. Consider $(\mathbb{R}^3, \xi_{std})$. First we show that any knot type can be represented by a Legendrian knot. This is now quite simple, just take any diagram for the knot (Figure 5), make the modifications shown in Figure 6 and then use the above procedure to recover a Legendrian knot (Figure 5).

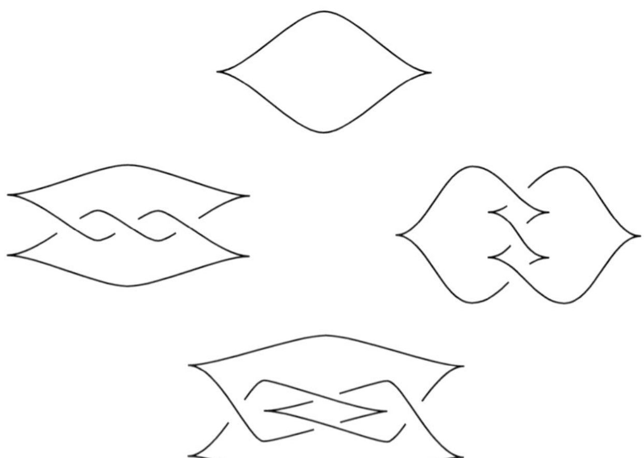

Fig. 4. Legendrian knots realizing the unknot, right and left trefoil knots and the figure eight knot.

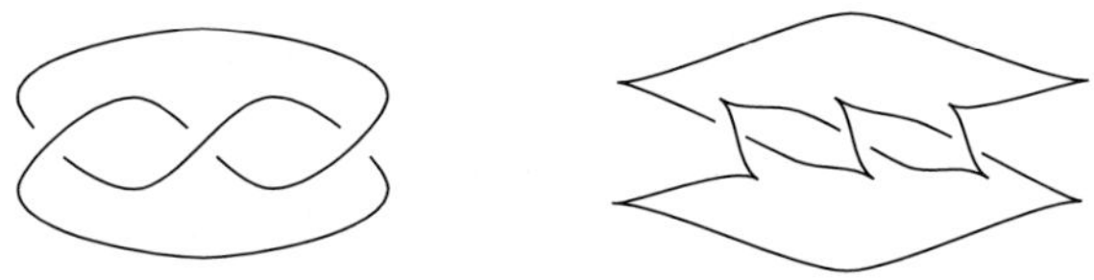

Fig. 5. Converting a knot diagram (left) into a Legendrian front (right).

Of course the problem with the construction of a Legendrian knot this way is that it is certainly not C^0 close to the original knot since the difference between y coordinates of the original knot and the Legendrian knot determined by the front projection can be quite far apart. This problem with the y coordinate can be fixed with the idea illustrated in Figure 7 that any arc may be C^0 approximated rel end points by a Legendrian arc.

Using Darboux's theorem (that all contact structures are locally the same as $(\mathbb{R}^3, \xi_{std})$) one may easily finish the proof for a general contact 3-manifold. □

Of course we could have used this last technique to show that any knot type has a Legendrian representative thus avoiding the first part of the proof. But, in practice, if one is trying to construct Legendrian representatives of a knot type (and not C^0 approximations of a specific knot) one uses Figure 6. This is because using Figure 7 introduces too many "zig-zags". We will see below it is best to avoid these as much as possible.

Just as there are Reidemeister moves for topological knot diagrams there is a set of "Reidemeister" moves for front diagrams too.

THEOREM 2.6. (See [76]). *Two front diagrams represent Legendrian isotopic Legendrian knots if and only if they are related by regular homotopy and a sequence of moves shown in Figure 8.*

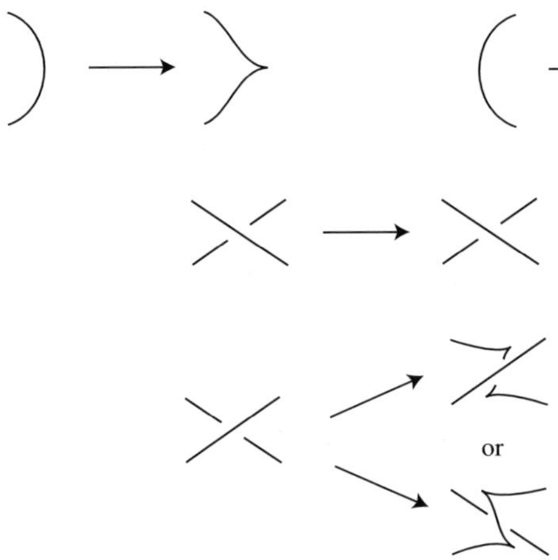

Fig. 6. Realizing a knot type as a Legendrian knot.

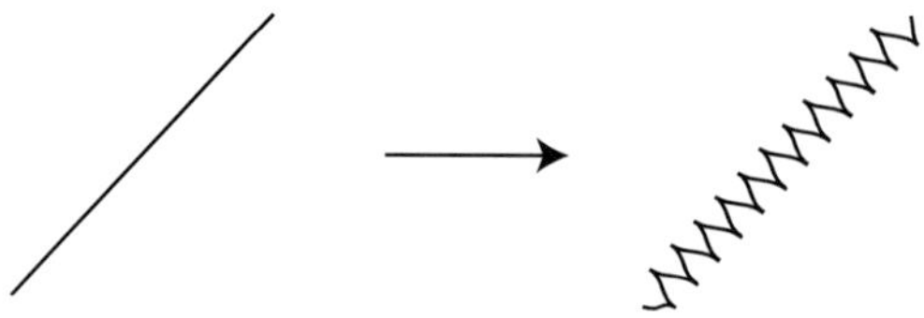

Fig. 7. C^0 approximating an arc by a Legendrian arc.

EXAMPLE 2.7. If Figure 9 we show that two different front diagrams for a Legendrian unknot are Legendrian isotopic.

To illustrate the difficulty in using these move one might want to try to show the two Legendrian figure eight knots in Figure 10 are Legendrian isotopic.

We now discuss the *Lagrangian projection* of a Legendrian knot. Let

$$\pi : \mathbb{R}^3 \to \mathbb{R}^2 : (x, y, z) \mapsto (x, y).$$

Then the Lagrangian projection of a Legendrian knot L is $\pi(L)$. The terminology "Lagrangian projection" comes from the fact that $d\alpha|_{xy\text{-}plane}$, which is a symplectic form, vanishes when restricted to $\pi(L)$. This is very important when considering Legendrian knots in $\mathbb{R}^{2n+1}$ but is irrelevant in our discussion. However, we keep the terminology for the sake of consistency. If we again parameterize L by ϕ (all notation is as above) then

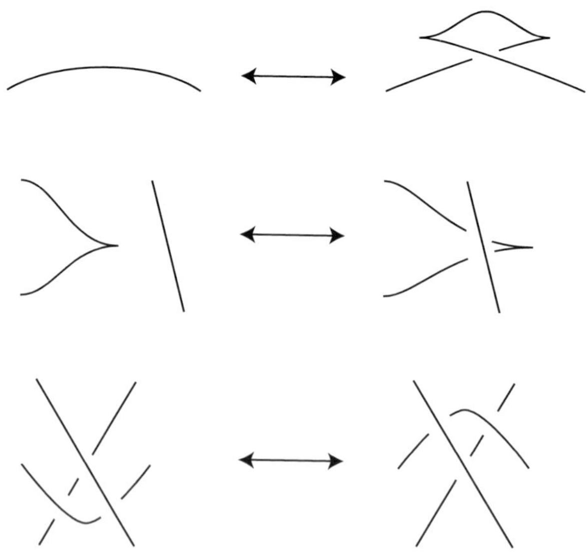

Fig. 8. Legendrian Reidemeister moves. (Also need the corresponding figures rotated 180° about all three coordinate axes.)

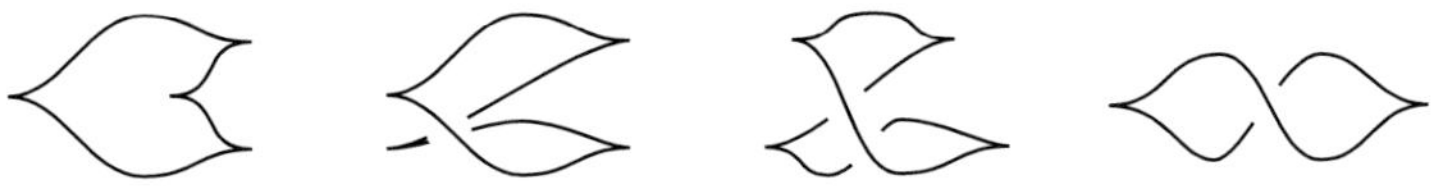

Fig. 9. Various fronts of the same Legendrian unknot.

Fig. 10. Two fronts of the same Legendrian figure eight knot.

$\pi(L)$ is parameterized by $\phi_\pi(\theta) = (x(\theta), y(\theta))$. Unlike the front projection, the Lagrangian projection is always parameterized by an immersion, since if $x'(\theta) = y'(\theta) = 0$ for some θ then $z'(\theta) \neq 0$ so the tangent vector to L is pointing in the $\partial/\partial z$ direction which does not lie in ξ.

The Legendrian knot L can be recovered (up to translation in the z-direction) from the Lagrangian projection as follows: pick some number z_0 and define $z(0) = z_0$. Then define

$$z(\theta) = z_0 + \int_0^\theta y(\theta)x'(\theta)\mathrm{d}\theta. \tag{2.4}$$

Since a Legendrian knot satisfies (2.1) we see this equation can be written

$$z(\theta) = z_0 + \int_0^\theta z'(\theta)\mathrm{d}\theta,$$

which is a tautology. So the only ambiguity in recovering L is the choice of z_0.

Let us observe a few restrictions on immersions $S^1 \to \mathbb{R}^2$ that can be Lagrangian projections of a Legendrian knot. First let $g : S^1 \to \mathbb{R}^2 : \theta \mapsto (x(\theta), y(\theta))$ be any immersion. If we try to define $z(\theta)$ by (2.4) then we run into problems. Specifically, if we think of $\theta \in [0, 2\pi]$ then $z(\theta)$ will be a well defined function on S^1 only if $z(0) = z(2\pi)$. This condition can be written

$$\int_0^{2\pi} y(\theta)x'(\theta)\mathrm{d}\theta = 0$$

and of course is not satisfied for all immersions g. This is the only obstruction to lifting g to an immersion G: $S^1 \to \mathbb{R}^3$ whose image is tangent to ξ. The image of G will be a Legendrian knot if G is an embedding. The only way it can fail to be an embedding is if double points in the image of g lift to have the same z coordinate. Thus an immersion g lifts

Fig. 11. Using the Lagrangian projection to approximate topological arcs.

to a Legendrian knot (well defined up to isotopy) if

(1) $\int_0^{2\pi} y(\theta)x'(\theta)\mathrm{d}\theta = 0$ and

(2) $\int_{\theta_0}^{\theta_1} y(\theta)x'(\theta)\mathrm{d}\theta \neq 0$ for all $\theta_0 \neq \theta_1$ with $g(\theta_0) = g(\theta_1)$.

Unfortunately these conditions are not easy to interpret diagrammatically, making working with Lagrangian projections somewhat harder that working with front projections. None the less, Lagrangian projections are still quite useful as we will see below. We also note that Theorem 2.5 can be proven using Lagrangian projections. Specifically, given any topological arc $\gamma: [0,1] \to \mathbb{R}^3$ we want to use the Lagrangian projection to prove $\gamma([0,1])$ can be C^0 approximated by a Legendrian arc with the same end points. By considering Figure 11 one may clearly do this.

We also have a weak Reidemeister type theorem.

THEOREM 2.8. *Two Lagrangian diagrams represent Legendrian isotopic Legendrian knots only if their diagrams are related by a sequence of moves shown in Figure 12.*

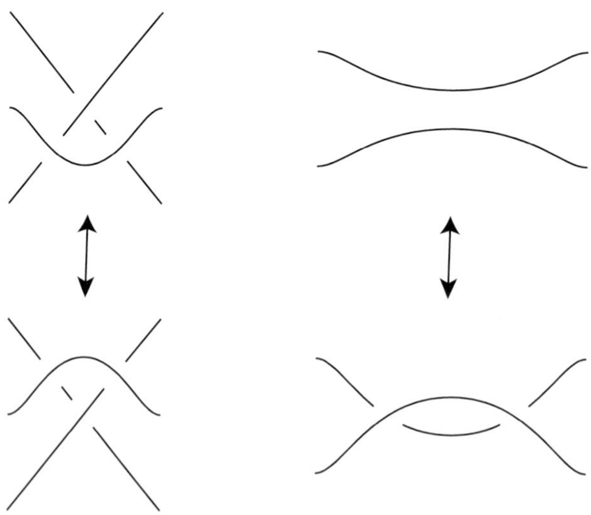

Fig. 12. Legendrian Reidemeister moves in the Lagrangian projection. (Also need the corresponding figures rotated 180° about all three coordinate axes.)

Unfortunately these moves are not sufficient to guarantee Legendrian isotopy due to the integral constraints discussed above.

In Section 4.5 we will discuss converting a front projection into a Lagrangian projection.

2.4. *Transverse knots*

A *transverse knot* T in a contact manifold (M^3, ξ) is an embedded S^1 that is always transverse to ξ:

$$T_xT \oplus \xi_x = T_xM, \quad x \in T.$$

If ξ is orientable (i.e., if it is defined globally by a 1-form) then we can fix an orientation on ξ. Recall our definition of contact structure requires a fixed orientation on M, thus if T is transverse to ξ we can orient T so that it always intersects ξ positively. When T is so oriented we call T a *positive transverse knot*. If we give T the opposite orientation then we call T a *negative transverse knot*. If we do not put positive/negative in front of the phrase "transverse knot" then positive is always implied. In particular transverse knots are always oriented (unless ξ is not orientable).

For the rest of this section we restrict attention to $(\mathbb{R}^3, \xi_{\text{std}})$. Transverse knots are usually studied via the analogy of the front projection: $\Pi : \mathbb{R}^3 \to \mathbb{R}^2 : (x, y, z) \mapsto (x, z)$. Unlike for Legendrian knots we cannot recover T from $\Pi(T)$, however we can reconstruct T up to isotopy through transverse knots (i.e., we can recover the transverse isotopy class of T). Thus it is reasonable to study T through $\Pi(T)$.

Let us examine $\Pi(T)$. Since the condition on a knot being transverse is an open condition in the space of embeddings $S^1 \to \mathbb{R}^3$ it is easy to check that for a generic transverse knot, $\Pi(T)$ will be the image of an immersion. This immersion satisfies two obvious constraints:

(1) the immersion has no vertical tangencies pointing down and

(2) there are no double points as shown in Figure 13.

To check these conditions let $\phi : S^1 \to \mathbb{R}^3 : \theta \mapsto (x(\theta), y(\theta), z(\theta))$ be a parameterization for T. The fact that T is a positive transverse knot implies that

$$z'(\theta) - y(\theta)x'(\theta) > 0.$$

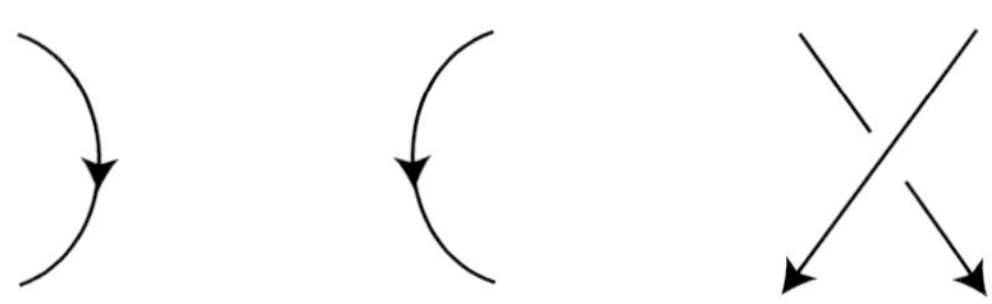

Fig. 13. Segments excluded from projections of transverse knots.

Now the projection of T is parameterized by $\phi_\Pi(\theta) = (x(\theta), z(\theta))$. At a vertical tangency pointing down we have $x' = 0$ and $z' < 0$, this contradicts the above equation thus establishing condition (1). To see condition (2) note that

$$y(\theta) < \frac{z'(\theta)}{x'(\theta)},$$

If $x'(\theta) > 0$ and the opposite inequality if $x'(\theta) < 0$.

Thus the y coordinate is bounded by the slope of the diagram in the xz-plane. (Once again recall the positive y axis points into the page.) One may now easily check that the picture on the right side of Figure 13 cannot be the projection of a transverse knot.

THEOREM 2.9. (See [36,76]). *Any diagram satisfying conditions* (1) *and* (2) *above can be lifted to a transversal knot in* $\mathbb{R}^3$ *well defined up to isotopy through transversal knots. Moreover, two diagrams will represent the same transverse isotopy class of transverse knots if any only if they are related by a sequence of move shown in Figure 14.*

A second way to study transversal knots is by using closed braids. Recall a closed braid is simply a knot (or link) in $\mathbb{R}^3$ (we will use cylindrical coordinates (r, ψ, z) now) that can be parameterized by a map $f : S^1 \to \mathbb{R}^3 : \theta \mapsto (r(\theta), \psi(\theta), z(\theta))$ for which $r(\theta) \neq 0$ and $\psi'(\theta) > 0$ for all θ. For more on braids and closed braids see [6].

To see the connection between braids and transverse knots we use the symmetric version of the standard contact structure $(\mathbb{R}^3, \xi_{sym})$. Since this contact structure is contactomorphic to the standard one we can transfer any question about the contact structure ξ_{std} to the contact structure ξ_{sym}. Now given a closed braid B we can isotopy it

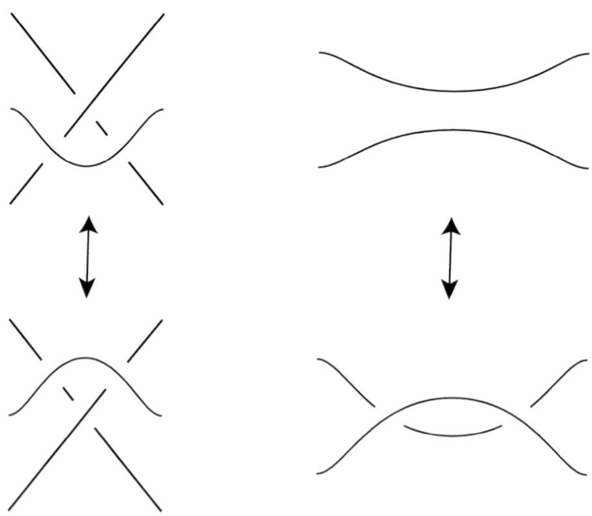

Fig. 14. Transverse Reidemeister moves. Add arrows in all ways that do not violate the conditions in Figure 13. (Also need the corresponding figures rotated 180° about all three coordinate axes.)

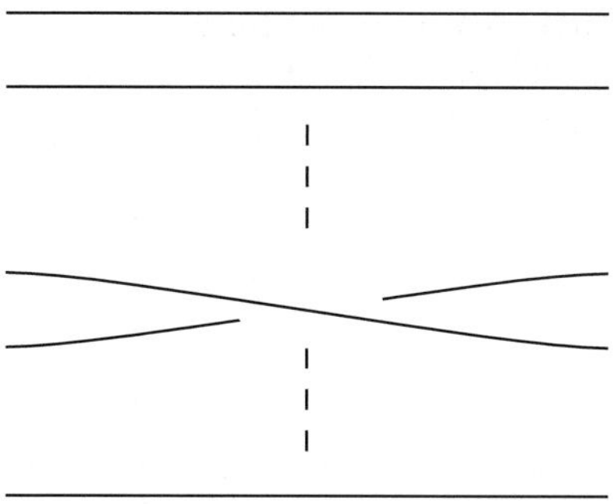

Fig. 15. A generator σ_i for the braid group B_n.

through closed braid so that it is far from the z-axis. Very far from the z-axis the planes that make up ξ_{sym} are almost vertical, that is close to the planes spanned by $\partial/\partial z$ and $\partial/\partial r$. Thus the closed braid type B represents a transversal knot.

Bennequin proved the opposite assertion:

THEOREM 2.10. (Bennequin 1983, [5]). *Any transverse knot in* $(\mathbb{R}^3, \xi_{sym})$ *is transversely isotopic to a closed braid.*

The proof of this theorem is quite similar to the proof that knots can be braided [6]. One just needs the check that the "braiding process" can be done in a transverse way. For details see [5,70].

Recall, fixing n points p_i, in a disk D^2, an n-braid is an embedding of n arcs $\gamma_i : [0,1] \to D_2 \times [0,1]$ so that $\gamma_i(t) \in D^2 \times \{t\}$ and the endpoints of the γ_i corresponding to 0 (resp. 1) as a set map to $\{p_i\}_{i=1}^n$ in $D^2 \times \{0\}$ (resp. $D^2 \times \{1\}$). The set of all n-braids B_n form a group. It is easy to see the group is generated by σ_i, $i = 1, \ldots n-1$ where σ_i is the n braid with the i and $i+1$ strands interchanging in a right handed fashion and the rest unchanged, see Figure 15.

The group B_n naturally includes in B_{n+1}. Given a braid b in B_n the *positive stabilization* of b is $b\sigma_n$ in B_{n+1}.

THEOREM 2.11. (Orevkov and Shevchishin 2003, [70]; Wrinkle, [80]). *Two braids represent the same transverse knot if and only if they are related by positive stabilization and conjugation in the braid group.*

2.5. *Types of classification*

In trying to classify Legendrian or transverse knots one could mean many different things. We concentrate on Legendrian knots here but an analogous discussion holds for ransverse knots too.

One can classify Legendrian knots up to *isotopy through Legendrian knots*. That is L_0 and L_1 are Legendrian isotopic if there is a continuous family L_t, $t \in [0,1]$, of Legendrian knots starting at L_0 and ending at L_1. One can also classify Legendrian knots up to *ambient contact isotopy*. Here we mean L_0 and L_1 are ambient contact isotopic if there is a one parameter family $\phi_t : M \to M, t \in [0,1]$, (here (M, ξ) is the ambient contact manifold) of contactomorphisms of M such that ϕ_0 is the identity map and $\phi_1(L_0) = L_1$.

THEOREM 2.12. *The classification of Legendrian knots up to Legendrian isotopy is equivalent to the classification up to contact isotopy. The analogous statement is true for transverse knots.*

PROOF. The implication of contact isotopy to Legendrian isotopy is obvious. For the other implication assume $L_t, t \in [0,1]$ is an isotopy through Legendrian knots in the contact manifold (M, ξ). There is a family of diffeomorphisms $\phi_t : M \to M, t \in [0,1]$ such that $\phi_t(L_0) = L_t$. Moreover, it is easy to arrange that $\phi_t^*(\xi|_{L_t}) = \xi|_{L_0}$. Let $\xi_t = \phi_t^*(\xi)$. This is a one parameter family of contact structures with $\xi_t = \xi_0$ along L_0. Thus Gray's Theorem, [1,59], implies there is a family of diffeomorphisms ψ_t such that $\psi_t^*(\xi_t) = \xi_0$ and ψ_t is the identity on L_0. Now set $f_t = \phi_t \circ \psi_t$. Note

$$f_t^*(\xi_0) = \psi_t^*(\xi_t) = \xi_0.$$

Thus f_t are all contactomorphisms of $\xi_0 = \xi$. Moreover $f_t(L_0) = \phi_t(L_0) = L_t$. □

In S^3 with the standard contact structure there is another type of classification that is equivalent to these.

THEOREM 2.13. *In (S^3, ξ_{std}) (or $(\mathbb{R}^3, \xi_{std})$) two Legendrian knots are Legendrian isotopic if and only if their compliments are contactomorphic.*

This theorem is not necessarily true in other contact manifolds.

PROOF. A Legendrian knot L has a canonical neighborhood $N(L)$. Denote by $M(L)$ the closure of the complement of $N(L)$. If two Legendrian knots L_0 and L_1 in (S^3, ξ_{std}) have contactomorphic complements, then let $\psi : M(L_0) \to M(L_1)$ be the contactomorphism. We can extend ψ over $N(L_0)$ so that it takes $N(L_0)$ to $N(L_1)$ and is a contactomorphism of (S^3, ξ_{std}). In [25], Eliashberg has shown that there is a unique tight contact structure on S^3. That coupled with Gray's Theorem implies there is a family of contactomorphisms $\psi_t : S^3 \to S^3, t \in [0,1]$ with ψ_0 the identity map and $\psi_1 = \psi$. Thus $L_t = \psi_t(L_0)$ is a Legendrian isotopy from L_0 to L_1. □

2.6. *Invariants of Legendrian and transversal knots*

Though it is not always essential we will always consider oriented knots.

2.6.1. *Classical invariants of Legendrian knots.* The most obvious invariant of a Legendrian knot is its underlying topological knot type, since any Legendrian isotopy between two Legendrian knots is, among other things, a topological isotopy of the underlying knots. Given a Legendrian knot L we will denote its underlying topological knot type by $k(L)$. We will use $\mathcal{K}$ to denote a topological knot type (i.e., the set of topological knots isotopic to a fixed knot) and the set of all Legendrian knots L with $k(L) \in \mathcal{K}$ will be denoted by $\mathcal{L}(\mathcal{K})$.

The next invariant of a Legendrian knot L is the *Thurston–Bennequin invariant* which intuitively measures the "twisting of ξ around L". More rigorously this invariant is defined by a trivialization of the normal bundle ν of L. A fixed identification of ν with $L \times \mathbb{R}^2$ is called a *trivialization* of ν or a *framing* of L. A Legendrian knot has a canonical framing: since ξ_x and ν_x intersect transversally ($T_xL \subset \xi_x$) one gets a line bundle $l_x = \xi_x \cap \nu_{x,}$ for $x \in L$. The line bundle l gives a framing of ν over L. This framing is the *Thurston–Bennequin framing* of L and is denoted *tbf* (L).

If the normal bundle has a preassigned framing $\mathcal{F}$ then we can assign a number to the Thurston–Bennequin framing of L. This number $tw(L, \mathcal{F})$ is just the twisting of l with respect to $\mathcal{F}$ and is called the twisting of L with respect to $\mathcal{F}$. If L is null homologous then L has a framing given by a Seifert surface. The twisting of L with respect to this Seifert framing will be called the *Thurston–Bennequin invariant* of L and is denoted $tb(L)$. One may alternately define $tb(L)$ as follows: let v be a non-zero vector field along L in $\nu \cap \xi$ and let L' be a copy of L obtained by pushing L slightly in the direction of v. Now define $tb(L)$ as the linking of L with L', i.e., $tb(L) = lk(L, L')$.

REMARK 2.14. If v' is a non-zero vector field along L transverse to ξ and L'' is obtained from L by pushing L slightly in the direction of v' then $tb(L) = lk(L, L'')$.

The last "classical" invariant of a Legendrian knot L is the *rotation number* of L. This invariant will only be defined for null homologous knots, so assume that $L = \partial\Sigma$ where Σ is an embedded orientable surface. The contact planes when restricted to Σ, $\xi|_\Sigma$, form a trivial 2 dimensional bundle (any orientable two plane bundle is trivial over a surface with boundary). This trivialization of $\xi|_\Sigma$ induces a trivialization $\xi|_L = L \times \mathbb{R}^2$. Recall L has an orientation, so let v be a non-zero vector field tangent to L pointing in the direction of the orientation on L. The vector field v is in $\xi|_L = L \times \mathbb{R}^2$ and thus using this trivialization we can think of v as a path of non-zero vectors in $\mathbb{R}^2$, as such it has a winding number. This winding number $r(L)$ is the rotation number of L. Note the rotation number depends on the orientation of L and changes sign if the orientation is reversed.

REMARK 2.15. One can also define $r(L)$ as the obstruction to extending v to a non-zero vector field in $\xi|_\Sigma$.

The invariants $k(L)$, $tb(L)$ and $r(L)$ will be called the *classical invariants* of L.

2.6.2. *Computation of the classical invariants via projection.* We now consider Legendrian knots L in the standard contact structure on $\mathbb{R}^3$ and interpret the classical invariants in the front and Lagrangian projection of L. We begin with the rotation number. Let $w = \partial/\partial y$. This is a non-zero section of ξ and thus can be used to trivialize $\xi|_L$ independent of finding a Seifert surface for L. Now to compute the rotation number of L we just need to see how many times a non-zero tangent vector field v to L, thought of as living in $\mathbb{R}^2$ via the trivialization given by w, winds around the origin of $\mathbb{R}^2$. This is equivalent to making a signed count of how many times v and w point in the same direction, we call this an intersection of v and w. The "sign of the intersection" is determined by whether v passes w counterclockwise $(+1)$ or clockwise (-1). Now in the front projection v will be pointing in the direction of $\pm w = \pm\partial/\partial y$ at the cusps. One may easily check that the intersection will be positive when going down a cusp and negative when going up a cusp. Moreover, since this counts the number of times v "intersects" $\pm w$ we need to divide by two in order to get $r(L)$. Thus in the front projection

$$r(L) = \frac{1}{2}(D - U), \tag{2.5}$$

where U is the number of up cusps in the front projection and D is the number of down cusps. In the Lagrangian projection w projects to $\partial/\partial y$ and thus the rotation number is easily seen to be the winding number of the tangent vectors to $\pi(L)$

$$r(L) = \text{winding}(\pi(L)). \tag{2.6}$$

For the Thurston–Bennequin invariant. Let $v = \partial/\partial z$ then for any Legendrian knot L, v is a vector field transverse to ξ along L (as in Remark 2.14). Thus $tb(L)$ is the linking of L with a copy L' of L obtained by shifting slightly in the z direction. So in the front projection L and L' are as in Figure 16. Now the linking number of L and L' is just one half the signed count of the intersections between them. Where an intersection is positive if it is right handed and negative if it is left handed (recall L and hence L' are oriented). See Figure 17. One can see that at each right (left) handed self crossing of the projection of L there will be two right (left) handed crossings of L and L' and at a right or left cusp of L there will be a left handed crossing of L and L'. Thus in the

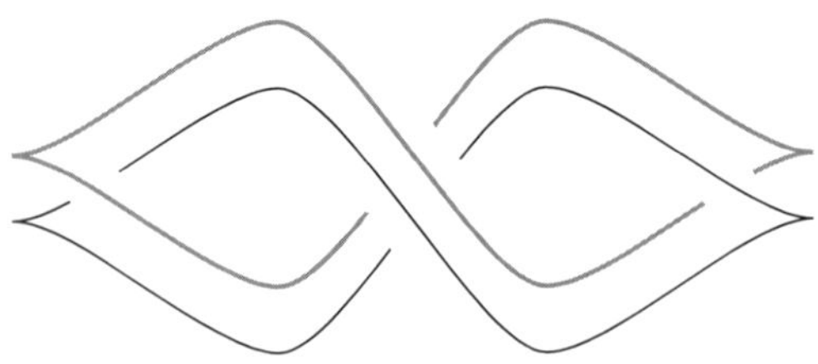

Fig. 16. Knots L (black) and L' (grey) used to compute $tb(L)$.

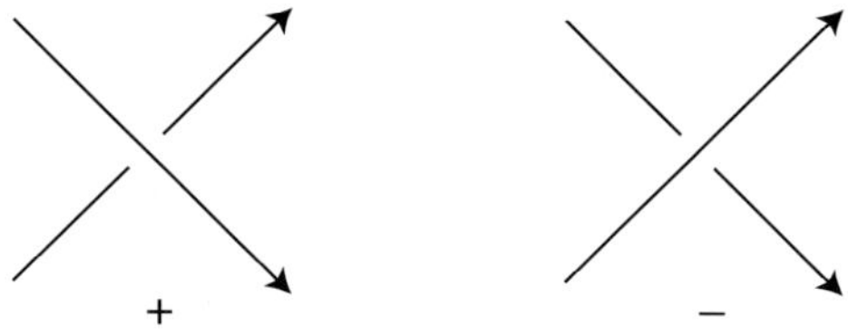

Fig. 17. Right handed crossing (on left) and a left handed crossing (on right). Note this figure is to indicate that right (left) crossings contribute $+$ $(-)$ to the write of a knot. The picture on the right cannot occur in a front projection, but non the less left handed crossings can.

front projection

$$tb(L) = \text{writhe}(\Pi(L)) - \frac{1}{2}(\text{number of cusps in } \Pi(L)). \tag{2.7}$$

In the Lagrangian projection L and L' project to the same diagram, but we have

$$tb(L) = \text{writhe}(\pi(L)). \tag{2.8}$$

One may prove this formula by first recalling $tb(L)$ is the linking number of L and L'. In the Lagrangian projection think about trying to pull L' straight up. If during this process L' never intersects L then their linking is 0. Moreover, each crossing in the diagram for L contributes ± 1 to the linking number.

2.6.3. *Classical invariants of transverse knots.* For transverse knots T there are only two classical invariants, the topological knot type $k(T)$ and the *self-linking number* (this is sometimes called the Bennequin number). We denote transverse knots in a fixed topological knot type $\mathcal{K}$ by $\mathcal{T}(\mathcal{K})$. To define the self-linking number of T we assume it is homologically trivial. (There is a "relative" version of the self-linking number if this is not true but we will not discuss this point here.) Thus there is an orientable surface Σ such that $\partial\Sigma = T$. As above we know that $\xi|_\Sigma$ is trivial so we can find a non-zero vector field v over Σ in ξ. Let T' be a copy of T obtained by pushing T slightly in the direction of v. The self-linking number $sl(T)$ of T is the linking of T' with T.

The self-linking number also has an interpretation in terms of a relative Euler class. To see this let v be a non-zero vector in $\xi \cap T\Sigma$ along T that points out of Σ If this v were to extend over Σ then we could use it to get T' above and the self-linking would be 0. If v does not extend then the self-linking is not 0.

REMARK 2.16. The self-linking number $sl(T)$ is precisely the obstruction to extending v over Σ to a non-zero vector field in ξ.

2.6.4. *Computations of the self-linking number.* To compute $sl(T)$ for a transverse knot in the standard contact structure on $\mathbb{R}^3$ note that the vector $v = \partial/\partial y$ is always in ξ_{std} and thus can be used to trivialize ξ_{std} independent of a Seifert surface for T. Now let T' be

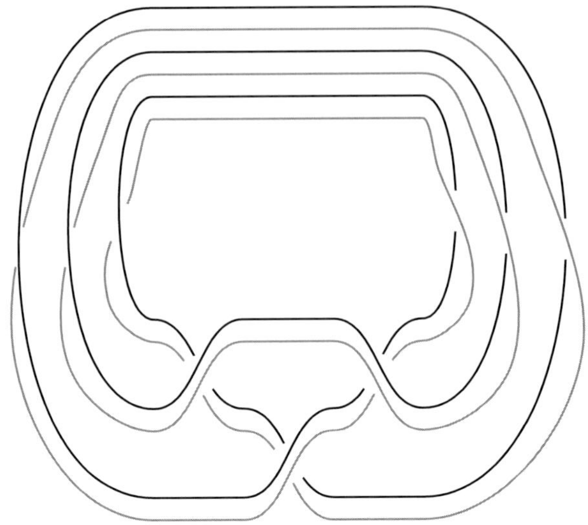

Fig. 18. The braid T in black and its push-off by v in grey.

a copy of T obtained by pushing T along v. The self-linking number of T is the linking number of T and T'. If we consider T and T' in the "front projection" we see that

$$sl(T) = \text{writhe}(\Pi(L)). \tag{2.9}$$

The argument for this formula is exactly like the one for (2.8).

We would also like to see a formula for $sl(T)$ in terms of a braid representation for T. To do this recall we must use the cylindrically symmetric contact structure ξ_{sym} on $\mathbb{R}_3$. For our non-zero vector field we take $v = \partial/\partial r$ on $\{x \geq \epsilon\}$, $v = -\partial/\partial r$ for $\{x \leq -\epsilon\}$ and on $\{-\epsilon < x < \epsilon\}$ we have interpolate between these two choices by rotating clockwise in the contact planes. If we think of all the braiding as occurring in $\{x \geq \epsilon\}$ the we see each generator $\sigma^{\pm}$ contributes ± 1 to the self-linking number (see Figure 18).

Moreover each strand as it passes in and out of $\{x \leq \epsilon\}$ contributes -1 to the self linking number. Thus we have

$$sl(T) = a(T) - n(T), \tag{2.10}$$

where $n(T)$ is the number of strands in the braid b representing T and $a(T)$ is the algebraic length of b when written in terms of the generators σ_i (algebraic length is the sum of the exponents on the generators). This useful equality was first observed by Bennequin [5].

2.7. *Stabilizations*

Given a Legendrian knot L there is a simple way to get another Legendrian knot in the same topological knot type: stabilization. We describe *stabilization* in the standard contact

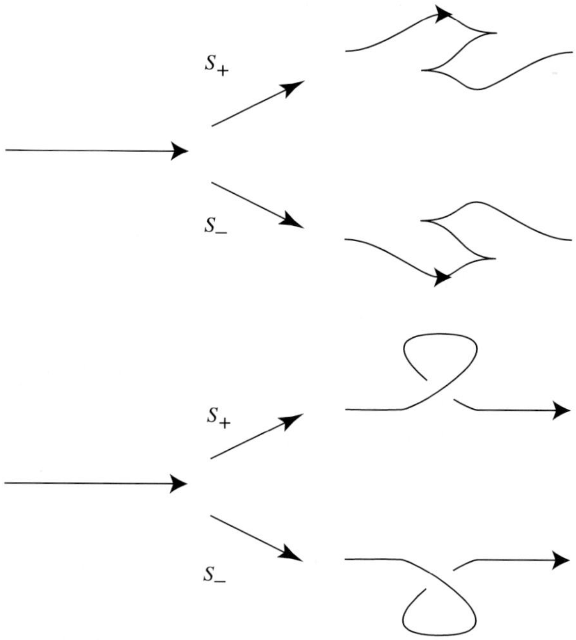

Fig. 19. Stabilizations in the front projection (top) and the Lagrangian projection (bottom).

structure on $\mathbb{R}^3$. If a strand of L in the front projection of L is as on the left hand side of Figure 19 then the stabilizations of L is obtained by removing the stand and replacing it with one of the zig-zags shown on the right hand side of Figure 19. If down cusps are added then the stabilization is called positive, and denoted $S_+(L)$, and if up cusps are added then the stabilization is called negative, and denoted $S_-(L)$. Since stabilizations are done locally, this actually defines, via Darboux's Theorem, stabilizations for Legendrian knots in any manifolds. Note

$$tb(S_\pm(L)) = tb(L) - 1, \tag{2.11}$$

and

$$r(S_\pm(L)) = r(L) \pm 1. \tag{2.12}$$

It is important to observe that stabilization is a well-defined operation, that is, it does not depend at what point the stabilization is done. This is true in general [36] but in $\mathbb{R}^3$ it is somewhat easier to prove [45]. One only needs to check that the "zig-zags" used to define the stabilization can be move past cusps and crossings in the front projection. Using Theorem 2.6 this is a fun exercise.

One important fact concerning stabilizations is the following.

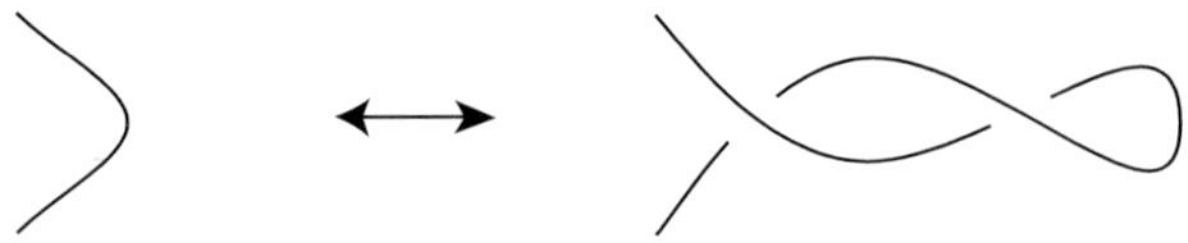

Fig. 20. Transverse stabilization.

THEOREM 2.17. (Fuchs and Tabachnikov 1997, [45]). *Given two Legendrian knots L_1 and L_2 in $(\mathbb{R}^3, \xi_{\text{std}})$ that are topologically isotopic then after each has been stabilized some number of times they will be Legendrian isotopic.*

So the topological classification of knots is equivalent to the stable classification of Legendrian knots. To prove this theorem one checks that the topological Reidemeister moves can be performed via Legendrian Reidemeister moves after sufficiently many stabilizations. It should not be difficult to prove this theorem in any contact 3-manifold, but such a proof does not exist in the literature.

Note using stabilization it is easy to get Legendrian knots with arbitrarily negative Thurston–Bennequin invariants. In any tight contact structure there is an upper bound on the Thurston–Bennequin invariant (see Section 3). Thus while it is easy to stabilize a knot it is not necessarily easy to "destabilize" a knot. We say L destabilizes if there is a Legendrian knot L' such that $L = S_{\pm}(L')$.

There is notion of transverse stabilization too. The stabilization of a transverse knot T is formed by taking an arc in the front projection as shown on the left hand side of Figure 20 and replacing it with the arc on the right hand side. Denote the resulting knot by $S(T)$. Note there is only one type of stabilization here and

$$sl(S(T)) = sl(T) - 2. \tag{2.13}$$

Once again it is not hard to show that stabilization is well defined for transverse knots. We also have

THEOREM 2.18. (Fuchs and Tabachnikov 1997, [45]). *Given two transverse knots T_1 and T_2 in $(\mathbb{R}^3, \xi_{\text{std}})$ that are topologically isotopic then after each has been stabilized some number of times they will be transversely isotopic.*

There is a notion of destabilization for transverse knots that is exactly analogous to destabilization of Legendrian knots.

2.8. *Surfaces and the classical invariants*

In this section we see how to compute the classical invariants of a Legendrian and transverse knot in terms of the characteristic foliation on a surface bounded by the knot.

We will also see how convex surface theory [48,51] can be used to understand the classical invariants of Legendrian knots.

Let T be a transverse knot bounding the surface Σ. Orient Σ so that T is its oriented boundary. With this orientation the characteristic foliation Σ_ξ points out of $\partial\Sigma = T$ (recall the characteristic foliation is oriented). Perturb Σ so that Σ_ξ is generic. In particular, we can assume that all the singularities are isolated elliptic or hyperbolic points. Moreover, each singularity has a sign depending on whether the orientation of ξ and $T\Sigma$ agree at the singularity. Let $e_\pm$ be the number of $\pm$ elliptic singularities in Σ_ξ and let $h_\pm$ be the number of $\pm$ hyperbolic singularities. We now have

$$-sl(T) = (e_+ - h_+) - (e_- - h_-). \tag{2.14}$$

This formula follows easily by interpreting $sl(T)$ as a relative Euler class (that is let v be a vector field along T tangent to Σ_T and contained in ξ and pointing into Σ_T, then $sl(T)$ is the obstruction to extending v to a non-zero vector field on Σ_T).

By looking at the characteristic foliation one may frequently see how to destabilize a transverse knot. In particular, if one sees Figure 21 in the characteristic foliation of a surface with a transverse knot T in its boundary, then one can let T' be the knot shown in the picture and it is clear that $sl(T') = sl(T) + 2$. With a little work one can, in fact, see that $S(T') = T$. Thus when one finds a negative hyperbolic point on a surface whose unstable manifolds separate off a disk with one elliptic point (necessarily positive), then the knot can be destabilized. Such a disk is called a *transverse bypass*, first appeared in [32] and was used in the classification of some transverse torus knots in [31].

Now consider a Legendrian knot L with Seifert surface Σ. It is relatively easy to see we may isotopy Σ relative to its boundary so that the singularities in the characteristic foliation along the boundary alternate in sign. When this is arranged it is easy to see $|tb(L)|$ is simply half the number of singularities along the boundary. We now observe:

LEMMA 2.19. (Eliashberg and Fraser 1998, [28]). *Assume the singularities in the characteristic foliation along $\partial\Sigma = L$ alternate in sign. If $tb(L) \leq 0$ then positive/negative singularities along the boundary will be sources/sinks (along the boundary). If $tb(L) > 0$*

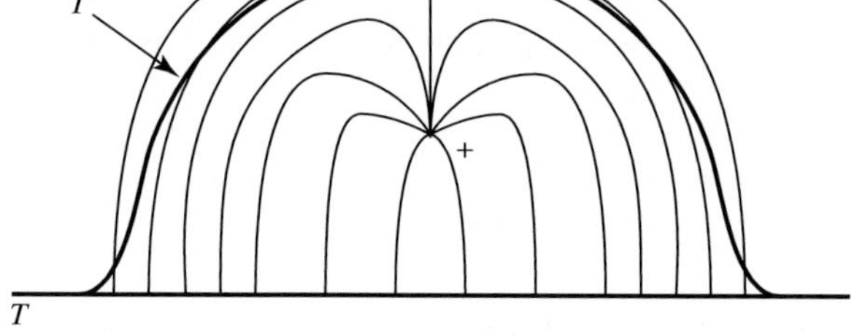

Fig. 21. Using the characteristic foliation to recognize a destabilization.

the positive/negative singularities along the boundary will be sinks/sources (along the boundary).

The idea for this lemma is simply to examine how the characteristic foliation inherits an orientation from the contact planes and the Seifert surface. A singularity along the boundary is positive or negative depending on whether the contact planes are twisting past Σ in a right handed or left handed fashion.

Recall that elliptic singularities in the characteristic foliation are sources (sinks) when the singularity is positive (negative). Thus when $tb(L) > 0$ we see from this lemma that all the singularities must be hyperbolic! So the characteristic foliation can be standardized along the boundary when $tb(L) > 0$.

When $tb(L) \leq 0$ it takes a little more work to normalize to characteristic foliation near the boundary. In addition, the foliation can be normalized in various ways. In particular, one can arrange that all the singularities are hyperbolic, or that all the singularities are elliptic, or that they alternate between hyperbolic and elliptic [28].

If $tb(L) \leq 0$ then we can isotope Σ relative to L so that it is convex. Denote the dividing curves of Σ by Γ and the $\pm$ regions of $\Sigma \backslash \Gamma$ by $\Sigma_\pm$ Then we have

$$tb(L) = -\frac{1}{2}(L \cap \Gamma), \tag{2.15}$$

and

$$r(L) = \chi(\Sigma_+) - \chi(\Sigma_-). \tag{2.16}$$

Call a properly embedded arc $a \subset \Gamma$ in Σ boundary parallel if the closure of one of the components of $\Sigma \backslash a$ is a disk that contains no components of Γ in its interior. Such a boundary parallel dividing curve is frequently called a *bypass*, because it allows one to "bypass" some twisting as the following lemma shows. Bypasses were first introduced in [51].

LEMMA 2.20. *If the dividing curves Γ of Σ contain a boundary parallel arc and $tb(L) < -1$ or Σ has genus greater than 0, then L can be destabilized.*

PROOF. The Legendrian Realization Principle [48,51] says that we can alter the characteristic foliation to any other singular foliation as long as Γ "divides" this foliation. Thus we may realize the foliation shown in Figure 22 near the boundary parallel arc. We then let L' be the Legendrian knot indicated on the right hand side of Figure 22 (after the corners are smoothed). It is not hard to show that $L = S_\pm(L')$. □

REMARK 2.21. A bypass is technically the disk cobounded by L and L' but using this lemma we see that this is essentially equivalent to the definition above.

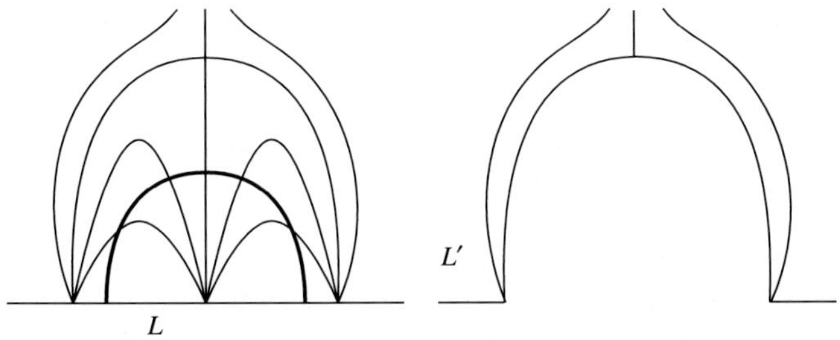

Fig. 22. Characteristic foliation associated to a boundary parallel dividing curve (left). The destabilization L' of L (right).

2.9. *Relation between Legendrian and transversal knots*

Let L be a Legendrian knot in a contact manifold (M, ξ). Let $A = S^1 \times [-1, 1]$ be an embedded annulus in M such that $S^1 \times \{0\} = L$ and A is transverse to ξ. If A is sufficiently "thin" then the characteristic foliation on A is as shown in Figure 23. Given such an A the curve $L_+ = S^1 \times \{+1/2\}$ (resp. $L_- = S^1 \times \{-1/2\}$) is a positively (resp. negatively) transverse knot. The knots $L_\pm$ are called the positive (resp. negative) transverse push-off of L. It is easy to check that any two transversal push-offs are transversely isotopic.

WARNING. *The definition of the positive and negative push-off is not standard in the literature. Some authors (e.g., Bennequin* [5]) *reverse the naming of positive and negative push-offs.*

In the standard contact structure on $\mathbb{R}^3$ we can see this transverse push-off in the front projection. Let L be an oriented Legendrian knot and consider its front projection. Away from the cusps, any strand oriented to the right should be pushed in the $-y$-direction slightly to get L_+ any strand oriented to the left should be pushed in the y direction slightly. Thus way from the cusps the front projection of L and L_+ agree. A careful analysis near the cusps shows that they change as shown in Figure 24.

Given a Legendrian knot L we can construct a positive and a negative transversal knot $T_\pm(L)$ the relation between their classical invariants is described in the following lemma.

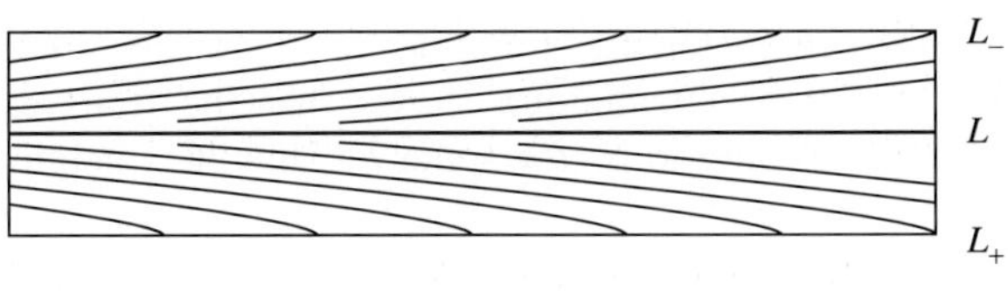

Fig. 23. Annulus involved in the transverse push-off of L.

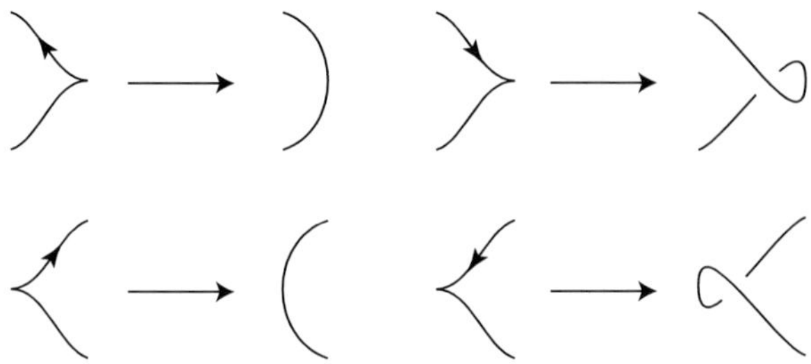

Fig. 24. Positive transverse push-off in the front projection.

LEMMA 2.22. *The invariants of Legendrian knots and their transverse push-offs are related by*

$$sl(T_\pm(L)) = tb(L) \mp r(L). \tag{2.17}$$

PROOF. Given L a Legendrian knot and Σ a Seifert surface for L we let τ be a non-zero vector in $\xi|_L$ that extends to a non-zero vector in $\xi|_\Sigma$. Thus if v is a non-zero vector field tangent to L then $r(L)$ is the twisting of v relative to τ in ξ which we denote: $r(l) = t(v, \tau, \xi|_L)$. If ν is the normal bundle to L, s is the outward pointing normal to Σ and w is a vector field in $\xi|_L$ transverse to v then $tb(L) = t(w, s, \nu)$.

Let v_+, w_+, τ_+ be the vector fields in $\xi|_{L+}$ coming naturally from the definition of L_+. Note we can think of s as the same as s_+ since L and L_+ are the same topological knots. Now keeping in mind $v = \xi|_{L+}$ we have

$$\begin{aligned} sl(L_+) &= t(\tau_+, s, \nu) = t(\tau_+, w_+, \nu) + t(w_+, s, \nu) \\ &= t(\tau, w, \xi) + t(w, s, \nu) = -r(L) + tb(L). \end{aligned}$$

The equation for L_- is similar except that $\nu = -\xi$ so $t(\tau_+, w_+, \nu) = -t(\tau, w, \xi)$.

For Legendrian knots in $(\mathbb{R}^3, \xi_{std})$ there is a direct diagrammatic proof of these formulas using the formulas derived in the previous sections and Figure 24. □

There is also a Legendrian push-off of a transversal knot. Let T be a transverse knot. We want to construct a standard model for a neighborhood of T. To this end consider $(\mathbb{R}^3, \xi_{sym})$. Recall $\xi_{sym} = \ker(dz + r^2 d\phi)$. Now let M be $\mathbb{R}^3$ modulo the action $z \mapsto z + 1$. The contact structure ξ_{sym} clearly induces a contact structure on M (also denoted ξ_{sym}). Now $M = S^1 \times \mathbb{R}^2$ and $T_a = \{r = a\} \subset M$ is a torus that bounds the solid torus $S_a = \{r \leq a\} \subset M$. The characteristic foliation of T_a is by lines of slope a^2. A standard application of Moser's technique shows that the transverse knots T has a neighborhood N contactomorphic to S_a for some a. Thus in a neighborhood of T there is a torus T_b with $b < a$ and $b^2 = 1/n$ for some positive integer n. The characteristic foliation on T_b is by lines of slope $b^2 = 1/n$. Let T_l be a leaf in the foliation of T_b. It is easy to see that T_l is a Legendrian knot topologically isotopic to T.

The Legendrian knot T_l is called a *Legendrian push-off* of T. By considering the obvious annulus between T and T_l it is easy to see that the positive transverse push-off of T_l is T.

Notice that T_l is called *a* Legendrian push-off and not *the* Legendrian push-off. The reason for this is that there are many other Legendrian push-offs. Note that we can find a $c < b < a$ such that $c^2 = 1/(n+1)$. The the Legendrian push-off obtained using T_c, denote it T'_l, is related to T_l by a negative stabilization [36]: $T'_l = S_-(T_l)$. We say two Legendrian knots are *negatively stably isotopic* if they are Legendrian isotopic after each has been negatively stabilized some number of times. Thus the Legendrian push-off of a transverse knot gives a well defined negative stable isotopy class of Legendrian knots.

THEOREM 2.23. (In $(\mathbb{R}^3, \xi_{\text{std}})$, Epstein, Fuchs and Meyer 2001, [30]; in a general contact manifold, Etnyre and Honda 2001, [36]). *Two Legendrian knots are negatively stably isotopic if and only if their transverse push-offs are transversely isotopic.*

It is quite easy to show the "only if" part of this theorem. The "if" part is much more difficult.

Since we know all transverse knots are the transverse push-off of some Legendrian knot, the classification of transversal knots is equivalent to the classification of Legendrian knots up to negative stabilization.

3. Tightness and bounds on invariants

In the standard contact structure on $\mathbb{R}^3$, or more generally in a tight contact structure, there are many bounds on the classical invariants of a Legendrian or transversal knot. In fact, the existence of some of these bounds is at the heart of the nature of tightness. In this section we discuss these inequalities.

3.1. *Bennequin's inequality*

Part of the origins of modern contact geometry can be found in the following theorem.

THEOREM 3.1. (Eliashberg 1992, [25]). *Let (M, ξ) be a tight contact 3-manifold. Let L be a Legendrian knot in M with Seifert surface Σ_L and T be a transverse knot with Seifert surface Σ_T. Then*

$$sl(T) \leq -\chi(\Sigma_T), \tag{3.1}$$

and

$$tb(L) + |r(L)| \leq -\chi(\Sigma_L). \tag{3.2}$$

SKETCH OF PROOF. Clearly the second inequality follows from the first and (2.17). To prove the first inequality recall the discussion in Section 2.8, if the foliation on Σ is generic then

$$-sl(T) = (e_+ - h_+) - (e_- - h_-).$$

Moreover one may easily show

$$\chi(\Sigma_T) = (e_+ + e_-) - (h_+ + h_-).$$

Thus

$$sl(T) + \chi(\Sigma_T) = 2(e_- - h_-).$$

So we can prove Inequality (3.1) by showing that we may isotop Σ_T (rel boundary) so that $e_- = 0$. This is done by showing that the tightness of ξ implies that each negative elliptic point is connected to a negative hyperbolic point and then canceling this pair. This argument is discussed in [25,33]. □

This bound is not sharp for many knot types. This is easily seen once other inequalities are established, so we defer this discussion until later in this section.

The inequalities in Theorem 3.1 are called the Bennequin inequalities. However, the theorem as stated is due to Eliashberg [25]. What Bennequin actually showed was that for any transverse knot in $(\mathbb{R}^3, \xi_{\text{std}})$ (3.1) holds [5]. One reason for this is that the notion of tight vs. overtwisted was unknown at the time. Bennequin was trying to show that $\mathbb{R}^3$ has more than one contact structure. He did this by showing that in the standard contact structure (3.1) is true while it is not in the contact structure ξ_{ot} defined in Example 2.2. Specifically the Legendrian unknot $L = \{z = 0, r = \pi\}$ in $(\mathbb{R}^3, \xi_{\text{ot}})$ is easily seen to have $tb(L) = 0$.

This was one of the first indications of the existence of a tight vs. overtwisted dichotomy. In fact, tightness can be characterized in terms of knots.

THEOREM 3.2. *A contact structure ξ is overtwisted if and only if there is a Legendrian unknot with Thurston–Bennequin invariant equal to 0 if and only if there is a transverse unknot with self-linking number equal to 0.*

Using Theorem 3.1 the only non-trivial (but still easy) part of this theorem is that overtwisted implies the existence of a transverse unknot with self-linking 0.

Since 3.2 gives an upper bound on the Thurston–Bennequin invariant of a knot L in a tight contact structure we can make the following definition

$$\overline{tb}(\mathcal{K}) = \max\{tb(L) | L \in \mathcal{L}(\mathcal{K})\}.$$

So $\overline{tb}(\mathcal{K})$ is the maximal Thurston–Bennequin invariant for Legendrian knots in the knot type $\mathcal{K}$. Clearly this is an invariant of the topological knot type.

3.2. *Slice genus*

There has been a refinement of Bennequin's inequality for knots in Stein fillable contact structures. For a discussion of Stein fillable contact structures see [33,50]. The important point here is that the standard tight contact structure on S^3 is Stein fillable by B^4 (with its standard complex structure).

THEOREM 3.3. (Akbulut and Matveyev 1997, [2]; Lisca and Matic 1998, [58]). *Let X be a Stein manifold with boundary M and ξ the induced contact structure. If L is a Legendrian knot in M then*

$$tb(L) + |r(L)| \leq -\chi(\Sigma), \tag{3.3}$$

where Σ is an embedded oriented surface in X with $\partial\Sigma = L$.

Given a knot type $\mathcal{K}$ in $M = \partial X$, let $g_s(\mathcal{K})$ denote the minimal genus of an embedded orientable surface Σ in X with $\partial\Sigma = \mathcal{K}$. This is called the *slice genus* of $\mathcal{K}$ in (X, M). Equation (3.3) can be interpreted in terms of the slice genus:

$$\frac{1}{2}(tb(L) + |r(L)|) + 1 \leq g_s(\mathcal{K})$$

for all $L \in \mathcal{L}(\mathcal{K})$. Clearly the genus of a knot is an upper bound on the slice genus, thus this inequality implies the Bennequin inequality.

For Legendrian knots in the standard contact structure in S^3 this theorem is due to Rudolph [73]. In full generality is appears in [2,58]. Currently, the most direct proof of this theorem involves a standard adjunction type inequality in Seiberg-Witten theory and an embedding theorem for Stein manifolds proved in [57].

While much of the deep mathematical content to the above theorem is Seiberg-Witten theory, the language of Legendrian knots provides a user friendly "front end" for the Seiberg-Witten theory. In particular, in many examples it is easy to simply draw pictures to see knots cannot be slice or to get bounds on their slice genus without ever having to explicitly invoke Seiberg-Witten theory. We will see applications of this in Section 7.4. Here we illustrate the usefulness of this theorem with a simple example. Let $\mathcal{K}$ be a knot type and $\overline{\mathcal{K}}$ be its mirror image. It is well known [71] that $\mathcal{K}\#\overline{\mathcal{K}}$ is slice. Thus the maximal Thurston–Bennequin invariant satisfies $\overline{tb}(\mathcal{K}\#\overline{\mathcal{K}}) \leq -1$. But later we will see (Section 5.4) that $\overline{tb}(\mathcal{K}\#\overline{\mathcal{K}}) = \overline{tb}(\mathcal{K}) + \overline{tb}(\overline{\mathcal{K}}) + 1$. Thus

$$\overline{tb}(\overline{\mathcal{K}}) \leq -\overline{tb}(\mathcal{K}) - 2.$$

By drawing pictures (see Section 5.2) one may easily see that for p, q positive and relatively prime, a (p,q)-torus knot $\mathcal{K}(_{p,q})$ has $\overline{tb}(\mathcal{K}_{(p,q)}) = pq - p - q$. Thus

$$\overline{tb}(K_{(-p,q)}) = \overline{tb}(\overline{\mathcal{K}}_{(p,q)}) \leq p + q - pq - 2,$$

and we see the Bennequin inequality is not sharp. (Later we will see that $\overline{tb}(\mathcal{K}_{(-p,q)}) = -pq$.)

3.3. *Other inequalities in* $(\mathbb{R}^3, \xi_{\mathrm{std}})$

We begin by recalling the definition of several knot polynomials. Consider a projection of an oriented knot K. To this projection we associate the HOMELY (Laurent) polynomial (aka, two variable Jones polynomial) $P_K(v,z)$ which is defined by the skien relation

$$\frac{1}{v} P_{K_+} - v P_{K_-} = z P_{K_0},$$

where K is either K_+ or K_- and K_+, K_-, K_0 are related to each other as shown in the top row of Figure 25.

Let D be a regular projection of a (non-oriented) topological knot K. To this projection we associate a two variable (Laurent) polynomial $L_K(a,x)$ defined by

$$L_U = \frac{a - a^{-1}}{x} + 1,$$

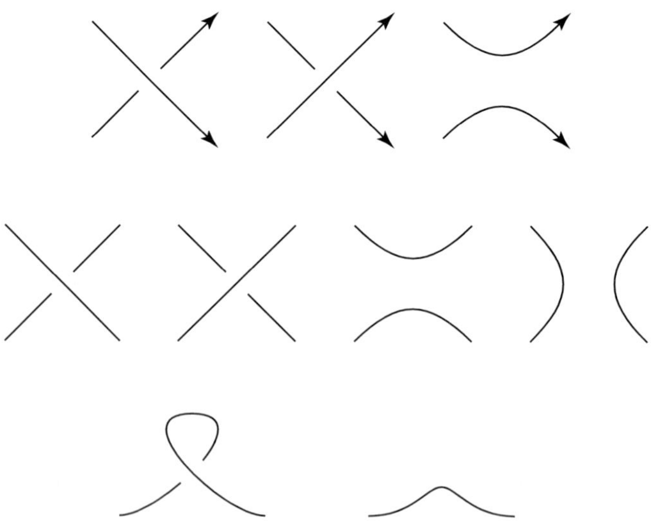

Fig. 25. The relation between K_+, K_-, K_0 (top), K_+, K_-, K_0 K_∞(middle) and K_r, K (bottom).

$$L_{K_+} - L_{K_-} = x(L_{K_0} - L_{K_\infty}),$$

$$L_{K_r} = aL_K,$$

where K_+, K_-, K_0, K_∞ are as related in the middle row of Figure 25, K_r, K are as related in the bottom row of Figure 25 and U is the diagram for the unknot with no crossings. The polynomial L is an invariant of D through regular homotopies. The polynomial $F_K(a,x) = a^{\text{writhe}(D)}L(a,x)$ is the Kauffman polynomial of K. It is an invariant of the topological knot K.

THEOREM 3.4. *Let $\mathcal{K}$ be a topological knot type. For any $L \in \mathcal{L}(\mathcal{K})$ we have*
(1) $tb(L) + |r(L)| \le d_{P_K}$ *and*
(2) $tb(K) \le d_{F_K}$.
where d_{P_K} is the lowest degree in v of $P_K(v,z)$ and d_{F_K} is the lowest degree in a of $F_K(a,x)$.

The first inequality, observed by Fuchs and Tabachnikov [45], follows from work of Franks and Williams [40] and Morton [64] who showed that for an n braid with algebraic length a:

$$a - n \le d_{P_K},$$

and the fact we observed in Section 2.6 that $sl = a - n$. Thus this inequality gives a bound on the self-linking numbers of transverse knots. The relation between the invariants of a Legendrian knot and its transverse push-off complete the proof of the first inequality in the theorem.

The second inequality is originally due to Rudolph [72]. Both inequalities can be proved using state models for the polynomials, see [78]. Both inequalities have been extended to Legendrian knots in J^1S^1, the one jet space of S^1, see [14]. For a more detailed discussion of the history of these inequalities see [39].

EXAMPLE 3.5. For the left handed trefoil we have

$$P_K(v,z) = v^{-3}\left(\frac{v - v^{-1}}{z}\right)(2v - v^{-1} - vz^2)$$

and

$$F_K(a,x) = a^{-3}\left(1 + \frac{a - a^{-1}}{x}\right)(2a - a^{-1} + z - a^{-2}z + az^2 - a^{-1}z^2).$$

Thus we get the following bounds on Legendrian left handed trefoils L

$$tb(L) + |r(L)| \le d_{P_K} = -5$$

and

$$tb(L) \leq -6.$$

We will see below that the maximal tb for left handed trefoils is -6 thus the second inequality seems sharper, but the trefoils with $tb = -6$ must have $r = \pm 1$. Thus both inequalities are sharp in this case. Note that in this case the Bennequin inequality is not sharp, as it only gives an upper bound of 1.

While no known bound on tb is always sharp the Kauffman polynomial F_K frequently seems to be very good. For example

THEOREM 3.6. (Ng 2001, [69]). *If $\mathcal{K}$ is a 2-bridge knot, then the Kauffman bound is sharp: $\overline{tb}(\mathcal{K}) = d_{F_{\mathcal{K}}}$. Moreover, the Kauffman bound is sharp for all knot types with eight or fewer crossings except the* $(4, -3)$ *torus knot* (8_{19} *in* [71]).

4. New invariants

In this section we discuss two invariants of Legendrian knots that have been defined in recent years. They are considerably more complicated than the "classical invariants" of Legendrian knots and currently we do not know exactly what they tell us about Legendrian knots. But they do show us that the classical invariants do not completely determine the Legendrian isotopy type of Legendrian knots.

4.1. *Contact homology (aka Chekanov–Eliashberg DGA)*

In [27] Eliashberg first outlined the general theory of contact homology. It is a sophisticated algebraic tool for keeping track of "Reeb cords" and "pseudo-holomorphic curves" in symplecizations of contact manifolds. Subsequently the more elaborate framework of Symplectic Field Theory has been developed in [29]. The foundational parts of these grand theories (especially the parts of interest here) are currently being developed. In certain circumstances one can translate the general ideas into simpler terms and rigorously establish a theory based in these simpler terms. The differential graded algebra (DGA) of Chekanov [11] is such a theory for Legendrian knots in $\mathbb{R}^3$. We will concentrate on this simpler, but still deep, theory here. For a discussion of how the theory described here fits into the "big picture" see [38]. In [75] a similar theory was established for Legendrian knots in circle bundles (with contact structures transverse to the circle fibers).

Let L be such a Legendrian knot. To L we want to assign a graded algebra and a differential on the algebra. There are various levels of sophistication one can use in defining the algebra and its differential. We begin with the simplest description and discuss generalizations in Section 4.4.

The algebra. Denote the double points in $\pi(L)$ (recall this is the Lagrangian projection of L) by $\mathcal{C}$. We assume $\mathcal{C}$ is a finite set of transverse double points (for a generic Legendrian this is clearly true). Let $\mathcal{A}_{\mathbb{Z}_2}$ be the free associative unital algebra over $\mathbb{Z}_2$ generated by $\mathcal{C}$. If we want to make the generators explicit we write $\mathcal{A}_{\mathbb{Z}_2}(\mathcal{C})$.

The grading. To each crossing $c \in \mathcal{C}$ there are two points c^+ and c^- in $L \subset \mathbb{R}^3$ that project to c. We let c^+ be the point with larger z coordinate. Choose a map $\gamma_c : [0, 1] \to L$ that parametrizes an arc running from c^+ to c^-. (Note there are two such arcs, we can choose either.) Let P be the projectivized unit circle in the xy-plane (i.e., identify antipodal points). We get a map $g_c : [0, 1] \to P$ by $g_c(t) = \gamma_c'(t)/|\gamma_c'(t)|$. Since c is a transverse double point $g_c(0) \neq g_c(1)$. We extend g_c to a map $g_c : S^1 \to P$ by rotating $g_c(1)$ clockwise until it agrees with $g_c(0)$. Now define the grading on c to be

$$|c| = \text{degree } (g_c).$$

See Figure 26. Note the grading is not well defined! If we chose the other arc from c^+ to c^- in L then we could have gotten a different number. Note the union of the two arcs is all of L. From this it is easy to see that the difference between the two possible gradings on c is the twice the rotation number of L (recall the rotation number is the degree of the Gauss map for the immersion $\pi(L)$). Thus $|c|$ is well defined modulo $2r(L)$. Thus $\mathcal{A}$ is a graded algebra with grading in $\mathbb{Z}_{2r(L)}$.

The differential. We will define the differential ∂ on $\mathcal{A}_{\mathbb{Z}_2}$ by defining it on the generators and then extending by the signed Leibniz rule:

$$\partial ab = (\partial a)b + (-1)^{|a|} a \partial b.$$

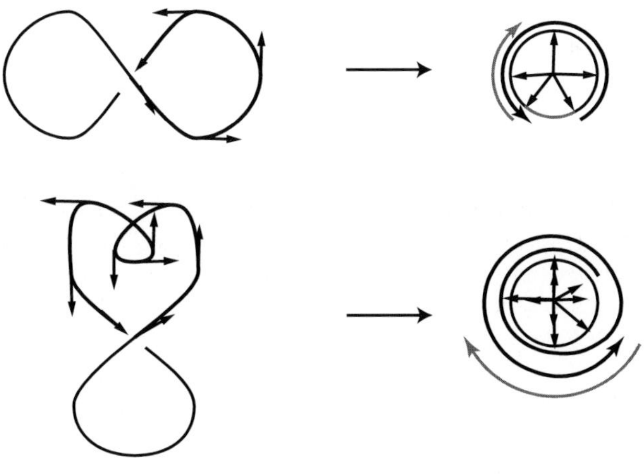

Fig. 26. On the right are two crossings, the thicker arc is γ_c. On the right is the image of the Gauss map for the arc. The grey arrow indicates the completion of g_c to a closed loop. When the Gauss map is projectivized we see the top example has grading 1 and the bottom example has grading 3.

(Note that since we are working over $\mathbb{Z}_2$ we do not really need the $(-1)^{|a|}$ in this formula, but when working over other rings/fields the sign is important so we keep it in the formula.) The neighborhood of a crossing c in $\mathbb{R}^2$ is divided into four quadrants by $\pi(L)$ two are labeled $(+)$ quadrants and two are labeled $(-)$ quadrants as shown in Figure 27. Let $a \in \mathcal{C}$ be a generator of $\mathcal{A}$ and let $b_1 \ldots b_k$ be a word in the "letters" $\mathcal{C}$. Let P_{k+1} be a $k+1$ sided polygon with vertices labeled counterclockwise $v_0, \ldots, v_k$ and set

$$\mathcal{M}^a_{b_1 \ldots b_k} = \{u : (P_{k+1}, \partial P_{k+1}) \to (\mathbb{R}^2, \pi(L)) | u \text{ satisfies 1. } - \text{ 3. below}\}/\text{reparam.}$$

where $\mathbb{R}^2$ denotes the xy-plane and the conditions are

(1) u is an immersion;
(2) $u(v_0) = a$ and near a the image of a small neighborhood of v_0 under u covers a $+$ quadrant.
(3) $u(v_i) = b_i$, $i = 1, \ldots, k$, and near b_i the image of a small neighborhood of v_i under u covers a $-$ quadrant.

We can now define

$$\partial a = \sum_{b_1 \ldots b_k} (\#_2 \mathcal{M}) b_1 b_2 \ldots b_k,$$

where the sum is taken over all words in the letters $\mathcal{C}$ and $\#_2$ denotes the modulo two count of elements in $\mathcal{M}$. It is not hard to show that $\mathcal{M}$ is finite and the sum is finite [11], thus the differential is well-defined.

LEMMA 4.1. (Chekanov 2002, [11]). *The map ∂ is a differential:*

$$\partial \circ \partial = 0,$$

and lowers degree by one.

EXAMPLE 4.2. Here we compute the graded algebra and the differential for the Legendrian unknot with $tb = -2$ (and $r = \pm 1$). The Lagrangian projection is shown on the right hand side of Figure 28.

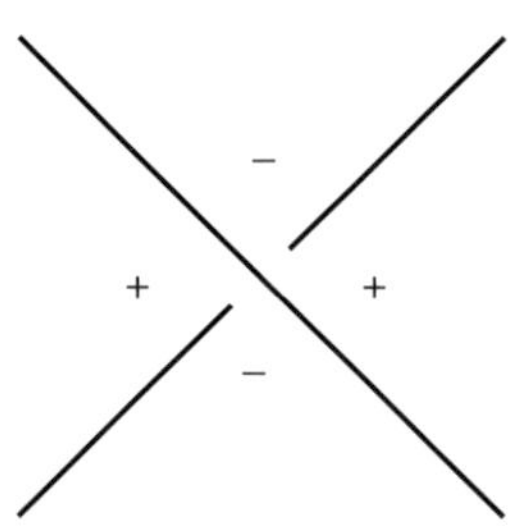

Fig. 27. Quadrants near a crossing in $\pi(L)$.

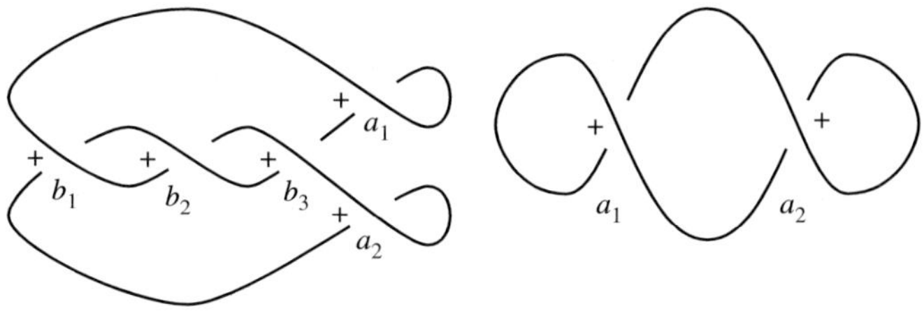

Fig. 28. The Lagrangian projection of a Legendrian trefoil and unknot. One corner of each crossings has been labeled with a $+$ to help determine the signs of all the quadrants near the crossing.

We have two generators of the algebra a_1, a_2. Each has grading 1. Moreover, $\partial a_i = 1$ for $i = 1, 2$.

EXAMPLE 4.3. Consider the Legendrian right handed trefoil with $tb = 1$. The Lagrangian projection is shown on the left of Figure 28. The algebra has five generators a_1, a_2, b_1, b_2, b_3. Their gradings are

$$|a_i| = 1, \qquad |b_i| = 0.$$

One easily computes

$$\partial a_1 = 1 + b_1 + b_3 + b_1 b_2 b_3, \qquad \partial a_2 = 1 + b_1 + b_3 + b_3 b_2 b_1, \qquad \partial b_i = 0.$$

To discuss the invariance of $(\mathcal{A}_{\mathbb{Z}_2}, \partial)$ under Legendrian isotopy we need a few preliminary definitions. An automorphism of $\mathcal{A}_{\mathbb{Z}_2}(c_1, \ldots, c_n)$ of the form

$$\phi(c_i) = \begin{cases} c_i, & i \neq j \\ \pm c_j + u, & u \in \mathcal{A}(c_1, \ldots, c_{j-1}, c_{j+1}, \ldots, c_n), \qquad i = j \end{cases} \tag{4.1}$$

for some fixed j is called *elementary*. A composition of such automorphism is called a *tame automorphism*. A tame isomorphism is an identification of the generators of two algebras followed by a tame automorphism. (Since our algebras are all graded our automorphisms are also assumed to be graded automorphisms).

An *index i stabilization* of a DGA $(\mathcal{A}_{\mathbb{Z}_2}(c_1, \ldots, c_n), \partial)$ is the graded algebra $\mathcal{A}_{\mathbb{Z}_2}(c_1, \ldots, c_n, a, b)$ where $|a| = |b| - 1 = i$ and the differential on this algebra agrees with the original differential on the c_i's, $\partial a = 0$, and $\partial b = a$. If the index of a stabilization is unimportant then it is simply referred to as a stabilization instead of an index i stabilization.

Two DGA's are *stably tame isomorphic* if after stabilizing each of the DGA's some number of times they become tame isomorphic.

THEOREM 4.4. (Chekanov 2002, [11]). *The DGA* $(\mathcal{A}_{\mathbb{Z}_2}, \partial)$ *associated to a Legendrian knot changes by stable tame isomorphisms under Legendrian isotopy, moreover the homology*

$$CH_*(L) = \frac{\ker \partial}{\text{image } \partial}$$

is unchanged under Legendrian isotopy.

EXAMPLE 4.5. The contact homology for the Legendrian unknot in Example 4.2 is

$$CH_*(L) = 0,$$

since $\partial a_1 = 1$. Indeed $\ker(\partial)$ is spanned by $a_1 - a_2$ but $\partial(a_1(a_1 - a_2)) = a_1 - a_2$.

REMARK 4.6. In general it is easy to show that anytime $\partial a = 1$ for some element in the algebra then $CH_* = 0$.

EXAMPLE 4.7. The contact homology for the Legendrian trefoil in Example 4.3 is

$$CH_*(L) = CH_0(L) = \langle b_1, b_2, b_3 | b_1 b_2 b_3 = b_3 b_2 b_1 = 0 \rangle.$$

PROPOSITION 4.8. (Chekanov 2002, [11]). *The DGA associated to a stabilized knot is stably tame isomorphic to the "trivial" algebra, that is the algebra generated by one element a such that* $\partial a = 1$. *In particular, the contact homology is* 0.

The main observation in the proof of this proposition is that a stabilization in the Lagrangian projection is achieved by adding a small loop (Figure 19). Since this loop is small a generator a has been added to the algebra for which $\partial a = 1$ by the following lemma.

LEMMA 4.9. (Chekanov 2002, [11]). *If* $u : P_{k+1} \to \mathbb{R}^2$ *is a polygon in* $\mathcal{M}^a_{b_1 \ldots b_k}$ *then*

$$h(a) - \sum_{i=1}^{k} h(b_i) = \int_{P_{k+1}} u^*(dx \wedge dy)$$

where $h(c)$ *is the difference in the z-coordinates of the two points* c^+, c^- *in L lying above a double point c.*

Thus we easily see the contact homology is 0 as in Example 4.5. With a little more work one can see the algebra is as claimed in the proposition.

4.2. *Linearization*

Since our algebra is non-commutative it is quite hard in general to look at the homology of two DGA's and determine if they are the same or not. In this section we describe two methods to extract more "computable" invariants out of the DGA associated to a Legendrian knot.

Given a DGA $(\mathcal{A}, \partial)$ we can filter $\mathcal{A}$ by "word length". That is, define $\mathcal{A}_n$ to be the vector space generated by words in the generators of $\mathcal{A}$ of length less than or equal to n. Note $\mathcal{A}_n$ is not an algebra, but merely a vector space. We call the DGA *augmented* if for each generator $a \in \mathcal{A}$, ∂a contains no constant term. Said another way the "the differential is non-decreasing on word length". It is easy to see that if $(\mathcal{A}, \partial)$ is an augmented DGA then ∂ induces a differential ∂_n on $\mathcal{A}_n$ by applying ∂ to a generator of $\mathcal{A}_n$ and projecting the result to $\mathcal{A}_n$. If $(\mathcal{A}, \partial)$ is the DGA associated to a Legendrian knot L and it is augmented then we call the homology of $(\mathcal{A}_n, \partial_n)$ the *order n contact homology* of $(\mathcal{A}, \partial)$ denote it $L_nCH_*(L)$. The first order contact homology is frequently referred to as the *linear (or linearized) contact homology*. The Poincaré polynomial

$$P_n(\lambda) = \sum_{i \in \mathbb{Z}_{2r(L)}} \dim(L_nCH_i(L))\lambda^i$$

of $L_nCH_*(L)$ is called the *order n Chekanov-Poincaré polynomial* of $(\mathcal{A}, \partial)$. It is interesting to observe [11] that one can compute the Thurston–Bennequin invariant of a Legendrian knot by evaluating $P_1(\lambda)$ at -1:

$$tb(L) = P_1(-1).$$

IMPORTANT FACTS. First, not all DGA's are augmented, and second, the order n contact homology and order n Chekanov-Poincaré polynomial of $(\mathcal{A}, \partial)$ is not an invariant of the stable tame isomorphism class of $(\mathcal{A}, \partial)$!

These two facts seem to indicate these notions are useless for Legendrian knots and that our notation is not so good. It turns out we can still make use of these order n approximations to the contact homology. If $\mathcal{A}$ is generated by $a_1, \ldots, a_k$ then set

$$G(\mathcal{A}) = \{g \text{ a graded automorphism of } \mathcal{A} \text{ of the form } a_i \mapsto a_i + c_i \text{ where } c_i \in \mathbb{Z}_2.\}$$

Every element $g \in G(\mathcal{A})$ gives a tame isomorphism of DGA's from $(\mathcal{A}, \partial)$ to $(\mathcal{A}, \partial^g)$ where $\partial^g = g\partial g^{-1}$. Now set

$$G_a(\mathcal{A}) = \{g \in G(\mathcal{A}) : (\mathcal{A}, \partial^g) \text{ is augmented}\}.$$

THEOREM 4.10. (Chekanov 2002, [11]). *The sets*

$$I_n(L) = \{\textit{Chekanov-Poincaré polynomial of order } n \textit{ for } (A, \partial^g) : g \in G_a(\mathcal{A})\}$$

and

$$L_nCH_*(L) = \{L_nCH_*(\mathcal{A}, \partial^g) : g \in G_a(\mathcal{A})\}$$

are invariants of the Legendrian knot L.

It is relatively easy to show these sets do not change under tame isomorphisms. Thus to prove this theorem one only needs to check what happens to G_a, I_n and L_nCH_* under stabilization.

A convenient way to find $G_a(A)$ is via augmentations. Let $(\mathcal{A}, \partial)$ be a DGA. A map

$$\epsilon : \mathcal{A} \to \mathbb{Z}_2$$

is called an *augmentation* of $\mathcal{A}$ if $\epsilon(1) = 1$, $\epsilon \circ \partial = 0$ and ϵ vanishes on any element of non-zero degree.

LEMMA 4.11. *The augmentations of* (A, ∂) *are in one-to-one correspondence with* $G_a(\mathcal{A})$.

This lemma is simple to prove once one observes that given an augmentation ϵ the tame automorphism $g(a_i) = a_i + \epsilon(a_i)$ is in $G_a(\mathcal{A})$.

EXAMPLE 4.12. In this example we consider the two "Chekanov–Eliashberg knots". These were the first two Legendrian knots with the same tb, r and knot type that were shown to be non-Legendrian isotopic. The Lagrangian projection of these knots is shown in Figure 29. The generators for L are a_i, $i = 1, \ldots 9$. Their gradings are

$$|a_i| = 1,\ i = 1\ldots 4, \quad |a_5| = 2, \quad |a_6| = -2, \quad |a_i| = 0,\ i = 7\ldots 9$$

and the boundary map is

$$\partial a_1 = 1 + a_7 + a_7a_6a_5, \quad \partial a_2 = 1 + a_9 + a_5a_6a_9, \quad \partial a_3 = 1 + a_8a_7,$$

$$\partial a_4 = 1 + a_9a_8, \quad \partial a_i = 0,\ i \geq 5.$$

The generators for L' are b_i, $i = 1, \ldots 9$. Their gradings are

$$|b_i| = 1,\ i = 1\ldots 4, \quad |b_i| = 0,\ i = 5\ldots 9$$

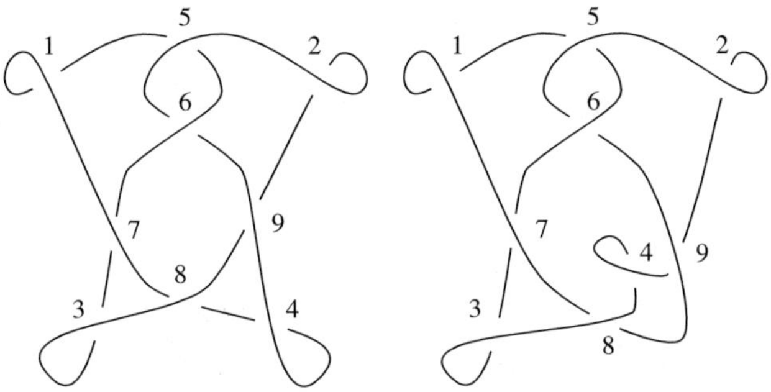

Fig. 29. The knot L (left) and L' (right).

and the boundary map is

$$\partial b_1 = 1 + b_7 + b_7 b_6 b_5 + b_5 + b_9 b_8 b_5, \quad \partial b_2 = 1 + b_9 + b_5 b_6 b_9,$$

$$\partial b_3 = 1 + b_8 b_7, \quad \partial b_4 = 1 + b_9 b_8, \quad \partial b_i = 0, \ i \geq 5.$$

Though these differentials look different it is difficult at this stage to see that the algebras are not stability tame isomorphic. We now compute the linearized contact homology of L. We begin by looking for augmentations. These must be maps $\epsilon : \mathcal{A} \to \mathbb{Z}_2$ that send a_i to 0 for $i \leq 6$ and send a_i to $a_i + c_i$ for $i \geq 7$ where c_i is 0 or 1. The equation $\epsilon \circ \partial = 0$ implies

$$1 + c_7 + c_7 c_6 c_5 = 0, \quad 1 + c_9 + c_5 c_6 c_9 = 0, \quad 1 + c_8 c_7 = 0, \quad 1 + c_9 c_8 = 0.$$

The only solution to these equations is $c_7 = c_8 = c_9 = 1$. The differential ∂_ϵ associated to this augmentation is

$$\partial_\epsilon a_1 = a_7 + a_6 a_5 + a_7 a_6 a_5, \quad \partial_\epsilon a_2 = a_9 + a_5 a_6 + a_5 a_6 a_9,$$

$$\partial_\epsilon a_3 = a_8 + a_7 + a_8 a_7, \quad \partial_\epsilon a_4 = a_9 + a_8 + a_9 a_8, \quad \partial_\epsilon a_i = 0, \ i \geq 5.$$

Thus one easily computes the linearized homology is generated by $a_1 + a_2 + a_3 + a_4, a_5, a_6$. It has dimension one in grading -2, 1, 2 so $P_1(\lambda) = \lambda^{-2} + \lambda + \lambda^2$.

For $\mathcal{A}(L')$ one finds three augmentations $c_i = 0, i = 1 \ldots 4$, $c_7 = c_8 = c_9 = 1$ and $c_5 c_6 = 0$. No matter which augmentation is used one finds the linearized homology generated by $b_1 + b_2 + b_3 + b_4, b_5, b_6$. Thus it has dimension one in grading 1 and dimension two in grading 0.

Hence one can distinguish L and L' using the linearized contact homology.

4.3. *The characteristic algebra*

Let $(\mathcal{A}, \partial)$ be the DGA generated by $a_1, \ldots, a_k$. Let I be the two sided idea in $\mathcal{A}$ generated by $\langle \partial a_i \rangle_{i=1}^k$. The characteristic algebra of $(\mathcal{A}, \partial)$ is

$$\mathcal{C}(\mathcal{A}) = \mathcal{A}/I.$$

Two characteristic algebras $\mathcal{A}_1/I_1$ and $\mathcal{A}_2/I_2$ are *tame isomorphic* if after adding some generators to $\mathcal{A}_i$ and the same generators to I_i, $i = 1, 2$, then there is a tame isomorphism of $\mathcal{A}_1$ to $\mathcal{A}_2$ taking I_1 to I_2. It is important to note that the definition of tame isomorphism for characteristic algebras actually requires the pair $\mathcal{A}, I$ not just $\mathcal{A}/I$. So one should probably define the characteristic algebra to be the pair $(\mathcal{A}, I)$ but we will hold with tradition and define $\mathcal{C}(\mathcal{A}) = \mathcal{A}/I$, keeping in mind the need to remember both $\mathcal{A}$ and I. We say $\mathcal{A}_1/I_1$ and $\mathcal{A}_2/I_2$ are *equivalent* if after adding some (possibly different) number of generators to both $\mathcal{A}_1$ and $\mathcal{A}_2$ (but not the ideals I_i) the algebras $\mathcal{A}_1/I_1$ and $\mathcal{A}_2/I_2$ are tame isomorphic.

THEOREM 4.13. (Ng 2001, [68]). *Legendrian isotopic knots have equivalent characteristic algebras.*

We will illustrate the efficacy of the characteristic algebra by addressing Ng's solution to the "Legendrian mirror problem" [45]. That is, given a Legendrian knot L is it isotopic to its image under the map $(x, y, z) \mapsto (x, -y, -z)$? The answer is sometimes NO.

EXAMPLE 4.14. Consider the Legendrian knot L whose Lagrangian projection is shown in Figure 30. The algebra has 11 generators with gradings $|a_1| = |a_2| = |a_7| = |a_9| = |a_{10}| = 1$, $|a_3| = |a_4| = 0$ and $|a_5| = |a_6| = |a_8| = |a_{11}| = -1$. The differential is

$$\partial a_1 = 1 + a_{10}a_5a_3 \qquad \partial a_6 = a_{11}a_8$$

$$\partial a_2 = 1 + a_3(1 + a_6a_{10} + a_{11}a_7) \qquad \partial a_7 = a_8a_{10}$$

$$\partial a_4 = a_{11} + (1 + a_6a_{11} + a_{11}a_7)a_3 \qquad \partial a_9 = 1 + a_{10}a_{11}$$

$$\partial a_5 = \partial a_8 = \partial a_{11} = \partial a_3 = 0.$$

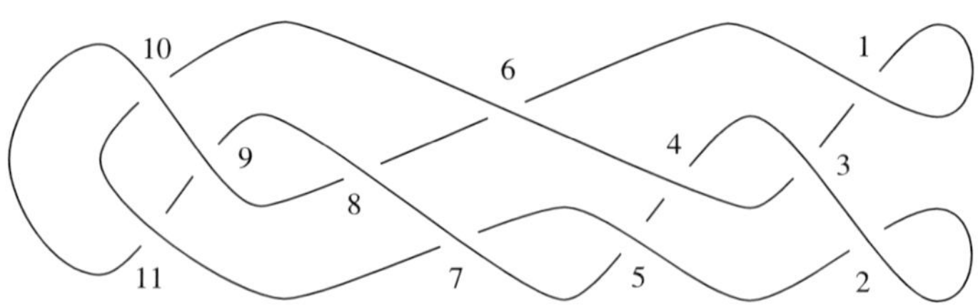

Fig. 30. The Legendrian knots L.

Let I be the ideal generated by these relations and set $\mathcal{C}(\mathcal{A}) = \mathcal{A}/I$. The characteristic algebra $\mathcal{C}(\mathcal{A}')$, the algebra associated to the Legendrian mirror of L, is the same except all the terms in I are reversed.

One may show [69] that

$$\mathcal{C}(\mathcal{A}) = \langle a_1, \ldots, a_{10}\rangle/\langle 1 + a_{10}a_5a_3, 1 + a_3a_{10}a_5, 1 + a_{10}^2a_5^2, 1 + a_{10}a_5 + a_6a_{10} + a_{10}a_5^2a_7\rangle$$

We note that $\mathcal{C}(\mathcal{A})$ does not have elements v and w of grading -1 and 1, respectively, that satisfy $vw = 1$. The reason for this is if there were such elements then we could consider the further quotient algebra $\mathcal{C}'$ obtained by setting $a_3 = 1, a_1 = a_2 = a_6 = a_7 = a_9 = 0$. This algebra has presentation $\langle a_5, a_{10}\rangle/\langle 1 + a_{10}a_5\rangle$. Moreover the elements v and w would map to such elements in $\mathcal{C}'$. But in $\mathcal{C}'$ one may easily see that such elements do not exist.

Now in $\mathcal{C}(\mathcal{A}')$ we see the elements $v = a_{10}$ and $w = a_5a_3$ satisfy $vw = 1$. Thus L is not Legendrian isotopic to its mirror.

Let γ_i be the number of generators of degree i of the DGA $(\mathcal{A}, \partial)$. The *degree distribution* of $\mathcal{A}$ is the map

$$\gamma : \mathbb{Z}_{2r(L)} \to \mathbb{Z}_{\geq 0} : i \mapsto \gamma_i$$

where $r(L)$ is the rotation number of L.

THEOREM 4.15. (Ng 2001, [68]). *The first and second order Poincaré–Chekanov polynomials for all possible augmentations of the DGA $(\mathcal{A}, \partial)$ are determined by the characteristic algebra $\mathcal{C}(\mathcal{A})$ and the degree distribution of $\mathcal{A}$.*

Ng has found examples that the characteristic algebra distinguishes that have the same linearized contact homology. Thus the characteristic algebra is a strictly stronger invariant that the linearized contact homology.

4.4. *Lifting the DGA to $\mathbb{Z}[t, t^{-1}]$*

It is fairly simple to describe the lift of the DGA to $\mathbb{Z}[t, t^{-1}]$. Once again we start with the Lagrangian projection $\pi(L)$ of the Legendrian knot L and assume the set of double point $\mathcal{C}$ is finite and consists of transverse double points. Now let $\mathcal{A}$ be the free associative unital algebra over $\mathbb{Z}[t, t^{-1}]$ generated by $\mathcal{C}$. We grade the generators as above and set $|t| = 2r(L)$. The grading in this case is over $\mathbb{Z}$ (and not just $\mathbb{Z}_{2r(L)}$). Recall to define the grading on $c \in \mathcal{C}$ we chose paths γ_c from the upper point $c^+ \in L$ above c to the lower point $c^- \in L$. To define the differential we want to assign to each $u \in \mathcal{M}^a_{b_1\ldots b_k}$ (for the notation

Fig. 31. Signs $s(v_i)$ associated to a corner.

see Section 4.1) an integer n_u and a sign s_u. We begin with the integer. Given $u \in \mathcal{M}^a_{b_1 \dots b_k}$ note that the image of the boundary $u(\partial P_{k+1})$ is a union of arcs in $\pi(L)$ that can all be uniquely lifted to arcs α_i in L. Let $\Gamma_u = \cup \alpha_i \cup \gamma_a \cup \gamma_{b_i}$. This is a closed loop in L. We orient Γ_u so that all the arcs α_i are oriented as the boundary of the polygon $u(P_{k+1})$. Now Γ_u is an oriented loop in the oriented $L = S^1$. As such it corresponds to an element in $H_1(L, \mathbb{Z}) = \mathbb{Z}$. This integer is denoted n_u. Now for the sign. To each corner v_i of P_{k+1} we assign a sign $s(v_i)$ as shown in Figure 31. The sign is u is defined to be

$$s_u = \prod_{i=0}^{k} s(v_i).$$

Now the boundary map is defined on a generator $a \in C$ by

$$\partial a = \sum_{b_1 \dots b_k} \sum_{u \in \mathcal{M}^a_{b_1 \dots b_k}} s_u b_1 b_2 \dots b_k t^{n_u},$$

where the first sum is taken over all words in the letters $\mathcal{C}$.

Once again we have

THEOREM 4.16. (Etnyre, Ng and Sabloff 2002, [38]). *The map ∂ is a differential*

$$\partial \circ \partial = 0.$$

The stable tame isomorphism class and homology of $(\mathcal{A}, \partial)$ are invariant under Legendrian isotopy.

EXAMPLE 4.17. Here we compute the contact homology of the Legendrian figure eight knot with $\mathrm{tb} = -3$. From Figure 32 we see the algebra has seven generators with $|a_2| = |a_4| = |a_5| = |a_7| = 1$, $|a_1| = |a_3| = 0$ and $|a_6| = -1$. We compute the differential

$$\partial a_1 = a_6 - a_6 a_3 - t a_6 a_3 a_5 a_6, \quad \partial a_2 = t^{-1} + a_1 a_3 - a_6 a_3 a_4,$$

$$\partial a_4 = 1 - a_3 + t a_5 a_6 a_3, \quad \partial a_7 = t^{-1} + a_3 - t a_3 a_6 a_3 a_5, \quad \partial a_3 = \partial a_5 = \partial a_6 = 0.$$

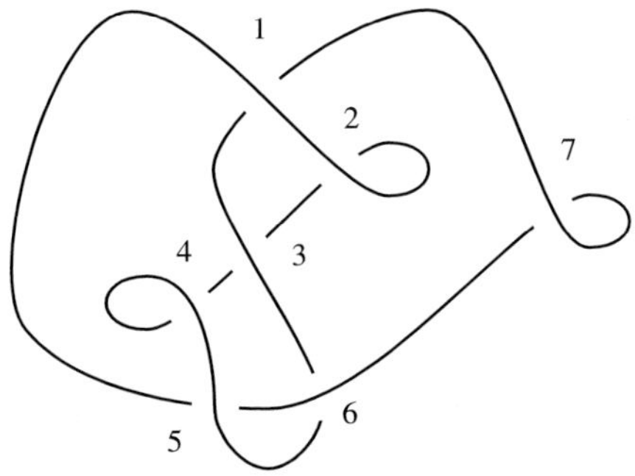

Fig. 32. The Legendrian knots L.

It is not currently known if this refined contact homology is a stronger invariant of Legendrian knots than the original contact homology. One can still define the order n contact homology, and in particular the linearization (though augmentations are much harder to find), and the characteristic algebra of this enhanced contact homology.

One potential benefit of including orientations when we count disks in $\mathcal{M}^a_{b_1\ldots b_k}$ is that we can abelianize. In particular we can define $\mathcal{A}_{\mathbb{Q}}$ to be the supercommutative associative algebra with unit over $\mathbb{Q}[t, t^{-1}]$ generated by $\mathcal{C}$, the crossings in a Lagrangian projection of a Legendrian knot. We grade the elements of $\mathcal{A}_{\mathbb{Q}}$ as we did before and the differential is also defined as above.

THEOREM 4.18. (Etnyre, Ng and Sabloff 2002, [38]). *The stable tame isomorphism class and homology of* $(\mathcal{A}_{\mathbb{Q}}, \partial)$ *are invariants of the Legendrian isotopy class of the Legendrian knot* L.

The homology of $(\mathcal{A}_{\mathbb{Q}}, \partial)$ is called the *abelianized contact homology* of L. It is surprising but not known if one can abelianize the contact homology over $\mathbb{Z}$ or $\mathbb{Z}_2$. In the proof of this theorem it is simple to show the stable tame isomorphism class of the differential algebra is an invariant of L but to show its homology is invariant one writes down a chain map that involves division by integers. Thus if the abelianized contact homology over $\mathbb{Z}$ or $\mathbb{Z}_2$ is invariant the proof will have to be considerably different from the proof of the above theorem.

4.5. *DGA's in the front projection*

Though it is more natural to define Chekanov's DGA in the Lagrangian projection it is somewhat difficult to work with Lagrangian projections of Legendrian knots. In this section we discuss Ng's description of the DGA in the front projection [69]. This front description is based on the following observation.

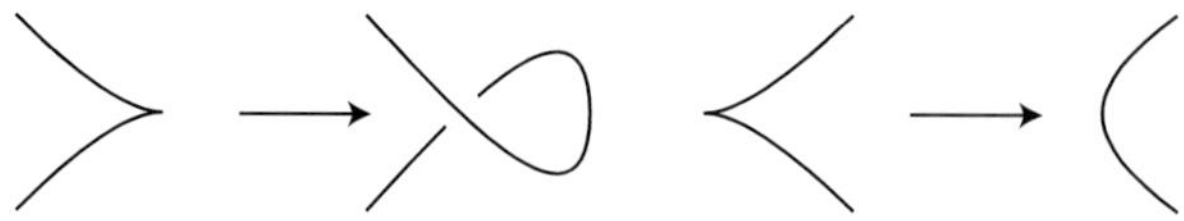

Fig. 33. Transition from a front projection to a Lagrangian projection.

THEOREM 4.19. *Given the front projection of a Legendrian knot L one can obtain a diagram isotopic to the Lagrangian projection by altering the cusps in the front diagram as shown in Figure 33.*

SKETCH OF PROOF. The idea is to stretch any Lagrangian projection in the x-direction so that it is very long and flat. Then it is easy to arrange that except near crossings and cusps each strand of the projection has constant z-value. Now, in the complement of the crossings and the cusps, tilt the strands so that they each have a distinct slope ϵz where z is the z coordinate of the line before it is tilted. Form the crossings by arcs whose slopes lie between the slopes of the arcs involved and form the left cusps in a similar fashion (Figure 34). It is easy to see the Lagrangian projection of these constant slope lines. With a moments thought the crossings and cusps project as indicated in the statement of the theorem. For full details see [69]. □

With this theorem in mind it is simple to translate the elements used in the DGA into the front projection.

The algebra. Let $\mathcal{C}$ be the double points and right cusp points in the projection. Let $\mathcal{A}$ be the free associative unital algebra over $\mathbb{Z}_2$ generated by $\mathcal{C}$. (One can of course translate the algebra over $\mathbb{Z}[t, t^{-1}]$ too, but we leave this to the reader, or see [38].)

The grading. Each right cusp point has grading 1. To each double point c choose a capping path γ_c as we did in Section 4.1 for double points in the Lagrangian projection.

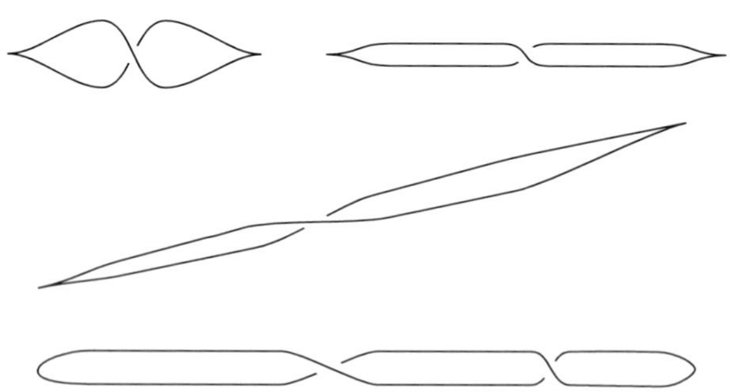

Fig. 34. Top left: the front projection of a Legendrian unknot. Top right: the stretched projection. Middle: the tilted projection that is easy to convert into the Lagrangian projection (bottom).

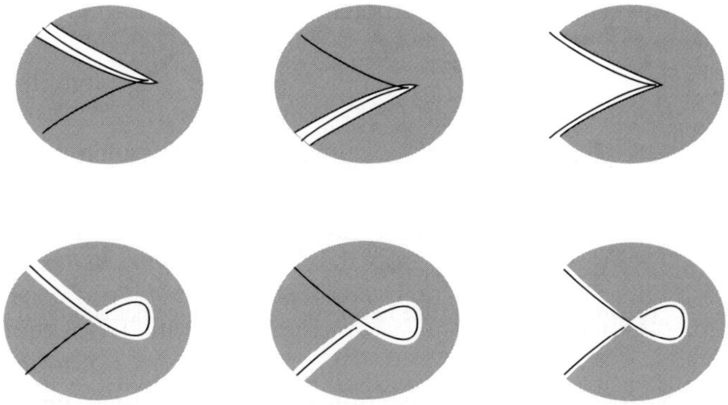

Fig. 35. The ways a disk can have a negative corner at a right cusp (top) and the corresponding disk in the Lagrangian projection (bottom). The gray is the image of the disk. The left two diagrams indicate a branch point at the cusp. In this case we count the cusp as a corner. The right hand picture indicates a corner at the cusp. In this case we count the "corner" as a double corner. That is we count it twice when reading off the work that goes into the differential.

The grading of the crossing is

$$|c| = D_c - U_c,$$

where D_c (respectively, U_c) is the number of cusps traversed downward (respectively upward) along γ_c.

The differential. Label the quadrants near a double point as in Figure 27. We will think of a right cusps as a corner (or quadrant) and put a $+$ there. Left cusps are not thought of as corners. The differential is exactly as in Section 4.1 for crossings, except that it may have a corner or branch point at a right vertex (see Figure 35). Branch points are counted as corners and a corner at a right cusp counts as two corners. The reason for this can be seen by transforming the front projection into the Lagrangian projection as described in Theorem 4.19. Moreover a polygon can pass by a right cusp as shown in Figure 36. The differential of a right cusp is 1 plus the differential defined above for a crossing.

One may easily use Theorem 4.19 to see that this DGA is exactly the one you would get using the Lagrangian projection of L.

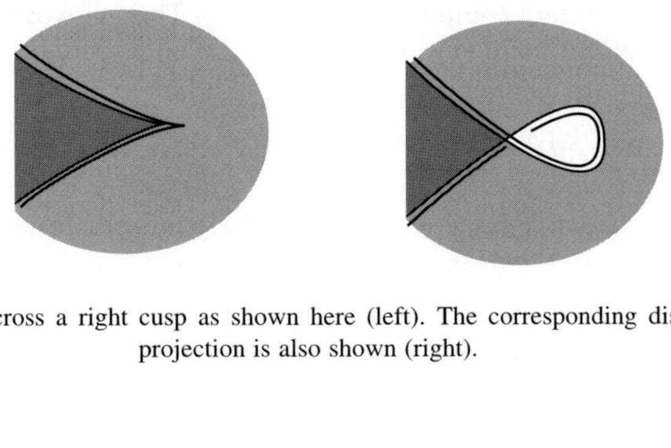

Fig. 36. A polygon can cross a right cusp as shown here (left). The corresponding disk in the Lagrangian projection is also shown (right).

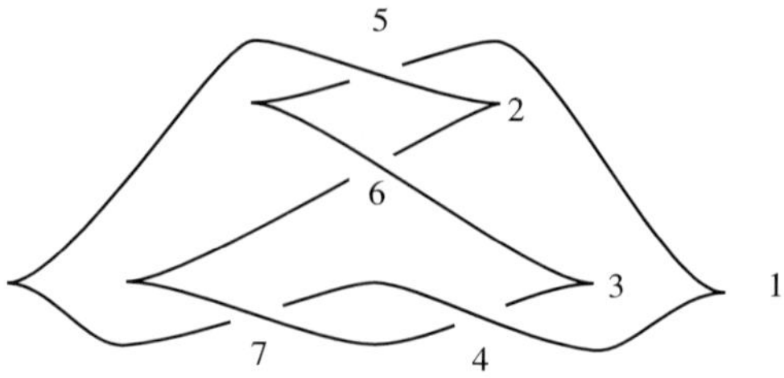

Fig. 37. The front projection of the Legendrian figure eight knot with tb $= -3$.

EXAMPLE 4.20. Consider the front projection of the Legendrian figure eight knot in Figure 37. There are seven generators for the algebra, three from cusps and four from crossings. Their gradings are $|a_i| = 1, i = 1, \dots 4$, $|a_i| = 0, i = 5, 6, 7$ and $|a_7| = -1$. The boundary map is

$$\partial a_1 = 1 + a_6a_6a_4a_6a_7 + (1 + a_6a_5)a_3a_6a_7 + a_6a_2(1 + a_6 + a_7a_4a_6)a_7,$$

$$\partial a_2 = 1 + a_5a_6, \quad \partial a_3 = 1 + a_6 + a_6a_7a_4, \quad \partial a_4 = \partial a_5 = \partial a_6 = \partial a_7 = 0.$$

REMARK 4.21. Note that many of the disks in the above computation are difficult to find. This is because of the right cusps. They allow disks to be quite complicated. To eliminate this problem one can use the Legendrian Reidemeister moves (Theorem 2.6) to move them all to the far right of the diagram. This will usually increase the number of crossings in the diagram, but the computation of the boundary map will usually be considerably easier.

4.6. *Decomposition invariants*

Here we define another new invariant of Legendrian knots. This invariant was also discovered by Chekanov [12,13] and can be used to solve Arnold's famous four cusp conjecture (see Section 7.3).

Let F be the front projection of a Legendrian knot L. Let $C(F)$ denote the cusp points in F and let $S(F)$ be the smooth components of $F \setminus C(F)$. These are called the set of "strands" in F. A *Maslov potential* for F is a map

$$\mu : S(F) \to \mathbb{Z}_{2r(L)}$$

whose value on the upper stand at a cusp is one larger than its value on the lower strand. Note the Maslov potential is well defined up to adding a constant. A double point in F is called *Maslov* if the value of the Maslov potential evaluated on the two strands crossing there are the same (see Figure 38).

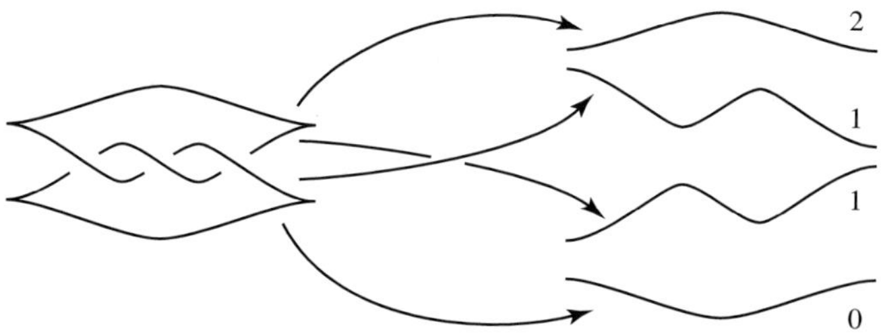

Fig. 38. The four strands associated (right) associated with the front (left). The numbers are a Maslov potential.

Now let $D(F)$ be the double points in F and let $A(F)$ be the closure of the components in $F \setminus (D(F) \cup C(F))$. We call $A(F)$ the arcs in F. A *decomposition* of F is the collection of simple closed curves $D_F = \{F_1 \dots F_k\}$ such that each F_i is the union of arcs in F, each arc is used in only one F_i and all arcs are in some F_i. A crossing in F is called *switching* or *non-switching* as indicated in Figure 39. A decomposition is called *admissible* if

(1) Each F_i bonds a disk B_i;
(2) The "vertical slice" $B_{i,x} = B_i \cap \{x = \text{constant}\}$ of each disk is an interval, a cusp point or empty;
(3) Near a switching crossing the intervals $B_{i,x}$ and $B_{j,x}$ are disjoint or one is contained within the other, where F_i and F_j are two components of the decomposition coming together at the crossing;
(4) Each switching crossing is Maslov.

Note Item (1) rules out trivial decompositions (except for the unknot with $tb = -1$). Item (2) rules out the left two decompositions shown in Figure 40. (In fact, a stabilization in a diagram will prevent it from having any admissible decompositions because of Item (2).) Item (3) rules out things like that shown in Figure 40.

Let $\text{Adm}(F)$ be the set of admissible decompositions of F. Define the function

$$\theta : \text{Adm}(F) \rightarrow \mathbb{Z}$$

by $\theta(D_F) = \#(D_F) - S(D_F)$ where $S(D_F)$ is the number of switching crossings that occur in the decomposition D_F.

THEOREM 4.22. (Chekanov and Pushkar, [13]). *Let L, L' be two Legendrian knots in $\mathbb{R}^3$ with generic front projections F, F'. If L and L' are Legendrian isotopic then there is a one-to-one correspondence* $f : \text{Adm}(F) \rightarrow \text{Adm}(F')$ *such that* $\theta(f(D_F)) = \theta(D_F)$.

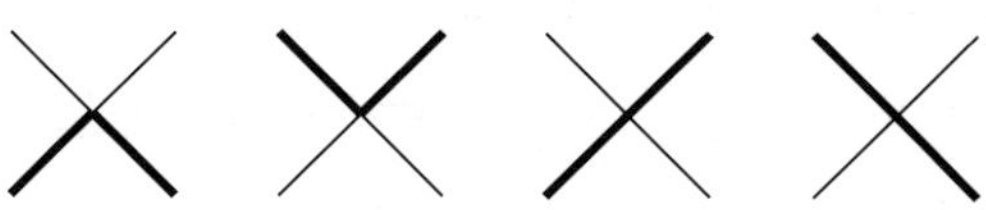

Fig. 39. The left two figures are switching crossings the right two are non-switching crossings. (Different line weights represent different components in the decomposition.)

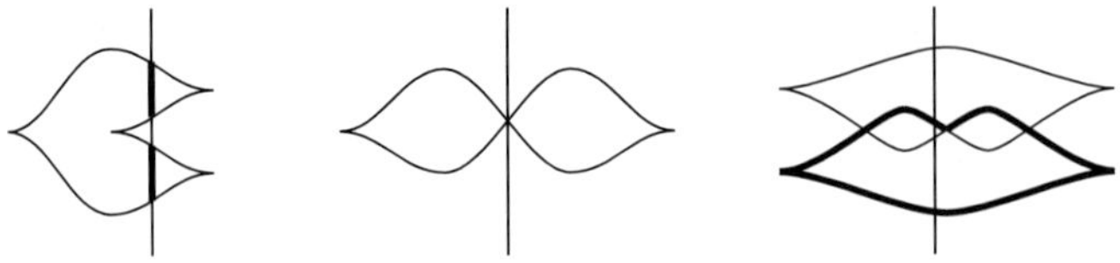

Fig. 40. Left two figures are not admissible by Item (2). The right figure is not admissible by Item (3).

Thus we have two new invariants of Legendrian knots! The cardinality of $\mathrm{Adm}(\Pi(L))$ and the map θ.

EXAMPLE 4.23. We can once again distinguish the Chekanov–Eliashberg knots. Here we draw the Chekanov examples somewhat differently (see Figure 41). In this projection of L one may easily see in any admissible decomposition crossings 2, 3, 4, 5 must be switching crossings (by property (1) and (2)). Moreover, crossings 1 and 6 cannot be switching since they are not Maslov. Thus one sees that this projection of L has $|\mathrm{Adm}(\Pi(L))| = 1$. Similarly in the projection of L' crossings 2, 3, 4, 5 must be switching and either crossings 1, 6 are both switching or both non-switching. Thus $|\mathrm{Adm}(\Pi(L'))| = 2$.

It is interesting to note that, just as the contact homology did, this invariant also vanishes on stabilized Legendrian knots since there are no admissible decompositions of a stabilized knot diagram.

The existence of admissible decompositions has an interesting relation to augmentations of the Chekanov–Eliashberg DGA. In particular Fuchs has shown the following theorem.

THEOREM 4.24. (Fuchs, [43]). *If the front projection of a Legendrian knot has an admissible decomposition then its DGA has an augmentation.*

Very recently Fuchs and Ishkhanov, and independently Sabloff, have found a converse to the above theorem.

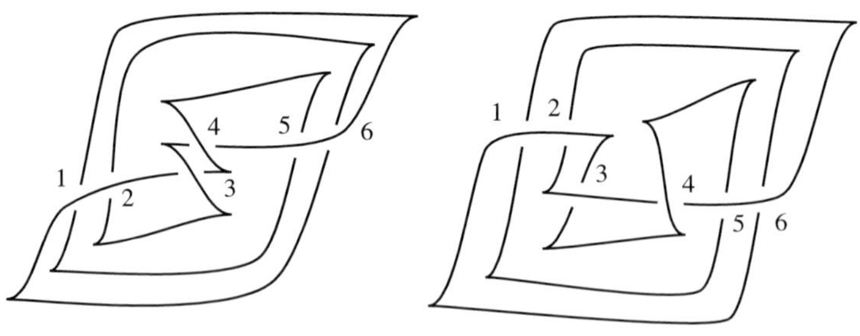

Fig. 41. Legendrian projections of the two "Chekanov knots" from Figure 29. L is on the left and L' is on the right.

THEOREM 4.25. (Fuchs and Ishkhanov [44]; Sabloff [74]). *If the DGA of a Legendrian knot L has a (graded) augmentation, then any front projection of L has a ruling.*

An interesting corollary of this result is the following.

COROLLARY 4.26. *If the DGA of a Legendrian knot L has a (graded) augmentation then its rotation number is zero.*

5. Classification results

In considering the classification of Legendrian knots it is convenient to consider the map

$$\Psi : \widetilde{\mathcal{L}} \to \widetilde{\mathcal{K}} \times \mathbb{Z} \times \mathbb{Z} : (L) \mapsto (k(L), tb(L), r(L)),$$

where $\widetilde{\mathcal{L}}$ is the space of *all* oriented Legendrian knots and $\mathcal{K}$ is the space of *all* oriented topological knots and $k(L)$ is the topological knot type underlying L. The main questions one would like to answer are: What is the image of Ψ? and is Ψ injective? The answer to the second question is NO as the Chekanov–Eliashberg examples in Figure 29 show. See also [41] for the first such examples.

We can refine the above questions by considering the map Ψ restricted to $\mathcal{L}(\mathcal{K})$ for some knot type $\mathcal{K}$. This map is denoted $\Psi_{\mathcal{K}}$. A knot type $\mathcal{K}$ is called *Legendrian simple* if $\Psi_{\mathcal{K}}$ is injective. So for Legendrian simple knot types a classification of Legendrian knots amounts to the identification of the range of $\Psi_{\mathcal{K}}$.

One can also consider the map

$$\Phi : \widetilde{\mathcal{L}} \to \widetilde{\mathcal{T}} \times \mathbb{Z},$$

where $\widetilde{\mathcal{T}}$ is the space of all transverse knots and $\Phi(L) = (L_+, tb(L))$. Recall L_+ is the (positive) transverse push-off of L. Note we did not include the rotation number in the range of Φ since $r(L) = tb(L) - \mathrm{sl}(L_+)$. This map is potentially a better invariant than Ψ. It turns out this map is not injective either.

EXAMPLE 5.1. Let $E(k,l)$ be the Legendrian knots shown in Figure 42. Note the topological knots underlying $E(k,l)$ are determined by $k+l$. In [30] it was shown that $E(k,l)$ is Legendrian isotopic to $E(k',l')$ if and only if the unordered pairs $\{k,l\}\{k',l'\}$ are the same. This proof involves computations of the linearized contact homology. However if l is odd then the transverse push-offs $E(k,l)_+$ and $E(k-1,l+1)_+$ are transversely isotopic.

Finally we note that all the new invariants described in Section 4 vanish on stabilized knots thus cannot distinguish non-Legendrian isotopic knots if the knots are stabilized. In Section 5.4 we will show that there are stabilized Legendrian knots that are not Legendrian isotopic. Thus contact homology and the decomposition invariants are not complete invariants.

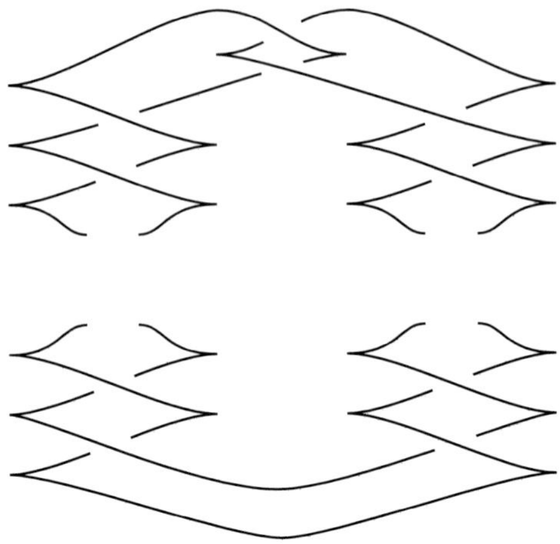

Fig. 42. The Legendrian knot $E(k, l)$. There are l crossings on the right hand side and K on the left.

A standard procedure for trying to classify Legendrian knots in a knot type $\mathcal{K}$ is to show

- All Legendrian knots in $\mathcal{L}(\mathcal{K})$ can be destabilized except those with maximal Thurston–Bennequin invariant.
- Classify Legendrian knots in $\mathcal{L}(\mathcal{K})$ with maximal Thurston–Bennequin invariant.
- Understand when Legendrian knots become the same under stabilization.

Recall, a Legendrian knot L destabilizes if there is a Legendrian knot L' such that $L = S_{\pm}(L')$. This procedure does not always work (see the sections on connected sums and cablings) but it is frequently a useful strategy.

5.1. *The unknot*

The main theorem concerning unknots is the following.

THEOREM 5.2. (Eliashberg and Fraser 1995, [28]). *In any tight contact three manifold, Legendrian unknots are determined by their Thurston–Bennequin invariant and rotation number. All Legendrian unknots are stabilizations of the unique one with $tb = -1$ and $r = 0$ (see Figure 4).*

The bound on the Thurston–Bennequin invariant follows from Bennequin's inequality (3.2). The theorem then follows from two lemmas.

LEMMA 5.3. *Any Legendrian unknot with $tb < -1$ destabilizes.*

PROOF. Let D be a disk that L bounds. If $tb(L) < -1$ then we may make D convex and it will have at least two dividing curves by (2.15). Since D is a disk one of these dividing curves must be boundary parallel (recall this means that it separates off a disk containing no other dividing curves). As discussed in Section 2.8 we may use this dividing curve to destabilize L. □

LEMMA 5.4. *There is a unique Legendrian unknot with $tb = -1$.*

PROOF. Let L_1 and L_2 be two Legendrian unknots with $tb = -1$. Let D_i be a disk that L_i bounds. We can make D_i convex. Using the Legendrian Realization Principle [48,51] we can arrange that the characteristic foliation on D_i has two elliptic point on ∂D_i and the foliation on the rest of D_i is by arcs. Thus we may Legendrian isotop L_i to an arbitrarily small neighborhood of an arc A_i. One may easily show that any two Legendrian arcs are Legendrian isotopic. Thus we can take $A_1 = A_2$. It is now an exercise that any two "thickenings" of $A = A_i$ to a Legendrian unknot are Legendrian isotopic. □

5.2. *Torus knots*

Let T be the boundary of a neighborhood of an unknot in M^3. A knot that can be isotoped to sit on T is called a torus knot.

THEOREM 5.5. (Etnyre and Honda 2001, [36]). *In any tight contact three manifold, Legendrian torus knots are determined up to Legendrian isotopy by their knot type, Thurston–Bennequin invariant and rotation number.*

To complete the classification of Legendrian torus knots we need to identify the invariants that are realized. For this we need to more carefully describe torus knots. Let V be an embedded unknoted solid torus in M. Let μ be the meridian to V and λ be the longitude. Now all torus knots can be isotoped onto ∂V so they can be expressed by $p\mu + q\lambda$ where $|p| > q > 0$ and $(p, q) = 1$. If $p > 0$ then the torus knot is called *positive* otherwise it is called *negative*.

THEOREM 5.6. (Etnyre and Honda 2001, [36]). *All Legendrian positive (p, q)-torus knots are stabilizations of the unique one with $tb = pq - p - q$ and $r = 0$ (see Figure 43).*

THEOREM 5.7. (Etnyre and Honda 2001, [36]). *All Legendrian negative (p, q)-torus knots are stabilizations of one with maximal tb which is equal to pq. Moreover, (p, q)-torus knots with maximal tb are classified by their rotation number and the set of realized rotation number is*

$$\left\{\pm(p + q + n2q) | 0 \le n \le \frac{2(p + q)}{q}\right\},$$

see Figure 44.

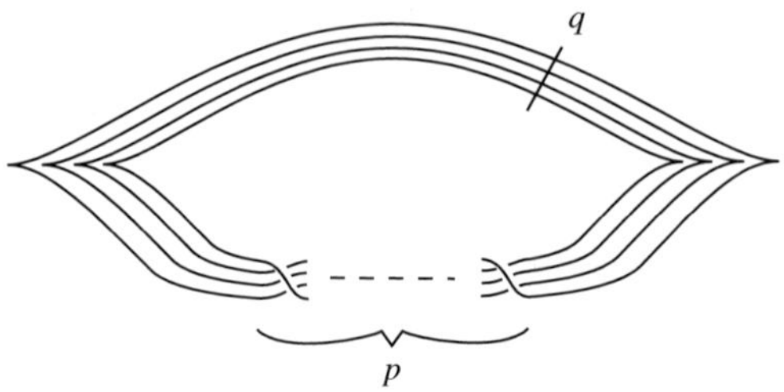

Fig. 43. Positive (p, q)-torus knot.

In Figure 45 we see the possible tb and r of Legendrian $(-9, 4)$-torus knots. The points indicate a unique Legendrian knot with corresponding invariants. The edges indicate either positive or negative stabilizations (depending on whether the edge has a positive or negative slope). This picture will be called the *mountain range* associated to the knot type. We have not discussed mountain ranges up till now because all previous knot types had a particularly simple mountain range.

Proof of Theorem 5.5. We restrict our attention to (S^3, ξ_{std}) for convenience. Let T be a standardly embedded convex torus in (S^3, ξ_{std}) with two dividing curves. We will need two facts [49,51] concerning this situation:

(1) The slope s of the dividing curves Γ_T on T is less than 0.
(2) Any other slope in $(-\infty, 0)$ can be realized by a standardly embedded convex torus parallel to (and disjoint from) T with two dividing curves. Moreover, slopes larger than s are realized by tori on one side of T and slopes smaller than s are realized on the other.

We also need the following topological fact: the framing on the normal bundle of a (p, q)-torus knot given by its Seifert surface differs from the one induced by the standard torus T on which it sits by pq.

Now consider a Legendrian positive (p, q)-torus knot L. The Bennequin inequality says that $tb(L) \leq pq - p - q$. Thus $tw(L, T) \leq -p - q < 0$, so we can make T convex without moving L. Throughout this section we will assume T has only two dividing curves. It is easy to deal with the more general case but this is left to the reader. Suppose the dividing

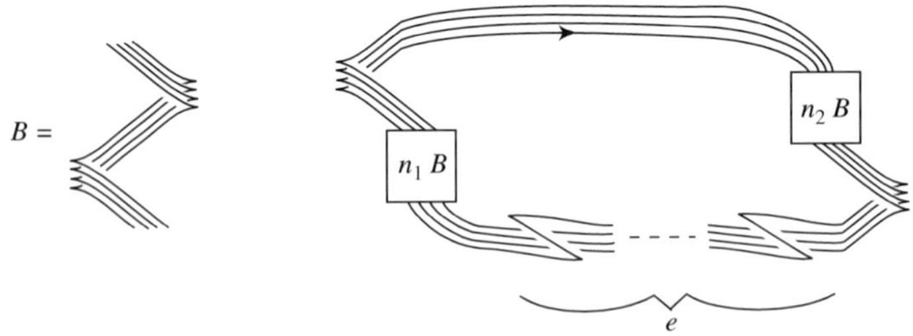

Fig. 44. Negative $(-p, q)$-torus knot. Here $p = (n_1 + n_2 + 1)q + e$. So for each such pair of positive integers n_1, n_2 there is a maximal tb, $(-p, q)$-torus knot.

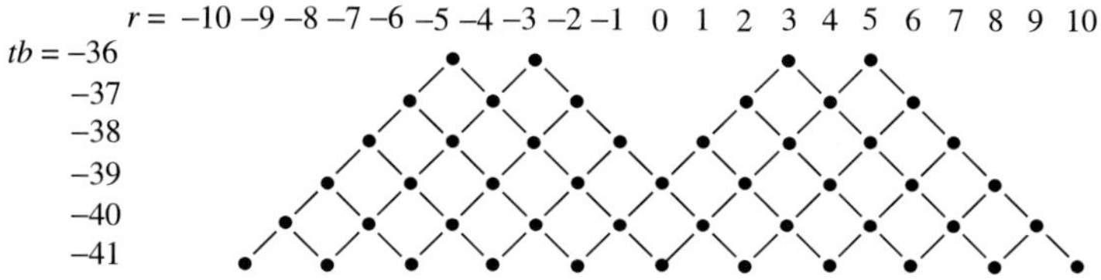

Fig. 45. Possible *tb* and *r* for the $(-9, 4)$-torus knot.

curves on T have slope $-n/m$, then $tw(L,T) = -(pm + qn)$. For all positive, relatively prime n, m this number is always less than or equal to $-(p+q)$ with equality only when $m = n = 1$. Thus if $tw(L,T) < -(p+q)$ (i.e., $tb(L) < pq - p - q$) then there is a convex torus T' whose dividing curves have slope -1 and cobounds with T a toric annulus $N = T^2 \times [0,1]$. Let $A = L \times [0,1]$ in N, here $L = L \times \{0\}$. We can assume that ∂A is Legendrian and A is convex. All the dividing curves on A are properly embedded arcs and they intersect $L \times \{0\} \subset \partial A$, $2(pm+qn)$ times and $L \times \{1\} \subset \partial A$, $2(p+q)$ times. From this one may easily conclude that there is a dividing curve on A that has boundary on $L = L \times \{0\}$ and separates off a disk from A. Thus as discussed in Section 2.8 we can use this dividing curve to destabilize L.

Using almost the same argument it is easy to show that a negative (p, q)-torus knot destabilized to pq. This is left as an exercise, or see [36]. The fact that this is the maximal *tb* for such a Legendrian knot is more difficult to establish. In particular, when q is odd it does not follow from any of the inequalities in Section 3 (see [43]). □

PROOF OF THEOREM 5.6. Let L_1 and L_2 be two Legendrian positive (p, q)-torus knots with maximal *tb*. From the proof of the previous theorem we know that L_i sits on a convex torus T_i with two dividing curves of slope -1, for $i = 1, 2$. Let V_i and V'_i be the two solid tori into which T_i breaks S^3. Using the Legendrian realization principle we may isotop T_i relative to L_i so that the characteristic foliation on T_1 is the same as the one on T_2. Let ϕ be a diffeomorphism T_1 to T_2 that preserves the characteristic foliation and takes L_1 to L_2. By the classification of contact structures on solid tori we know ϕ extends, as a contactomorphism, over V_1 and V'_1. Thus ϕ is a contactomorphism of S^3 to itself that takes L_1 to L_2. Applying Theorem 2.13 we see that L_1 and L_2 are Legendrian isotopic. □

The proof of Theorem 5.7 is similar to the above proof in spirit but somewhat more involved. We refer the reader to [36].

5.3. *Figure eight knot*

The main theorem concerning Legendrian figure eight knots is the following.

THEOREM 5.8. (Etnyre and Honda 2001, [36]). *In any tight contact three manifold, Legendrian figure eight knots are determined by their Thurston–Bennequin invariant and* *rotation number. All Legendrian figure eight knots are stabilizations of the unique one with $tb = -3$ and $r = 0$ see Figure 4.*

The bound on *tb* comes from the inequality in Theorem 3.4. Thus the following two lemmas establish the theorem.

LEMMA 5.9. *Any Legendrian figure eight knot with $tb < -3$ destabilizes.*

LEMMA 5.10. *There is a unique Legendrian figure eight knot with $tb = -3$.*

To prove any Legendrian figure eight knot L with $tb < -3$ destabilizes we consider a minimal genus Seifert surface Σ for L. So Σ is a punctured torus and we can make it convex. We would like to find a bypass for L on Σ so we can destabilize L. We might not always be able to do this. Thus to prove L destabilizes we need to isotop Σ until we do see a bypass for L on Σ. To accomplish this we use the fibration on the complement of the figure eight knot. Specifically, let $X = S^3 \backslash N$ where N is a standard neighborhood of a L. Then X is a Σ bundle over S^1. The monodromy for X is

$$\Psi = \begin{bmatrix} 2 & 1 \\ 1 & 1 \end{bmatrix}.$$

If we cut X open along Σ then we get $\Sigma \times [0, 1]$. Because of the monodromy map the dividing curves on $\Sigma \times \{0\}$ and $\Sigma \times \{1\}$ will be different. We can then take an annulus $A = S^1 \times [0, 1]$ where $S^1 \subset \Sigma$, make ∂A Legendrian and A convex. If we choose S^1 correctly we will be able to find a bypass on A for a boundary component of A. We can use this to alter the dividing curves on Σ, [51]. This will frequently produce a bypass on Σ for L. The rigorous argument is somewhat involved. The reader is referred to [36] for details.

To prove there is a unique Legendrian figure eight knot with $tb = -3$ we use the above ideas to show that the dividing curves on Σ in this situation can always be arranged to look a certain way and that there is a unique tight contact structure on X that is a subset of S^3 and has the given dividing curves. The uniqueness proof is then finished as in the proof of Theorem 5.6.

5.4. *Connected sums*

There is a standard way to take a contact connect sum of manifolds. If (M_i, ξ_i) are two contact three manifolds, then let B_i be a ball with convex boundary contained in a neighborhood of a point. The connect sum $M_1 \# M_2$ is obtained from $\overline{M_i \backslash B_i}$ by gluing their boundaries together by a contactomorphism. This construction preserves tightness [15]. If L_i is a Legendrian knot in M_i we can arrange that $L_i \cap B_i$ looks like the intersection of the x-axis with a unit ball about the origin in $(\mathbb{R}^3, \xi_{std})$. We can then define the *connect sum*

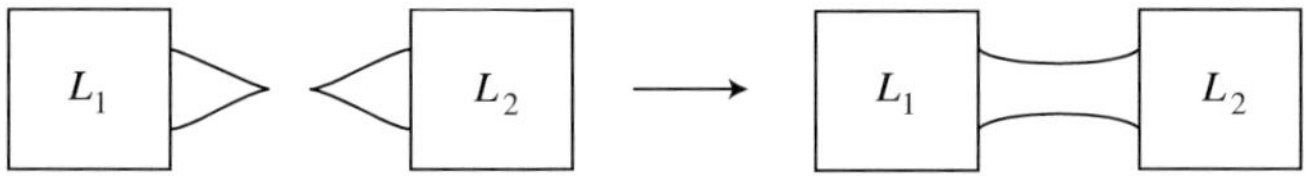

Fig. 46. The connect sum in the front projection.

$L_1\#L_2$ in $M_1\#M_2$ to be the closure of $L_1\backslash B_1 \cup L_2\backslash B_2$ (here we of course need to arrange that the gluing map for $M_1\#M_2$ sends $\partial(L_1 \cap B_1)$ to $\partial(L_2 \cap B_2)$).

In $(\mathbb{R}^3, \xi_{std})$ the connect sum has a diagrammatic interpretation in the front projection (note $\mathbb{R}^3\#\mathbb{R}^3$ is not $\mathbb{R}^3$ so we are somewhat abusing notation, but we may think of one of the $\mathbb{R}^3$'s as S^3 with its standard contact structure). This diagrammatic connect sum is shown in Figure 46. It is a good exercise to show that this is the same a the connect sum defined above and that it is well defined. Moreover using this diagrammatic interpretation it is easy to see

$$tb(L_1\#L_2) = tb(L_1) + tb(L_2) + 1$$

and

$$r(L_1\#L_2) = r(L_1) + r(L_2).$$

While all the theorems concerning the connect sum work with the more general definition we restrict attention here to the diagrammatic version in $\mathbb{R}^3$. The following theorem says that the prime decomposition of Legendrian knots is unique up to shifting stabilization from one summand to the other and topological symmetries.

THEOREM 5.11. (Etnyre and Honda 2003, [35]). *Let $\mathcal{K} = \mathcal{K}_1\#...\#\mathcal{K}_n$ be topological connected sum knot type in $\mathbb{R}^3$ with $\mathcal{K}_i$ prime. The map*

$$\frac{\mathcal{L}(\mathcal{K}_1) \times ... \times \mathcal{L}(\mathcal{K}_n)}{\sim} \to \mathcal{L}(\mathcal{K}_1\#...\#\mathcal{K}_n)$$

given by connect sum is a one-to-one correspondence where $\sim$ is generated by

(1) $(..., S_\pm(L_i), ..., L_j, ...) \sim (..., L_i, ..., S_\pm(L_j), ...)$ *and*
(2) $(L_1, ..., L_n) \sim (L_{\sigma(1)}, ..., L_{\sigma(n)})$ *where σ is permutation of* $1, ..., n$ *such that* $\mathcal{K}_i = \mathcal{K}_{\sigma(i)}$.

The idea for the proof of this theorem is quite simple. One just uses the spheres implicated in the definition of connect sum. Make these spheres convex, cut the manifold along the spheres and glue in standard contact balls with a standard Legendrian arc in them. One then analyses what happens if one decomposes using a different convex sphere. The details can be found in [35].

While the proof of this theorem is fairly straightforward it has some very interesting consequences.

COROLLARY 5.12. *Given any integers n and m there is a topological knot type $\mathcal{K}$ and Legendrian knots $L_1, \dots, L_n \in \mathcal{L}(\mathcal{K})$ with the same tb and r that are not Legendrian isotopic. Moreover, they stay non-isotopic even after m stabilizations of any type.*

These are the first examples that cannot be distinguished by the Chekanov–Eliashberg DGA or the decomposition invariants. In fact, there is no known invariant distinguishing many of them.

Recall Theorem 2.17 says that any two Legendrian knots in the same knot type become Legendrian isotopic after some number of positive and negative stabilizations. There are many theorems in topology concerning some type of equivalence under stabilization (e.g., for Heegaard splittings of three manifolds, smooth structures on 4 manifolds,...), but this is the first time one can show that more than one stabilization is required in the equivalence.

We indicate the proof of this corollary in the case $n = 2$ and $m = 0$. To this end consider the $(-5, 2)$ torus knots $\mathcal{K}$. In $\mathcal{L}(\mathcal{K})$ there are four Legendrian knots with maximal *tb* (this maximum is -10) having rotation numbers -3, -1, 1, 3. Denote them by K_r where r indicates their rotation number. Now set $L_1 = K_{-3}\#K_3$ and $L_2 = K_{-1}\#K_1$. Clearly both L_1 and L_2 have $tb = -19$ and $r = 0$. However they are not Legendrian isotopic since their summands are not related by the equivalence relation in Theorem 5.11. Using the fact that $S_+(K_{-3}) = S_-(K_{-1})$ and $S_-(K_3) = S_+(K_1)$ one may easily check that $S_+(L_1) = S_+(L_2)$ and $S_-(L_1) = S_-(L_2)$. Thus our examples only satisfy $m = 0$ in the corollary. The reason for this is that the valleys in the mountain range for $\mathcal{K}$ (see Section 5.2 for the terminology) are only one deep. Thus to find examples with $m > 0$ we just need to find torus knots having very deep valleys. This is a simple exercise, or see [35].

5.5. *Cables*

We now discuss Legendrian knots in cabled knot types. To do this we need some preliminary definitions. We say a solid torus S represents a knot type $\mathcal{K}$ if its core curve is in $\mathcal{K}$. Define the width of a knot type as

$$w(\mathcal{K}) = \sup \frac{1}{\text{slope}(\Gamma_{\partial S})}, \tag{5.1}$$

where the supremum is taken over all solid tori with convex boundary representing $\mathcal{K}$, and $\Gamma_{\partial S}$ are the dividing curves on the boundary of S. This is clearly an invariant of topological knots. Since the standard neighborhood of a Legendrian knot with $tb = n$ has dividing curves on its boundary with slope $1/n$ we can see that

$$\overline{tb}(\mathcal{K}) \leq w(\mathcal{K}) \leq \overline{tb}(\mathcal{K}) + 1.$$

A knot type $\mathcal{K}$ is called *uniformly thick* if

(1) $w(\mathcal{K}) = \overline{tb}(\mathcal{K})$ and
(2) any solid torus S representing $\mathcal{K}$ is contained in the interior of a solid torus that is the standard neighborhood for a Legendrian knot in $\mathcal{L}(\mathcal{K})$ with maximal *tb*.

For the unknot one may easily check that $w = 0$ while $\overline{tb} = -1$, thus the unknot is not uniformly thick. We will see below that there are many knot types that are uniformly thick, but first we indicate the importance of being uniformly thick.

Let γ be a (p, q) curve on ∂S, where $S = S^1 \times D^2$. Given a knot type K let S' be a solid torus representing $\mathcal{K}$ and choose a diffeomorphisms $S \to S'$ sending $S^1 \times \{pt\}$ to a longitude of S' and $\{pt\} \times \partial D^2$ to a meridian. Then the knot type $\mathcal{K}_{(p,q)}$ determined by γ under this diffeomorphism is called the *(p, q)-cable* of $\mathcal{K}$.

THEOREM 5.13. (Etnyre and Honda, [34]). *Let $\mathcal{K}$ be a Legendrian simple, uniformly thick knot type. Then the knot type $\mathcal{K}_{(p,q)}$ is also Legendrian simple.*

Thus we cannot say the cables of Legendrian simple knots are Legendrian simple, but this is true if the knot type is uniformly thick.

THEOREM 5.14. (Etnyre and Honda, [34]). *We have the following:*

(1) *Negative torus knots are uniformly thick.*

(2) *If $\mathcal{K}$ is uniformly thick and $(p/q) < w(\mathcal{K})$ then $\mathcal{K}_{(p,q)}$ is uniformly thick.*

(3) *If $\mathcal{K}_1$ and $\mathcal{K}_2$ are uniformly thick then $\mathcal{K}_1 \# \mathcal{K}_2$ is uniformly thick.*

The proofs of these last two theorems are in principle quite similar to the proofs in the proceeding sections, especially the section on torus knots.

There are knot types that are not uniformly thick. As mentioned above the unknot is not uniformly thick, but it acts somewhat as if it is. For example, its cables (torus knots) are Legendrian simple. It would appear that all positive torus knots are not Legendrian simple, but this seems difficult to prove in general. We have the following:

THEOREM 5.15. (Etnyre and Honda, [34]). *If $\mathcal{K}'$ is the $(2, 3)$-cable of the $(2, 3)$-torus knot, then $\mathcal{L}(\mathcal{K}')$ is classified as in Figure 47. This entails the following:*

(1) *There exist exactly two maximal Thurston–Bennequin representatives $K_\pm \in \mathcal{L}(\mathcal{K}')$. They satisfy $tb(K_\pm) = 6$, $r(K_\pm) = \pm 1$.*

(2) *There exist exactly two non-destabilizable representatives $L_\pm \in \mathcal{L}(\mathcal{K}')$ which have non-maximal Thurston–Bennequin invariant. They satisfy $tb(L_\pm) = 5$ and $r(L_\pm) = \pm 2$.*

(3) *Every $L \in \mathcal{L}(\mathcal{K}')$ is a stabilization of one of K_+, K_-, L_+, or L_-.*

(4) *$S_+(K_-) = S_-(K_+)$, $S_-(L_-) = S_-^2(K_-)$, and $S_+(L_+) = S_+^2(K_+)$.*

(5) *$S_+^k(L_-)$ is not (Legendrian) isotopic to $S_+^k S_-(K_-)$ and $S_-^k(L_+)$ is not isotopic to $S_-^k S_+(K_+)$, for all positive integers k. Also, $S_-^2(L_-)$ is not isotopic to $S_+^2(L_+)$.*

This example is quite interesting since it is the first known example where there are Legendrian knots which do not have the maximal tb for the knot type and yet they do not destabilize. Moreover, this is the first knot type where there are Legendrian representatives with the same invariants that do not become Legendrian isotopic after some number of

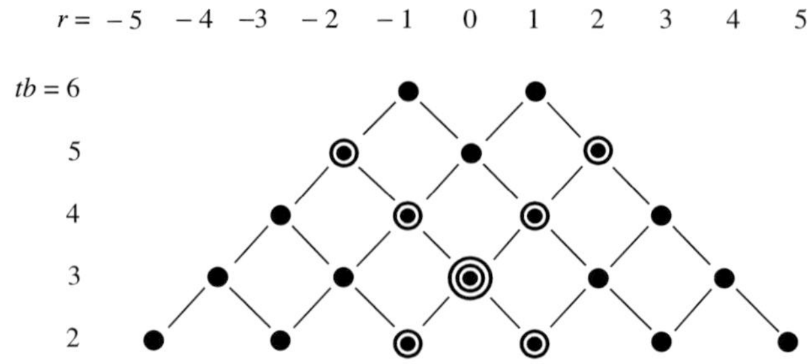

Fig. 47. Classification of Legendrian (2,3) -cables of (2,3)-torus knots. Concentric circles indicate multiplicities, i.e., the number of distinct isotopy classes with a given r and tb.

stabilizations of a fixed sign. See the section on transverse knots for more discussion of this knot type.

5.6. *Links*

There has not been much study of Legendrian links. One of the first result about links appeared in [63] and addressed the question: given a link type is there a Legendrian realization such that each component has maximal Thurston–Bennequin invariant in its knot type? The answer is NO. For a link $L = \cup L_i$ define its Thurston–Bennequin invariant to be the sum of the Thurston–Bennequin invariants of its components $tb(L) = \Sigma tb(L_i)$. Then the inequality for the Thurston–Bennequin invariant involving the HOMFLY polynomial in Theorem 3.4 still holds. For the Borromean rings B and the Whitehead link W a simple computations of the HOMFLY polynomial yield

$$tb(B) \leq -4$$

and

$$tb(W) \leq -5.$$

Since B had three components and W has two we clearly cannot realize them with components consisting of unknots with maximal Thurston–Bennequin invariant. These bounds are realized, see Figure 48. Also in this figure it is shown that the mirror of the Whitehead link can be realized by Legendrian unknots of $tb = -1$, once again illustrating the "cirality" of Legendrian knots. In general, very little is known concerning realizing Legendrian links with given Legendrian knot components and the restrictions on the Thurston–Bennequin invariants.

The N-copy of a Legendrian knot has also been studied. Given a Legendrian knot L in $(\mathbb{R}^3, \xi_{std})$, let L_N be the Legendrian link obtained by taking N copies of L each shifted slightly in the z direction. See Figure 49. The link L_N is a Legendrian link in the topological link type of the $(Ntb(L), N)$-cable of L. If L had maximal tb for Legendrian knots in its knot

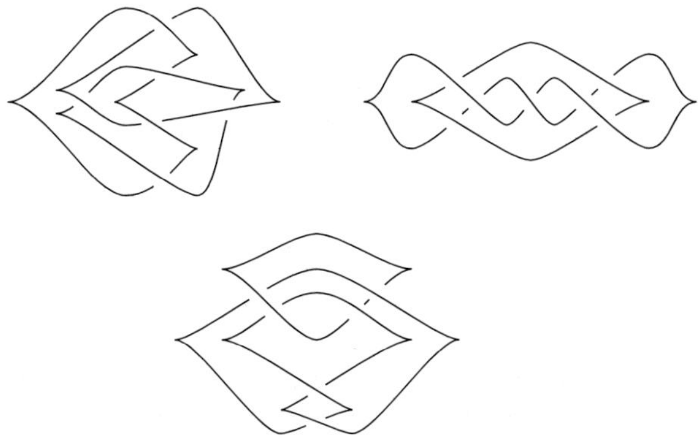

Fig. 48. Legendrian Borromean rings with $tb = -4$ (top left), Legendrian Whitehead link with $tb = -5$ (top right) and Legendrian Whitehead mirror realized by two $tb = -1$ unknots (bottom).

type then L_N is a Legendrian realization of its link type with each component having maximal *tb* in its knot type.

Topologically one can always realize any arbitrary permutation of the components of this link via isotopy. Can this be done for the components of the Legendrian N-copy. We have the following result.

THEOREM 5.16. (Michatchev 2001, [62]). *If L is a Legendrian unknot, then only cyclic permutations of the components of L_N are possible via Legendrian isotopy.*

One may easily check that cyclic permutations are possible. To show other permutations are not possible one must use an enhanced version of contact homology see [62].

One can study other cables as well. The following result follows easily from [36].

THEOREM 5.17. *Let $p > q > 0$ be relatively prime. There is a unique Legendrian link with maximal Thurston–Bennequin invariant in the (np, nq)-torus link type. Moreover, each component of the link has maximal tb in its knot type and any permutation of its components can be realized by a Legendrian isotopy.*

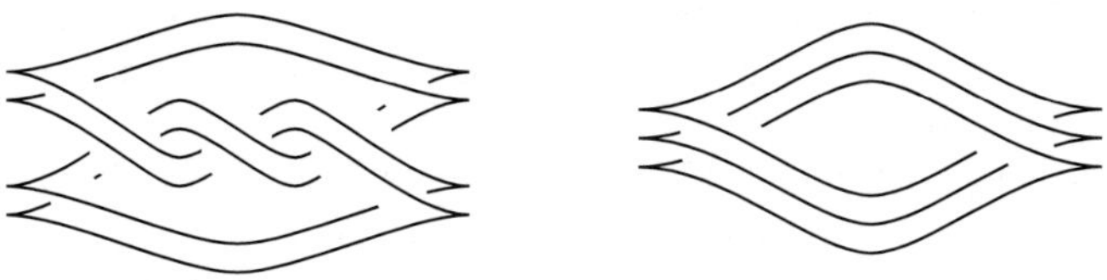

Fig. 49. A 2-copy of the right handed trefoil (left) and a 3-copy of the unknot (right).

5.7. *The homotopy type of the space of Legendrian knots*

Recall $\mathcal{K}$ denotes a topological knot type. That is $\mathcal{K}$ is a space of topological knots isotopic to a fixed knot. Also recall $\mathcal{L}(\mathcal{K})$ denotes the set of all Legendrian knots in $\mathcal{K}$. All the classification results above can be thought of in terms of identifying the kernel of the map

$$i_* : \pi_0(\mathcal{L}(\mathcal{K})) \to \pi_0(\mathcal{K}),$$

where $i : \mathcal{L}(\mathcal{K}) \to \mathcal{K}$ is the natural inclusion map. In general, studying the homotopy groups of $\mathcal{L}(\mathcal{K})$ is an interesting problem. To this end, fix a component $\mathcal{L}$ of $\mathcal{L}(\mathcal{K})$ and consider the inclusion $i : \mathcal{L} \to \mathcal{K}$. Recently Kálmán [54,55] has shown the induced map on π_1 need not be injective. Specifically, if $\mathcal{L}$ is the set of maximal Thurston–Bennequin invariant, Legendrian right handed trefoil knots then consider the loop Ω in $\mathcal{L}$ illustrated in Figure 50.

THEOREM 5.18. (Kálmán [54,55]). *The loop Ω^6 is non-trivial in $\pi_1(\mathcal{L})$ but $i(\Omega^6)$ is trivial in $\pi_1(K)$.*

The key ingredient in proving this theorem is the following result.

THEOREM 5.19. (Kálmán [54]). *If L is a fixed (generic) Legendrian knot in $\mathcal{L}$ then there is a multiplicative homomorphism*

$$\mu : \pi_1(\mathcal{L}, L) \to Aut(CH(L))$$

defined by continuation on the Chekanov–Eliashberg contact homology.

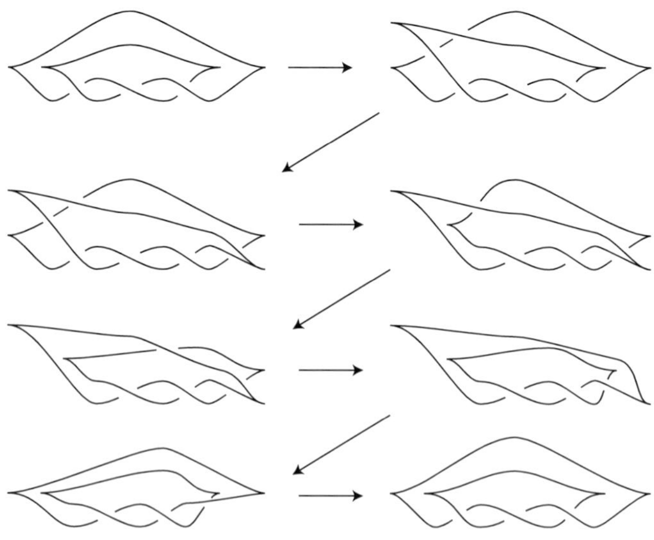

Fig. 50. A loop of Legendrian trefoils.

Now showing that Ω^6 is a non-trivial loop amounts to a computation that μ of this loop is non-trivial. This is carried out in [54]. Moreover, in [55] non-trivial loops are found for all maximal Thurston–Bennequin invariant positive torus knots.

5.8. *Transverse knots*

Recall a knot type K is transversely simple if transverse knots in the knot type are determined by their self-linking numbers. By Theorem 2.23 we know that the negative stable classification of Legendrian knots implies the transverse classification. Thus from the results above we know the following knot types are transversely simple

(1) the unknot ([26]),
(2) torus knots (positive torus knots in [31], all torus knots in [36,60,61]), and
(3) the figure eight knot ([36]).

The theorem on Legendrian connect sum implies that the connect sum of Legendrian simple knot types need not be Legendrian simple, but this is not the case for transverse knots.

THEOREM 5.20. (Etnyre and Honda 2003, [35]). *If $\mathcal{K}_1$ and $\mathcal{K}_2$ are transversely simple then $\mathcal{K}_1 \# \mathcal{K}_2$ is also transversely simple.*

This theorem easily follows from Theorem 5.11 on Legendrian connect sums and the observation if a knot type is transversely simple then its $tb - r$ mountain range (see Figure 45) is connected via negative stabilizations. The proof is an easy exercise or see [35].

Once again there is an interesting relation between transverse knots and braids. We call a knot type $\mathcal{K}$ *exchange reducible* if any braid representing $\mathcal{K}$ can be reduced to a unique minimal braid index braid for $\mathcal{K}$ by a sequence of braid destabilizations and exchange moves. An exchange move is shown in Figure 51.

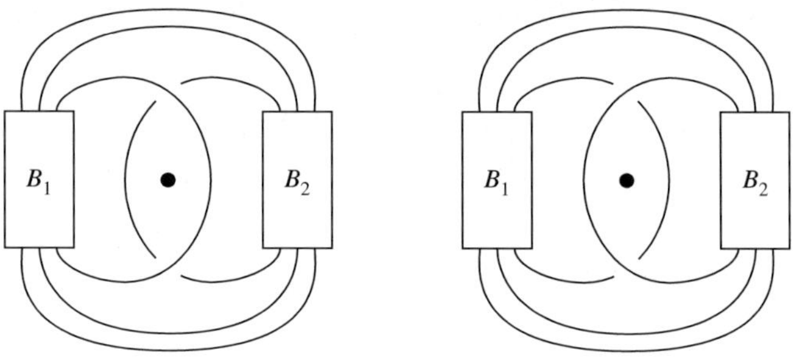

Fig. 51. Exchange move.

THEOREM 5.21. (Birman and Wrinkle 2000, [10]). *If $\mathcal{K}$ is exchange reducible then it is transversely simple.*

To prove this theorem one needs to observe that there are two types of braid destabilizations. A positive one and a negative one. The positive one can be realized by a transverse isotopy and the negative one corresponds to a transverse destabilization of the transverse knot. Moreover, an exchange move can also be done by a transverse isotopy of the knot. Thus being exchange reducible implies that any transverse knot destabilizes to a unique maximal self-linking number transverse knot and thus the knot type is transversely simple. One can easily reprove that the unknot is transversely simple with this theorem [9], see also [60,61] for a discussion of torus knots and cables from this perspective.

The search for non-transversely simple knot types was long and difficult, due in part to the lack of non-classical invariants for transverse knots. But we now know there are non-transversely simple knot types.

THEOREM 5.22. (Birman and Menasco, [7]). *The transverse knots represented by the closures of the braids $\sigma_1^3\sigma_2^4\sigma_1^2\sigma_2^{-1}$ and $\sigma_1^3\sigma_2^{-1}\sigma_1^2\sigma_2^4$ are not transversely isotopic, but have the same self-linking number.*

Many other examples arise from Birman and Menasco's analysis, see [7,8], but they are all three braids. They prove the knots are not transversely isotopic by classifying all closed three braids and understanding how to pass from one three braid to another representing the same knot type. (They can also show that there must exist non-transversely simple knots types that are not three braids, but the specific knot types cannot identified at this time.)

The classification of Legendrian knots in the knot type $\mathcal{K}$ of the (2, 3)-cable of the (2, 3)-torus knot in Theorem 5.15 implies that $\mathcal{K}$ is not transversely simple. In particular we get the first classification of transverse knots in a non-transversely simple knot type.

THEOREM 5.23. (Etnyre and Honda, [34]). *Let $\mathcal{K}$ be the $(2,3)$-cable of the $(2,3)$-torus knot. Each odd integer less than or equal to seven is the self-linking number of a transverse knot in $\mathcal{T}(\mathcal{K})$. Moreover, the self-linking number determines the transverse knot if $sl \neq 3$ and there are precisely two transverse knots with $sl = 3$.*

One should be able to construct many more such examples of cables of positive torus knot using the ideas from [34].

5.9. *Knots in overtwisted contact structures*

Overtwisted contact structures have been classified by Eliashberg in [23] and are determined by their homotopy type of plane field. Since [23] overtwisted structures have not been studied much. In particular, Legendrian knots in overtwisted contact structures have been somewhat ignored. There are however a few things we do know.

If L is a Legendrian knot in an overtwisted contact manifold (M, ξ) then call L *loose* if the contact structure on $M \backslash L$ is overtwisted. While it has been claimed by several authors that if a Legendrian knot L violates the Bennequin inequality then it is loose, this is not actually true as can be seen from the example below (this was first observed in [18]). Concerning the classification of loose knots we have the following theorem.

THEOREM 5.24. (Eliashberg and Fraser 1995, [28] and Dymara 2001, [19]). *Fix an overtwisted contact structure on S^3, then two loose Legendrian knots in the same knot type and having the same tb and r are contactomorphic. Moreover, if there is a fixed overtwisted disk in the complement of both Legendrian knots then they are Legendrian isotopic.*

The proof of this theorem relies on the classification of overtwisted contact structures in [23]. In particular, tb and r determine the homotopy type of the contact structure on the complement of L. Thus two loose knots with the same invariants are related by a contactomorphism of the ambient contact manifold. One gets a global contact isotopy if there is a fixed overtwisted disk in the complement of both Legendrian knot.

In [28] it was asked if non-loose knots exist in overtwisted contact structures. In fact they do, as shown in [19]. The example given there goes as follows. The standard tight contact structure on S^3 can be thought to be the planes orthogonal (in the standard round metric) to the Hopf fibration. Now let ξ be the contact structure obtained by a "half-Lutz twist." To understand this write the Hopf fibration as $\pi : S^3 \to S^2$. Let $A \subset S^2$ be an embedded annulus $A = S^1 \times [0, 1]$. Let $N = \pi^{-1}(A)$. Thus $N = A \times S^1$ where the S^1's are the Hopf fibers. The standard contact structure intersects the tori $T_t = S^1 \times \{t\} \times S^1$ in a linear foliation. On N, alter the contact structure by adding an extra half twist, that is, leave the contact structure fixed at the boundary of N but as t moves from 0 to 1 make the slopes of the foliations induced on the T_t change more rapidly so that the angles swept out by the foliations are π more that then angles swept out by the standard contact structure. Call this new contact structure ξ. Note there will be a t such that the characteristic foliation on T_t is parallel to the fibers in the Hopf fibration. Let γ be a leaf in this characteristic foliation. In [19] it was shown that the overtwisted contact structure ξ is tight when restricted to $S^3 \backslash \gamma$. Intuitively this is because γ intersects all the obvious overtwisted disks. To prove the complement is tight one needs to examine the universal cover, $\mathbb{R}^3$, of the complement and prove that the contact structure pulls back to the standard contact structure on $\mathbb{R}^3$. Note γ is a Legendrian unknot and one may easily compute (in a manner similar to the computation for torus knots) that $tb = 1$ and $r = 0$. It is easy to find another Legendrian γ' with $tb = 1$ and $r = 0$ so that the complement is still overtwisted (just take the connected sum the boundaries of two disjoint overtwisted disks). Now γ and γ' cannot be Legendrian isotopic since their complements are not contactomorphic.

From this example it seems the classification of Legendrian unknots in (S^3, ξ) is more complicated than the classification of Legendrian unknots in tight contact structures. It would be quite interesting (and probably not too hard) to completely classify Legendrian unknots in this contact structure, or more generally all overtwisted contact structures on S^3. It seems studying Legendrian unknots in overtwisted contact structures might be very interesting. But this must wait for future work.

6. Higher dimensions

A contact structure on a $2n + 1$ dimensional manifold M is a hyperplane field ξ that can be locally given as the kernel of a 1-form α such that $\alpha \wedge (d\alpha)^n \neq 0$. As in dimension three one should think of such a plane field as "maximally non-integrable". The standard contact structure ξ on $\mathbb{R}^{2n+1}$ is given as the kernel of

$$\alpha = \mathrm{d}z - \sum_{j=1}^{n} y_j \mathrm{d}x_j, \tag{6.1}$$

where $x_1, y_1, \dots, x_n, y_n, z$ are Euclidean coordinates on $\mathbb{R}^{2n+1}$. In this section we restrict attention to $(\mathbb{R}^{2n+1}, \xi)$. A *Legendrian knot* (or Legendrian submanifold) of $\mathbb{R}^{2n+1}$ is an n dimensional submanifold $L \subset \mathbb{R}^{2m+1}$ that is tangent to ξ at each point. We will frequently use L to denote the Legendrian submanifold and the domain of a Legendrian embedding. We also recall that the standard symplectic structure on $\mathbb{C}^n$ is given by

$$\omega = \sum_{j=1}^{n} \mathrm{d}x_j \wedge \mathrm{d}y_j, \tag{6.2}$$

and a Lagrangian submanifold is an n dimensional submanifold $L \subset \mathbb{C}^n$ for which $\omega(v_1, v_2) = 0$ for all vectors $v_1, v_2 \in TL$.

6.1. *Legendrian knots in* $\mathbb{R}^{2n+1}$

As in 3 dimensions there are two standard projections that are useful in studying Legendrian knots. The first is the Lagrangian projection (sometimes called the complex projection) which projects out the z coordinate:

$$\pi : \mathbb{R}^{2n+1} \to \mathbb{C}^n : (x_1, y_1, \dots, x_n, y_n, z) \mapsto (x_1, y_1, \dots, x_n, y_n). \tag{6.3}$$

The second is called the *front projection* and it projects out the y_j's:

$$\Pi : \mathbb{R}^{2n+1} \to \mathbb{R}^{n+1} : (x_1, y_1, \dots, x_n, y_n, z) \mapsto (x_1, \dots, x_n, z). \tag{6.4}$$

We begin with the Lagrangian projection. If L is an embedded Legendrian knot then $\pi(L)$ is an immersed Lagrangian submanifold of $\mathbb{C}^n$. If L is generic, that is in a C^∞ dense subset of all Lagrangian embeddings, then $\pi(L)$ will have a finite number of isolated double points. The embedding of L in $\mathbb{R}^{2n+1}$ can be recovered (up to rigid translation in the z direction) from $\pi(L)$ as in the $n = 1$ case described above. Specifically, pick a point $p \in \pi(L)$ and choose any z coordinate for p then the z coordinate of any other point p' is determined by

$$\sum_{j=1}^{n} \int_{\gamma} y_j \mathrm{d}x_j, \tag{6.5}$$

where γ is any path in $\pi(L)$ from p to p'. Furthermore, given any Lagrangian immersion in $\mathbb{C}^n$ with isolated double points, if the integral in 6.5 is independent of the path γ then we obtain a Legendrian immersion in $\mathbb{R}^{2n+1}$. Note that we will get an embedding as long as the above integral is not zero for paths connecting the double points. The integral in 6.5 will be independent of γ when we have exact Lagrangian immersions $i : L \to \mathbb{R}^{2n}$, i.e., ones for which $i^*(\sum_{j=1}^{n} x_j \mathrm{d}y_j)$ is exact. In particular, if $H^1(L) = 0$, as is the case for S^n if $n \neq 1$, then all Lagrangian immersions are exact. Thus an isotopy L_t' of exact Lagrangian manifolds in $\mathbb{C}^n$ with transverse double points will lift to an isotopy L_t of Legendrian manifolds in $\mathbb{R}^{2n+1}$ (again there is no guarantee that the manifolds will be embedded).

EXAMPLE 6.1. Here we consider the most basic example of an exact Lagrangian immersion of a sphere in $\mathbb{C}^n$, thus providing a Legendrian sphere in $\mathbb{R}^{2n+1}$. Let $S^n = \{(x, y) \in \mathbb{R}^n \times \mathbb{R} : |x|^2 + y^2 = 1\}$ and then define

$$f(x, y) : S^n \to \mathbb{C}^n : (x, y) \mapsto ((1 + iy)x). \tag{6.6}$$

We claim that the image of f is an exact Lagrangian sphere in $\mathbb{C}^n$ with one double point that lifts to an embedded sphere in $\mathbb{R}^{2n+1}$. When $n = 1$ the image is a figure eight in the plane with a double point at the origin.

The Reidemeister type theorem here is

LEMMA 6.2. *If two Legendrian knots in* $\mathbb{R}^{2n+1}$, $n > 1$, *are Legendrian isotopic then their Lagrangian projections are related by regular homotopies and isolated double point moves (see the right hand side of Figure 12).*

Though the Lagrangian projection of Legendrian knots has many nice properties, for example, Lemma 6.2, it has some drawbacks as well. Specifically, given a submanifold L^n of $\mathbb{C}^n$ it is not clear whether or not it is an exact Lagrangian submanifold or lifts to an embedded Legendrian knot. Thus it is sometimes useful to consider the front projection of a Legendrian knot. The front projection $\Pi(L)$ is a codimension one subvariety of $\mathbb{R}^{n+1}$. Living in a lower dimensional space makes it somewhat easier to visualize. The drawback to the front projection is that $\Pi(L)$ will have certain singularities. In fact, any singularity with a well-defined tangent space one would expect from a projection of

$L^n \subset \mathbb{R}^{2n+1}$ to $\mathbb{R}^{n+1}$ can occur except tangential double points (these would correspond to double points of L in $\mathbb{R}^{2n+1}$). But any map from an n dimensional space to $\mathbb{R}^{n+1}$ with no vertical tangencies and the appropriate type of singularities will lift to a Legendrian knot in $\mathbb{R}^{2n+1}$.

To see that any map into $\mathbb{R}^{n+1}$ with the appropriate singularities can be lifted to a Legendrian knot we observe that at cusp edges (and all other allowable singularities) there is a well defined tangent plane to the image of the map. Thus we can recover the y_i coordinate of the Legendrian knot by looking at the slope of the tangent plane in the $x_i z$-plane (since for the lift to be Legendrian its tangent planes have to satisfy $\mathrm{d}z = \sum_{i=1}^n y_i \mathrm{d}x_i$).

If $\phi : L \to \mathbb{R}^{2n+1}$ is a Legendrian embedding then (after a C^∞ small perturbation to another Legendrian) off of a codimension one subset $\Sigma \subset L$ the map $\Pi \circ \phi : L \to \mathbb{R}^{n+1}$ will be an immersion. For a generic point p in Σ we can find a neighborhood N of p in L, with coordinates $(x_1, \ldots, x_n)$, and a coordinate neighborhood N' of $\Pi \circ \phi(p)$ such that $\Pi_F \circ \phi$ is expressed in these coordinates by $(x_1, \ldots, x_n) \mapsto (3x_1^2, x_2, \ldots, x_n, \pm 2x_1^3)$. We say such a point is on a *cusp edge* (see Figure 52). Note that the image of $\Pi \circ \phi$ cannot have tangent planes containing the z-direction (since $\phi(L)$ is Legendrian). It is the cusp edges that allow our front projections to avoid these vertical tangencies. For a generic 1 dimensional Legendrian knot only cusp edges occur in the front projection. In higher dimensions there can be other singularities as well, however, they all occur in codimension larger than one. So the top dimensional strata of Σ consists entirely of cusp edges. In particular, later it will be important that any path on L can be perturbed so that it only intersects Σ in cusp edges. For a thorough discussion of allowable singularities see [4].

For Legendrian knots L in $\mathbb{R}^{2n+1}$ we have two "classical" invariants. First, Tabachnikov [77] has shown how to generalize the Thurston–Bennequin invariant to higher dimension. Specifically, let L' be a copy of L obtained by pushing L slightly in the z-direction, then

$$tb(L) = \text{linking}(L, L'). \tag{6.7}$$

It is interesting to note that when n is even $tb(L) = \frac{1}{2}\chi(L)$ and thus is a trivial invariant of the Legendrian isotopy class of L. When n is odd $tb(L)$ is a non-trivial invariant (see Section 6.3).

It is more difficult to generalize the rotation number of a Legendrian knot in three manifolds to higher dimensions. To each point $x \in L$ we get a Lagrangian plane $d\pi_x(T_x L)$ in $\mathbb{C}^n$. Thus we get a map L to $\mathrm{Lag}(\mathbb{C}^n)$ the space of Lagrangians in $\mathbb{C}^n$. The homotopy class

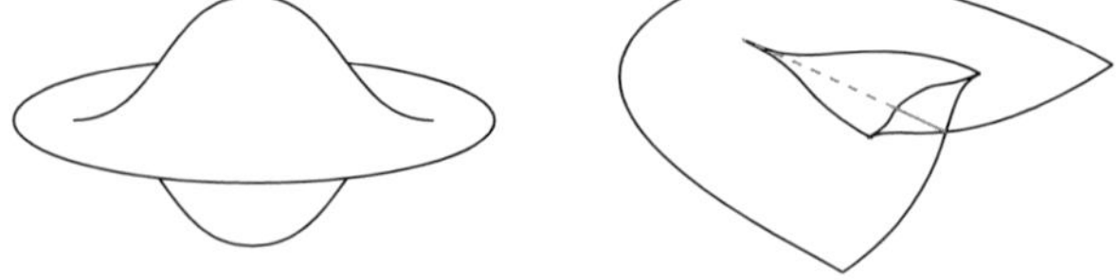

Fig. 52. Front projection with a cusp edge (right) and a "swallow-tail" singularity (left).

of this map is an invariant of L. In dimension three, $\pi_1(\mathrm{Lag}(\mathbb{C})) = \mathbb{Z}$ and thus to Legendrian knots in three manifolds we have a rotation number. In higher dimensions and when L is not just a sphere this homotopy class can be much more complicated than an integer. Again it is interesting to note that when n is even and L is an n sphere then this rotation class is always trivial. Thus when n is even there are no classical invariants of Legendrian knots.

6.2. *Generalizations of the Chekanov–Eliashberg DGA*

Contact homology naturally generalizes to higher dimensions. Here we define the DGA associated to a Legendrian knot in $\mathbb{R}^{2n+1}$.

The algebra. Denote the double points of the Lagrangian projection, $\pi(L)$, by $\mathcal{C}$. We assume $\mathcal{C}$ is a finite set of transverse double points. Let $\mathcal{A}$ be the free associative unital algebra over $\mathbb{Z}_2$ generated by $\mathcal{C}$.

The grading. To each crossing $c \in \mathcal{C}$ there are two points c^+ and c^- in $L \subset \mathbb{R}^{2n+1}$ that project to c. We denote the point with larger z-coordinate by c^+. Choose a map $\gamma_c : [0,1] \to L$ that parametrizes an arc running from c^+ to c^-. (Note there could be more that one path.) For each point $\gamma(\theta) \in L$ we have a Lagrangian plane $d\pi_{\gamma(\theta)}(T_{\gamma(\theta)}L)$ in $\mathbb{C}^n$. Thus γ gives us a path $\hat{\gamma}$ in $\mathrm{Lag}(\mathbb{C}^n)$. Since c is a transverse double point $\hat{\gamma}(0)$ is transverse to $\hat{\gamma}(1)$. Thus we can find a complex structure J on $\mathbb{R}^{2n}$ that (1) induces the same orientation on $\mathbb{C}^n$ as the standard complex structure and (2) $J(\hat{\gamma}(1)) = \hat{\gamma}(0)$. Now set $\bar{\gamma}(\theta) = e^{J\theta}\hat{\gamma}(1)$. Note that $\hat{\gamma}$ followed by $\bar{\gamma}$ is a closed loop in $\mathrm{Lag}(\mathbb{C}^n)$. Moreover $\pi_1(\mathrm{Lag}(\mathbb{C}^n)) = \mathbb{Z}$ thus to this closed loop we get an integer, $\mathrm{cz}(c)$, the Conley-Zehnder invariant of c. The grading on c is

$$|c| = \mathrm{cz}(c) - 1.$$

We note that c in general depends on the path γ with which we started. To take care of this ambiguity note that to each circle immersed in L we get an integer through the above procedure. Let n be the greatest common divisor of all these integers. It is easy to convince oneself that $|c|$ is well defined modulo n.

The differential. We will define the differential ∂ on $\mathcal{A}$ by defining it on the generators of $\mathcal{A}$ and then extending by the signed Leibniz rule:

$$\partial ab = (\partial a)b + (-1)^{|a|} a \partial b.$$

Let $a \in \mathcal{C}$ be a generator of $\mathcal{A}$ and let $b_1 \ldots b_k$ be a word in the "letters" $\mathcal{C}$. Let P_{k+1} be a $k+1$ sided polygon in $\mathbb{C}$ with vertices labeled counterclockwise $v_0, \ldots, v_k$. We will consider maps $u : (P_{k+1}, \partial P_{k+1}) \to (\mathbb{C}^n, \pi(L))$ such that $u|_{\partial P_{k+1} \setminus \{v_i\}}$ lifts to a map to $L \subset \mathbb{R}^{2n+1}$. Call a vertex v_i mapping to the double point c is *positive* (resp. *negative*) if the lift of the arc just clockwise of v_i in ∂P_{k+1} lifts to an arc approaching c^+ (resp. c^-) and the arc just counterclockwise of v_i lifts to an arc approaching c^- (resp. c^+), where $c^\pm$ are as in the definition of grading. Set

$$\mathcal{M}^a_{b_1 \ldots b_k} = \{u : (P_{k+1}, \partial P_{k+1}) \to (\mathbb{C}^n, \pi(L)) \text{ such that } u \text{ satisfies 1.–4. below}\}/\sim$$

where $\sim$ is holomorphic reparameterization and the conditions are

(1) $u|_{\partial P_{k+1}\setminus\{v_i\}}$ lifts to a map to $L \subset \mathbb{R}^{2n+1}$,
(2) $u(v_0) = a$ and v_0 is positive,
(3) $u(v_i) = b_i$, $i = 1, \ldots, k$, and v_i is negative,
(4) u is holomorphic.

We can now define

$$\partial a = \sum_{b_1 \ldots b_k} (\#_2 \mathcal{M}) b_1 b_2 \ldots b_k,$$

where the sum is taken over all words in the letters $\mathcal{C}$ for which $\dim(\mathcal{M}^a_{b_1,\ldots,b_n}) = 0$ and $\#_2$ denotes the modulo two count of elements in $\mathcal{M}$.

THEOREM 6.3. (Ekholm, Etnyre and Sullivan, [20,21]). *With the notation above:*
(1) *The map ∂ is a well defined differential that reduces the grading by 1.*
(2) *The stable tame isomorphism class of $(\mathcal{A}, \partial)$ is an invariant of L.*
(3) *The homology of $(\mathcal{A}, \partial)$ is an invariant of L.*

We will compute some simple examples below.

6.3. *Examples*

The DGA of a Legendrian knot L is defined in terms of the Lagrangian projection, but it will be much easier to construct examples in the front projection. Thus we would like to be able to recognize the double points in the Lagrangian projection by looking at the front projection. We begin by noting that generically we can assume that the double points in the Lagrangian projection do not occur along singularities in the front projection. So we consider a point $p \in \Pi(L)$ in the front projection that is not singular. Since $\Pi(L)$ is embedded near p and the tangent plane does not contain the vector $\partial/\partial z$ we can find an open set U in $\mathbb{R}^n$ and a function $f : U \longrightarrow \mathbb{R}$ such that near p, $\Pi(L)$ is given by the graph of f.

Let c^+ and c^- be two points in $\Pi(L)$. If they correspond to double point in the Lagrangian projection then they must have the same x_i coordinates, which we denote x, and different z coordinates. We assume the z coordinate of c^+ is larger. Let $f^\pm$ be the functions described above for $c^\pm$. We can assume that f^+ and f^- have the same domain. Let $f = f^+ - f^-$. The following are equivalent:

(1) c^+ and c^- correspond to a double point in $\pi(L)$;
(2) $T_{c^+}\Pi(L) = T_{c^-}\Pi(L)$;
(3) $df^+_x = df^-_x$;
(4) $df_x = 0$.

Thus we have a double point if x is a critical point of f. Moreover, the double point is transverse if and only if x is a non-degenerate double point. Thus to a transverse double point we have a Morse index, $\mathrm{Ind}_x(f)$. Suppose c^+, c^- correspond to a double point. Let γ be a path from c^+ to c^- that is transverse to the singular set of $\Pi(L)$. Thus it intersects the singular set along the cusp edges in a finite number of points. Let D be the number of times γ intersects the cusps while its z coordinate is decreasing and U the number of times with its z coordinate increasing.

LEMMA 6.4. (Ekholm, Etnyre and Sullivan, [20]). *With the notation as above. If c^+ and c^- correspond to a transverse double point c then*

$$\mathrm{cz}(c) = \mathrm{Ind}_x(f) + D - U. \tag{6.8}$$

We are now ready for some examples.

EXAMPLE 6.5. In Figure 53 the front projections of two Legendrian knots are shown. Though the pictures are of Legendrian S^2's in $\mathbb{R}^5$ there are clearly analogous Legendrian S^n's in $\mathbb{R}^{2n+1}$. We use contact homology to distinguish these examples in all dimensions. When n is even they cannot be distinguished by any classical invariants. When n is odd they have different Thurston–Bennequin invariants. Note the Legendrian L on the left has only one double point (the "axis" of the flying saucer) c. Using the above lemma we see that $\mathrm{cz}(c) = n + 1$ so $|c| = n$. Since ∂ is degree -1 we see that $\partial c = 0$. So the contact homology is generated by one element in grading n.

For L' on the right of the figure, there is a double point c corresponding the the "axis" again, but now one must go down three cusp edges from c^+ to c^- thus $|c| = n + 2$. If the picture is drawn symmetrically, there are an S^{n-1}'s worth of double points. By deforming the picture slightly so the cusps are tilted we can break the S^{n-1}'s worth of double points into two double point a and b. They will have gradings $|a| = n$ and $|b| = 1$. We can again conclude that $\partial c = 0$ but we do not know if $\partial b = 0$ or 1, and when $n = 2$ if $\partial a = b$ or 0. But in any case the contact homology of L' will be different from that of L.

We now describe a procedure for changing a given Legendrian knot. Suppose L is a Legendrian knot in $\mathbb{R}^{2n+1}$ containing two disks L_u and L_l that project to the same disk under the Lagrangian projection. This is a degenerate situation, but it is easy to isotop any Legendrian knot so that it has such disks. Assume L_u is above L_l. Let M be a k-manifold embedded in L_l. There is an ambient isotopy of $\mathbb{R}^{n+1}$ that is supported near $\Pi(M) \times \mathbb{R}$, where the $\mathbb{R}$ factor is the z-axis, that moves $\Pi(M)$ up just past $\Pi(L_u)$. In the front projection replace $\Pi(L_l)$ with the disk obtained from $\Pi(L_l)$ by applying the ambient isotopy. This new

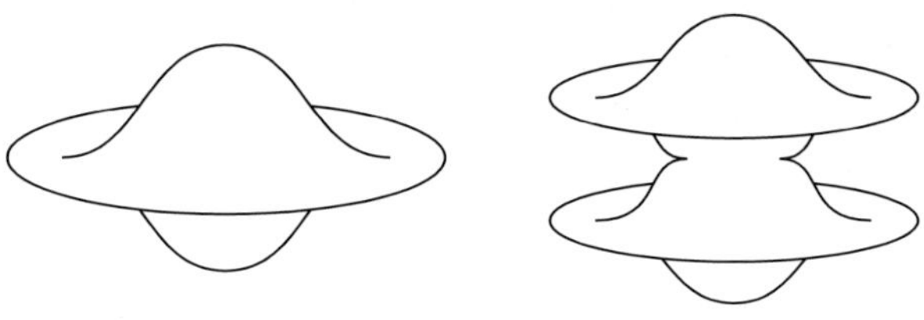

Fig. 53. The front projection of two Legendrian spheres.

front diagram describes a new Legendrian knot L_M called the *stabilization of L along M*. One may show

PROPOSITION 6.6. (Ekholm, Etnyre and Sullivan, [20]). *If L_M is the stabilization of L along M and the notation is as above then*

(1) *The rotation class of L_M is the same as the rotation class of L.*

(2) *If D, U are the number of down, up, cusps along a generic path from $\Pi(L_u)$ to $\Pi(L_l)$ in $\Pi(L)$ then*

$$tb(L_M) = \begin{cases} tb(L), & n \text{ even}, \\ tb(L) + (-1)^{(D-U)} 2\chi(M), & n \text{ odd}, \end{cases}$$

(3) *If the stabilization of the Legendrian knot L takes place in a small neighborhood of a cusp edge of $\Pi(L)$ then $CH_*(L) = 0$.*

REMARK 6.7. Note using this proposition it is easy to see that when n is odd and not 1 then we can realize any any integer, of the appropriate parity, as the tb of some Legendrian knot.

It is important that stabilization does not always force the contact homology to be trivial.

EXAMPLE 6.8. Let L be the Legendrian knot shown in Figure 54. It is clearly Legendrian isotopic to the standard "flying saucer" in Figure 53. Let L_1 be the stabilization of L along the point p indicated in the figure. Let $L_k = L_{k-1} \# L_1$ where # is defined by replacing an arc connecting cusp edges of the two front projection by the tube shown on the right of Figure 54. One can show [20] that the grading -1 linearized contact homology of L_k has dimension k:

$$dim(L_1 CH_{-1}(L_k)) = k.$$

Thus none of the L_k's are Legendrian isotopic. These examples generalize to all dimensions but when n is odd the knots L_k are already distinguished by their tb. Thus when

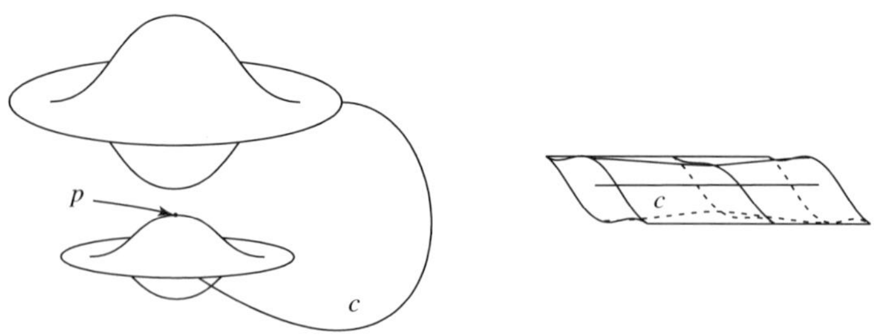

Fig. 54. The Legendrian knot L (left). To form L a neighborhood of the arc c should be replaced with the tube shown on the right.

n is odd to get an infinite family of examples of distinct Legendrian knots with the same classical invariants one must alter the above construction. For details see [20].

For other examples of Legendrian knots in high dimensions and other constructions see [20,22].

7. Applications

There are numerous applications of Legendrian knot theory to topology and contact geometry. We describe a few of these applications below.

7.1. *Legendrian surgery*

Let (M,ξ) be a contact manifold and L a Legendrian knot in M. There is a standard neighborhood N of L with convex boundary. The dividing curves $\Gamma_{\partial N}$ consist of two parallel curves. We can choose a framing on N so that the dividing curves have slope ∞ and of course the meridian has slope 0. Let M' be the manifold obtained by gluing $S^1 \times D^2$ to $\overline{M \setminus N}$ by the map $A : \partial(S^1 \times D^2) \to \partial\overline{M \setminus N}$ given by

$$A = \begin{pmatrix} 1 & 1 \\ -1 & 0 \end{pmatrix}.$$

Here we are using the framing on ∂N and product structure on $\partial(S^1 \times D^2)$ to write the map as a matrix.

The manifold M' has a contact structure defined on the complement of $S^1 \times D^2$ but $\partial(S^1 \times D^2)$ has an induced convex characteristic foliation. Using the classification of tight contact structures on the solid torus there is a unique tight contact structure on $S^1 \times D^2$ inducing this characteristic foliation. Thus we can get a well defined contact structure ξ' on M'. We say the contact manifold (M', ξ') is obtained form (M, ξ) by *Legendrian surgery* on L. There is an obvious generalization to Legendrian surgery on a Legendrian link in (M, ξ). The following is the fundamental result concerning Legendrian surgery.

THEOREM 7.1. *If (M, ξ) is a fillable contact structure then any contact manifold obtained by Legendrian surgery from (M, ξ) is also fillable.*

Recall that a fillable contact structure is tight [24]. Moreover, the standard contact structure on S^3 is fillable. Thus we can construct many tight contact structures on three manifolds by doing Legendrian surgery on links in S^3.

EXAMPLE 7.2. Legendrian surgery on a Legendrian unknot in S^3 with $tb = -k$ yields tight contact structures on the lens spaces $L(k + 1, 1)$.

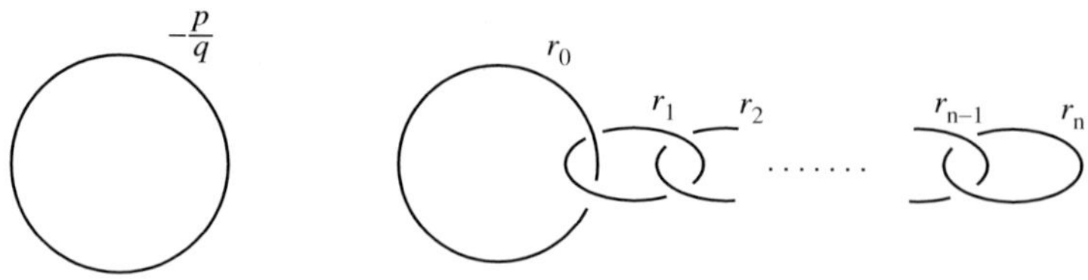

Fig. 55. Two surgery presentation of the lens space $L(p, q)$.

To get other lens spaces $L(p, q)$ consider the continued fractions expansion of $-p/q$:

$$-\frac{p}{q} = r_0 - \frac{1}{r_1 - \cdots \frac{1}{r_n}}.$$

Then the lens space $L(p, q)$ as a surgery presentation as shown in Figure 55, [71]. It is easy to see that all the $r_i < -1$ thus we can realize the right hand side of Figure 55 by Legendrian unknots with $\mathrm{tb} = r_i + 1$. Thus Legendrian surgery on this link will yield a tight contact structure on $L(p, q)$. In is interesting to note that all tight contact structures on $L(p, q)$ come from Legendrian surgery on a link in S^3 [49,51].

For an extensive discussion of the use of Legendrian surgery to construct tight contact structures see [50].

One of the main questions in contact geometry is the following.

QUESTION. Does Legendrian surgery on a tight contact structure produce a tight contact structure?

The answer is NO if the manifold is allowed to have boundary [53] (*cf.* [16]), but it is still possible that for closed contact three manifolds this answer is YES.

7.2. *Invariants of contact structures*

Now that we have seen how to construct contact structures using Legendrian knots we examine how to use Legendrian knots to distinguishing contact structures on three manifolds. For examples consider the contact structures

$$\xi_n = \ker(\sin(nz)dx + \cos(nz)dy),$$

on $T^3 = T^2 \times S^1$ where x, y are coordinates on T^2, and z is the coordinate on S^1. It is easy to see all these contact structures on T^3 are tight (when pulled back to the universal cover of T^3 the contact structures become the standard contact structure on $\mathbb{R}^3$, [49,51,56]) and certainly look different. But how can they actually be distinguished? They are in the same homotopy class of plane field so there is no "algebraic" way to distinguish them. Consider $\gamma = \{(x, y, z)|x = y = 0\}$. This is a Legendrian knot in (T^3, ξ_n) and the twisting of the contact planes relative to the natural product framing $\mathcal{F}$ is $tw(\gamma, \mathcal{F}) = -n$. The twisting with respect to $\mathcal{F}$ we will denote $tb(\gamma)$, this is a slight abuse of notation but should not be confusing. It turns out that this is the maximal twisting for any Legendrian knot in this knot type.

THEOREM 7.3. *Let $\mathcal{K}$ be the topological knot type containing γ. Then for the contact structure ξ_n we have*

$$\overline{tb}(\mathcal{K}) = -n.$$

Thus all the contact structures ξ_n on T^3 are distinct.

Similar arguments have been used in distinguish some contact structures on other T^2 bundles over S^1 and on some S^1 bundles over surfaces [49,51,52]. This illustrates that Legendrian knots are a very subtle invariant of contact structures. In fact, it is possible that contact structures are completely determined by the Legendrian knots that exist within them.

7.3. *Plane curves*

Chekanov and Pushkar have recently solved Arnold's famous "four vertex conjecture" [3] using invariants of Legendrian knots [13]. Arnold conjecture is about generic wave fronts. A *generic wave front* is a curve in $\mathbb{R}^2$ that is immersed at all but a finite number of points at which there is a semi cubic cusp. Thus a curve looks very much like the front projection of a Legendrian knot discussed in Section 2.3 except there might be vertical tangents. Consider the manifold $M = \mathbb{R}^2 \times S^1$ with the contact structure

$$\xi = \ker((\cos\theta)\mathrm{d}x + (\sin\theta)\mathrm{d}y),$$

where x, y are coordinates on $\mathbb{R}^2$ and $\theta(\text{mod } 2\pi)$ is the coordinate on S^1. (If we think of M as the unit cotangent bundle of $\mathbb{R}^2$ then ξ is a natural contact structure induced from the standard complex structure on $T^*\mathbb{R}^2$.) Now if γ is a generic (oriented) wave front in $\mathbb{R}^2$ then we can lift γ to γ_l in M by sending $p \in \gamma$ to (p, θ) in M where θ is the $\pi/2$ plus the angle $T_p\gamma$ makes with the x-axis in $\mathbb{R}^2$. Clearly γ_l is a Legendrian curve in M. In addition generic Legendrian curves in M will project to generic wave fronts in $\mathbb{R}^2$. Thus we may clearly study wave fronts by studying their corresponding Legendrian lifts.

Suppose $\gamma_t, t \in [0, 1]$ is a family of generic (oriented) wave fronts in $\mathbb{R}^2$ with γ_0 and γ_1 oppositely oriented smooth circular wave fronts and for all t, γ_t having no self tangencies at which the orientations on the two intersecting strands agree. Such a family is usually called an "eversion of a smooth circular wave front with no dangerous self tangencies". Note this last condition guarantees that the Legendrian lift of each γ_t is an embedded Legendrian knot in M. Thus such a family of wave fronts gives a Legendrian isotopy of the Legendrian lifts.

CONJECTURE 1. (Arnold's Four Cusp Conjecture, [3]). *An eversion of a smooth circular wave front with no dangerous self tangencies must contain some curves with at least four cusps.*

In [3] Arnold verified this conjecture in certain cases. Using an invariant for Legendrian knots in M similar to the decomposition invariant described in Section 4.6, Chekanov and Pushkar have verified the conjecture in all cases [13].

7.4. *Knot concordance*

Using the inequality

$$\frac{1}{2}(tb(L) + |r(L)|) + 1 \leq g_s(\mathcal{K})$$

for all $L \in \mathcal{L}(\mathcal{K})$ from Section 3.2 we can easily make many observations about (non) slice knots and, in particular, find many examples of elements of infinite order in the knot concordance group. We be begin with the simple observation.

LEMMA 7.4. *If there is a Legendrian knot $L \in \mathcal{L}(\mathcal{K})$ with $tb(L) + |r(L)| + 1 \geq 0$ then $\mathcal{K}$ is not smoothly slice. Moreover $\mathcal{K}$ is of infinite order in the knot concordance group.*

The first observation follows directly from the above inequality. The second follows from the inequality and the fact that $tb(L\#L') = tb(L) + tb(L') + 1$.

We now apply these ideas to study Whitehead doubles of knots. If the neighborhood N of a knot K is replaced by the solid torus $D^2 \times S^1$ shown in Figure 56, so that $\{\text{pt}\} \times S^1$ with $\text{pt} \in \partial D^2$, is identified with a longitude for K on ∂N then the image of C is called the 0-*Whitehead double* of a knot K and is denoted $W_0(K)$. The *n-twisted Whitehead double* $W_n(K)$ is obtained in an analogous way after putting n-full right handed twist in the solid torus $D^2 \times S^1$. One may compute that the Alexander polynomial of $W_n(K)$ is 0 for any knot type. Freedman [42] has used this to show that the zero twisted Whitehead double of any knot is *topologically locally flatly* slice. There is a well-known conjecture:

CONJECTURE 2. *A knot is smoothly slice if and only if its zero twisted Whitehead double is.*

In relation to this conjecture we can show the following theorem is true.

THEOREM 7.5. (Rudolph 1995, [73]). *Given any knot type $\mathcal{K}$ for which $\overline{tb}(\mathcal{K}) \geq 0$ then all iterated zero twisted Whitehead doubles of K are not smoothly slice.*

To prove this theorem we consider the a Legendrian version of the Whitehead double. Let L be a Legendrian knot in $(\mathbb{R}^3, \xi_{\text{std}})$ and L' a parallel copy pushed up (in the z-direction) slightly from L. The Legendrian Whitehead double $LW(L)$ of L is obtained by

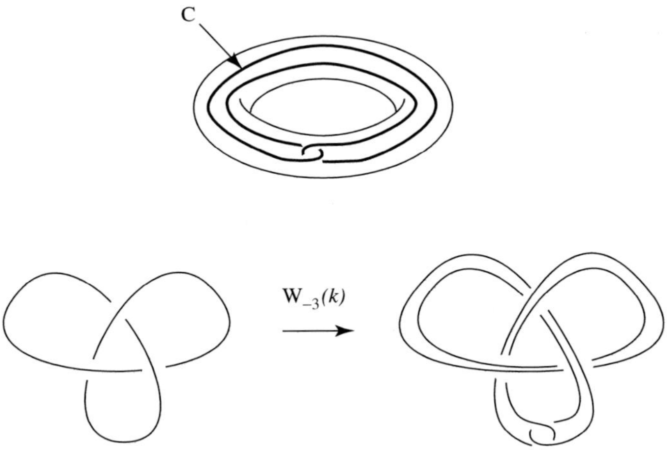

Fig. 56. The solid torus used in the Whitehead double (top) and the -3 twisted double of the left handed trefoil knot.

replacing a horizontal stand of L and L' by a clasp as shown in Figure 57. One may easily compute that as a topological knot $LW(L)$ is $W_{tb(L)}(L)$ and that

$$tb(LW(L)) = 1 \text{ and } r(LW(L)) = 0.$$

Thus given $\mathcal{K}$ let L be an Legendrian knot in $\mathcal{L}(\mathcal{K})$ realizing maximal tb. Forming $LW(L)$ we see that $W_{\overline{tb}(\mathcal{K})}(\mathcal{K})$ is of infinite order in the concordance group. By considering the Legendrian doubles of stabilizations of L we see that $W_n(\mathcal{K})$ is of infinite order for all $n \leq \overline{tb}(\mathcal{K})$. In particular if $\overline{tb}(\mathcal{K}) \geq 0$ then the zero twisted Whitehead double is not slice.

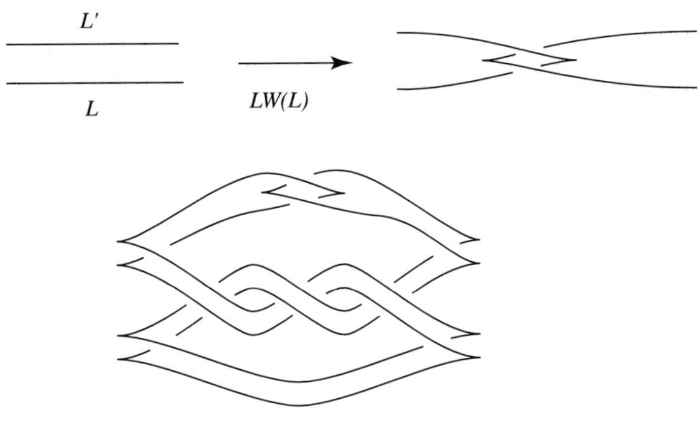

Fig. 57. The change to $L \cup L'$ used to make $LW(L)$. Legendrian Whitehead double of the trefoil (bottom).

7.5. *Invariants of classical knots*

It is quite easy to construct invariants of topological knots out of Legendrian knot theory. For example the maximal Thurston–Bennequin invariant is a topological knot invariant

$$\overline{tb}(\mathcal{K}) = \max\{tb(L) | L \in \mathcal{L}(\mathcal{K})\}.$$

As we have seen in Section 3.2 this invariant is closely related to the slice genus of a knot. Moreover it is very sensitive to mirroring.

EXAMPLE 7.6. Let $\mathcal{K}$ be a (p, q) torus knot, with $p, q > 0$. Its mirror $\bar{\mathcal{K}}$ is a $(-p, q)$ torus knot. From Section 5.2 we know $\overline{tb}(\mathcal{K}) = pq - p - q$ and $\overline{tb}(\bar{\mathcal{K}}) = -pq$.

It also seems that $\overline{tb}$ is sensitive to many topological operations on a knot, like mutation. This prospect seems very interesting but has not been pursued.

Analogous to the definition of finite type invariants of topological knots one can define finite type invariants of Legendrian knots.

THEOREM 7.7. (Fuchs and Tabachnikov 1997, [45]). *Any finite type Legendrian knot invariant for Legendrian knots in* $(\mathbb{R}^3, \xi_{std})$ *with the same tb and r is also finite type invariant of the underlying framed topological knot type.*

Put another way, two distinct Legendrian knots with the same topological knot type, *tb* and *r* cannot be distinguished by a finite order invariant. This theorem has been generalized to other contact manifold, [79].

One might try to use this theorem to try to construct infinite order invariants of topological knots out of Legendrian knot theory. For example let $\mathcal{L}_{\mathrm{m}}(\mathcal{K})$ be all the Legendrian knots with maximal *tb* and set $CH(\mathcal{K}) = \{CH_*(L) | L \in \mathcal{L}\}$. So $CH(\mathcal{K})$ is the set of all contact homologies for Legendrian knots in $\mathcal{L}(\mathcal{K})$ with maximal tb. Since $CH_*(L)$ is not a finite type invariant of L it would be surprising if $CH(\mathcal{K})$ is a finite type invariant of $\mathcal{K}$. More generally we ask the following question.

QUESTION. Can one extract non-finite type invariants of topological knots $\mathcal{K}$ out of the DGA associated to Legendrian knots in $\mathcal{L}(\mathcal{K})$?

7.6. *Contact homology and topological knot invariants*

In this section we discuss invariants of topological knots in $\mathbb{R}^3$ that come from from high dimensional Legendrian knot theory. Specifically, to a topological knot in $\mathbb{R}^3$ we will construct a Legendrian torus in $\mathbb{R}^3 \times S^2$. The Legendrian isotopy class of this torus will be an invariant of the knot. Thus any invariant of the Legendrian torus will be an invariant of the topological knot.

Let $W = \{(x, v) \in T^*\mathbb{R}^3 | v \in T_x^*M \text{ and } |v| = 1\}$ be the unit cotangent bundle of $\mathbb{R}^3$. It is easy to see there is a natural contact structure ξ on W. Specifically let x_i be coordinates on $\mathbb{R}^3$ and y_i be coordinates on the fibers of $T^*\mathbb{R}^3$. The restriction of $\lambda = \Sigma y_i dx_i$ to W is a contact form on W. Let $\xi = \ker \alpha$.

Given a topological knot K in $\mathbb{R}^3$ let $T_K = \{(x, v) \in W | v(T_xK) = 0\}$ be the unit conormal bundle. One may easily check that T_K is a Legendrian torus in W and that a topological isotopy of K will produce a Legendrian isotopy of the associated torus. Thus T_K is an invariant of K.

As discussed above there is no "classical invariants" of T_K, but one can use the contact homology of T_K to come up with an invariant of K. It is actually somewhat difficult to compute the contact homology (not to mention contact homology has not been shown to be well defined in this case yet), but following this idea Ng [65] has combinatorially defined the invariants of topological knots and braids. Moreover, in [65] it is shown that these contact homology invariants can distinguish knots with the same Alexander polynomial and signature. In [66] a topological interpretation is given to a small piece of the contact homology invariant. Here we describe some of the geometric idea behind the braid invariants and refer the reader to [65–67] for details on the knot invariants and the combinatorial proofs of invariance.

Let U be the unknot in $\mathbb{R}^3$ represented as a 1-braid. So T_U is a Legendrian torus in W. A Legendrian submanifold always has a neighborhood contactomorphic to the 1-jet space of the submanifold [47,59]. That is, given a manifold M its one jet space is $J^1(M) = T^*M \times \mathbb{R}$ with the contact form $\alpha = \mathrm{d}z + \lambda$ (here z is the coordinate in the $\mathbb{R}$ direction and λ is the tautological 1-form on T^*M). It is easy to see that the zero section Z in $J^1(M)$ is a Legendrian submanifold. Thus the Legendrian torus T_U has a neighborhood contactomorphic to $J^1(T^2)$.

Any braid B in $\mathbb{R}^3$ can be isotoped into an arbitrarily small neighborhood of the unknot U. Moreover, any isotopy of the braid can be done in this small neighborhood. Thus the Legendrian manifold T_B in W can be thought to sit inside $J^1(T^2)$ and the Legendrian isotopy class of T_B as a subset of $J^1(T^2)$ depends only on the isotopy class of B as a braid.

There is a projection $\pi : J^1(T^2) \to T^*T^2$ that has properties identical to the Lagrangian projection discussed in Section 6.1. Thus $\pi(T_B)$ is a Lagrangian submanifold of T^*T^2 and generically has a finite number of transverse double points. We can now define the contact homology of T_B as in Section 6.2. It is the contact homology of T_B that is the invariant of B.

We now describe (at least heuristically) how to compute $CH_*(T_B)$. The 1-jet space also has a "front projection" $\Pi : J^1(T^2) \to T^2 \times \mathbb{R}$. This front projection has all the properties described in Section 6.1 for the standard front projection. If B is an n-braid we can find n functions $f_i : [0, 1] \to R^2$ that parameterize B in the sense that their graphs in $[0, 1] \times \mathbb{R}^2/(0, x) \sim (1, x)$ give B. With such a representation of B we can express $\Pi(T_B)$ as the union of the graphs of the functions $F_i(t, \theta) = f_i(t) \cdot (\cos 2\pi\theta, \sin 2\pi\theta) : [0, 1]^2 \to \mathbb{R}$, where $[0, 1]^2$ glued to form T^2. As described in Section 6.3 the double point the Lagrangian projection of T_B correspond to the critical points of $F_i - F_j$, and these correspond to the critical points of $|f_i - f_j|$. The function $|f_i - f_j|$ must have a maximum and a minimum, we label these b_{ij}, a_{ij}, respectively. Though there may be other critical points for $|f_i - f_j|$ we can eliminate them up to stable tame isomorphism. There are no cusps in the front projection of T_B so the grading on the double points is one less than the Morse index of $F_i - F_j$, thus $|a_{ij}| = 0$ and $|b_{ij}| = 1$.

We now describe the differential. Clearly we have

$$\partial a_{ij} = 0.$$

The differential on the b_{ij}s is somewhat more complicated.

$$\partial b_{ij} = a_{ij} - \phi_B(a_{ij}). \tag{7.1}$$

Where ϕ_B is defined as follows: for a generator σ_k of the braid group we define

$$\phi_{\sigma_k} : \begin{cases} a_{ki} \mapsto -a_{k+1,i} - a_{k+1,k}a_{ki} & i \neq k, k+1 \\ a_{ik} \mapsto -a_{i,k+1} - a_{ik}a_{k,k+1} & i \neq k, k+1 \\ a_{k+1,i} \mapsto a_{ki} & i \neq k, k+1 \\ a_{i,k+1} \mapsto a_{ik} & i \neq k, k+q \\ a_{k,k+1} \mapsto a_{k+1,k} & \\ a_{k+1,k} \mapsto a_{k,k+1} & \\ a_{ij} \mapsto a_{ij} & i,j \neq k, k+1 \end{cases} \tag{7.2}$$

If $B = \sigma_{i_1} \cdots \sigma_{i_m}$ then define $\phi_B = \phi\sigma_{i_m} o \cdots o \phi_{\sigma i_1}$. This differential ∂ is computed by looking for holomorphic disks in $T^*(T^2)$ with boundary on the projection of T_B. This is done by considering "gradient flow trees" for all the functions $F_i - F_j$. See [65]. In addition, see [65], for a description of the knot invariant.

References

[1] B. Aebisher et al., *Symplectic Geometry*, Progress in Mathematics, Vol. 124, Birkhäuser, Basel (1994).

[2] S. Akbulut and R. Matveyev, *Exotic structures and adjunction inequality*, Turk. J. Math. **21** (1) (1997), 47–53.

[3] V.I. Arnold, *Sur les propriétés topologiques des proections Lagrangiennes eu geométrie symplectique des caustiques*, Preprint no. 9320 CEREMADE, Université Paris, Dauphine (1993).

[4] V. Arnold and A. Givental, *Symplectic Geometry, Dynamical Systems IV*, V.I. Arnold and S.P. Novikov, eds, Springer-Verlag, Berlin (1990).

[5] D. Bennequin, *Entrelacements et équations de Pfaff*, Astérisque **107–108** (1983), 87–161.

[6] J. Birman, *Braids, Links, and Mapping Class Groups*, Annals of Mathematical Studies, 82, Princeton University Press, Princeton (1975).

[7] J. Birman and W. Menasco, *Stabilization in the Braid Group II: Transversal Simplicity of Knots* (www.arxiv.org/abs/math.GT/0310280).

[8] J. Birman and W. Menasco, *Studying links via closed braids III. Classifying links which are closed 3-braids*, Pac. J. Math. **161** (1) (1993), 25–113.

[9] J. Birman and W. Menasco, *Studying links via closed braids V. The unlink*, Trans. Am. Math. Soc. **329** (2) (1992), 585–606.

[10] J. Birman and N. Wrinkle, *On transversally simple knots*, J. Differential Geom. **55** (2) (2000), 325–354.

[11] Yu. Chekanov, *Differential algebra of Legendrian links*, Invent. Math. **150** (3) (2002), 441–483.

[12] Yu. Chekanov, *New invariants of Legendrian knots*, European Congress of Mathematics, Vol. II, (2000), 525–534, Barcelona.

[13] Y. Chekanov and P. Pushkar, *Arnold's Four Cusp Conjecture and Invariants of Legendrian Knots*, in preparation.

[14] S. Chmutov and V. Goryunov, *Polynomial invariants of Legendrian links and their fronts*, KNOTS '96 (Tokyo), World Science Publishing, River Edge, NJ (1997), 239–256.

[15] V. Colin, *Chirurgies d'indice un et isotopies de sphères dans les variétés de contact tendues*, C. R. Acad. Sci. Paris Sér. I Math. **324** (1997), 659–663.

[16] V. Colin, *Chirurgies de Dehn admissibles dans les variétés de contact tendues*, Ann. Inst. Fourier (Grenoble) **51** (2001), 1419–1435.

[17] L. Conlon, *Differentialble Manifolds a First Course*, Birkhäuser, Basel (1993).

[18] K. Dymara, *Legendrian Knots in Overtwisted Contact Structures* (www.arxiv.org/abs/math.GT/0410122).

[19] K. Dymara, *Legendrian knots in overtwisted contact structures on* S^3, Ann. Global Anal. Geom. **19** (3) (2001), 293–305.

[20] T. Ekholm, J. Etnyre and M. Sullivan, *Non-Isotopic Legendrian Submanifolds in* $\mathbb{R}^{2n+1}$ (www.arxiv.org/abs/math.SG/0210124).

[21] T. Ekholm, J. Etnyre and M. Sullivan, *The Contact Homology of Legendrian Submanifolds in* $\mathbb{R}^{2n+1}$, preprint 2003.

[22] T. Ekholm, J. Etnyre and M. Sullivan, *Orientations in Legendrian Contact Homology and Exact Lagrangian Immersions* (www.arxiv.org/abs/math.SG/0408411).

[23] Y. Eliashberg, *Classification of overtwisted contact structures on 3-manifolds*, Invent. Math. **98** (1989), 623–637.

[24] Y. Eliashberg, *Filling by Holomorphic Discs and Its Applications, Geometry of Low-Dimensional Manifolds*, London Math. Soc. Lecture Note Ser., 151, Vol. II, S.K. Donaldson and C.B. Thomas, eds, Cambridge University Press, Cambridge, (1990), 45–67.

[25] Y. Eliashberg, *Contact 3-manifolds twenty years since J. Martinet's work*, Ann. Inst. Fourier **42** (1992), 165–192.

[26] Y. Eliashberg, *Legendrian and Transversal Knots in Tight Contact 3-Manifolds*, Topological Methods in Modern Mathematics, Publish or Perish, Houston, TX (1993), 171-193.

[27] Y. Eliashberg, *INvariants in Contact Topology*, Proceedings of the International Congress of Mathematicians, Vol. II (Berlin, Doc. Math. 1998, Extra Vol. II (1998), 327–338.

[28] Y. Eliashberg and M. Fraser, *Classification of Topologically Trivial Legendrian Knots*, Geometry, Topology, and Dynamics (Montreal, PQ, 1995), 17–51, CRM Proceedings of the Lecture Notes, 15, American Mathematical Society, Providence, RI (1998).

[29] Y. Eliashberg, A. Givental and H. Hofer, *Introduction to symplectic field theory, GAFA (Tel Aviv, 1999)*, Geom. Funct. Anal. 2000, Special Volume, Part II (2000), 560–673.

[30] J. Epstein, D. Fuchs and M. Meyer, *Chekanov–Eliashberg Invariants and Transverse Approximations of Legendrian Knots*, Pac. J. Math. **201** (1) (2001), 89–106.

[31] J. Etnyre, *Transversal torus knots*, Geom. Topol. **3** (1999), 253–268.

[32] J. Etnyre, *Tight contact structures on lens spaces*, Commun. Contemp. Math. **2** (4) (2000), 559–577.

[33] J. Etnyre, *Introductory Lectures on Contact Geometry*, Topology and Geometry of Manifolds (Athens, GA, 2001), 81–107, Proceedings of the Symposium on Pure Mathematics, 71, Amerian Mathematical Society, Providence, RI (2003).

[34] J. Etnyre and K. Honda, *Legendrian Cables and Transverse Simplicity*, preprint 2003 (www.arxiv.org/abs/math.SG/0306330).

[35] J. Etnyre and K. Honda, *On connected sums and Legendrian knots*, Adv. Math. **179** (1) (2003), 59–74.

[36] J. Etnyre and K. Honda, *Knots and contact geometry I: torus knots and the figure eight*, J. Sympl. Geom. **1** (1) (2001), 63–120.

[37] J. Etnyre and L. Ng, *Problems in Low Dimensional Contact Topology*, Topology and Geometry of Manifolds (Athens, GA, 2001), 337–357, Proceedings of the Symposium on Pure Mathematics, 71, American Mathematical Society, Providence, RI (2003).
[38] J. Etnyre, L. Ng and J. Sabloff, *Invariants of Legendrian knots and coherent orientations*, J. Sympl. Geom. **1** (2) (2002), 321–367.
[39] E. Ferrand, *On Legendrian knots and polynomial invariants*, Proc. Am. Math. Soc. **130** (4) (2002), 1169–1176.
[40] J. Franks and R. Williams, *Braids and the Jones Polynomial*, Trans. Am. Math. Soc. **303** (1) (1987), 97–108.
[41] M. Fraser, *Example of nonisotopic Legendrian curves not distinguished by the invariants tb and r*, Int. Math. Res. Not. **19** (1996), 923–928.
[42] M. Freedman, *The topology of four-dimensional manifolds*, J. Differential Geom. **17** (1982), 357–453.
[43] D. Fuchs, *Chekanov–Eliashberg invariants of Legendrian knots: existance of augmentations*, J. Geom. Phys. **47** (1) (2003), 43–65.
[44] D. Fuchs and T. Ishkhanov, *Invariants of Legendrian knots and decompositions of front diagrams*, Mosc. Math. J. **4** (3) (2004), 707–717.
[45] D. Fuchs and S. Tabachnikov, *Invariants of Legendrian and transverse knots in the standard contact space*, Topology **36** (5) (1997), 1025–1053.
[46] H. Geiges, *A brief history of contact geometry and topology*, Expo. Math. **19** (1) (2001), 25–53.
[47] H. Geiges, *Handbook of Differential Geometry*, Contact Geometry, Vol. 2, to appear.
[48] E. Giroux, *Convexité en topologie de contact*, Comment. Math. Helv. **66** (4) (1991), 637–677.
[49] E. Giroux, *Structures de contact en dimension trois et bifurcations des feuilletages de surfaces*, Invent. Math. **141** (2000), 615–689.
[50] R. Gompf, *Handlebody construction of Stein surfaces*, Ann. Math. **148** (1998), 619–693.
[51] K. Honda, *On the classification of tight contact structures I*, Geom. Topol. **4** (2000), 309–368.
[52] K. Honda, *On the classification of tight contact structures II*, J. Differential Geom. **55** (1) (2000), 83–143.
[53] K. Honda, *Gluing tight contact structures*, Duke Math. J. **115** (3) (2002), 435–478.
[54] T. Kálmán, *One Parameter Families of Legendrian Torus Knots* (www.arxiv.org/abs/math.GT/0409005).
[55] T. Kálmán, *Contact Homology and One Parameter Families of Legendrian Knots* (www.arxiv.org/abs/math.GT/0407347).
[56] Y. Kanda, *The classification of tight contact structures on the 3-torus*, Commun. Anal. Geom. **5** (1997), 413–438.
[57] P. Lisca and G. Matić, *Tight contact structures and Seiberg-Witten invariants*, Invent. Math. **129** (3) (1997), 509–525.
[58] P. Lisca and G. Matić, *Stein 4-manifolds with boundary and contact structures*, Symplectic, Contact and Low-Dimensional Topology (Athens, GA, 1996), Topology Appl. **88** (1-2) (1998), 55–66.
[59] D. McDuff and D. Salamon, *Introduction to Symplectic Topology*, Oxford University Press, Oxford (1995).
[60] W. Menasco, *On iterated torus knots and transversal knots*, Geom. Topol. **5** (2001), 651–682.
[61] W. Menasco, *Erratum to: On Iterated Torus Knots and Transversal Knots*, in preparation.
[62] K. Michatchev, *Invariants of Legendrian and transversal links*, PhD thesis, Stanford, (2001).
[63] K. Mohnke, *Legendrian links of topological unknots*, Contemporary Mathematics, 279, American Mathematical Society, Providence, RI (2001), 209–211.
[64] H. Morton, *Seifert circles and knot polynomials*, Math. Proc. Cambridge Philos. Soc. **99** (1) (1986), 107–109.
[65] L. Ng, *Knot and Braid Invariants from Contact Homology* (www.arxiv.org/abs/math.GT/0302099).
[66] L. Ng, *Knot and Braid Invariants from Contact Homology II* (www.arxiv.org/abs/math.GT/0303343).
[67] L. Ng, *Framed Knot Contact Homology* (www.arxiv.org/abs/math.GT/0407071).
[68] L. Ng, *Computable Legendrian invariants*, Topology **42** (1) (2003), 55–82.
[69] L. Ng, *Maximal Thurston–Bennequin number of two-bridge links*, Algebr. Geom. Topol. **1** (2001), 427–434.
[70] S. Orevkov and V. Shevchishin, *Markov theorem for transverse links*, J. Knot Theory Ramifications **12** (7) (2003), 905–913.
[71] D. Rolfsen, *Knots and Links*, Mathematics Lecture Series 7 (1976), Publish or Perish Inc.

[72] L. Rudolph, *A congruence between link polynomials*, Math. Proc. Cambridge Philos. Soc. **107** (2) (1990), 319–327.

[73] L. Rudolph, *An obstruction to sliceness via contact geometry and "classical" gauge theory*, Invent. Math. **119** (1995), 155–163.

[74] J. Sabloff, *Augmentations and Rulings of Legendrian Knots* (www.arxiv.org/abs/math.SG/0409032).

[75] J. Sabloff, *Invariants of Legendrian knots in circle bundles*, Commun. Contemp. Math. **5** (4) (2003), 569–627.

[76] J. Swiatkowski, *On the isotopy of Legendrian knots*, Ann. Glob. Anal. Geom. **10** (1992), 195–207.

[77] S. Tabachnikov, *An invariant of a submanifold that is transversal to a distribution* (Russian), Uspekhi Mat. Nauk **43** 3(261) (1988), 193–194 translation in Russian Math. Surveys **43**(3) (1988), 225-226.

[78] S. Tabachnikov, *Estimates for the Bennequin number of Legendrian links from state models for knot polynomials*, Math. Res. Lett. **4** (1) (1997), 143–156.

[79] V. Tchernov, *Vassiliev invariants of Legendrian, transverse, and framed knots in contact three-manifolds*, Topology **42** (1) (2003), 1–33.

[80] N. Wrinkle, *The Markov Theorem for Transverse Knots* (www.arxiv.org/abs/math.GT/0202055).

CHAPTER 4

Knot Spinning

Greg Friedman

Department of Mathematics, Yale University, 10 Hillhouse Ave/P.O. Box 208283, New Haven, CT 06520-8283, USA
E-mail: friedman@math.yale.edu

Contents

Abstract

We survey spinning constructions for higher-dimensional knots. These methods for creating new knots from knots of lower dimension can be described geometrically, and so spun knots often can be visualized directly (or at least schematically). We review the basics of higher-dimensional knot theory and then turn to specific constructions, including simple spinning, superspinning, twist spinning, frame spinning, rolling, and deform spinning. We also discuss some new hybrid constructions, such as frame twist spinning, frame deform spinning, and frame rolling. Throughout, we provide examples of important historical applications and results.

HANDBOOK OF KNOT THEORY
Edited by William Menasco and Morwen Thistlethwaite

1. Introduction

This exposition is intended to provide some introduction to higher dimensional knots – embeddings of S^{n-2} in S^n – through spinning constructions. Once our shoe laces, those archetypal hand tools of knot theory, have been turned into spheres, how can we construct and visualize concrete examples of such knots?

There are many important ways to construct higher dimensional knots. If we are interested in *algebraic knots*, we can look at the links of singularities of complex algebraic varieties in $\mathbb{C}^n$ [7,32]. We can also construct knots by *surgery theory* (see [29] for a recent survey). There are powerful and complex tools for studying the knots that arise in these manners, but such construction methods frequently do not allow one to "see" the knot. Often these knots can be described only in terms of their algebraic invariants. We want to be able to visualize our knots, at least as far as it is possible to do so with our 3-dimensional brains. This brings us to a series of constructions known as *knot spinnings*. Many extensions have been made to Artin's original spinning technique, which dates back to 1925, but the various spinning constructions all have the appeal of being completely geometric in nature and thus highly visual. On top of providing a myriad of examples of importance in knot theory, these constructions provide an excellent introduction to thinking about higher dimensional knots and higher dimensional topology in general.

Unfortunately, there do not seem to be many general references for knot theory in high dimensions, but we list a few sources that might be of interest for a beginner to the subject. More advanced references can be found in these sources.

Colin Adams's popular treatment of knot theory in *The Knot Book* [1] contains a chapter on visualizing high-dimensional knots, as does Charles Livingston's introductory text *Knot Theory*. Dale Rolfsen's classic introduction to knot theory, *Knots and Links* [34], touches on the high-dimensional theory throughout, including spinning constructions on pages 85–87 and 96–99 and a "Higher Dimensional Sampler" in Chapter 11. The paper "A survey of multidimensional knots" by Kervaire and Weber in [25] provides a survey through 1977. Andrew Ranicki's book *High-Dimensional Knot Theory: Algebraic Surgery in Codimension 2* [33] provides a more modern and extensive look at the theory from the point of view of algebraic surgery theory, while the article by Jerome Levine and Kent Orr [29] provides a more compact survey of high-dimensional knot theory via surgery.

In addition, three recent books deal exclusively with knotted 2-spheres (and other surfaces) in $\mathbb{R}^4$ and S^4. These are *Braid and Knot Theory in Dimension Four* [20] by Seiichi Kamada, *Knotted Surface and Their Diagrams* [5] by Scott Carter and Masahico Saito, and *Surfaces in 4-Space* [6] by Carter et al. All the three books have a strong pictorial flavor, and each mentions knot spinning, the first book dealing with it more extensively in its Chapter 10 and the third book in Chapter 2.

I thank Joan Doran for drawing the included figures.

2. Some basics

2.1. *What is a knot?*

We begin with the precise definition of a knot.

Let S^n be the *n-dimensional sphere*, which we will be free to think of in several ways: as an abstract manifold, as the set of points in $\mathbb{R}^{n+1}$ unit distance from the origin, or as $\mathbb{R}^n$ compactified by adding a point at infinity. More generally, we will use S^n to denote any object piecewise linearly (PL) homeomorphic to the sphere. Similarly, we will use B^n to denote any object PL homeomorphic to the unit ball in $\mathbb{R}^n$, the set of points with distance ≤ 1 from the origin. The boundary of B^n, denoted ∂B^n, is PL homeomorphic to S^{n-1}.

With these conventions, a *knot of dimension n* is a PL locally flat embedding $K : S^{n-2} \hookrightarrow S^n$ or $K : S^{n-2} \hookrightarrow \mathbb{R}^n$. Recall that the PL condition simply means that there exist triangulations of S^{n-2} and S^n (or $\mathbb{R}^n$) with respect to which K is simplicial, while K is locally flat if each point $K(x)$ in the image of K has a neighborhood U such that $(U, U \cap K(S^{n-2}))$ is PL homeomorphic to the standard coordinate pair $(\mathbb{R}^n, \mathbb{R}^{n-2})$. There is no real theoretical difference between letting S^n or $\mathbb{R}^n$ serve as the codomain of the knot since we are free to rechoose the point at infinity of S^n so that the image of the knot will lie in $\mathbb{R}^n \subset S^n$ (technically, we are replacing the knot with an *equivalent* one; see below). Sticking with spheres has some technical advantages, and we will principally use spheres as the ambient space, though occasionally it will suit us to use $\mathbb{R}^n$ instead (Figure 1).

The requirement that a knot be PL locally flat is a common restriction, designed to avoid singularities of the embedding. For example, if we only required the embedding to be continuous, "infinite knottedness" might occur (Figure 2). Requiring the knot to be piecewise linear prevents this level of unpleasantness, but local-flatness is also necessary to prevent other kinds of pathologies, such as local knotting, that may occur when one is not working with differentiable maps – note that local-flatness certainly holds for differentiable embeddings by the Tubular Neighborhood Theorem. See [15] for a discussion of non-locally flat knots.

In the classical dimension, $S^1 \hookrightarrow S^3$, PL knots and smooth knots are equivalent. In fact, for any n, any codimension two PL locally flat embedding can be made differentiable (see Corollary 6.8 in [37]). However, in high dimensions the smooth structure on the embedded sphere may not be the standard one. By working in the PL category, we allow these knots but do not concern ourselves with any eccentricities in their smooth structures. All this being said, most of the constructions we will discuss work equally well in the differentiable category and generate knots with the standard smooth structure provided we

Fig. 1. Some smooth knots $S^1 \subset \mathbb{R}^3$. A smooth knot can always be given a PL structure by employing a suitable triangulation of $\mathbb{R}^3$ that contains the knot as a subcomplex.

Fig. 2. A "wild" knot that is not locally flat at the point of infinite "knottedness".

start with a knot with the standard smooth structure. Knot spinning can also be done on non-locally flat PL knots provided some minor extra care is employed (see Section 4.3 of [15]) or with topological non-locally flat knots provided that the embedding is flat at some point. In the sequel, we will stick with PL locally flat knots for convenience and consistency.

2.2. *Knot equivalence*

Now, if you have a knotted string lying on your desk and you pick it up and move it someplace else, we would like to think of it as the same knot. Thus we should really consider equivalence classes of knots. We call two knots $K_0, K_1 : S^{n-2} \to S^n$ *equivalent* if there is an orientation-preserving PL homeomorphism $f : S^n \to S^n$ such that $fK_0(S^{n-2}) = K_1(S^{n-2})$. In other words, f should take the image of K_0 to the image of K_1. In particular, this will be true if there is a PL ambient isotopy of S^n taking $K_0(S^{n-2})$ to $K_1(S^{n-2})$. In fact, this stronger condition is sometimes used as the definition of knot equivalence. Since we work in the PL locally flat category, these two conditions are equivalent (see Proposition 1.10 in [3]), but the analogous equivalence does not hold in the smooth category due to the failure of the Alexander trick.

It is a standard abuse, in which we shall engage freely, to use the word "knot" and the same symbol, K, to refer to the equivalence class of the knot K or even to the image of K.

We refer to $K : S^{n-2} \hookrightarrow S^n$ as *n-dimensional* or an *n-knot*. This is not a universal notation; it is perhaps more standard to refer to such a knot as an $n - 2$ knot. We also refer to the knots $K : S^1 \hookrightarrow S^3$ as *classical knots*.

One also sometimes speaks of *oriented equivalence* for which S^{n-2} is given a fixed orientation and it is required that the orientation-preserving PL homeomorphism $f : S^n \to S^n$ taking $K_0(S^{n-2})$ to $K_1(S^{n-2})$ also preserves the orientation of these subspaces. However, we will not impose this stricter condition except when stated explicitly.

2.3. *The unknot and toroidal decompositions of S^n*

The *unknot* in dimension n is the equivalence class of the "standard embedding" $S^{n-2} \subset S^n$. In other words, if $S^n = \{\vec{x} \in \mathbb{R}^{n+1} \,|\, |\vec{x}| = 1\}$, then the unknot can be represented

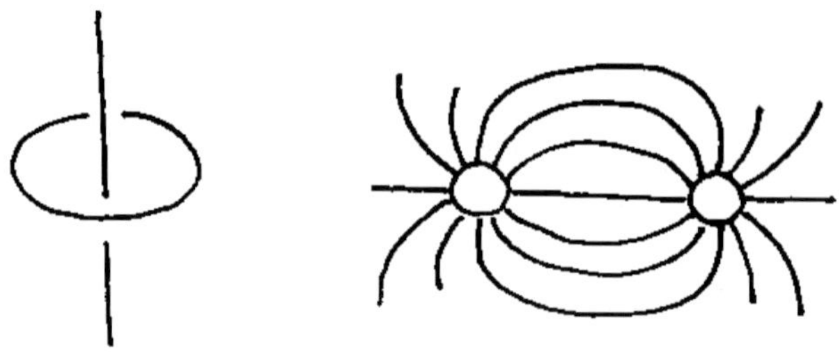

Fig. 3. The decomposition of S^3 (thought of as $\mathbb{R}^3$ plus a "point at infinity") into two solid tori $S^1 \times B^2$ and $B^2 \times S^1$. The left-hand picture shows the circular cores of the tori (the vertical line becomes a circle as it wraps through the point at infinity). The right-hand picture shows a slice along the $y-z$ plane: the two disks are a slice of one solid torus (cut a donut in half and then view it on end), while the arcs represent slices of the meridional disks of the other solid torus.

as $\{\vec{x} \in \mathbb{R}^{n+1} | |\vec{x}| = 1, x_{n+1} = x_n = 0\}$. The classical 3-dimensional unknot is equivalent to the unit circle in the $x-y$ plane in (the compactified) $\mathbb{R}^3$.

By setting more coordinates equal to 0, we can define standard embeddings of any sphere into any other sphere of higher dimension. This leads to nice decompositions of S^n into two generalized solid tori: the standard embedding of S^m into S^n, $m < n$, has a tubular neighborhood PL homeomorphic to $S^m \times B^{n-m}$, and the complement of the interior of this neighborhood is PL homeomorphic to $B^{m+1} \times S^{n-m-1}$. Thus S^n can be decomposed as a union of $S^m \times B^{n-m}$ and $B^{m+1} \times S^{n-m-1}$, identified in the obvious way along their common boundary $S^m \times S^{n-m-1}$ (Figure 3):

$$S^n = S^m \times B^{n-m} \bigcup_{S^m \times S^{n-m-1}} B^{m+1} \times S^{n-m-1}$$

This follows, e.g., from the fact that S^n can be written as the join of $S^m * S^{n-m-1}$. The most familiar case is the standard genus one Heegard decomposition of S^3, in which a neighborhood of the unknot $S^1 \subset S^3$ is the solid torus $S^1 \times B^2$, whose complement is another solid torus, $B^2 \times S^1$.

2.4. *A useful excision*

We conclude this introductory section with one other construction that will be used repeatedly. Consider an n-dimensional knot K, and choose any point $x \in K \subset S^n$ (here – by our standard notational abuse – we use "K" to represent the image of the knot). Since K is PL locally flat, there is a neighborhood B^n_- of x in S^n such that $(B^n_-, B^{n-2}_-) := (B^n_-, B^n_- \cap K)$ is PL homeomorphic to an unknotted ball pair, i.e., it is PL homeomorphic to the standard ball pair $(B^n, B^n \cap \mathbb{R}^{n-2})$ in $\mathbb{R}^n$. Since the closure of the complement of a PL n-ball in S^n is also an n-ball, the closures of the complements $S^n - B^n_-$ and $K - B^{n-2}_-$ will each be balls, and we label this complementary pair by (B^n_K, B^{n-2}_K). The ball B^{n-2}_K may be knotted in B^n_K (Figure 4).

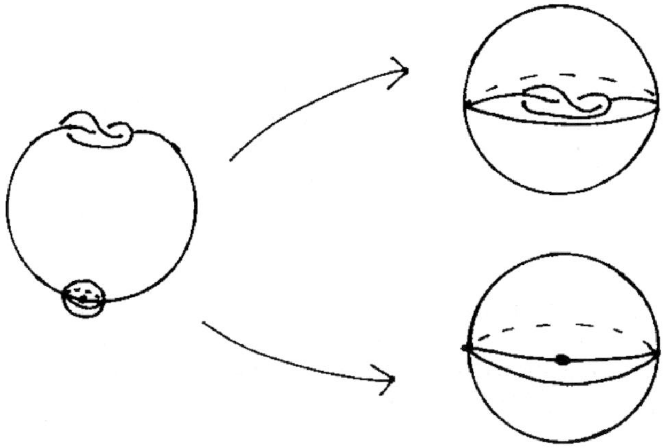

Fig. 4. Removing a trivial neighborhood from a knotted circle to obtain a knotted arc.

We also observe that the common boundary of the pairs (B^n_-, B^{n-2}_-) and (B^n_K, B^{n-2}_K) is the *unknotted* pair of spheres (S^{n-1}, S^{n-3}) (since it is PL homeomorphic to the boundary of the standard ball pair). In what follows, it will often be convenient to identify this with the *standard unknot*, which we have already discussed.

What if we choose the neighborhood of a different point to remove in this construction? It turns out that we get the same pair (B^n_K, B^{n-2}_K) up to PL homeomorphism. To see this, consider the ball neighborhoods of two different points. We can simply slide one ball to the other along the knot, which complementarily takes the complement of one neighborhood to the complement of the other. Note that while this idea has nice intuitive appeal, it does require some technical checking to ensure that such sliding is always allowed. However, this theory is well established, and we avoid going too far afield to visit the details here (see, e.g., Chapter 6 of [19]).

3. Basic spinnings

3.1. *Simple spinning*

As we know, there are an infinite number of knots $S^1 \hookrightarrow S^3$ and myriad examples can be created by anyone with a piece of string and some time on their hands (classifying these knots is another matter!). To get knots of higher dimensions requires a little bit more ingenuity. One method is to get high-dimensional knots from knots of lower dimension by spinning them.

The earliest spinning construction is due to Emil Artin in 1925 [2]. Artin used spinning to construct 4-dimensional knots from classical knots, but the same idea can be used to create an $n+1$ dimensional knot from any n-dimensional knot. This construction is generally referred to just as "spinning", but we will call it *simple spinning* to differentiate it from the more general constructions to follow.

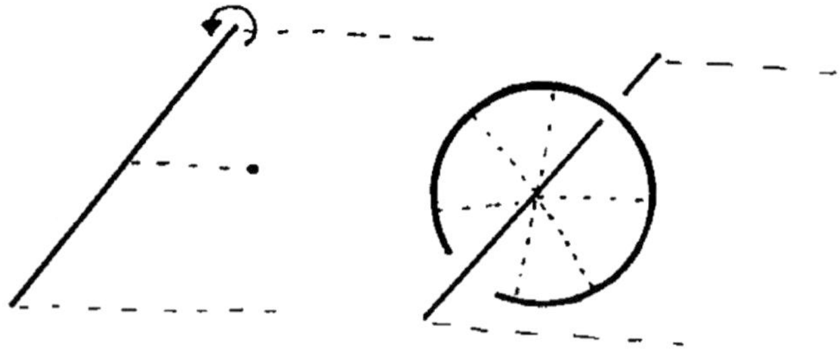

Fig. 5. Spinning a point in the half-plane around the axis.

In this section, it will be most convenient to consider knots in $\mathbb{R}^n$ instead of S^n (see Section 2), though of course we can easily transform from one type to the other by adding or removing a point at infinity.

To see the basic idea, consider the upper half-plane $H^2 = \{(x, y) \in \mathbb{R}^2 | y \geq 0\}$ and choose a point $(x_0, y_0) \in H^2$ with $y_0 > 0$. Now rotate H^2 around the x-axis in $\mathbb{R}^3$. The point will sweep out a circle (Figure 5). Analytically, the circle will be parametrized in $\mathbb{R}^3$ by the set of points $(x_0, y_0 \cos\theta, y_0 \sin\theta)$, as θ runs from 0 to 2π (assuming that we rotate counterclockwise as seen from the positive x-axis looking in the negative x direction).

To see how this applies to knots, let us consider a knot K in $\mathbb{R}^3$. Up to equivalence, we can arrange for K to lie in the upper half-space $H^3 = \{(x, y, z)|z \geq 0\}$ except for an unknotted arc that dips below the x–y plane $\mathbb{R}^2 = \{(x, y, z)|z = 0\}$ (Figure 6). Let us remove the interior of this unknotted arc; what remains is a knotted arc in H^3 with its endpoints (and only its endpoints) in $\mathbb{R}^2$. We can now rotate H^3 around $\mathbb{R}^2$ in $\mathbb{R}^4$ just as we rotated H^2 around $\mathbb{R}^1$ in $\mathbb{R}^3$. Analytically, we parametrize by θ, and each point (x, y, z) in the upper half-space sweeps out the circle $(x, y, z\cos\theta, z\sin\theta)$. Note that $\mathbb{R}^2$ remains fixed. By thinking about how the longitude lines swing around the globe with the north and south poles remaining fixed, we can imagine how the knotted arc gets spun into the image of a 2-sphere S^2. Thus, by spinning, we obtain a knotted S^2 in $\mathbb{R}^4$ (Figures 7 and 8).

You might be asking, what if we had chosen a different way to split our original knot into a knotted arc? It turns out that we get the same spun knot, essentially for the same reason by which we noted in Section 2.4 that B^n_K is independent of the choice of B^n_-. In fact, notice that if we start with our knot in S^3, then our construction to get a knotted arc in the upper half-space by removing an unknotted arc in the lower half-space is completely

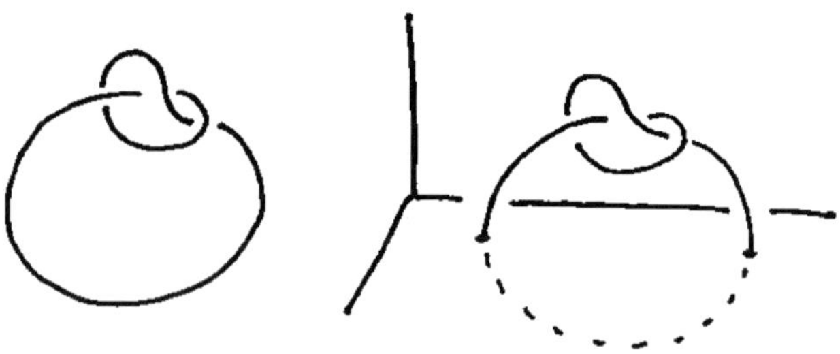

Fig. 6. Turning a knotted circle into a knotted arc in the upper half-space in order to spin it about the plane.

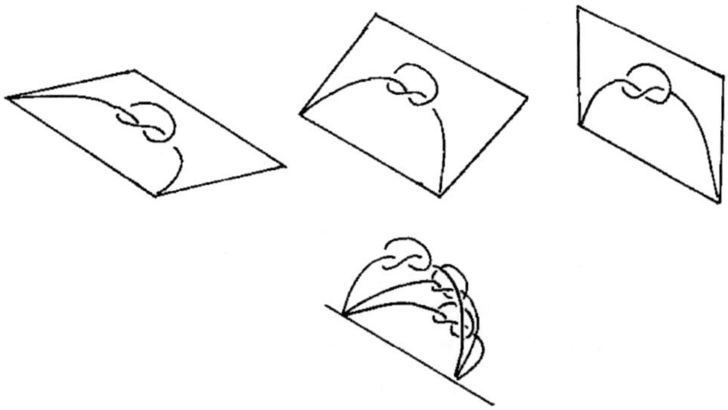

Fig. 7. A schematic of knot spinning.

equivalent to the construction of B_K^n by removing a small ball neighborhood of a point on the knot.

This simple spinning construction already has several important ramifications. For example, it can be shown very easily [34] that the fundamental group of the complement of this spun knot in $\mathbb{R}^4$ (its knot group) is isomorphic to the knot group of our original knot in $\mathbb{R}^3$. Based on known results about knots in $\mathbb{R}^3$, this implies the existence of an infinite number of inequivalent knots in $\mathbb{R}^4$. By contrast, Dennis Roseman showed in [35] that the spins of two distinct knots may be equivalent. For example, he showed that spinning the square knot yields the same 4-knot as does spinning the granny knot. We discuss a much stronger result along these lines at the end of Section 3.2.

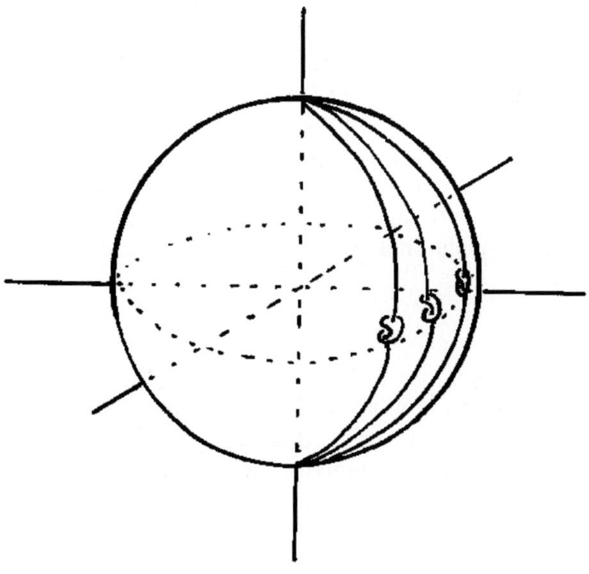

Fig. 8. A second schematic of knot spinning.

The construction for higher dimensions is similar. We begin with a knot $K: S^{n-2} \to \mathbb{R}^n$. Again, we can manipulate the knot within its equivalence class so that it lies mostly in the upper half-space $H^n = \{(x_1, \dots, x_n) \in \mathbb{R}^n | x_n \geq 0\}$ and so that the intersection of the knot with the lower half-space is unknotted. We then remove the interior of this unknotted ball to obtain the complementary knotted ball B^{n-2} in H^n. Its intersection with $\mathbb{R}^{n-1}$ is an unknotted S^{n-3}. Now we spin H^n around $\mathbb{R}^{n-1}$ into $\mathbb{R}^{n+1}$ so that each point $(x_1, \dots, x_n)$ sweeps out the circle $(x_1, \dots, x_{n-1}, x_n \cos \theta, x_n \sin \theta)$.

It is a little harder now to see that our knotted ball in the upper half-plane gets spun into a sphere S^{n-1}, but the idea of pivoting a longitude around its poles extends to higher dimensions. To see this, we employ the following coordinate analysis, which will also be useful when we need to describe more general spinnings, below.

Consider S^{n-1} as the unit sphere in $\mathbb{R}^n$, $S^{n-1} = \{\vec{x} \in \mathbb{R}^n | |\vec{x}| = 1\}$, and consider $\mathbb{R}^n$ as $\mathbb{R}^{n-2} \times \mathbb{R}^2$. Then we can define the latitude for a point $y \in S^{n-1}$ as its orthogonal projection onto the $\mathbb{R}^{n-2}$ factor and the longitude of y as the angular polar coordinate of its projection onto the $\mathbb{R}^2$ factor. Hence the latitude is always well defined, while the longitude is either undefined or a unique angle, dependent upon whether or not y lies in $\mathbb{R}^{n-2} \times 0$. Notice that in the case where the longitude is undefined, the point on the sphere is uniquely determined by its latitude (just as on a globe). To simplify the notation in abstract cases, we will simply refer to the latitude–longitude coordinates (z, θ), whether θ is defined or not. Then the point (z, θ) in S^{n-1} corresponds to the point in $\mathbb{R}^n$ determined by the rectangular coordinates $(z, r \cos \theta, r \sin \theta)$ for $z \in \mathbb{R}^{n-2}, r \geq 0, \theta \in [0, 2\pi)$, and such that $|z|^2 + r^2 = 1$ (note that this determines r, given z).

Now, consider the closure of the set of points in S^{n-1} with fixed longitude $\theta = 0$. These points can be written in rectangular coordinates as $(z, r, 0)$, with $|z|^2 + r^2 = 1$ and $r \geq 0$. This set is homeomorphic to a ball B^{n-2} (in fact, it is the graph of $r = \sqrt{1 - |z|^2}$). Its boundary is the $n-3$ sphere with $|z|^2 = 1$ and $r = 0$, and we call this boundary the *generalized pole* of S^{n-1}. Now for each point $(z, r, 0)$ in rectangular coordinates, we can spin to get the set of points $(z, r \cos \theta, r \sin \theta)$ as θ runs from 0 to 2π. The points $(z, 0, 0)$ of the generalized pole remain fixed, and the rest of the 0 longitude sweeps out the rest of the sphere. Analogously, as we spin a knot, the knotted ball sweeps out a knotted sphere. We leave it to the reader to formulate a precise analytical description.

Preservation of knot groups under simple spinning continues to hold in this higher dimensional setting, and by iterating the spin construction, we establish the existence of an infinite number of inequivalent knots in any dimension $n \geq 3$.

3.2. *Superspinning*

Having spun knots in circles, how about spinning around higher dimensional spheres? *Superspinning* of classical knots was introduced by Zeeman [42] and Epstein [8] separately in 1960. Zeeman used the spinning of classical knots about spheres to show that it is possible to embed two $n-2$ spheres in S^n such that each sphere is unknotted but the pair is linked, i.e., there is no homeomorphism of S^n sending one knot to the "northern hemisphere" and the other knot to the "southern hemisphere". Epstein strengthened these results to show that two $n-2$ spheres can be embedded in Euclidean n-space in each of the following ways: (i) neither can be shrunk to a point in the complement of the other; (ii) one can and one cannot be shrunk to a point in the complement of the other. In 1970, Sylvain Cappell [4] generalized this construction as a way to construct an $n+p$ dimensional knot from any n-dimensional knot by spinning it around a p-sphere $S^p, p \geq 1$. He called this method "superspinning", though it is sometimes referred to in the literature as p-spinning. Cappell utilized superspinning to demonstrate the existence of knots whose complements are homotopy equivalent but not homeomorphic.

This time, let us jump straight to the general construction, p-spinning for a knot $K : S^{n-2} \hookrightarrow S^n$.

Imagine $S^p \times B^n$ embedded in our standard unknotted way in S^{p+n} so that we can write

$$S^{p+n} = [S^p \times B^n] \bigcup_{S^p \times S^{n-1}} [B^{p+1} \times S^{n-1}]$$

(see Section 2). Here $\cup_{S^p \times S^{n-1}}$ indicates that we are gluing the two spaces along their common boundary $S^p \times S^{n-1}$. We can decompose the unknot S^{p+n-2} in S^{p+n} by its intersections with the pieces of this decomposition as

$$[S^p \times B^{n-2}] \bigcup_{S^p \times S^{n-3}} [B^{p+1} \times S^{n-3}]$$

$$\begin{array}{ccc} S^p \times B^{n-2} & \cup & B^{p+1} \times S^{n-3} \\ \Big\downarrow & & \Big\downarrow \\ S^p \times B^n & \cup & B^{p+1} \times S^{n-1} \end{array}$$

(Here we think of B^{n-2} as the unknotted subset of B^n given by setting the last two coordinates to 0.) This is one of our standard decompositions of S^{p+n-2}, but now we see it lying within a decomposition of the larger sphere S^{p+n}. We can write the pair of spaces more compactly as

$$S^p \times (B^n, B^{n-2}) \cup B^{p+1} \times (S^{n-1}, S^{n-3})$$

and we should think of this as the product of S^p with a trivial (unknotted) ball pair, "capped off" by another standard piece.

Try to picture this decomposition of the unknotted S^2 in S^4, taking $p = 1$, $n = 3$ and recalling that S^0 is a pair of points. In this case, S^2 decomposes into a neighborhood of the equator and neighborhoods of the north and south poles. The decomposition of S^4 consists of a neighborhood in S^4 of the equator of S^2 and its complement, which is $B^2 \times S^2$. So, written as pairs, we have the unknot (S^4, S^2) decomposed into $S^1 \times (B^3, B^1)$, which will play the important role in our spinning construction, and $B^2 \times (S^2, S^0)$, "the rest".

Now, within this construction, we can replace each trivial pair (B^n, B^{n-2}) in the product $S^p \times (B^n, B^{n-2})$ with the knotted ball pair (B^n_K, B^{n-2}_K), obtained from K as in Section 2.4. In other words, we construct

$$S^p \times (B^n_K, B^{n-2}_K) \cup B^{p+1} \times (S^{n-1}, S^{n-3})$$

and we define the superspun knot K^* to be the subset given by

$$[S^p \times B^{n-2}_K] \bigcup_{S^p \times S^{n-3}} [B^{p+1} \times S^{n-3}]$$

As K^* is PL homeomorphic to the standard decomposition of S^{n+p-2}, we see that K^* is a sphere of dimension $n+p-2$ knotted in S^{n+p}.

If $p = 1$, superspinning K gives us the same simple spun knot that we obtained in Section 3.1. Can you see why? Try thinking about 1-spinning classical knots.

It turns out that the knot group of a superspun knot is also the same as the knot group of the original knot, but in general, superspinning does not create the same knots as does iterated simple spinning.

Also as for simple spinning, $p > 1$ superspinning may take inequivalent knots to equivalent knots: Cameron Gordon showed in [17] that all superspun knots are amphicheiral, i.e., they are oriented equivalent to the knot obtained by reversing the orientation of both S^{n+p} and K^* (this is sometimes called $(-)$-amphicheirality). As a corollary, this generalizes Roseman's result, cited above, on the equivalence of the spun granny knot with the spun square knot, and it implies that the $p > 1$ superspins of inequivalent knots may be equivalent. Another result along these lines was obtained by Cherry Kearton, who showed in [24] that the superspins of two classical knots $K_1, K_2 \subset S^3$ are equivalent if and only if their knot groups are isomorphic. This is false, however, for the superspins of knots of higher dimension (see, e.g., [23]).

3.3. *Frame spinning*

Even more general than superspinning is *frame spinning*: why limit ourselves to spinning about spheres? How about other manifolds? Frame spinning was introduced by Dennis Roseman in 1989 [36], though the name is due to Alexander Suciu [38], who used frame spinning to construct new examples of inequivalent knots that have the same complement (this is a phenomenon that occurs only for knots above the classical dimension, though at most two higher dimensional knots can share a given complement; see, e.g., [28]).

To describe frame spinning, let us once again begin with an n-dimensional knot K. This time, however, our additional data comes in the form of an m-dimensional manifold M^m embedded in S^{n+m-2} with a framing ϕ. This last condition means that we in fact consider an embedding $\phi\colon M^m \times B^{n-2} \hookrightarrow S^{n+m-2}$. Furthermore, we assume that S^{n+m-2} is embedded in the standard, unknotted way into S^{n+m} with the standard framing as in the generalized torus decomposition. Putting these framings together, we get a pair of tubular neighborhoods of M^m in (S^{n+m}, S^{n+m-2}) of the form $N = M^m \times (B^n, B^{n-2})$, where each

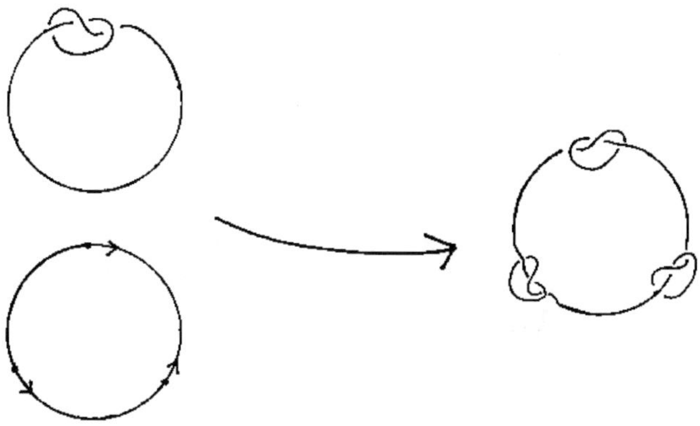

Fig. 9. The trefoil knot spun about the manifold M consisting of three disjoint points in S^1. Note that the framing at each point (indicated by an arrow that depicts the orientation of the framing) determines how to attach the knot.

(B^n, B^{n-2}) is an *unknotted* ball pair (although the exact embedding of N into S^{n+m} depends on our choice of framing ϕ) (Figure 9).

The idea now is to take all those unknotted ball pair fibers in N and replace them with our knotted ball pair (B^n_K, B^{n-2}_K) as we did for superspinning. In other words, having used the framing to identify the neighborhood pair N as $M^m \times (B^n, B^{n-2})$, we remove it and then replace it with $M^m \times (B^n_K, B^{n-2}_K)$, glued in along the same framing. Thus, our frame spun knot will be

$$(S^{n+m-2} - M^m \times B^{n-2}) \bigcup_{M^m \times S^{n-3}} M^m \times B^{n-2}_K$$

sitting inside the $n + m$ sphere

$$(S^{n+m} - M^m \times B^n) \bigcup_{M^m \times S^{n-1}} M^m \times B^n_K.$$

In the special case where M^m is the sphere S^m embedded in the standard way and with standard framing in S^{n+m-2}, we recover m-superspinning (why?).

If the manifold M has multiple components, or even components of different dimensions, then we can spin different knots (also possibly of different dimensions) around each component. It is possible to generalize this construction even further, but first we should study some other types of spinning.

4. Spinning with a twist

4.1. *Twist spinning*

Twist spinning, introduced by Zeeman in 1965 [43], was an early generalization of Artin's simple spinning construction. Again, we begin with an n-dimensional knot and

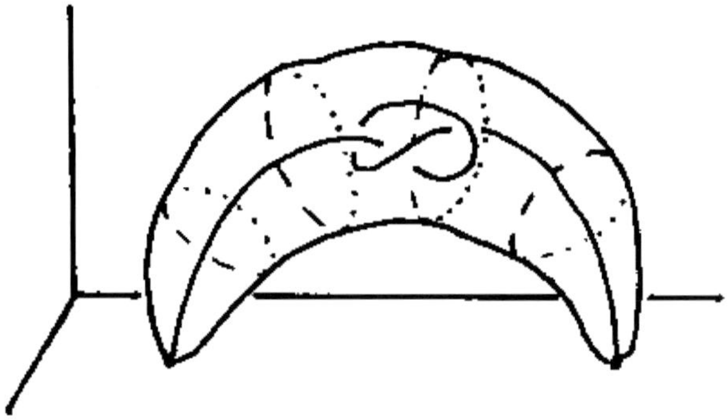

Fig. 10. We rotate a ball around the knotted arc.

obtain an $n + 1$ dimensional knot, but the difference between simple spinning and twist spinning can be illustrated celestially: As the moon orbits the Earth, it always keeps the same face towards the Earth. This is analogous to simple spinning in which the knot is rotated around the plane but always keeps "the same face" towards the plane serving as the axis of rotation. By comparison, twist spinning is like the Earth orbiting the sun: as the earth orbits, it also rotates around its own axis.

Before giving a general formula, let us consider heuristically the case of twist spinning a classical knot. As in the simple spinning construction, we replace the knot with a knotted arc in the upper half-space whose endpoints lie in the $x-y$ plane. We can also assume that the knotted part of the arc is contained within a ball whose intersection with the arc is its north and south poles (Figure 10). Now, as we rotate half-space around the plane as in simple spinning, we simultaneous spin this ball on its axis (Figure 11). It is only necessary that the end result lines up with the starting position, so we are free to spin the ball on its axis any integral number k times as we rotate H^3.

Let us be more specific. Given an n-knot K, then just as for superspinning about S^1 (which is equivalent to simple spinning), we decompose the $n + 1$ dimensional unknot as the two space pairs $S^1 \times (B^n, B^{n-2})$ and $B^2 \times (S^{n-1}, S^{n-3})$. To superspin, we simply removed $S^1 \times (B^n, B^{n-2})$ and glued in $S^1 \times (B^n_K, B^{n-2}_K)$, reattaching along the original boundary $S^1 \times (S^{n-1}, S^{n-3})$. In order to create the k-twist spin, however, we glue in the

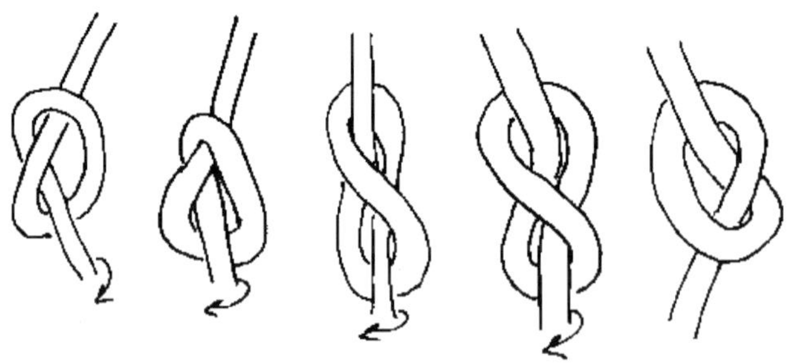

Fig. 11. A 180° twist of the trefoil knot (thickened for improved visualization).

following way: we represent points in the common boundary $S^1 \times S^{n-1}$ by (η, z, θ), where $\eta \in S^1$ and (z, θ) are latitude–longitude coordinates for S^{n-1} such that the unknotted $S^{n-3} \subset S^{n-1}$ is the generalized pole (see Section 3.1). If (η, z, θ) is such a point in the boundary of $B^2 \times S^{n-1}$, we attach that point to the boundary of $S^1 \times B^n_K$ by $(\eta, z, \theta) \rightarrow (\eta, z, \theta + k\eta)$. The addition here is the standard angle addition in the circle, which we can think of as $\mathbb{R}/2\pi\mathbb{Z}$. In this way, as we glue the pieces together, we introduce a k-fold twisting by rotations of the longitude coordinate. You should convince yourself that this procedure corresponds to our earlier heuristic description.

If your first instinct is to think that the twisting does not add anything to the spinning since the knotted arc lands back where it started, you should consider the following toy example. Recall from Section 3.1 our original toy example of simple spinning in which we rotated a point in the upper half-plane around the x-axis. This time, however, imagine two points in the upper half-plane. As we sweep the half-plane around the x-axis, let these two points rotate around each other in the plane k times, where k is any integer. At the end of the process, we have two curves in space that link each other k times (Figure 12). This is not quite the same procedure as twist spinning, but it should illustrate the idea that interesting things can happen if we deform as we spin.

Zeeman showed in [43] that a twist spun knot depends only on K and $|k|$, i.e., k-twist spinning and $-k$-twist spinning yield the same knot. Furthermore, he proved the slightly surprising fact that any 1-twist spun knot (and hence also any -1-twist spun knot) is unknotted! This is actually a corollary of the much stronger theorem in the same paper stating that any k-twist spun knot is a fibered knot with fiber, the punctured k-fold branched cover of S^n determined by the knot being spun. It has also been shown, by Cameron Gordon in his thesis [18], that if k and l are coprime, then a k-twist spin followed by an l-twist spin yields the unknot; this was generalized by Tristram (unpublished), who showed that any sequence of k_i-twist spins, where the k_i are coprime, is unknotted. A short proof of both statements can be found in [22]. Deborah Goldsmith and Louis Kauffman found another generalization by showing in [16] that, if $L_{k,l}(K)$ indicates the l-twist spin of the k-twist spin of K, then $L_{k,l}(K)$ is equivalent to one of $L_{0,g}(K)$ or $L_{g,g}(K)$, where $g = \text{g.c.d.}(k, l)$.

Unlike the constructions of Section 3, twist spinning does not preserve knot groups. In fact, if G is the knot group of K, then the group of the k-twist spin of K is isomorphic to $(\mathbb{Z} \times G)/\langle t^{-1}g^k \rangle$, where t is a generator of $\mathbb{Z}$ and $g \in G$ represents a meridian of the knot K [14]. Note that if $k = 0$, i.e., we simple spin, this group is just G. It is less easy to see, though it must be true by Zeeman's theorem, that if $k = \pm 1$, then this group is $\mathbb{Z}$, the knot group of the unknot! Actually, this is not hard to show algebraically, using the fact that any knot group is the normal closure of any element representing a meridian.

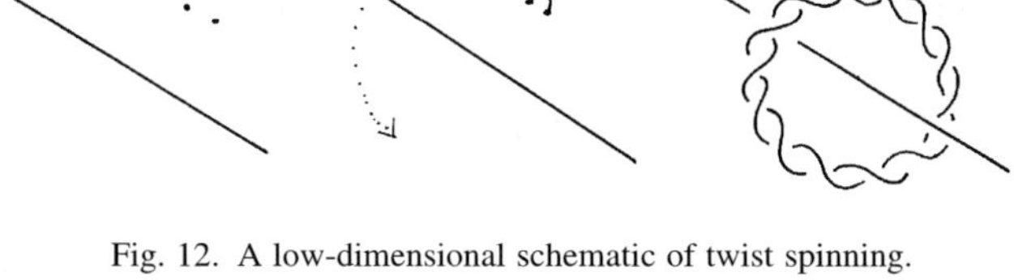

Fig. 12. A low-dimensional schematic of twist spinning.

4.2. *Frame twist spinning*

Now that we have seen how to add twisting to Artin's basic spinning construction, can we add some kind of twisting to our other spins? For superspinning about spheres of dimension greater than 1, the answer is no! This is because $\pi_n(S^1) = 0$ for all $n > 1$, which implies that we cannot use higher dimensional spheres to parametrize spinning. Any attempt at twisting can be deformed to give back ordinary superspinning. On the other hand, since $\pi_1(S^1) = \mathbb{Z}$, there are countably many maps $S^1 \to S^1$ that cannot be so deformed, and the element k in $\pi_1(S^1)$ corresponds to k-twist spinning.

However, where superspinning fails to be twistable, frame spinning does allow a twist if the manifold M^m admits a map $M^m \to S^1$ that cannot be deformed into the trivial map to a point. Just as for twist spinning, this map provides us with enough data to alter the gluing map of the construction by twisting the longitude coordinate of B^n_K as we glue. The gluing of the latitude coordinates is once again controlled by the framing ϕ of M.

So let us be specific. Recall that, in frame spinning, we used the framing of M^m in S^{n+m-2} along with the trivial framing of S^{n+m-2} in S^{n+m} to identify a neighborhood N of M^m in (S^{n+m}, S^{n+m-2}) with the product $M^m \times (B^n, B^{n-2})$. Then we replaced $M \times (B^n, B^{n-2})$ with $M \times (B^n_K, B^{n-2}_K)$ and glued it back in *along the same framing*. Suppose, however, that we are given a map $\tau : M^m \to S^1$. Then we can use this map to augment the gluing with a twist along the longitude. This is done as follows: we use the framings to assign coordinates (x, z, θ) to the boundary $M \times (S^{n-1}, S^{n-3})$ of the neighborhood N. Here $x \in M$ and (z, θ) are latitude/longitude coordinates on S^{n-1} such that the unknotted S^{n-3} is the generalized pole. The boundary of the complement of N in S^{n+m} possesses the same coordinates, as these two boundaries agree. Again, we cut out N and replace it with $M^m \times (B^n_K, B^{n-2}_K)$, which we glue back in, but instead of gluing the point (x, z, θ) in $\partial(S^{n+m} - N)$ right back to its counterpart in ∂N, we glue it by the attaching map $f : (x, z, \theta) \to (x, z, \theta + \tau(x))$, where again the addition in S^1 is the standard angle addition.

In other words, we form

$$[(S^{n+m}, S^{n+m-2}) - M^m \times (B^n, B^{n-2})] \bigcup_f [M^m \times (B^n_K, B^{n-2}_K)]$$

where $\cup_f$ indicates a gluing via the attaching map given above.

This construction was introduced by the author in his dissertation. He goes on to calculate various algebraic invariants of frame twist spun knots, including Alexander polynomials in [15] and the knot groups in [14]. If the map τ is homotopic to the trivial map, we recover Roseman's frame spinning. If $M = S^1$ and τ is the map that wraps the circle around itself k times, we recover Zeeman's k-twist spinning.

5. More general spinnings

5.1. *Deform spinning*

An even more general class of spinning constructions is known to exist. The first example, *roll spinning*, was introduced in a short paper by Ralph Fox in 1966 [13]. Unfortunately,

Fox provided only one example and did not include specific details of the construction, which has led to some contention over the exact definition of roll spinning. In 1979, Litherland [30] provided a formal definition and a generalization, *deform spinning*, of which both roll spinning and twist spinning are special cases. According to Masakazu Teragaito [39], Fox's original construction is actually an example of what Litherland calls *symmetry spinning*, but by now Litherland's definition of roll spinning is the one that has caught on.

Deform spinning is another construction that takes n-knots to $n+1$ knots. The tersest description of deform spinning comes from once again thinking of a simple spin as a special case of a frame spin, i.e., as

$$[(S^{n+1}, S^{n-1}) - S^1 \times (B^n, B^{n-2})] \bigcup_{\partial} [S^1 \times (B^n_K, B^{n-2}_K)]$$

where $\cup_\partial$ indicates gluing along the common boundary in the obvious (untwisted) fashion. Suppose now that we have a 1-parameter family f_ψ of *deformations* of B^n_K rel ∂B^n_K such that f_0 is the identity and $f_{2\pi}(B^{n-2}_K) = B^{n-2}_K$. The family f_ψ should also depend piecewise linearly on the parameter ψ. Litherland then describes the deform spin of K as

$$[(S^{n+1}, S^{n-1}) - S^1 \times (B^n, B^{n-2})] \bigcup_{\partial} (S^1 \times B^n_K, \bigcup_{\psi \in S^1} \psi \times f_\psi(B^{n-2}_K))$$

In other words, as we spin, we deform the knotted ball according to f_ψ. Note that in this description $S^1 \times B^n_K$ is the ordinary undeformed product, but we equally well could have used the deformation of the pair; Litherland demonstrates the equivalence of the two approaches and uses it to redefine the deform spin in terms of *crossed products* of spaces. However, for our purposes in the following sections, it is perhaps easier to maintain the original viewpoint. In this language, it is easy to observe that simple spinning corresponds to setting f_ψ equal to the identity for each ψ, while k-twist spinning corresponds to setting f_ψ equal to the rotation of the longitude coordinate of B^n_K by $k\psi$. (Technically, to get around the fact that we need to keep the boundary of B^n_K fixed, we rotate a smaller interior ball allowing the region between the two boundaries to become stretched around; however, it is clear that so long as the ball being rotated encompasses the knotted part of B^{n-2}_K, this does not affect the final construction of the deform spun knot, and we recover our original description of twist spinning.)

Litherland also shows that, thinking of the collection f_ψ as a PL map

$$f: \ B^n_K \times [0, 2\pi] \to B^n_K \times [0, 2\pi],$$

the type of the deform spun knot is dependent only upon the *pseudo-isotopy class of* f *rel* ∂B^n_K *as a map of pairs* (f and g are pseudo-isotopic rel ∂B^n_K as maps of pairs if there is a PL homeomorphism

$$H: \ (B^n_K, B^{n-2}_K) \times [0, 2\pi] \to (B^n_K, B^{n-2}_K) \times [0, 2\pi]$$

such that $H|_{B^n_K \times 0}$ and $H|_{\partial B^n_K \times [0,2\pi]}$ are the identity maps and $H|_{B^n_K \times 2\pi} = f_{2\pi} g^{-1}_{2\pi}$).

We can now define roll spinning of a classical knot $K: S^1 \hookrightarrow S^3$. We present a new geometric description of Litherland's construction that we hope will be valuable to our readers. This interpretation of roll spinning is due independently to Dennis Roseman.

Recall our definition of (B_K^3, B_K^1) by removing an unknotted ball neighborhood of a point of K. Since we are dealing with a classical knot, we can parametrize S^1 by angles ψ and consider $(B_{K,\psi}^3, B_{K,\psi}^1)$ built as the complement of the neighborhood of the point $K(\psi)$ of the knot. We have already noted that for different choices of ψ, the pairs $(B_{K,\psi}^3, B_{K,\psi}^1)$ are all PL homeomorphic. Nevertheless, starting from a fixed base, say $\psi = 0$, we can view the collection of homeomorphisms $f_\psi : (B_{K,0}^3, B_{K,0}^1) \to (B_{K,\psi}^3, B_{K,\psi}^1)$ as a 1-parameter family of deformations and use this to deform spin. This construction is roll spinning. k-roll spinning can be created by rolling the basepoint around the knot k times. A more technically precise formulation is given in [30] (see also [39]). Note that this construction depends on a choice of framing of K in order to control the roll (in the aeronautical sense) of the ball as it traverses the knot; thus one usually defines rolling with respect to some fixed standard framing of K, usually the one in which K and a longitude of the boundary torus of the framed neighborhood of K do not link. If we use a different framing, we will twist as we roll; this leads to twist roll spun knots or, more specifically, l-twist k-roll spun knots (Figure 13).

Litherland treats another example of deform spinning that applies only to knots which possess symmetries, i.e., periodic homeomorphisms $S^n \to S^n$ that take the knot to itself. This construction is called *symmetry spinning*. However, the construction is slightly technical, involving certain branched covers of the sphere, and so we omit a detailed description. The interested reader is referred to [21] or [30], in which Taizo Kanenobu utilizes symmetry spinning to obtain some counterexamples to the 4-dimensional Smith conjecture.

Unfortunately, roll spinning cannot be generalized to roll spinning of classical knots about higher dimensional spheres. As for twist spinning, this is because $\pi_n(S^1) = 0$ for $n > 1$, so any attempt to parametrize rolling by a sphere of dimension >1 will re-create superspinning. Similarly, since the nth homotopy group $\pi_n(S^1 \times S^1)$ is trivial for $n > 1$, super twist roll spinning is also nothing new. However, Litherland does discuss generalizations such as roll spinning higher dimensional knots about S^1. This utilizes an isotopy of S^{n-1} just as classical roll spinning utilizes the isotopy that rotates S^1. Rather than visit the technical details here, we will jump right to a more general construction.

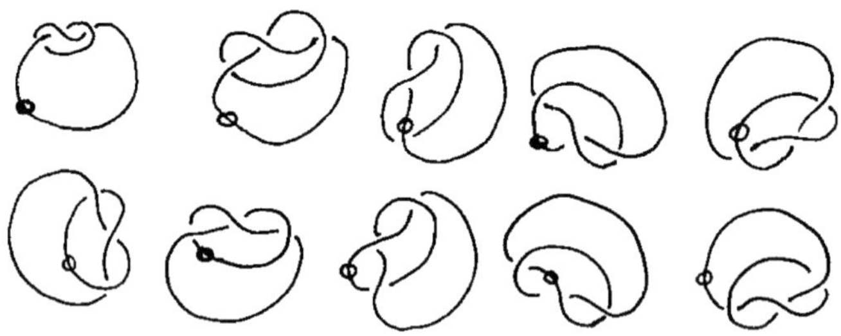

Fig. 13. Rolling the trefoil. The small circle in these figures represents B_-. Rather than moving B_-, it is more illustrative to hold it fixed and roll the knot around it! In fact, this roll of the trefoil is twisted with respect to the unlinked longitude. Furthermore, due to the symmetries of a torus knot, the rolling part of the deformation has no effect, only the twisting. Can you see why? Nonetheless, this diagram illustrates the idea of the procedure.

5.2. *Frame deform spinning*

Putting together frame spinning and deform spinning, we can introduce a new knot construction, *frame deform spinning* (this construction is implicit in a remark in Section 3 of [36]). By now the method may be obvious: we begin with an n-knot K and an m-manifold M^m embedded with framing in S^{n+m-2}, which itself sits unknotted and with the standard framing in S^{n+m}. We also posit a map f from M^m into the space of PL homeomorphisms of B^n_K rel ∂B^n_K, taking $x \in M^m$ to f_x. Then we can define the frame deform spin of K as

$$[(S^{n+m}, S^{n+m-2}) - M^m \times (B^n, B^{n-2})] \bigcup_{M^m \times S^{n-3}} (M^m \times B^n_K, \cup_{x \in M^m} f_x(B^{n-2}_K)).$$

If K is a classical knot and there is a non-trivial map $g : M^m \longrightarrow S^1$, we can compose g with the 1-parameter family of deformations used to define roll spinning to create *frame roll spinning*. We could also use a map $M \rightarrow S^1 \times S^1$ to define *frame twist roll spinning*.

An example of frame deform spinning due to Roseman provides a higher dimensional version of rolling. In classical rolling, we created a 1-parameter deformation by moving the point at which we cut out (B^3_-, B^1_-) around the knotted circle K. What if we similarly moved the excision point around a higher dimensional knot in order to roll spin a knotted S^p about S^p? The difficulty is that as we move the point around, we really need a frame on it to describe precisely how the ball B^n_- is situated in S^n so that we can say exactly how the complement is deformed under the motion. For non-deform spinning, the frame is irrelevant since all are equivalent by a homeomorphism and so yield the same spun knots. For roll spinning, however, the choice is relevant as our deformation depends directly on the global movement of the frame. In classic rolling, this frame was completely determined by our previous considerations – by the amount of twisting orthogonal to the knot and by the direction of the rolling (positive or negative). For higher dimensions, our previous considerations eliminate twisting and allow us to find a trivial normal framing, orthogonal to the knot, but the framing parallel to the knot will exist at all points only if the sphere is parallelizable. Thus this type of rolling is only possible if p equals 1, 3, or 7, yielding knots of dimension 4, 8, or 16. Very little is known as yet about this construction and the knots it generates.

6. Other constructions

We close by briefly mentioning two other known constructions of knots related to spinning.

The first, due to John Klein and Alexander Suciu [27], is called *diff-spinning*. It is a modified version of frame spinning in the smooth category in which the manifold M^m is altered by a diffeomorphism in the process of spinning. Klein and Suciu used diff-spinning to demonstrate the existence of inequivalent fibered knots whose homotopy Seifert pairings are isometric. It had been shown by Michael Farber in [9] that the homotopy Seifert pairing uniquely determines a fibered knot, provided the knot is sufficiently simple (i.e., its complement has the homotopy type of S^1 in sufficiently many dimensions). The construction of Klein and Suciu showed that Farber's result does not extend for all knots.

To define diff-spinning, note that the complement of the frame spin about M^m is diffeomorphic to the union of $B^{n+m-1} \times S^1$ with $M^m \times X$, where X is the complement of K (see [27] for a diagram illustrating this). Suppose now that we are given a self-diffeomorphism Φ of M^m that extends to a diffeomorphism $\bar{\Phi}$ of $B^{n+m-1} \supset S^{n+m-2} \supset M^m$. Then, roughly speaking, the diff-spin is formed by removing this complement and replacing it with the twisted product $(B^{n+m-1} \times_{\bar{\Phi}} S^1) \cup (M^m \times_{\Phi} X)$. If Φ satisfies a certain algebraic condition (see Section 5 of [27]), this space will also be the complement of a knot, the diff-spun knot.

Spinning in the smooth category is also considered in a low-dimensional context by Ronald Fintushel and Ronald J. Stern in [11]. Their construction, called *rim surgery*, begins with a surface Σ embedded in a smooth 4-manifold X. The procedure then is to spin a classical knot $K \subset S^3$ around a certain curve embedded in Σ. The technical details are essentially those we have seen before: a trivial neighborhood pair of the curve is removed, and it is replaced with a neighborhood pair whose fibers are (B^3_K, B^1_K). The result is a new surface embedding (X, Σ_K). Under certain assumptions, (X, Σ_K) will be homeomorphic to (X, Σ), but if the Alexander polynomial of K is non-trivial, (X, Σ_K) and (X, Σ) will not be diffeomorphic! More generally, (X, Σ_{K_1}) and (X, Σ_{K_2}) will be diffeomorphic only if the two knots K_1 and K_2 have the same Alexander polynomial. This is proven as an application of Seiberg–Witten theory. Generalizations of this construction and various applications have been considered by a variety of authors, including Fintushel and Stern [12], Stefano Vidussi [40,41], Sergey Finashin [10], and Hee Jung Kim [26], who considers a twist spinning analogue.

One last type of spinning, also introduced by Roseman in [36], is what he calls "spinning a knot about a projection". We shall refer to this as *projection spinning*. This clever construction involves many technical details, but, very roughly, the idea is to spin about an immersed manifold M, rather than an embedded one as we did for frame spinning. Let Σ denote the singular set of the embedding, i.e., the image of the points of M for which the immersion is not 1–1. Let $N(\Sigma)$ be a neighborhood of Σ in S^n. Outside of $N(\Sigma)$, the construction is the same as for frame spinning, i.e., each unknotted fiber pair in the tubular neighborhood of $M - N(\Sigma)$ is replaced with (B^n_K, B^{n-2}_K). However, $N(\Sigma)$ itself is broken up into sets that are homeomorphic to balls B^n and such that the intersection of B^n with the immersed M is a collection of hyperplanes. These balls with hyperplanes are then used as the data to create *multi-knots*, in which some knot K is blended together with itself in multiple directions. These ball neighborhoods are then replaced with the multi-knots. If M is embedded, we recover frame spinning. We refer the reader to [36] both for the technical definitions of projection spinning and for some nice graphical illustrations of the process.

In Remark 7 of [36], Roseman notes that it is further possible to deform projection spin, perhaps the ultimate in knot spinning constructions!

References

[1] C.C. Adams, *The Knot Book*, W.H. Freeman and Company, New York (1994).

[2] E. Artin, *Zur Isotopie zweidimensionalen Flächen im R_4*, Abh. Math. Semin. Univ. Hamburg **4** (1926), 174–177.

[3] G. Burde and H. Zieschang, *Knots*, de Gruyter Studies in Mathematics, 2nd edn., Vol. 5, Walter de Gruyter, Berlin (2003).

[4] S. Cappell, *Superspinning and Knot Complements*, Topology of Manifolds (Proc. Inst. Univ. of Georgia, Athens, GA, 1969), Markham, Chicago (1970), 358–383.

[5] J.S. Carter and M. Saito, *Knotted Surfaces and Their Diagrams*, Mathematics Surveys and Monographs, Vol. 55, American Mathematical Society, Providence, RI (1998).

[6] S. Carter, S. Kamada and M. Saito, *Surfaces in 4-Space*, Encyclopaedia of Mathematical Sciences: Low Dimensional Topology III, Vol. 142, Springer, Berlin (2004).

[7] A.H. Durfee, *Knot Invariants of Singularities*, Algebraic Geometry (Proc. Symp. Pure Math., Vol. 29, Humboldt State University, Arcata, CA, 1974), American Mathematical Society, Providence, RI (1975), 441–448.

[8] D.B.A. Epstein, *Linking spheres*, Proc. Camb. Philos. Soc. **56** (1960), 215–219.

[9] M.Š. Farber, *Isotopy types of knots of codimension two*, Trans. Am. Math. Soc. **261** (1980), 185–209.

[10] S. Finashin, *Knotting of algebraic curves in $\mathbb{C}P^2$*, Topology **41** (2002), 47–55.

[11] R. Fintushel and R.J. Stern, *Surfaces in 4-manifolds*, Math. Res. Lett. **4** (1997), 907–914.

[12] R. Fintushel and R.J. Stern, *Symplectic surfaces in a fixed homology class*, J. Diff. Geom. **52** (1999), 203–222.

[13] R.H. Fox, *Rolling*, Bull. Am. Math. Soc. **72** (1966), 162–164.

[14] G. Friedman, *Groups of locally-flat disk knots and non-locally-flat sphere knots*, J. Knot Theory Ramificat., 14 (2005) 189–215. See also http://www.math.yale.edu/~friedman.

[15] G. Friedman, *Alexander polynomials of non-locally-flat knots*, Indiana Univ. Math. J. **52** (2003), 1479–1578.

[16] D.L. Goldsmith and L.H. Kauffman, *Twist spinning revisited*, Trans. Am. Math. Soc. **239** (1978), 229–251.

[17] C.McA. Gordon, *A note on spun knots*, Proc. Am. Math. Soc. **58** (1976), 361–362.

[18] C. Gordon, *Knots and embeddings*, PhD thesis, University of Cambridge (1970).

[19] J.F.P. Hudson, *Piecewise Linear Topology*, W.A. Benjamin, Inc., New York (1969).

[20] S. Kamada, *Braid and Knot Theory in Dimension Four*, Mathematics Surveys and Monographs, Vol. 95, American Mathematical Society, Providence, RI (2002).

[21] T. Kanenobu, *Untwisted deform-spun knots: examples of symmetry-spun 2-knots*, *Transformation Groups (Osaka)*, Lecture Notes in Mathematics, Vol. 1375, K. Kawakubo, ed., Springer, Berlin (1987), 145–167.

[22] C. Kearton, *A theorem on twisted spun knots*, Bull. Lond. Math. Soc. **4** (1972), 47–48.

[23] C. Kearton, *Simple spun knots*, Topology **23** (1984), 91–95.

[24] C. Kearton, *Knots, groups, and spinning*, Glasgow Math. J. **33** (1991), 99–100.

[25] M. Kervaire and C. Weber, *A Survey of Multidimensional Knots, Knot Theory* (Proc. Semin., Plans-sur-Bex 1977) (Berlin), Lecture Notes in Mathematics, Vol. 685, Springer, Berlin (1978), 61–134.

[26] H.J. Kim, *Modifying surfaces in 4-manifolds by twist spinning*, see http://front.math.ucdavis.edu/math.GT/0411078.

[27] J.R. Klein and A.I. Suciu, *Inequivalent fibred knots whose homotopy Seifert pairings are isometric*, Math. Ann. **289** (1991), 683–701.

[28] R.K. Lashof and J.L. Shaneson, *Classification of knots in codimension two*, Bull. Am. Math. Soc. **75** (1969), 171–175.

[29] J. Levine and K. Orr, *A Survey of Applications of Surgery to Knot and Link Theory*, Annals of Mathematical Studies, Vol. 145, Princeton University Press, Princeton, NJ (2000), 345–364.

[30] R.A. Litherland, *Deforming twist-spun knots*, Trans. Am. Math. Soc. **250** (1979), 311–331.

[31] C. Livingston, *Knot Theory*, Carus Mathematical Monographs #24, The Mathematical Association of America, Washington, DC (1996).

[32] J. Milnor, *Singular Points of Complex Hypersurfaces*, Princeton University Press, Princeton, NJ (1968).

[33] A. Ranicki, *High-Dimensional Knot Theory: Algebraic Surgery in Codimension 2*, Springer, Berlin (1998).

[34] D. Rolfsen, *Knots and Links*, 2nd edn., Publish or Perish, Inc., Berkeley, CA (1976).
[35] D. Roseman, *The spun square knot is the spun granny knot*, Bol. Soc. Mat. Mexicana **20** (2) (1975), 49–55.
[36] D. Roseman, *Spinning knots about submanifolds; spinning knots about projections of knots*, Topol. Appl. **31** (1989), 225–241.
[37] C.P. Rourke and B.J. Sanderson, *Block bundles: I*, Ann. Math. **87** (2) (1968), 1–28.
[38] A.I. Suciu, *Inequivalent frame-spun knots with the same complement*, Comment. Math. Helv. **67** (1992), 47–63.
[39] M. Teragaito, *Roll-spun knots*, Math. Proc. Camb. Philos. Soc. **113** (1993), 91–96.
[40] S. Vidussi, *Lagrangian surfaces in a fixed homology class: existence of knotted Lagrangian tori*, J. Diff. Geom., in press. See also http://front.math.ucdavis.edu/math.GT/0311174.
[41] S. Vidussi, *Symplectic tori in homology E(1)'s*, Proc. Am. Math. Soc., in press. See also http://front.math.ucdavis.edu/math.GT/0401284.
[42] E.C. Zeeman, *Linking sphere*, Abh. Math. Semin. Univ. Hamburg **24** (1960), 149–153.
[43] E.C. Zeeman, *Twisting spun knots*, Trans. Am. Math. Soc. **115** (1965), 471–495.

CHAPTER 5

The Enumeration and Classification of Knots and Links

Jim Hoste

Pitzer College, Department of Mathematics, 1050 N Mills Avenue,
Claremont, CA 91711, USA
E-mail: jhoste@pitzer.edu

Contents

Abstract

The theoretical and practical aspects of link classification are described, with special emphasis on the mathematics involved in recent, large-scale link tabulations.

HANDBOOK OF KNOT THEORY
Edited by William Menasco and Morwen Thistlethwaite

1. Introduction

The first tables of knots were produced in the late 19th century at the very dawn of modern topology and knot theory. Inspired by Lord Kelvin's "Vortex Theory of the Atom" [59], the Scottish physicist P. G. Tait set out to systematically enumerate knots based on their crossing number. Joined later by the English Reverend T. P. Kirkman and America's first knot theorist, C. N. Little, the tabulating trio eventually produced, after untold hours of laborious handwork spread out over a period of about 25 years, a table of prime, alternating knots to 11 crossings and prime nonalternating knots to 10 crossings. A little more than a century later, in July 2003, S. Rankin, O. Flint and J. Schermann tabulated all 6,217,553,258 prime, alternating knots through 22 crossings in just over a day and a half of computer time! In between lies a story that touches nearly every aspect of the theory of knots – running from the beginning of the subject to the present day, and spanning the entire breadth of the topic in search of the powerful yet practical invariants needed to classify knots.

Much has been written about the history of knot theory and in particular the quest to tabulate knots and links. In this chapter, we will give only the most basic treatment of the subject, concentrating primarily on the mathematics involved in, and leading up to, the most recent large-scale tabulations. The reader seeking more details is urged to consult the original papers of Tait [55], Kirkman [33,34], and Little [37–40]; J. H. Conway [12]; M. B. Thistlethwaite [18,19,56]; and recent work by Rankin, Flint and Schermann [47–49]. These papers represent the core of the tabulating tradition. The early history of knot theory as well as knot tabulation are beautifully described in the historical articles of M. Epple [20–22], and are highly recommended. A survey of more recent work in knot tabulation may be found in [29]. In Section 2 we review some basic definitions and then outline, in Section 3, the theoretical aspects of knot and link classification. While a complete classification has been achieved, at least in theory, simple means of distinguishing *arbitrary* links with perhaps hundreds or thousands of crossings may never be forthcoming. Section 4, therefore, focuses on the practical ingredients of the most recent tabulations – work by Thistlethwaite, J. Hoste, and J. Weeks, and now Rankin, Flint and Schermann that have extended the tables into the 20-crossing realm. For lack of space, the details of many important pieces of work, perhaps most noticeably the groundbreaking and profound contributions of Conway [12] and the promising approach via braids of J. S. Birman and W. W. Menasco [6], are not discussed.

2. Definitions

We assume that the reader has a basic knowledge of knot theory, but we briefly describe the main definitions and important theorems. For a more detailed account see [51].

A *knot* is a smooth, unoriented, embedding of S^1 in S^3, where two such embeddings are considered equivalent if there is a homeomorphism h: $S^3 \to S^3$ which takes one embedded circle to the other. If the homeomorphism h preserves the orientation of S^3, then this is equivalent to saying the embeddings are related by an ambient isotopy. If not, then the two embeddings are related first by a reflection and then by ambient isotopy. No attempt at all will be made at this point to distinguish a knot from its mirror image, or *reflection*, even though such a pair might not be ambient isotopic. A knot which is ambient isotopic to its reflection is called *amphicheiral* or *achiral*. While many knots *obviously* appear to be chiral, it was not until 1914 that topology had developed sufficiently to allow a proof of this! (See M. Dehn [16].) Today, several invariants are known that can easily distinguish many knots from their mirror images (for example, the Jones polynomial [31]).

A *link* is a disjoint union of knots in S^3, again considered up to homeomorphisms taking one link to another. As before, the homeomorphism may or may not preserve the orientation of S^3. Each knot in the link is called a *component*. A link is called *trivial* if it is the boundary of a disjoint collection of smoothly embedded disks.

Of course, the equivalence relation on knots and links can be made finer by either allowing only ambient isotopy, or by orienting the links and requiring that the orientations be preserved, or both. Additionally, for links, we could order the components and require that the ordering be preserved. None of these refinements will be considered now. Instead, this chapter will discuss only the enumeration and classification of unoriented links under the coarsest equivalence defined above. We can then take the point of view of considering these refinements as *symmetries* enjoyed by a particular knot or link. (See [29] for symmetry data of all knots to 16 crossings.) Just as each integer may be decomposed into a product of primes, so also can each link be expressed in terms of simpler links. The most elementary decomposition occurs when a link is *split*. In this case there exists a smoothly embedded 2-sphere in the link complement which separates some components of the link from the rest. If a link is nonsplit it still might be made up of simpler links via the operation of *connected sum*, which is defined as follows. Given two links L_1 and L_2, in separate copies of S^3, first remove a ball B_i, $i = 1, 2$, from each copy of S^3 which meets each L_i in a diameter of B_i. Now form a new copy of S^3 by gluing together along their boundaries the complementary balls to B_1 and B_2, matching the orientations of the 3-manifolds, and matching the endpoints of what remains of the two links. There are two ways to do this, depending on how the two pairs of endpoints are matched. The newly formed link is the connected sum of L_1 and L_2 and is denoted $L_1 \sharp L_2$. A link is *prime* if it is nonsplit and not the connected sum of nontrivial links.

The operation of connected sum is not well defined. In the case of links having multiple components it clearly matters which components are chosen to be connected together. But additional, more subtle, problems exist too because of our choice of link equivalence. While we make no distinction between a knot K and its reflection $\overline{K}$, it may turn out that $K \sharp J$ and $\overline{K} \sharp J$ are not equivalent. Similarly, if K is an oriented knot and $-K$ is its *reverse* (obtained by reversing the orientation), then even though we consider K and $-K$ the same it may be the case that they are not ambient isotopic by an isotopy respecting orientation. This allows that, *even considered as unoriented knots*, $K \sharp J$ and $(-K) \sharp J$ might not be equivalent, where the connected sum was formed so as to respect the orientations of the summands. (*Nonreversible* knots, or sometimes called *noninvertible* knots, were first shown to exist by H. F. Trotter [60].)

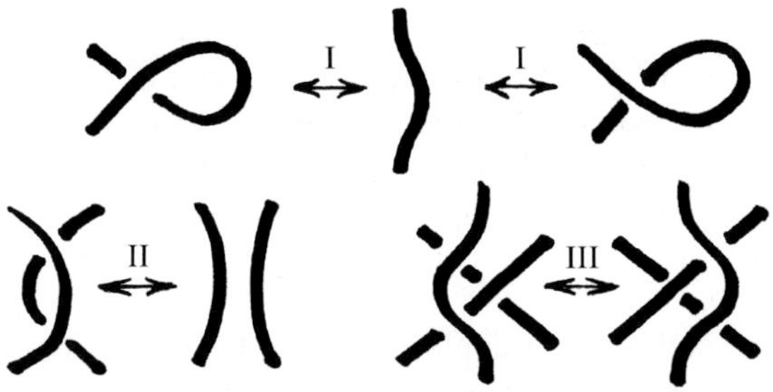

Fig. 1. Reidemeister moves.

Nevertheless, given a nonsplit, nonprime link we may decompose it as a connected sum and continue to decompose its summands if they remain nonprime until finally the process must end (because of the additivity of genus). Thus, every link can be expressed as a connected sum of prime links. In the case of knots, it was proven in 1949 by H. Schubert [53] that the decomposition of a knot into prime summands is unique (up to the ordering of the summands). The analogous result for nonsplit links was shown in 1958 by Y. Hashizume [26]. Because of these results the focus has traditionally been on prime knots and links, since the composite ones can all be built up out of the prime ones.

Every link can be projected into a plane so that the only singularities are a finite number of transverse double points. If at each double point we indicate an over and undercrossing strand in the obvious way then we call the resulting figure a link *diagram.* (See Figure 3.) The minimum number of crossings in any diagram of the same link is called the *crossing number* of the link. Crossings of oriented links can be labeled *right* or *left handed* as follows. Standing on the overcrossing strand of a right handed crossing and facing forward, the undercrossing strand will pass beneath from right to left. Atop a left handed crossing, the strand below runs from left to right.

Representing knots and links by diagrams is undoubtedly the oldest method in use. Many different local diagrammatic changes, some of which preserve the link, and some which do not, are central to the study of knot theory. Chief among these, and shown in Figure 1, are the *Reidemeister moves*, as well as the *flype* shown in Figure 2, which is itself always a combination of Reidemeister moves.

The importance of the Reidemeister moves lies in the fact that two link diagrams represent the same link (here "same" means ambient isotopic) if and only if they are related by a (finite) sequence of such moves[1]. Many important link invariants can be shown to exist by proving that some quantity derived from a link diagram is preserved by Reidemeister moves, and thus actually an invariant of the link type. This diagrammatic, or combinatorial, approach to knot and link theory has been, and continues to be, one of the cornerstones of the subject.

A crossing in a link diagram is *nugatory* if there is a circle in the projection plane that meets the diagram transversely only at that crossing. A nugatory crossing can clearly be

[1] The Reidemeister moves were known to J. C. Maxwell at least as early as 1868 (see [21].) The proof that they suffice to pass between equivalent diagrams was published by both Reidemeister [50], and J. W. Alexander and G. B. Briggs [2].

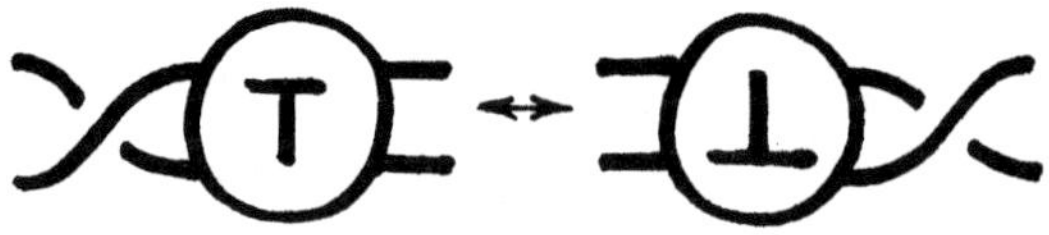

Fig. 2. A flype.

removed by a flype (or perhaps by a single Type I Reidemeister move). A diagram that has no nugatory crossings is called *reduced.*

3. Classifying knots and links

There are infinitely many different knots and links, and to date no practical or simple means to classify them has been found. The emphasis here, of course, must be placed on the words "practical or simple", although even the word "classify" deserves some clarification. In some sense, the theorem of Reidemeister, and Alexander and Briggs, classifies links according to their diagrams: every link can be represented by a diagram and two such diagrams represent the same link if and only if they are related by Reidemeister moves. But since no a priori bound on the number of such moves which might be required to pass between two given diagrams exists, we cannot algorithmically decide, by exploring a finite number of Reidemeister moves, if two diagrams are in fact equivalent.

Alternatively, W. Whitten has shown that prime knots with isomorphic fundamental groups have homeomorphic complements [65]. Coupled with the important theorem of C. McA. Gordon and J. Leucke that knots with homeomorphic complements are equivalent [24], we see that prime knots are classified by their fundamental groups. But as Alexander himself pointed out in 1927 [1], "unfortunately, the problem of determining when two such groups are isomorphic appears to involve most of the difficulties of the knot problem itself."

To address problems of this sort, we will say that links have been *classified* if we can solve the *recognition problem.* That is, is there an algorithm that can decide, in a finite amount of time, if any given pair of links are equivalent? Notice that given such an algorithm, we could then enumerate all links as follows. Since there are only finitely many link diagrams with a given crossing number, we could systematically list all diagrams arranged by crossing number. As each new diagram is produced, we could compare it to all the diagrams already on the list to see if it represents a new link. If it does, we add it to the list. If not, we discard it. But even this "solution" to the problem leaves us wanting, as ordering the links by crossing number is somewhat artificial, probably having no real bearing on the true topological nature of the links.

In theory, the link recognition problem has been solved. The main work was done by W. Haken, F. Waldhausen, and K. Johannson, with important contributions by G. Hemion, W. Jaco, P. B. Shalen, and S. Matveev. Their algorithm is too complicated to describe completely but we will give a *very brief* description. The interested reader should consult the recent book of S. Matveev [41], where the entire algorithm is described in detail. The algorithm compares the exteriors of the two links which have been additionally marked with the meridians of each component. (The *exterior* of a link is the closure of the complement of a regular neighborhood of the link. A *meridian* of a knot is the boundary of a disk meeting the knot transversely in one point.) Two links are equivalent if and only if there is a homeomorphism between their exteriors taking meridians to meridians. Because of the theorem of Gordon and Leucke, the meridians may be ignored in the case of knots, but in general the Haken algorithm does not rely on this simplification.

In order to determine if the two marked link exteriors are homeomorphic, the fundamental idea is to cut each exterior open along incompressible surfaces and continue to do so with what remains until, eventually, the process must end. By comparing the ending states and their markings, and the regluing instructions needed to return to the link exteriors, it can be decided if the original marked link exteriors are homeomorphic. In order to make the process algorithmic, the link exteriors are first triangulated and the theory of normal surfaces is used to find the incompressible surfaces. Each surface in the link exterior can be assumed to meet each tetrahedron of the triangulation in one of seven basic ways. Assigning seven variables to each tetrahedron to represent the number of local pieces which fit together to form the surface, we arrive at a finite number of matching equations in a finite number of variables. Only a finite number of fundamental solutions to these equations exists and these provide an algorithmic process for constructing the necessary incompressible surfaces in the link exterior. The entire procedure is quite complicated and may never be fully implemented on a computer.

While the Haken algorithm applies to all nonsplit links, the special case of the *unknot recognition* problem has received much attention. This is the problem of deciding if a given knot diagram represents the unknot. The complexity of this problem has been proven to be in class NP by J. Hass and J. C. Lagarias [25]. They also derive from the link recognition algorithm an upper bound on the number of Reidemeister moves m that might be needed to connect a knot diagram D having n crossings to the trivial diagram having zero crossings (assuming D represents the unknot). They prove that $m \leq 2^{cn}$, where $c = 10^{11}$. It should be noted that Hass and Lagarias made no attempt to find an optimal bound and in fact they believe it is quite likely that the actual upper bound on the required number of Reidemeister moves to connect two equivalent knot diagrams is polynomial in the (larger) number of crossings. Still, more work is clearly needed before the Haken link recognition algorithm becomes practical!

At least one computer program employing normal surface theory has been written. Developed by D. Letscher and B. Burton, and originally called *Normal* but now known as *Regina*, the software can carry out a variety of 3-manifold investigations based on normal surfaces [9]. Using *Regina* it may be possible using Haken's algorithm to identify unknots among diagrams with relatively few crossings.

An algorithmic solution to the unknot recognition problem, which is quite different from the Haken algorithm, has also been found by Birman and M. D. Hirsch [5]. Their approach makes use of braids. Given a knot diagram D with n Seifert circles and c crossings, it is known that D can be redrawn as the closure of a braid β on n strings with at most $c + (n-1)(n-2)$ crossings [61,66]. They prove that if D is the unknot, then some conjugate of β must be among a certain finite list of n-string braids that depends only on n and c. This list can be generated algorithmically and, making use of the solution to the conjugacy problem in the braid group B_n, each of its elements can be compared to D.

A computer program has been written by Birman, M. Rampichini, P. Boldi, and S. Vigna [8] to produce the list of braids. When combined with programs by other authors to solve the conjugacy problem in the braid group (see for example [36]), the entire process should be fully implemented.

Two other computer programs that can be used to attack the unknot recognition program, and which are frequently successful on diagrams with even hundreds of crossings are *SnapPea* by Weeks [62,64] and *Book Knot Simplifier* by M. Andreeva, I. Dynnikov, S. Koval, K. Polthier and I. Taimanov [4]. Working with a triangulation of the knot exterior, *SnapPea* looks for ways to reduce the number of tetrahedra in the triangulation. *Book Knot Simplifier* makes use of the fact that every link can be embedded in a book with three pages, that is, a union of three half-planes with common boundary. This way of viewing links seems to be particularly adept at computer manipulation. In particular, there are a large number of moves, which are easy to locate and apply, that change the 3-page presentation while preserving link type. The program searches for simplifying moves and seems reasonably successful at recognizing unknot diagrams.

Since the Haken link recognition algorithm is impractical, what can we *actually do* when confronted with two link diagrams wanting to know if they are the same link? Fortunately for us, many different knot and link invariants have been developed since the time of Tait, and if the links are different, perhaps some known invariant will distinguish them. Therefore, we begin by computing as many invariants as we can, beginning with the easiest to compute and moving to the more difficult. If all known invariants fail to tell the links apart, then perhaps they are the same, and we can launch ourselves into an attempt to relate the two diagrams by Reidemeister moves (or combinations of Reidemeister moves, such as flypes and other moves to be described later). Of course, the entire process is ad hoc, and may not lead to a definite answer. The harsh reality of knot theory is that we will probably never be able to decide if two arbitrary links are the same or not. Just imagine being given two link diagrams with a few million crossings each!

On the other hand, for certain classes of knots and links, spectacular classification results have been obtained. Wonderful examples include torus links [51], 2-bridge links [51], 3-string braids [7], alternating links and hyperbolic links. The last two classes have proven especially useful to the link tabulator wanting to classify all (prime) links up to a given crossing number. In preparation for the next section, where we describe how the current knot and link tables have actually been constructed, we briefly discuss both alternating and hyperbolic links.

A link is *alternating* if it is represented by a diagram whose crossings alternate, over–under–over–under and so on, as one travels around the components. When Tait first began his investigations he may have thought that all knots were alternating. (Indeed, it is nontrivial to prove that nonalternating knots exist! The first correct proof was given by R. H. Crowell in 1959 [14].) Tait made three conjectures about alternating knots, all of which can be stated for links. The first was that reduced, alternating diagrams have minimal crossing number; the second, that any two reduced alternating diagrams of the same link (here "same" means ambient isotopic) have the same *writhe* (the writhe of a diagram is the number of right handed crossings minus the number of left handed crossings); the third, that two alternating diagrams represent ambient isotopic links if and only if they are related by flypes[2]. (Since flypes preserve writhe, the third conjecture implies the second.) All of these conjectures were eventually proven not long after the discovery of the Jones polynomial, and its generalizations, in 1984. (See, for example, [27,31,32,43,44,46,57].) The proof of the Tait Flyping conjecture by Menasco and Thistlethwaite provides a classification of alternating links. Unlike the Reidemeister moves, the equivalence classes of diagrams related by flypes are all finite. Given two reduced alternating link diagrams, we can algorithmically generate the entire flype equivalence class of one and check to see if it contains the other. Moreover, this task is easily implemented on a computer making the comparison of alternating links reasonably practical. As satisfying as this result is, we must point out that the number of alternating knots and links is small compared to all links. In fact, the proportion of links which are alternating tends exponentially to zero with increasing crossing number [54]. Nevertheless, as we will see in the next section, the classification of alternating links provides the important first step in the construction of most link tables.

A truly remarkable situation exists for hyperbolic links. These are links for which the complement admits a complete Riemannian metric of constant curvature -1. The complement of every hyperbolic link can be decomposed in a canonical way into ideal polyhedra. This canonical triangulation depends only on the topology of the link complement. Once the canonical triangulation has been found for two links, the triangulations can be compared combinatorially to decide if the links have homeomorphic complements or not. The canonical triangulation (together with meridian data) provides a complete link invariant! For further details the reader should consult [23,52,63].

Unlike the general algorithm of Haken, computing the canonical triangulation of a hyperbolic link complement is much more practical. The program *SnapPea* does just this, as well as compute other invariants of hyperbolic manifolds such as volume, etc. *SnapPea* is so able to handle relatively small knots and links that it was used successfully by Hoste and Weeks in their tabulation with Thistlethwaite of all prime knots through 16 crossings. Unlike alternating links, the class of hyperbolic links is much "bigger." Of the 1,701,936 prime knots with 16 or less crossings, all but 32 are hyperbolic. According to Thurston, every knot is either a *torus* knot, a *satellite* knot, or a hyperbolic knot. The torus knots and links are those that can be embedded on a standard, unknotted torus, sitting inside S^3. Torus links are completely classified by how many times they wind in each direction around the torus. Moreover, the crossing number of a torus knot or link has been determined by K. Murasugi [35]. Thus, we can tell exactly how many torus links there are of a given crossing number. A satellite link is one that *orbits* a *companion* knot K in the sense that it lies inside a regular neighborhood of the companion. Since every knot is a satellite of the unknot we require the companion to be nontrivial. Note that composite knots are satellites of each of their summands. While few satellites exist at small crossing numbers, their numbers will grow tremendously as the number of crossings increases. If a satellite has *wrapping* number d (that is, the satellite meets every meridional disk in the regular neighborhood of the companion at least d times), then it seems plausible that the crossing number of the satellite is at least cd^2, where c is the crossing

[2] It may be that Tait never actually held this to be true. See [21].

number of the companion. Alas, this has not been proven, but it indicates why so few satellites exist to 16 crossings.

How *SnapPea* finds the canonical triangulation of a hyperbolic link complement is described nicely in [64]. However, it is worth pointing out that problems can arise. In order to find the triangulation, certain matching equations must be solved, the solutions of which *SnapPea* only approximates to a high degree of accuracy. Nevertheless, there is no guarantee that the level of accuracy is sufficient to insure that *SnapPea* finds the canonical triangulation, rather than some noncanonical triangulation. Thus, *SnapPea* may falsely declare two hyperbolic links different when in fact they are the same, if it had arrived at the noncanonical triangulation for one or both of the links. But fortunately, it cannot falsely declare two links to be the same. No matter what triangulations are found, canonical or not, if they match for two links, then the link complements are homeomorphic. And if further checking reveals that meridians are taken to meridians, then the links are equivalent. A program called *Snap* has been written by O. Goodman which specifically eliminates the round-off errors introduced by *SnapPea* and therefore cannot make the kind of mistake just described. But unlike *SnapPea*, *Snap* is less practical, taking much longer to find the canonical triangulation for a relatively small knot or link. See [13] for more information on *Snap*.

4. Producing link tables

Tabulating knots and links in order to discover the nature of physical matter – Gold is the trefoil! Lead is the figure eight! – was forgotten long ago. But somewhat ironically, tables of knots are now proving useful to scientists once again. For example, knot theory is playing an important role in recombinant DNA research, and researchers in that field need to identify knots and links that occur in their experiments. Important examples in knot theory have also come to light specifically because of the careful and methodical enumeration of knots and links. For example, nontrivial links with trivial Jones polynomial were first found in this way by Thistlethwaite [58], who also turned up the first examples of amphicheiral knots with odd crossing number [29]. No doubt other interesting and important examples will surface as the tables are extended even further. Having billions of links in the tables, as opposed to only a few hundred, provides a much richer and realistic data set for experimenting and testing conjectures. Thus, tables are an important part of our field and are likely to expand further as new algorithms and invariants are discovered and computers grow ever faster.

We outline here the basic plan that has been used to create the latest (and largest) tables of prime knots and links. The idea, already mentioned earlier, is simple: systematically list all diagrams to a given crossing number and then group them together according to link type. Because practical link recognition can only be carried out in an ad hoc way (except for alternating links) the second half of the program is clearly the more vexing part. But given the huge number of diagrams possible, even the first part must be undertaken with some care.

Once a table of prime knots and links has been found, these can then be connect-summed together in all possible ways to obtain the composite links. Thus, efforts have concentrated primarily on tabulating prime knots and links. It is worth pointing out though that one of the more obvious "theorems" of knot theory, that crossing number is additive under connected sum, has yet to be proven (or, incredibly, could it be false?). So while our table of prime knots and links might be arranged by crossing number, it is conceivable that using it to produce the composite links might not create them in order with respect to crossing number.

Our overall plan is thus the following. First, enumerate all prime alternating link diagrams to a given crossing number and group these together by flype equivalence class so as to create a table of prime alternating links. Three theorems are of particular importance here. The first Tait Conjecture assures us that a reduced alternating diagram is minimal in crossing number. Hence, we can be sure that the crossing numbers of the knots and links represented by the diagrams under consideration are exactly what they appear to be. The second theorem is an important result of Menasco that states that a reduced alternating diagram represents a prime link if and only if it is prime as a diagram [42]. That is, if no circle in the projection plane meets the diagram transversely in two points with crossings on either side, then in fact the link is prime. This result is an incredible boon to the tabulator, allowing easy recognition of composite alternating links. Finally, the already mentioned Tait Flyping Conjecture obviously plays a crucial role.

After the alternating links have been tabulated, we consider a diagram representing each alternating link, and change its crossings in all possible ways in order to create nonalternating diagrams. Most of these will reduce to fewer crossings, and some link diagrams may even represent split links or composite links. So clever methods will be needed to weed out the multitude of uninteresting and unwanted diagrams. Finally, ad hoc methods must be employed to either distinguish all that remains (by computing various invariants), or recognize repeats (by finding sequences of Reidemeister moves or, in the case of hyperbolic links, comparing canonical triangulations). The entire process begins with a scheme to encode diagrams.

4.1. *Encoding link diagrams*

There are a variety of ways to encode knot and link diagrams. One scheme, which was introduced by Tait (and was similar to ideas of C. F. Gauss and J. B. Listing), and further refined by C. H. Dowker and Thistlethwaite, has proven to be quite useful. Referred to here as the DT sequence of a link diagram (after Dowker and Thistlethwaite), it has been used successfully by a number of people in compiling modern tables [3,18,19,29,56]. The primary advantage of the DT sequence seems to be its brevity. With over 6 billion knots and links now in the tables, using as little computer memory as possible has obvious value. On the other hand, DT sequences are not easily transformed directly under the types of operations that need to be applied to diagrams in the course of a tabulation. Instead, it is usually necessary to derive additional, attendant information about the diagram, such as the signs of the crossings, or how the faces of the diagram (the complementary regions) meet one another, and so on, before operating on the diagram. Using this derived information assists in encoding new diagrams obtained from old ones via changes such as flypes, or other Reidemeister moves. Not only storage size but also computing speed are of great importance in any computer-assisted tabulation. Therefore, it usually makes sense to sacrifice brevity in favor of a more detailed and redundant encoding scheme if it can increase the speed at which diagrams can be manipulated. Indeed, the recent tabulation of alternating knots through 22 crossings by Rankin, Flint and Schermann uses an encoding scheme that is far more complicated than the DT sequence, takes more storage space per link, yet seems to be more finely tuned to the task of enumeration. Because of its historical importance, as well as continuing utility in modern tabulations and knot manipulation software, we will describe the DT sequence in some detail. Before doing so, however, we briefly mention other encoding schemes of importance.

A quite different method of notation was introduced by Conway [12], building on ideas used by Kirkman. Conway's scheme is quite efficient for links of low crossing number and in fact reflects deep structural properties of links. But it draws on a large set of symbols arranged according to a rather large set of rules, both of which grow with crossing number and for this reason does not lend itself well to computer programming.

However, for small knots and links Conway's system is so efficient that it allowed him to tabulate (by hand) all prime knots to 11 crossings and all prime links to 10 crossings in a few hours! Conway found 11 omissions and one duplication in Little's table of 11 crossing alternating knots. In the late 1970's A. Caudron [11] used an alternative version of Conway's notation to retabulate all prime knots to 11 crossings, discovering, in the process, four omissions in Conway's list of 11 crossing nonalternating knots. Continuing with the methodology of Kirkman and Conway, S. Jablan [30] completed the tabulation of 12 crossing alternating links in 1997.

Every link may be represented as a closed braid and so braid notation is an obvious choice for encoding diagrams. However, since transforming an arbitrary diagram into a closed braid usually involves an increase in the number of crossings, braids are perhaps not the best choice for a table organized by crossing number. But perhaps organizing a table instead with respect to one of the indices appropriate for braids, such as number of strings, would be better. At any rate, no major tabulation has yet to be undertaken based on braids. Perhaps, the work of Birman and Menasco [6], coupled with the unknot recognition algorithm of Birman and Hirsch [5], or similarly inspired algorithms, will one day lead to a major tabulation using braids. Instead, some effort has been made to systematically find braid representatives for already tabulated knots and links.

To encode a diagram with a DT sequence, first consider an arbitrary knot diagram with n crossings, and therefore $2n$ edges. (The edges are the components of the associated projection minus the double points.) Place a basepoint on one of the edges and also choose an orientation of the knot. We may now label the crossings with consecutive integers $1,2,3,\ldots,2n$ as we travel around the projection starting from the basepoint. Each crossing receives two labels and it is a consequence of the Jordan Curve Theorem that the labels at each crossing have opposite parity. (A slight variation is to label the edges rather than the crossings. In this case, the edge labels are paired, one with another, by seeing which two labels lead into each crossing.) The pairing of labels at each crossing gives a permutation, σ of the set $\{1,2,3,\ldots,2n\}$. The sequence of even numbers, $S = \{\sigma(1), \sigma(3),\ldots,\sigma(2n-1)\}$ is sufficient to denote σ. The final step in producing the DT sequence is to consider how the diagram differs from an alternating diagram. If the diagram is alternating then S is used to denote the diagram. If not, then some set of crossings of the diagram may be changed to produce an alternating diagram. The labels of S corresponding to these crossings are then negated to finally give a signed sequence of the even integers from 2 to $2n$. Since there are two possible alternating diagrams possible (each the reflection of the other), we see that S is only defined up to negation of all its entries. Since our aim is only to tabulate knots up to reflection, this presents no difficulty. It makes sense to either choose S to have the fewest number of minus signs, or to begin with a positive integer. The process is illustrated in Figure 3.

Since there are $4n$ possible choices of basepoint and direction, it is quite possible that all $4n$ DT sequences will be different. Choosing the lexicographically smallest one provides the *minimal* DT sequence for the diagram. A computer routine that finds the minimal DT sequence equivalent to a given one is easy to write.

Two problems arise regarding DT sequences. The first is that for an arbitrary diagram it may not be possible to recover the diagram from the sequence. However, it is proven in [18] that if the diagram represents a prime knot, and has no nugatory crossings, then it is determined by its sequence, although only up to reflection and isotopy in S^2. Again this presents no difficulty since these distinctions will not be made among knots in the table. The second difficulty, and one that presents a particularly annoying computing problem, is that most signed sequences of the numbers $\{2,4,\ldots,2n\}$ are not *realizable*, that is, do not correspond to any knot. A moment's thought reveals that what DT sequences really record are (certain) 4-valent graphs and of course, not all graphs are planar. While it is not difficult to decide if a "DT sequence" really is one, it is time consuming to do so. However, using the basic tabulation scheme of J. A. Calvo [10] described later, we will see that it will never be necessary to test arbitrary DT sequences to decide if they in fact represent knots.

Notice that the DT sequence of a knot with n crossings may be stored on a computer by using $2n$ bytes, since two bytes are usually used to store an integer. However, we can easily halve this to n bytes by using characters rather than integers, since one byte is typically used to store a character. A particularly nice scheme is to use "a" and "A" for 2 and -2, "b" and "B" for 4 and -4, etc. This allows for up to 26 crossings, but could obviously be

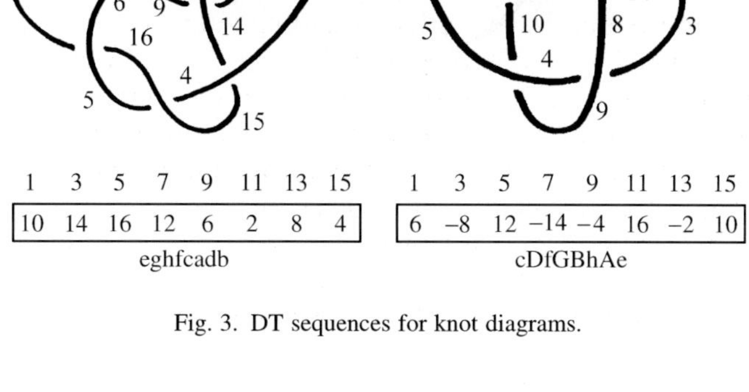

1	3	5	7	9	11	13	15
10	14	16	12	6	2	8	4

eghfcadb

1	3	5	7	9	11	13	15
6	−8	12	−14	−4	16	−2	10

cDfGBhAe

Fig. 3. DT sequences for knot diagrams.

extended considerably by using the full ASCII character set of 256 characters (and an agreed upon pairing of those characters.) Thus, DT sequences provide a fairly compact notation for diagrams.

The scheme can be extended to links by choosing an orientation and basepoint for each component and then numbering the crossings (or edges) consecutively, beginning at each basepoint in turn. In general, this may not produce the desired even–odd pairing at each crossing, but it is not hard to see that some choice of basepoints will. Restricting ourselves to these labelings we once again get a sequence of even integers that may be further refined by introducing minus signs to designate where the diagram differs from alternation. However, one more piece of information is needed in the DT sequence, namely which labels lie on the same component. This can be denoted by inserting vertical bars (or some other character) into the sequence to separate components. An example is given in Figure 4.

Again, as with knots, many signed sequences with bars do not represent actual links. Moreover, in general, it may not be possible to recover the diagram from the sequence without some serious ambiguity. But it is proven in [17] that if a DT sequence encodes a reduced diagram of a prime, nonsplit link, then it determines the diagram up to reflection and isotopy in S^2. Since we only wish to tabulate unoriented, prime, nonsplit links, up to reflection, we once again can avoid any possible ambiguity arising from the encoding scheme.

For the class of nonsplit links, the number of possible choices of basepoints and orientations can be reduced considerably. After the first basepoint and orientation are chosen there are a number of possible ways to then determine uniquely all the remaining basepoints and orientations. One example is as follows. After choosing the first basepoint and orientation of that component, change crossings if necessary to make the diagram begin with an overcrossing and also alternate. Number the crossings of the first component

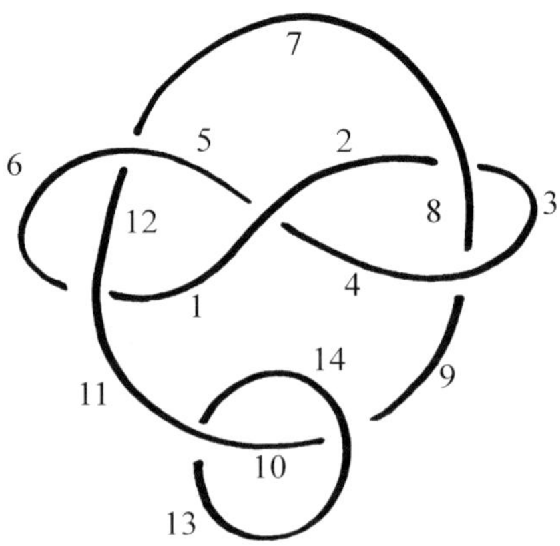

1 3 5 7 9 11 13
4 8 12 |2 −14 6|−10
bdf | aGc | E

Fig. 4. A DT sequence for a link diagram.

as usual. Because the link is nonsplit, the first component must cross other components and therefore some crossings of the first component have so far received only one label. Among these, pick the one, c, with the smallest label. Orient the other component at crossing c so that c is right handed, and place the second basepoint on that component so that c receives the smallest possible label such that the two labels at c have opposite parities. If any components remain we treat them in exactly the same way, continuing until the entire diagram has been labeled. This gives the sequence of even integers, and changing crossings to return to the original diagram indicates where minus signs must be inserted. Finally, bars are used to separate the components. The example in Figure 4 obeys this scheme. By restricting ourselves to this algorithm (or a similar one) we once again have $4n$ possible ways to encode a link diagram. To choose the lexicographically smallest DT sequence for the diagram, some consideration must be made of the bars. A nice choice is to regard the bar as zero (in fact, this is a nice way to store a DT sequence in the computer), and then use ordinary lexicographic order.

4.2. *Generating all alternating diagrams*

Each (unsigned) DT sequence corresponds to an alternating link diagram (actually, to a projection). Thus, to generate all possible alternating diagrams with n crossings we simply need to consider all possible DT sequences of length n. This is exactly the original approach taken by Tait as well as the computer equipped tabulators of the late 20th century [3,15,17,29]. But this approach has several major drawbacks that make it more and more impractical as n grows to 16, 17, and beyond. The main problem is that most DT sequences do not encode prime, nonsplit links. Thus, considerable time is spent testing each DT sequence to see if it is valid, with the greatest amount of time spent on deciding if the sequence is realizable. Even though some clever tricks can be introduced to avoid testing all possible DT sequences, huge amounts of time are still wasted considering useless DT sequences.

Instead, a significant savings can be achieved by inductively generating the n-crossing alternating diagrams from the k-crossing diagrams, where $k < n$. The basic idea is due to Calvo and K. Millett [10], and has been successfully used by the author to tabulate all alternating knots to 18 crossings, and by Thistlethwaite to tabulate all alternating knots and links to 19 crossings. A more refined version, which we will briefly describe later, has been used by Rankin, Flint and Schermann to tabulate all alternating knots to 22 crossings.

Suppose that D is a reduced, prime, alternating link diagram of n crossings. By *smoothing*, or *nullifying*, a crossing we may transform D to a diagram, D', of one less crossing. The two possible ways to smooth a crossing are illustrated in Figure 5.

Depending on whether the crossing was a *pure* crossing (between two strands of the same component) or a *mixed* crossing (between two strands of different components),

Fig. 5. Smoothing a crossing.

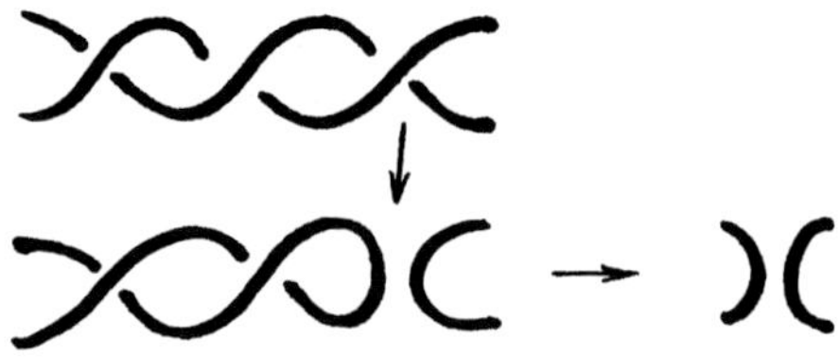

Fig. 6. Smoothing and reducing.

and which of the two possible smoothings is performed, the number of components of D and D' may be equal or differ by one. But regardless, D' represents an alternating link of at least one less crossing and therefore must already be in our census, provided it is prime. While D was assumed to be reduced, this need not be the case for D'. After the smoothing, nugatory crossings may be present that can then be eliminated. Figure 6 illustrates a possible scenario.

The important result of Calvo is that there is always at least one crossing in D that can be smoothed so that the resulting diagram D' will reduce, after eliminating nugatory crossings, to a reduced, prime, alternating link diagram. Thus, we may inductively build the collection of all prime, reduced, alternating link diagrams of n crossings by starting from the collection of all prime, reduced, alternating link diagrams with fewer crossings and splicing in twisted bands as in Figure 6 in all possible ways. Note that with this approach there is no reason to tabulate knots separately from links.

Using Calvo's algortihm, no testing of DT sequences for realizability, primality, or nugatory crossings is ever required! On the other hand, there is tremendous redundancy among the diagrams that are produced, especially if no attempt is made to account for the flype-structure of a diagram. However, Calvo also describes the general flype structure of a reduced prime alternating diagram. Each crossing c that can be involved in a nontrivial flype (and not all crossings can be) generates a unique *flype cycle* as shown in Figure 7.

Each flype tangle on the circuit (represented by a disk in the figure) is minimal in the sense that flipping it over cannot be achieved as a sequence of "smaller" flypes. A diagram is in *flype minimal* position if all crossings that generate the same flype cycle are grouped together in a single twisted band between two flype minimal tangles. (The diagram in Figure 7 is not flype minimal.) For a nonsplit, prime, alternating link L we may form a graph using flype minimal diagrams of L as vertices and connecting two vertices if they are related by a minimal flype (not the composition of smaller flypes). Calvo shows that this graph is always an f-dimensional torus lattice, where f is the number of flype cycles in the

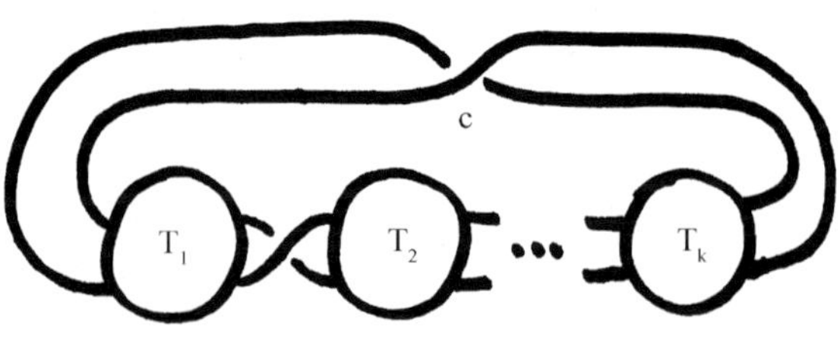

Fig. 7. Crossing c and its flype cycle.

link. Understanding the flype structure of a prime, alternating diagram, can greatly increase the efficiency of the overall program to inductively create all prime, alternating links.

Calvo's program has been carried out by Thistlethwaite for all prime, alternating links to 19 crossings, and independently by Hoste (for knots only), to 18 crossings. Their results agree with each other, and with the independent tabulation of Rankin, Flint and Schermann of prime alternating knots to 22 crossings.

In the remainder of this section we will briefly describe the Rankin, Schermann, and Smith tabulation, which is essentially a refinement of Calvo's approach with special effort having been taken to avoid redundant work thereby increasing efficiency. Their approach is extremely technical and the interested reader should consult their papers for details [47–49].

Rankin, Flint and Schermann consider four diagrammatic operations which they call D, ROTS, T, and OTS and which are pictured in Figure 8. These operations are applied to prime, alternating knot diagrams. Since the input diagrams are alternating, we have drawn only projections in the figure, not bothering to indicate the possible arrangements of crossings. However, given a choice of crossings in the input, the output must have its crossings chosen so as to remain alternating. The input diagrams are also unoriented, but in the case of the D operator, an orientation must be introduced in order to correctly apply D. Similarly, T is only applied if the orientation of the input matches that shown in the figure.

The basic idea, as with Calvo, is to inductively build up the $n + 1$ crossing knots from the n crossing knots. Given all prime, alternating n crossing knots, D and ROTS are first applied to build a collection of $n + 1$ crossing, prime, alternating knots. After this is done, T and OTS are repeatedly applied to the collection until no new knots appear.

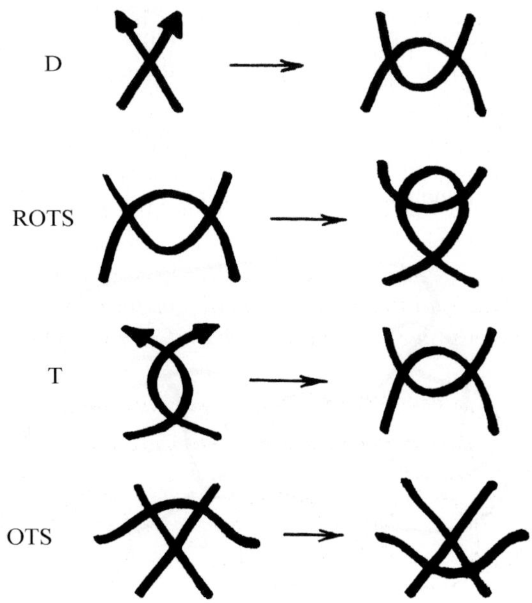

Fig. 8. The D, ROTS, T, and OTS operators.

In order to make the process more efficient by avoiding the creation of redundant diagrams, Rankin, Flint and Schermann make a careful analysis of the flype cycle structure of a knot and introduce a refined knot encoding scheme that contains this information. Their method is reminiscent of Conway's idea of inserting tangles at 4-valent vertices of graphs. Consider the knot shown in Figure 9. In several locations there are sets of crossings obtained by twisting a pair of parallel strands. Rankin, Flint and Schermann call these *groups* of crossings, and a *k-group* is a maximal such set of k crossings. For each k, the k-groups are labeled k_1, $k_2, \ldots k_{i_k}$ and a type of Gauss code is then recorded as one traverses the knot. Starting from some basepoint, and traveling in some direction, the labels of the groups are recorded as they are passed. Furthermore, each group is either *positive* or *negative* depending on whether the two strands in the group are oriented parallel or anti-parallel. Minus signs are then inserted into the Gauss code accordingly. The *group code* for the diagram in Figure 9, as it is called by Rankin, Flint and Schermann, is given in the figure.

As can be seen in Figure 7, all crossings that share the same flype cycle can be brought together by flyping into a (maximal) k-group. Assuming that this has been done for every flype cycle, Rankin, Flint and Schermann then show that it suffices to operate only on these flype minimal diagrams. They furthermore show that the four operations need only be applied as follows. First D is applied to any one crossing in each negative k-group, $k \geq 1$, and in each positive 2-group in each n crossing knot. After all possible applications of D have been made, the ROTS operator is then applied to each negative 2 or 3 group in every n crossing diagram. Rankin, Flint and Schermann remark that at this point they have usually produced about 98% of the $n + 1$ crossing diagrams. For example, of the 40,619,385 prime, alternating 19 crossing knots, 39,722,121 were found after only applying D and ROTS.

Next T is applied to each positive 2-group in all of the $n + 1$ knots that have been created so far, and this is continued until no new knots are added. The OTS operator is then applied until it also produces no new knots. The operations of T and OTS are then repeated alternately until no new knots result. At this point the construction of all $n + 1$ prime, alternating knots is complete.

An important feature of their work, which is too complicated to explain here, is that they employ a more sophisticated data structure than simply the group code explained earlier. Instead, they record the flype cycle information for each diagram in what they call the

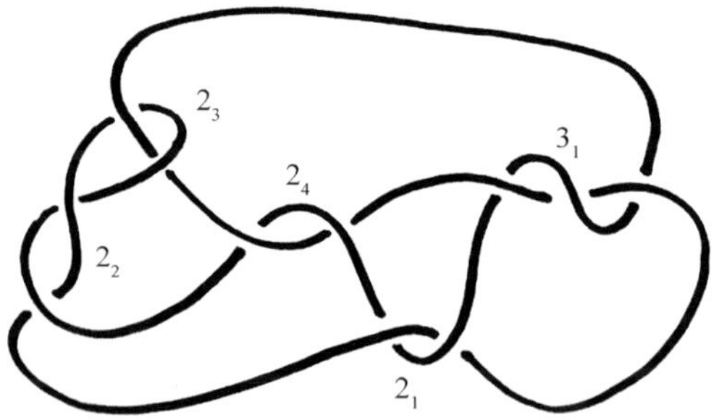

$$-2_1, -2_2, 2_3, -2_2, 2_4, -2_1, 3_1, 2_3, 2_4, 3_1$$

Fig. 9. A diagram with k-group encoding.

master group code and among all such encodings choose one, called the *master array* to represent the knot. Two diagrams are flype equivalent if and only if they have identical master arrays.

4.3. *Generating the nonalternating diagrams*

If a prime, alternating diagram D has n crossings, then we may produce 2^n nonalternating diagrams by switching crossings in all possible ways. Of course, half of these are unnecessary since we will consider any link and its mirror image as the same. Most of these diagrams will reduce to fewer crossings, and possibly even represent split or composite links. One way to avoid generating unwanted diagrams is to first group all the crossings of D into subsets where within each subset all the crossings must maintain the same state relative to each other lest an immediate reduction to fewer crossings is possible. For example, the set of crossings associated to a twisted band, or in the language of Rankin, Flint and Schermann, a k-group, is such a set. If all the crossings of D can be partitioned into j subsets of this kind then only 2^{j-1} vs 2^{n-1} nonalternating diagrams need be considered. Assuming that the nonalternating links of up to $n-1$ crossings have already been tabulated, we may first eliminate any diagram that reduces to fewer crossings. Note that while many diagrams will reduce to fewer crossings, relatively few will be unknots. Thus, attempting to apply one of the unknot recognition algorithms discussed in the last section would probably be unwarranted at this point. On the other hand, perhaps employing the ideas of I. Dynnikov and the 3-page book simplification moves could prove useful. At present, the only large-scale tabulations of nonalternating links have been carried out by Thistlethwaite (knots and links), and Hoste and Weeks (knots only), and in both cases a variety of Reidemeister moves were employed in an effort to eliminate diagrams that reduce to fewer crossings. In the author's case, the equivalence class generated by flypes and 2-passes was generated for each diagram and each of these diagrams was searched for (i, j)-pass moves which would reduce crossing number. An (i, j)-pass move is illustrated in Figure 10. This move "picks up" a bridge with i overcrossings and "lays it down" in a new location with j overcrossings. A 2-pass is a $(2, 2)$-pass. Thistlethwaite uses these moves plus many other esoteric moves (for example, one derived from the "Perko pair" equivalence [45]) to search for reductions. A more comprehensive description of Thistlethwaite's moves may be found in [29].

It is important to note that while lots of theoretical algorithms might be brought to bear on this stage of the problem the sheer number of link diagrams under consideration

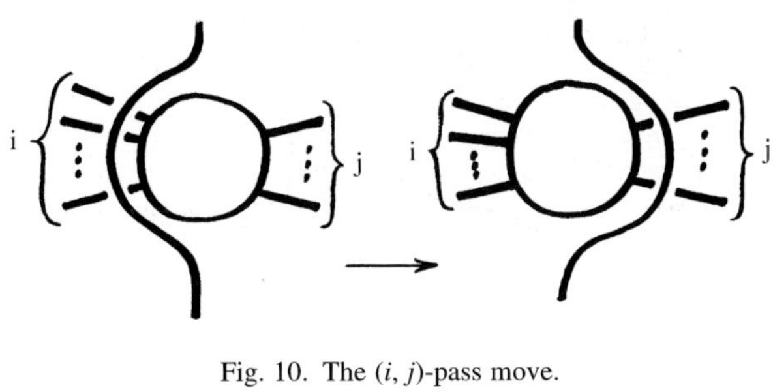

Fig. 10. The (i, j)-pass move.

requires great economy. Clever programming and ad hoc tricks might bring about greater gains than sophisticated algorithms derived from the most powerful theorems in topology. A certain side of tabulation remains an art form.

Experience shows that through about 17 crossings, flypes and pass moves appear to suffice to reduce the list of nonalternating links down to about 110–120% of its eventual size. At this point, the computation of various link invariants can be undertaken. Through 17 crossings, at least for knots, the computation of hyperbolic structure with *SnapPea* worked well. Very few knots were identified as nonhyperbolic, and very few hyperbolic knots had volumes so suspiciously close that further inspection (with other invariants) was warranted. The situation with links is not quite as favorable – many more pairs are too close to call with *SnapPea* and many more nonhyperbolic links exist. The winnowing out of duplicates from Thistlethwaite's nonalternating lists of links through 19 crossings still awaits completion.

5. Conclusion

Presently, over 6 billion knots and links have been tabulated, some with crossing number as high as 22. The numbers of prime, unoriented, alternating links per crossing number and component number are given in Table 1. The corresponding numbers for nonalternating

Table 1. Number of prime, unoriented, alternating links per crossing number n and number of components

n	Number of components								
	1	2	3	4	5	6	7	8	9
0	1								
1									
2		1							
3	1								
4	1	1							
5	2	1							
6	3	3	2						
7	7	6	1						
8	18	14	6	1					
9	41	42	12	1					
10	123	121	43	9	1				
11	367	384	146	17	1				
12	1288	1408	500	100	11	1			
13	4878	5100	2074	341	23	1			
14	19536	21854	8206	1556	181	13	1		
15	85263	92234	37222	7193	653	29	1		
16	379799	427079	172678	33216	3885	301	16	1	
17	1769979	2005800	829904	173549	19122	1129	36	1	
18	8400285	9716848	4194015	876173	105539	8428	471	19	1
19	40619385	48184018	21207695	4749914	599433	43513	1813	43	1
20	199631939								
21	990623857								
22	4976016485								

Table 2. Number of prime, unoriented, nonalternating links per crossing number n and number of components

n	Number of components			
	1	2	3	4
0				
1				
2				
3				
4				
5				
6			1	
7		2		
8	3	2	4	2
9	8	18	9	0
10	42			
11	185			
12	888			
13	5110			
14	27436			
15	168030			
16	1008906			

links are given in Table 2. As this article is being written, Rankin, Flint and Schermann are preparing to tabulate the 23-crossing prime alternating knots, and furthermore to turn their attention to alternating links. Thistlethwaite's alternating link tables to 19 crossings have yet to be confirmed, and his lists of nonalternating links still await the final removal of duplicates.

Tables of prime knots through 16 crossings are widely available in the software package *Knotscape* [28] written by Hoste and Thistlethwaite. This software not only contains the tables but will compute various knot invariants and locate knots in the tables. A major revision of *Knotscape* that will also handle links and include knot and link tables to 17 crossings is currently underway. As of this writing, the tables of Rankin, Flint and Schermann are not yet publicly available.

How far will this current burst of tabulation take us? To 25 or 30 crossings? Will a mega-tabulation distributed over thousands of machines via the internet be organized? Will a knot with trivial Jones polynomial, or some other surprising example be found this way? Perhaps, the greatest gains in tabulation will result from improvements in computers, but no doubt theoretical advances will be made as well, allowing more efficient algorithms, and providing better invariants. This is an exciting time in knot theory!

Acknowledgements

Special thanks are due to Joel Hass, Sergei Matveev, Morwen Thistlethwaite, and Jeff Weeks for their helpful comments.

References

[1] J.W. Alexander, *Topological invariants of knots and links*, Trans. Am. Math. Soc. **30** (2) (1928), 275–306.

[2] J.W. Alexander and G.B. Briggs, *On types of knotted curves*, Ann. Math. **28** (1927), 562–586.

[3] B. Arnold, M. Au, C. Candy, K. Erdener, J. Fan, R. Flynn, J. Hoste, R.J. Muir and D. Wu, *Tabulating alternating knots through 14 crossings*, J. Knot Theory Ramificat. **3** (4) (1994), 433–437.

[4] M. Andreeva, I. Dynnikov, S. Koval, K. Polthier and I. Taimanov, *Book Knot Simplifier*, computer software, http://www-sfb288.math.tuberlin.de/vgp/javaview/services/knots/.

[5] J.S. Birman and M.D. Hirsch, *A new algorithm for recognizing the unknot*, Geomet. Topol. **2** (9) (1998), 175–220.

[6] J.S. Birman and W.W. Menasco, *A Calculus on Links in the 3-Sphere*, Knots 90 (Osaka, 1990), de Gruyter, Berlin (1992), 625–631.

[7] J.S. Birman and W.W. Menasco, *Studying links via closed braids. III. Classifying links which are closed 3-braids*, Pacific J. Math. **161** (1) (1993), 25–113.

[8] J.S. Briman, M. Rampichini, P. Boldi and S. Vigna, *Towards an implementation of the B–H algorithm for recognizing the unknot*, J. Knot Theory Ramificat. **11** (4) (2002), 601–645.

[9] B. Burton, and D. Letscher, *Regina*, computer software, http://regina.sourceforge.net.

[10] J.A. Calvo, *Knot enumeration through flypes and twisted splices*, J. Knot Theory Ramificat. **6** (6) (1997), 785–797.

[11] A. Caudron, *Classification des noeuds et des enlacements*, Prepublication Math. d'Orsay, Orsay, France: Universite Paris-Sud (1981).

[12] J.H. Conway, *An Enumeration of Knots and Links and Some of Their Algebraic Properties*, Computational Problems in Abstract Algebra (Proc. Conf., Oxford, 1967), Pergamon, Oxford (1967), 329–358.

[13] D. Coulson, O. Goodman, C. Hodgson and W. Neumann, *Computing arithmetic invariants of 3-manifolds*, Experiment. Math. **9** (1) (2000), 127–152.

[14] R.H. Crowell, *Nonalternating links, III*, J. Math. **3** (1959), 101–120.

[15] O.T. Dasbach and S. Hougardy, *Does the Jones polynomial detect unknottedness?*, Exp. Math. **6** (1) (1997), 51–56.

[16] M. Dehn, *Die beiden Kleeblattschlingen*, Math. Ann **75** (1914), 402–413.

[17] H. Doll and J. Hoste, *A tabulation of oriented links*, Math. Computat. **57** (196) (1991), 747–761.

[18] C.H. Dowker and M.B. Thistlethwaite, *On the classification of knots*, C.R. Math. Rep. Acad. Sci. Canada **4** (2) (1982), 129–131.

[19] C.H. Dowker and M.B. Thistlethwaite, *Classification of knot projections*, Topol. Appl. **16** (1983), 19–31.

[20] M. Epple, *Branch points of algebraic functions and the beginnings of modern knot theory*, Historia Math. **22** (1995), 371–401.

[21] M. Epple, *Topology, matter space. I. Topological notions in 19th-century natural philosophy*, Arch. Hist. Exact Sci. **52** (4) (1998), 297–392.

[22] M. Epple, *Die Entstehung der Knotentheorie: Kontexte und Konstruktionen einer modernen mathematischen Theorie*, Vieweg, Wiesbaden (1999).

[23] D.B.A. Epstein and R.C. Penner, *Euclidean decompositions of noncompact hyperbolic manifolds*, J. Diff. Geom. **27** (1988), 67–80.

[24] C.McA. Gordon and J. Luecke, *Knots are determined by their complements*, J. Am. Math. Soc. **2** (1989), 371–415.

[25] J. Hass and J. C. Lagarias, *The number of Reidemeister moves needed for unknotting*, J. Amer. Math. Soc. **14** (2) (2001), 399–428.

[26] Y. Hashizume, *On the uniqueness of the decomposition of a link*, Osaka Math. J. **10** (1958), 283–300.

[27] P. Freyd, D. Yetter, J. Hoste, W.B.R. Lickorish, K. Millett and A. Ocneanu, *A new polynomial invariant of knots and links*, Bull. Amer. Math. Soc. **12** (1985), 239–249.

[28] J. Hoste and M.B. Thistlethwaite, *Knotscape*, computer software, http://www.math.utk.edu/~morwen.

[29] J. Hoste, M.B. Thistlethwaite and J. Weeks, *The first 1,701,936 knots*, Math. Intelligencer **20** (4) (1998), 33–48.

[30] S. Jablan, *Geometry of links, XII Yugoslav Geometric Seminar (Novi Sad, 1998)*, Novi Sad J. Math. **29** (3) (1999), 121–139.

[31] V.F.R. Jones, *A polynomial invariant for knots via Von Neumann algebras*, Bull. Amer. Math. Soc. **12** (1985), 103–111.

[32] L.H. Kauffman, *State models and the Jones polynomial*, Topology **26** (1987), 395–407.

[33] T.P. Kirkman, *The enumeration, description and construction of knots of fewer than ten crossings*, Trans. Roy. Soc. Edinburgh **32** (1885), 281–309.

[34] T.P. Kirkman, *The 364 unifilar knots of ten crossings enumerated and defined*, Trans. Roy. Soc. Edinburgh **32** (1885), 483–506.

[35] K. Murasugi, *On the braid index of alternating links*, Trans. Amer. Math. Soc. **326** (1) (1991), 237–260.

[36] J.C. Cha, K.H. Ko, S.J. Lee, J.W. Han and J.H. Cheon, *An efficient implementation of braid groups, Asiacrypt 2001*, Lecture Notes in Computer Science 2248 (2001), 157–174.

[37] C.N. Little, *On knots, with a census of order ten*, Trans. Connecticut Acad. Sci. **18** (1885), 374–378.

[38] C.N. Little, *Non alternate ± knots of orders eight and nine*, Trans. Roy. Soc. Edinburgh **35** (1889), 663–664.

[39] C.N. Little, *Alternate ± knots of order 11*, Trans. Roy. Soc. Edinburgh **36** (1890), 253–255.

[40] C.N. Little, *Non-alternate ± knots*, Trans. Roy. Soc. Edinburgh **39** (1900), 771–778.

[41] S. Matveev, *Algorithmic Topology and Classification of 3-Manifolds*, Algorithms and Computation in Mathematics, Vol. 9, Springer, Berlin (2003).

[42] W.W. Menasco, *Closed incompressible surfaces in alternating knot and link complements*, Topology **23** (1) (1984), 37–44.

[43] W.W. Menasco and M.B. Thistlethwaite, *The classification of alternating links*, Ann. Math. **138** (1993), 113–171.

[44] K. Murasugi, *The Jones polynomial and classical conjectures in knot theory*, Topology **26** (1987), 187–194.

[45] K. Perko, *On the classification of knots*, Proc. Am. Math. Soc. **45** (1974), 262–266.

[46] J.H. Przytycki and P. Traczyk, *Invariants of links of Conway type*, Kobe J. Math. **4** (1987), 115–139.

[47] S. Rankin, O. Flint and J. Schermann, *Enumerating the prime alternating knot, part I*, J. Knot Theory Ramificat. **13** (1) (2004), 57–100.

[48] S. Rankin, O. Flint and J. Schermann, *Enumerating the prime alternating knot, part II*, J. Knot Theory Ramificat. **13** (1) (2004), 101–150.

[49] S. Rankin and O. Smith, *Enumerating the prime alternating links*, J. Knot Theory Ramificat. **13** (1) (2004), 151–173.

[50] K. Reidemeister, *Elementare Begründung der Knotentheorie*, Abhandlungen aus dem Mathematischen Seminar der Hamburgischen Universität **5** (1926), 24–32.

[51] D. Rolfsen, *Knots and Links*, Mathematics Lecture Series, 7. Publish or Perish, Houston, TX (1990).

[52] M. Sakuma and J. Weeks, *The generalized tilt formula*, Geometriae Dedicata **55** (1995), 115–123.

[53] H. Schubert, *Die eindeutige Zerlegbarkeit eines Knotens in Primknoten*, Sitzungsber. Akad. Wiss. math.-nat. Kl. 3.Ahh. (1949), 57–104.

[54] C. Sundberg and M.B. Thistlethwaite, *The rate of growth of the number of prime alternating links and tangles*, Pacific J. Math. **182** (2) (1998), 329–358.

[55] P.G. Tait, *On Knots I, II, III*, Scientific Papers, Vol. 1, Cambridge University Press, London (1898), 273–347.

[56] M.B. Thistlethwaite, *Knot Tabulations and Related Topics, Aspects of Topology in Memory of Hugh Dowker 1912–1982*, London Math., Soc. Lecture Note Series 93, I.M. James and E.H. Kronheimer, eds, Cambridge Press, Cambridge, England (1985), 1–76.

[57] M.B. Thistlethwaite, *A spanning tree expansion of the Jones polynomial*, Topology **26** (3) (1987), 297–309.

[58] M.B. Thistlethwaite, *Links with trivial Jones polynomial*, J. Knot Theory Ramificat. **10** (4) (2001), 641–643.

[59] W. Thomson, *On vortex motion*, Trans. Roy. Soc. Edinburgh **25** (1867), 217–260.

[60] H.F. Trotter, *Noninvertible knots exist*, Topology **2** (1964), 275–280.

[61] P. Vogel, *Representation of links by braids: a new algorithm*, Comm. Math. Helv. **65** (1990), 104–113.

[62] J. Weeks, *SnapPea: a computer program for creating and studying hyperbolic 3-manifolds*, computer software, http://www.geometrygames.org.

[63] J. Weeks, *Convex hulls and isometries of cusped hyperbolic 3-manifolds*, Topol. Appl. **52** (1993), 127–149.

[64] J. Weeks, *Computation of Hyperbolic Structures in Knot Theory*, Handbook of Knot Theory, W.W. Menasco and M.B. Thistlethwaite, eds, Elsevier, Amsterdam (2005), 461–480.

[65] W. Whitten, *Knot complements and groups*, Topology **26** (1987), 41–44.

[66] S. Yamada, *The minimal number of Seifert circles equals the braid index of a link*, Invent. Math. **89** (1987), 347–356.

CHAPTER 6

Knot Diagrammatics

Louis H. Kauffman

Department of Mathematics, Statistics and Computer Science, University of Illinois, 851 South Morgan Street, Chicago, IL 60607-7045, USA

E-mail: kauffman@uic.edu

Contents

HANDBOOK OF KNOT THEORY
Edited by William Menasco and Morwen Thistlethwaite

Abstract

This paper is a survey of knot theory and invariants of knots and links from the point of view of categories of diagrams. The topics range from foundations of knot theory to virtual knot theory and topological quantum field theory.

1. Introduction

This paper is an exploration of the theme of knot diagrams. I have deliberately focused on basics in a number of interrelated domains. In most cases some fundamentals are done in a new or in a more concise way. Some parts of the paper are expository to fill in the context. This exposition is an outgrowth of the series of lectures [63] that I gave in Tokyo at the *Knots 96* conference in the summer of 1996. The present exposition goes considerably farther than those lectures, and includes material on virtual knot theory and on links that are undetectable by the Jones polynomial.

The paper is divided into seven sections (counting from Section 2 onward). Section 2, on the Reidemeister moves, gives a proof of Reidemeister's basic theorem (that the three Reidemeister moves on diagrams generate ambient isotopy of links in three-space). A discussion on graph embeddings extends Reidemeister's theorem to graphs and proves the appropriate moves for topological and rigid vertices. We hope that this section fills in some gaps in the literature. Section 3 discusses Vassiliev invariants and invariants of rigid vertex graphs. This section is expository, with discussions of the four-term relations, Lie algebra weights, relationships with the Witten functional intergal, and combinatorial constructions for some Vassiliev invariants. The discussion raises some well-known problems about Vassiliev invariants. The section on the functional integral introduces a useful abstract tensor notation that helps in understanding how the Lie algebra weight systems are related to the functional integral. Sections 4 and 5 are based on a reformulation of the Reidemeister moves so that they work with diagrams arranged generically transverse to a special direction in the plane. We point out how the technique by which we proved Reidemeister's theorem (it is actually Reidemeister's original technique) generalizes to give these moves as well. The moves with respect to a vertical are intimately related to quantum link invariants and to Hopf algebras. Section 4 is a quick exposition of quantum link invariants, their relationship with Vassiliev invariants, classical Yang–Baxter equation and infinitesimal braiding relations. Again, this provides the context to raise many interesting questions. Section 5 is a very concise introduction to the work of the author, David Radford and Steve Sawin on invariants of three-manifolds from finite-dimensional Hopf algebras. We touch on the question of the relationship of this work to the Kuperberg invariant. Section 6 is a discussion of the Temperley–Lieb algebra. Here we give a neat proof of the relation structure in the Temperley–Lieb monoid via piecewise linear diagrams. The last part of this section explains the relationship of the Temperley–Lieb monoid to parenthesis structures and shows how this point of view can be used to relate parentheses to the pentagon and the Stasheff polyhedron. Section 7 discusses virtual knot theory. Section 8 discusses the construction of links that while linked, have the same Jones polynomial as the unlink.

2. Reidemeister moves

Reidemeister [121] discovered a simple set of moves on link diagrams that captures the concept of ambient isotopy of knots in three-dimensional space. There are three basic Reidemeister moves. Reidemeister's theorem states that two diagrams represent ambient

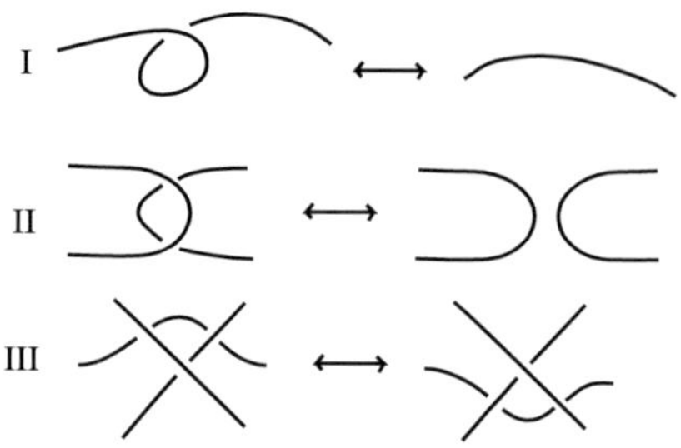

Fig. 1. Reidemeister moves.

isotopic knots (or links) if and only if there is a sequence of Reidemeister moves taking one diagram to the other. The Reidemeister moves are illustrated in Figure 1.

Reidemeister's three moves are interpreted as performed on a larger diagram in which the small diagram shown is a literal part. Each move is performed without disturbing the rest of the diagram. Note that this means that each move occurs, up to topological deformation, just as it is shown in the diagrams in Figure 1. There are no extra lines in the local diagrams. For example, the equivalence (A) in Figure 2 is *not* an instance of a single first Reidemeister move. Taken literally, it factors into a move II followed by a move I.

Diagrams are always subject to topological deformations in the plane that preserve the structure of the crossings. These deformations could be designated as "Move Zero" (see Figure 2).

A few exercises with the Reidemeister moves are in order. First of all, view the diagram in Figure 3. It is unknotted and you can have a good time finding a sequence of Reidemeister moves that will do the trick. Diagrams of this type are produced by tracing a curve and always producing an undercrossing at each return crossing. This type of knot is called a *standard unknot*. Of course we see clearly that a standard unknot is unknotted by just *pulling* on it, since it has the same structure as a coil of a rope that is wound down onto a flat surface.

Can one recognize unknots by simply looking for sequences of Reidemeister moves that undo them? This would be easy if it were not for the fact that there are examples of

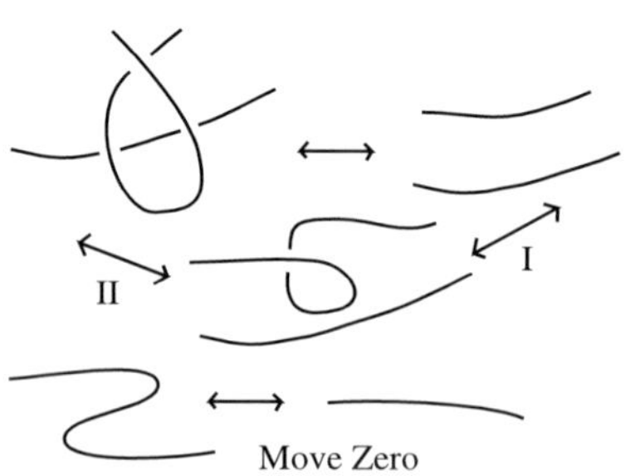

Fig. 2. Factorable move, move zero.

Fig. 3. Standard unknot.

unknots that require some moves that increase the number of crossings before they can be subsequently decreased. Such an demonic example is illustrated in Figure 4.

It is generally not so easy to recognize unknots. However, here is a tip: look for *macro moves* of the type shown in Figure 5. In a macro move, we identify an arc that passes entirely under some piece of the diagram (or entirely over) and shift this part of the arc, keeping it under (or over) during the shift. In Figure 5, we illustrate a macro move on an arc that passes under a piece of the diagram that is indicated by arcs going into a circular region. A more general macro move is possible where the moving arc moves underneath one layer of diagram, and at the same time, over another layer of diagram. Macro moves often allow a reduction in the number of crossings even though the number of crossings will increase during a sequence of Reidemeister moves that generates the macro move.

As shown in Figure 5, the macro move includes as a special case both the second and the third Reidemeister moves, and it is not hard to verify that a macro move can be generated by a sequence of type II and type III Reidemeister moves. It is easy to see that the type I

Fig. 4. A demon.

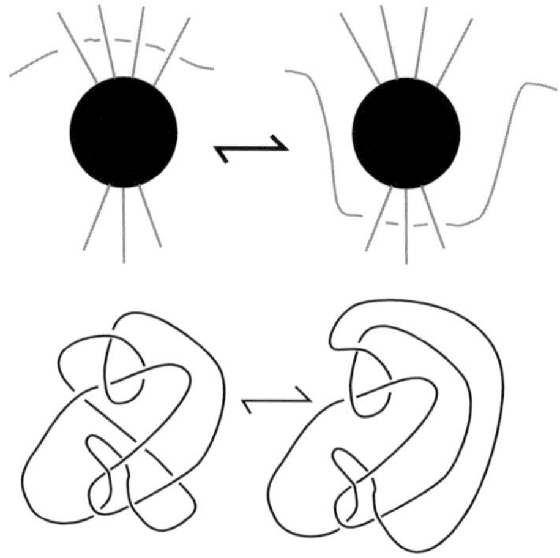

Fig. 5. Macro move.

moves can be left to the end of any deformation. The demon of Figure 4 is easily demolished by macro moves, and from the point of view of macro moves the diagram never gets more complicated.

Let us say that a knot can be *reduced* by a set of moves if it can be transformed by these moves to the unknotted circle diagram through diagrams that never have more crossings than the original diagram. Then we have shown that there are diagrams representing the unknot that cannot be reduced by the Reidemeister moves. On the other hand, I do not know whether unknotted diagrams can always be reduced by the macro moves in conjunction with the first Reidemeister move. If this were true it would give a combinatorial way to recognize the unknot.

Remark. In fact, there is a combinatorial way to recognize the unknot based on diagrams and moves. In [31] Dynnikov finds just such a result, using piecewise linear knot diagrams with all 90° angles in the diagrams, and all arcs in the diagram either horizontal or vertical. The interested reader should consult his lucid paper.

2.1. *Reidemeister's theorem*

We now indicate how Reidemeister proved his theorem.

An embedding of a knot or link in three-dimensional space is said to be *piecewise linear* if it consists of a collection of straight line segments joined end to end. Reidemeister started with a *single* move in three-dimensional space for piecewise linear knots and links. Consider a point in the complement of the link, and an edge in the link such that the surface of the triangle formed by the end points of that edge and the new point is not pierced by any other edge in the link. Then one can replace the given edge on the link by the other two edges of the triangle, obtaining a new link that is ambient isotopic to the original link. Conversely, one can remove two consecutive edges in the link and replace them by a new

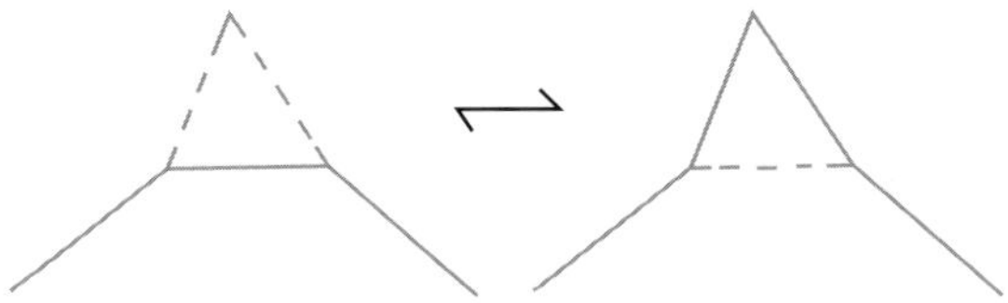

Fig. 6. Triangle move.

edge that goes directly from initial to final points, whenever the triangle spanned by the two consecutive edges is not pierced by any other edge of the link. This triangle replacement constitutes Reidemeister's three-dimensional move (see Figure 6). It can be shown that two piecewise linear knots or links are ambient isotopic in three-dimensional space if and only if there is a sequence of Reidemeister triangle moves from one to the other. This will not be proved here. At the time when Reidemeister wrote his book, equivalence via three-dimensional triangle moves was taken as the definition of topological equivalence of links.

It can also be shown that tame knots and links have piecewise linear representatives in their ambient isotopy class. It is sufficient for our purposes to work with piecewise linear knots and links. Reidemeister's planar moves then follow from an analysis of the shadows projected into the plane by Reidemeister triangle moves in space. Figure 7 gives a hint of this analysis. The result is a reformulation of the three-dimensional problems of knot theory to a combinatorial game in the plane.

To go beyond the hint in Figure 7 to a complete proof that Reidemeister's planar moves suffice involves preliminary remarks about subdivision. The simplest subdivision that one wants to be able to perform on a piecewise linear link is the placement of a new vertex at an interior point of an edge – so that this edge becomes two edges in the subdivided link. Figure 8 shows how to accomplish this subdivision via triangle moves.

Any triangle move can be factored into a sequence of smaller triangle moves corresponding to a simplicial subdivision of that triangle. This is obvious, since the triangles in the subdivision of the large triangle that is unpierced by the link are themselves unpierced by the link.

To understand how the Reidemeister triangle move behaves on diagrams it is sufficient to consider a projection of the link in which the triangle is projected to a non-singular triangle in the plane. Of course, there may be many arcs of the link also projected upon the interior of the projected triangle. However, by using subdivision, we can assume that the cases of the extra arcs are as shown in Figure 9. In this figure we have also shown how each of these cases can be accomplished by (combinations of) the three Reidemeister moves. This proves that a projection of a single triangle move can be accomplished by a sequence of Reidemeister diagram moves.

A piecewise linear isotopy consists in a finite sequence of triangle moves. There exists a direction in three-dimensional space that makes a non-zero angle with each of these triangles and is in general position with the link diagram. Projecting to the plane along this direction makes it possible to perform the entire ambient isotopy in the language of

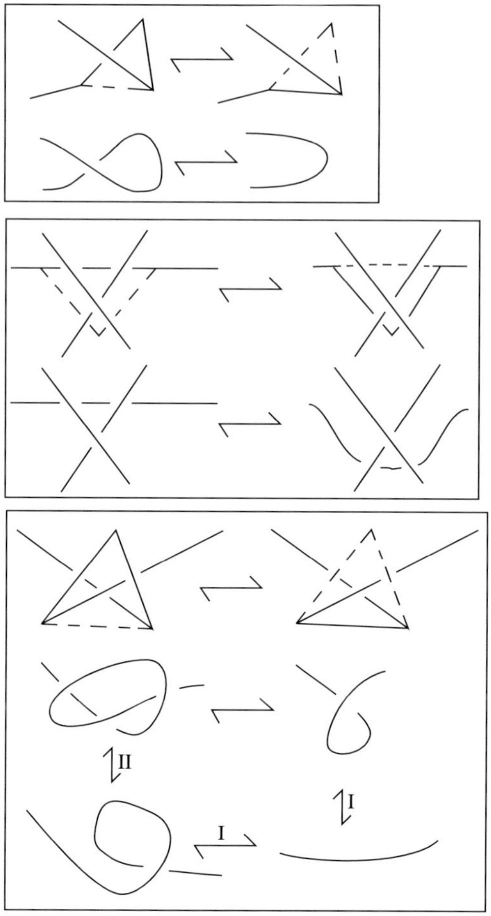

Fig. 7. Shadows.

projected triangle moves. Now apply the results of the previous paragraph and we conclude

REIDEMEISTER'S THEOREM. *If two links are piecewise linearly equivalent (ambient isotopic), then there is a sequence of Reidemeister diagram moves taking a projection of one link to a projection of the other.*

Note that the proof tells us that the two diagrams can be obtained from one spatial projection direction for the entire spatial isotopy. It is obvious that diagrams related by Reidemeister moves represent ambient isotopic links. Reidemeister's Theorem gives a complete combinatorial description of the topology of knots and links in three-dimensional space.

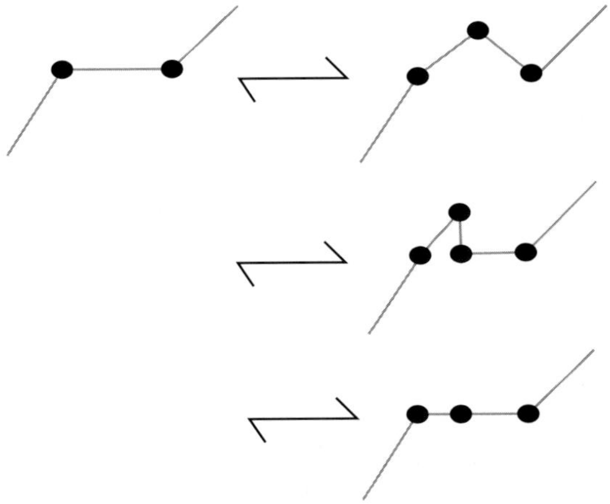

Fig. 8. Subdivision of an edge.

2.2. *Graph embeddings*

Let G be a (multi-)graph. That is, G is a finite abstract graph with, possibly, a multiplicity of edges between any two of its vertices. Now consider the embeddings of G in Euclidean three-space R^3. In the category of topological embeddings, any edge of G can acquire local

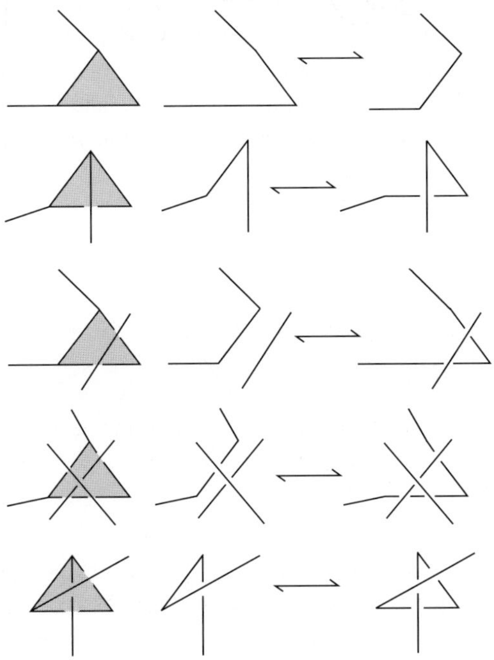

Fig. 9. Projections of triangle moves.

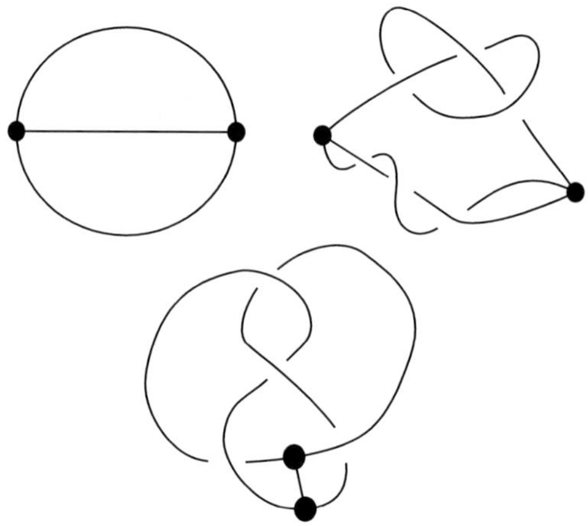

Fig. 10. Graph embedding.

knotting as shown in Figure 10. On top of this there is the possibility of global knotting that results from the structure of the graph as a whole.

Topological or piecewise linear ambient isotopy of graph embeddings is complicated by the fact that arbitrary braiding can be created or destroyed at a vertex, as illustrated in Figure 11.

For this reason, it is useful to consider ways to restrict the allowed movement in the neighborhood of a vertex. One way to accomplish this is to decree that each vertex will come equipped with a specific cyclic order of the edges meeting the vertex. This cyclic order can be instantiated on the boundary of a disk, and the graph replaced by a configuration of disks with cyclic orders of marked points along their boundaries. The edges of the original graph are replaced by edges that go from one disk to another

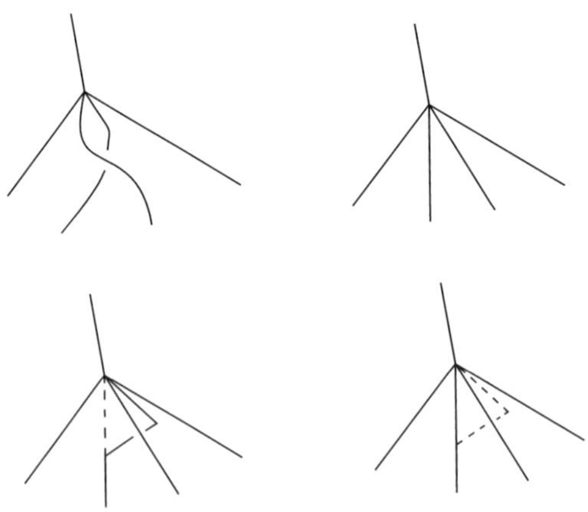

Fig. 11. Braiding at a vertex.

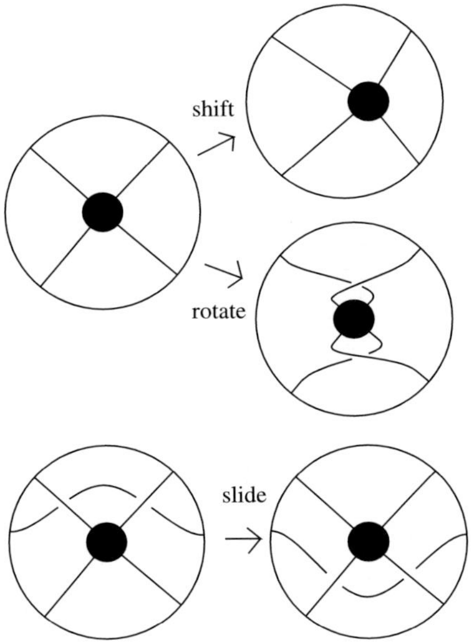

Fig. 12. Rigid vertex graphs and affine motions.

terminating in the marked points. Call such an arrangement a *rigid vertex graph G*. If G is a rigid vertex graph, then we consider embeddings of G where the disks are embedded metrically while the (graphical) edges are embedded topologically. A *rigid vertex isotopy* of one RV (RV stands for rigid vertex) embedding G to another G' is a combination of ambient isotopies of the embedded edges of the graph (the *strings* of the graph) relative to their endpoints on the disks, coupled with affine motions of the disks (carrying along the strings in ambient isotopy). An affine motion of a disk is a combination of parallel translations of the disk along a given direction in three-space and rotations of the disk about an axis through its center. We can think of a given disk as embedded inside a standard three-ball with the strings from the disk emanating straight to the boundary of the three-ball. Each basic affine motion is assumed to leave the points on the boundary of the containing three-ball fixed. Thus the types of affine motion are as illustrated in Figure 12.

We will give versions of the Reidemeister moves for both topological isotopy and rigid vertex isotopy of embedded graphs. In the topological case the extra moves are illustrated in Figure 13. Here we have indicated the elementary braiding at a vertex and slide moves that take an edge underneath a vertex. The proof that these moves suffice is a generalizaton of our original proof of the Reidemeister moves. That is, we model the graph embeddings by piecewise linear embeddings. This may entail subdividing the edges of the original graph so that those edges can have enough flexibility to sustain a given topological conformation. Thus, when we speak of a piecewise linear embedding of a given graph, we mean a piecewise linear embedding of a graph that is obtained from the given graph by subdividing some of its edges.

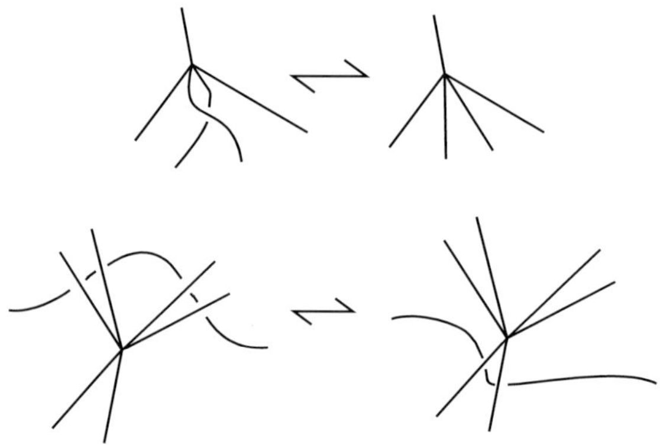

Fig. 13. Extra moves for topological isotopy of graphs.

Piecewise ambient isotopy of graph embeddings is defined exactly as in the case of piecewise linear isotopy for knots and links. The same projection arguments apply and the extra moves are obtained from the three-dimensional triangle move as illustrated in Figure 14. This completes the proof of our assertion about the topological Reidemeister moves for graphs.

Consider rigid vertex isotopy of rigid vertex (RV) graphs. We will assume that the topological moves are performed in the piecewise linear setting. Thus subdivisions of the edges of the graph can be produced. Basic translational affine moves of the embedded disks can have piecewise linear starting and ending states by drawing straight lines from the marked points on the disk boundaries to the corresponding points in the containing balls. Rotatory moves with the center of a disk as axis can also have piecewise linear

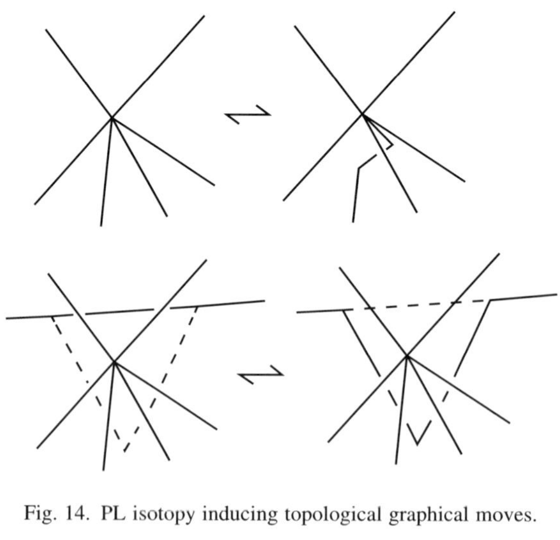

Fig. 14. PL isotopy inducing topological graphical moves.

starting and ending states by taking the braiding that is induced by the rotation and suitably subdividing it. These remarks show that RV isotopy can be achieved in the PL category.

The next point to consider is the result of projection of an RV isotopy on the corresponding diagrams. A sequence of elementary RV isotopies from a graph G to a graph G' has associated with it a direction of projection so that each PL triangle move has its triangle projected to a non-singular triangle in the plane and each affine move has its disk projected to a non-singular disk in the plane. In the case of the affine moves we can assume that the before to after appearance of the disk and its corresponding containing ball will represent either a topological identity map (albeit an affine shift) or a rotation about the disk axis by π radians. (Higher multiples of π can be regarded as iterates of a π rotation.) Therefore, the basic π rotation can be schematized as shown in Figure 15. The figure illustrates the moves that we need to add to the Reidemeister moves to obtain a planar diagram version of RV isotopy. The remaining moves in Figure 15 follow from the same projection arguments that we have used earlier in this section. This completes the construction of the diagrammatic calculus for RV isotopy.

Note that the generating moves for rigid vertex graph isotopy are almost the same as the generating moves for topological graph isotopy, except that the braiding at the vertex in the rigid vertex case comes from twisting the disk as a whole. This circumstance makes the construction of invariants of rigid vertex graphs much easier. We will discuss constructions of such invariants in the next section. In Section 4 we will return to the Reidemeister moves and reformulate them once again for the sake of quantum link invariants.

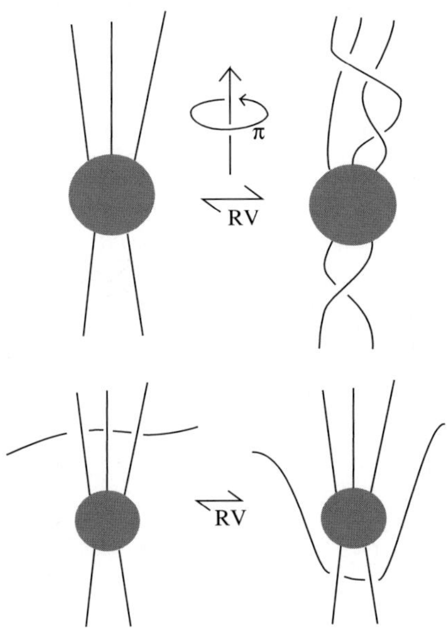

Fig. 15. Diagrammatic rigid vertex isotopy.

3. Vassiliev invariants and invariants of rigid vertex graphs

If $V(K)$ is a (Laurent polynomial valued, or more generally – commutative ring valued) invariant of knots, then it can be naturally extended to an invariant of rigid vertex graphs by defining the invariant of graphs in terms of the knot invariant via an "unfolding" of the vertex. That is, we can regard the vertex as a "black box" and replace it by any tangle of our choice. Rigid vertex motions of the graph preserve the contents of the black box, and hence entail ambient isotopies of the link obtained by replacing the black box by its contents. Invariants of knots and links that are evaluated on these replacements are then automatically rigid vertex invariants of the corresponding graphs. If we set up a collection of multiple replacements at the vertices with standard conventions for the insertions of the tangles, then a summation over all possible replacements can lead to a graph invariant with new coefficients corresponding to the different replacements. In this way each invariant of knots and links implicates a large collection of graph invariants (see [56,57]).

The simplest tangle replacements for a four-valent vertex are the two crossings, positive and negative, and the oriented smoothing. Let $V(K)$ be any invariant of knots and links. Extend V to the category of rigid vertex embeddings of four-valent graphs by the formula (see Figure 16)

$$V(K_*) = aV(K_+) + bV(K_-) + cV(K_0).$$

Here K_* indicates an embedding with a transversal four-valent vertex. This formula means that we define $V(G)$ for an embedded four-valent graph G by taking the sum

$$V(G) = \sum_S a^{i_+(S)} b^{i_-(S)} c^{i_0(S)} V(S)$$

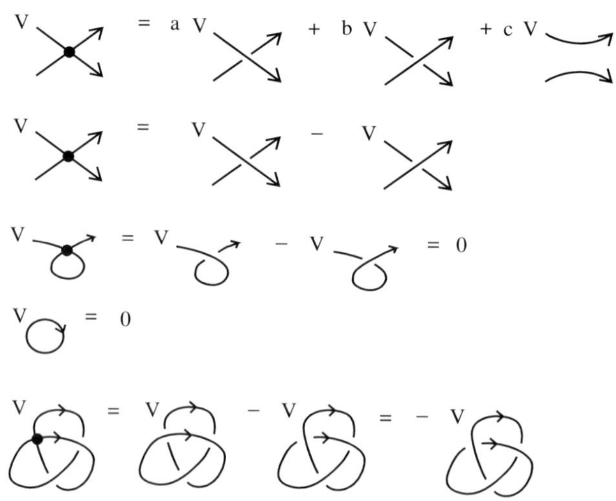

Fig. 16. Graphical vertex formulas.

with the summation over all knots and links S obtained from G by replacing a node of G with either a crossing of positive or negative type, or with a smoothing (denoted 0). Here $i_+(S)$ denotes the number of positive crossings in the replacement, $i_-(S)$ the number of negative crossings in the replacement, and $i_0(S)$ the number of smoothings in the replacement. It is not hard to see that if $V(K)$ is an ambient isotopy invariant of knots, then, this extension is a rigid vertex isotopy invariant of graphs. In rigid vertex isotopy the cyclic order at the vertex is preserved, so that the vertex behaves like a rigid disk with flexible strings attached to it at specific points (see previous section).

There is a rich class of graph invariants that can be studied in this manner. The Vassiliev invariants [4,10,142] constitute the important special case of these graph invariants where $a = +1$, $b = -1$ and $c = 0$. Thus $V(G)$ is a Vassiliev invariant if

$$V(K_*) = V(K_+) - V(K_-).$$

Call this formula the *exchange identity* for the Vassiliev invariant V. V is said to be of finite type k if $V(G) = 0$ whenever $|G| > k$ where $|G|$ denotes the number of four-valent nodes in the graph G. The notion of finite type is of paramount significance in studying these invariants. One reason for this is the following basic lemma.

LEMMA. *If a graph G has exactly k nodes, then the value of a Vassiliev invariant v_k of type k* on *G, $v_k(G)$, is independent of the embedding of G.*

PROOF. The different embeddings of G can be represented by link diagrams with some of the four-valent vertices in the diagram corresponding to the nodes of G. It suffices to show that the value of $v_k(G)$ is unchanged under switching of a crossing. However, the exchange identity for v_k shows that this difference is equal to the evaluation of v_k on a graph with $k + 1$ nodes and hence is equal to zero. This completes the proof. □

The upshot of this lemma is that Vassiliev invariants of type k are intimately involved with certain abstract evaluations of graphs with k nodes. In fact, there are restrictions (the four-term relations) on these evaluations demanded by the topology (we shall articulate these restrictions shortly) and it follows from results of Kontsevich [4] that such abstract evaluations actually determine the invariants. The invariants derived from classical Lie algebras are all built from Vassiliev invariants of finite type. All this is directly related to Witten's functional integral [145].

DEFINITION. Let v_k be a Vassiliev invariant of type k. The *top row* of v_k is the set of values that v_k assigns to the set of (abstract) four-valent graphs with k nodes. If we concentrate on Vassiliev invariants of knots, then these graphs are all obtained by marking $2k$ points on a circle, and choosing a pairing of the $2k$ points. The pairing can be indicated by drawing a circle and connecting the paired points with arcs. Such a diagram is called a *chord diagram*. Some examples are indicated in Figure 17.

Note that a top row diagram cannot contain any isolated pairings since this would correspond to a difference of local curls on the corresponding knot diagram (and these curls, being isotopic, yield the same Vassiliev invariants).

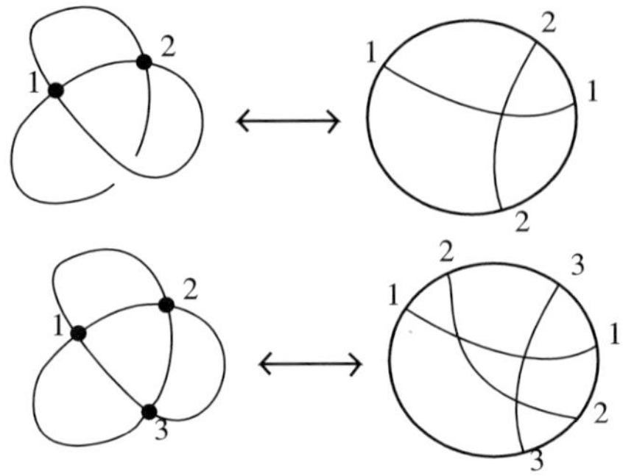

Fig. 17. Chord diagrams.

THE FOUR-TERM RELATION. (Compare [132].) Consider a single embedded graphical node in relation to another embedded arc, as illustrated in Figure 18. The arc underlies the lines incident to the node at four points and can be slid out and isotoped over the top so that it overlies the four nodes. One can also switch the crossings one-by-one to exchange the arc until it overlies the node. Each of these four switchings gives rise to an equation, and the left-hand sides of these equations will add up to zero, producing a relation corresponding to the right-hand sides. Each term in the right-hand side refers to the value of the Vassiliev invariant on a graph with two nodes that are neighbors to each other (see Figure 18).

There is a corresponding four-term relation for chord diagrams. This is the four-term relation for the top row. In chord diagrams the relation takes the form shown at the bottom of Figure 18. Here we have illustrated only those parts of the chord diagram that are relevant to the two nodes in question (indicated by two pairs of points on the circle of the chord diagram). The form of the relation shows the points on the chord diagram that are immediate neighbors. These are actually neighbors on any chord diagram that realizes this form. Otherwise there can be many other pairings present in the situation.

As an example, consider the possible chord diagrams for a Vassiliev invariant of type 3. There are two possible diagrams as shown in Figure 19. One of these has the projected pattern of the trefoil knot and we shall call it the *trefoil graph*. These diagrams satisfy the four-term relation. This shows that one diagram must have twice the evaluation of the other. Hence it suffices to know the evaluation of one of these two diagrams to know the top row of a Vassiliev invariant of type 3. We can take this generator to be the trefoil graph.

Now one more exercise: consider any Vassiliev invariant v and let us determine its value on the trefoil graph as in Figure 20.

The value of this invariant on the trefoil graph is equal to the difference between its values on the trefoil knot and its mirror image. Therefore, any Vassiliev invariant that assigns a non-zero value to the trefoil graph can tell the difference between the trefoil knot and its mirror image.

EXAMPLE. This example shows how the original Jones polynomial is composed of Vassiliev invariants of finite type. Let $V_K(t)$ denote the original Jones polynomial [44].

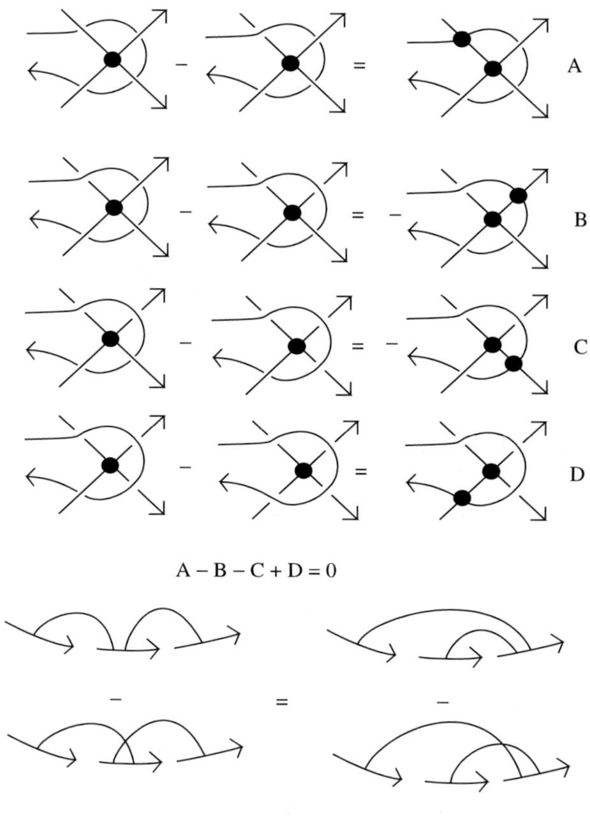

Fig. 18. The four-term relation.

Recall the oriented state expansion for the Jones polynomial [58] with the basic formulas (δ is the loop value).

$$V_{K_+} = -t^{1/2}V_{K_0} - tV_{K_\infty},$$
$$V_{K_-} = -t^{-1/2}V_{K_0} - t^{-1}V_{K_\infty},$$
$$\delta = -(t^{1/2} + t^{-1/2}).$$

Let $t = e^x$. Then

$$V_{K_+} = -\mathrm{e}^{x/2}V_{K_0} - \mathrm{e}^{x}V_{K_\infty},$$
$$V_{K_-} = -\mathrm{e}^{-x/2}V_{K_0} - \mathrm{e}^{-x}V_{K_\infty},$$
$$\delta = -(\mathrm{e}^{x/2} + \mathrm{e}^{-x/2}).$$

Thus

$$V_{K_*} = V_{K_+} - V_{K_-} = -2\,\sinh(x/2)V_{K_0} - 2\,\sinh(x)V_{K_\infty}.$$

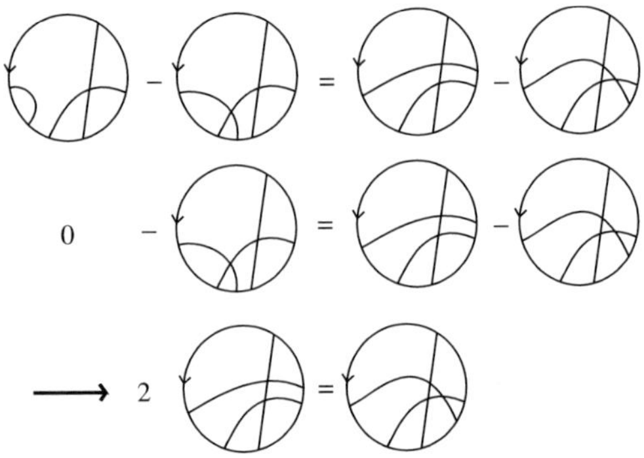

Fig. 19. Four-term relation for type three invariant.

Thus x divides V_{K_*}, and therefore, x^k divides V_G whenever G is a graph with at least k nodes. Letting

$$V_G(\mathrm{e}^x) = \sum_{k=0}^{\infty} v_k(G)x^k,$$

we see that this condition implies that $v_k(G)$ vanishes whenever G has more than k nodes. Hence *the coefficients of the powers of x in the expansion of $V_K(\mathrm{e}^x)$ are Vassiliev invariants*

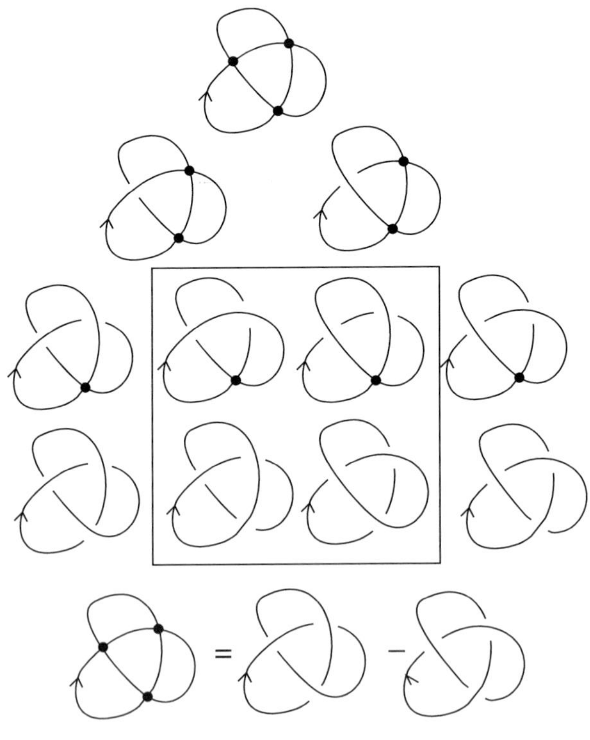

Fig. 20. Trefoil graph.

of finite type! This result was first observed by Birman and Lin [10] by a different argument.

Let us look a little deeper and see the structure of the top row for the Vassiliev invariants related to the Jones polynomial. By our previous remarks the top row evaluations correspond to the leading terms in the power series expansion. Since

$$\delta = -(\mathrm{e}^{x/2} + \mathrm{e}^{-x/2}) = -2 + [\text{higher}],$$

$$-\mathrm{e}^{x/2} + \mathrm{e}^{-x/2} = -x + [\text{higher}],$$

$$-\mathrm{e}^{x} + \mathrm{e}^{-x} = -2x + [\text{higher}],$$

it follows that the top rows for the Jones polynomial are computed by the recursion formulas

$$v(K_*) = -v(K_0) - 2V(K_\infty),$$

$$v([\text{loop}]) = -2.$$

The reader can easily check that this recursion formula for the top rows of the Jones polynomial implies that v_3 takes the value 24 on the trefoil graph and hence it is the Vassiliev invariant of type 3 in the Jones polynomial that first detects the difference between the trefoil knot and its mirror image.

This example gives a good picture of the general phenomenon of how the Vassiliev invariants become building blocks for other invariants. In the case of the Jones polynomial, we already know how to construct the invariant and so it is possible to get a lot of information about these particular Vassiliev invariants by looking directly at the Jones polynomial. This, in turn, gives insight into the structure of the Jones polynomial itself.

3.1. *Lie algebra weights*

Consider the diagrammatic relation shown in Figure 21. Call it (after Bar-Natan [4]) the STU relation.

LEMMA. *STU implies the four-term relation.*

PROOF. View Figure 22. □

STU is the smile of the Cheshire cat. That smile generalizes the idea of a Lie algebra. Take a (matrix) Lie algebra with generators T^a. Then

$$T^aT^b - T^bT^a = if_{abc}T^c$$

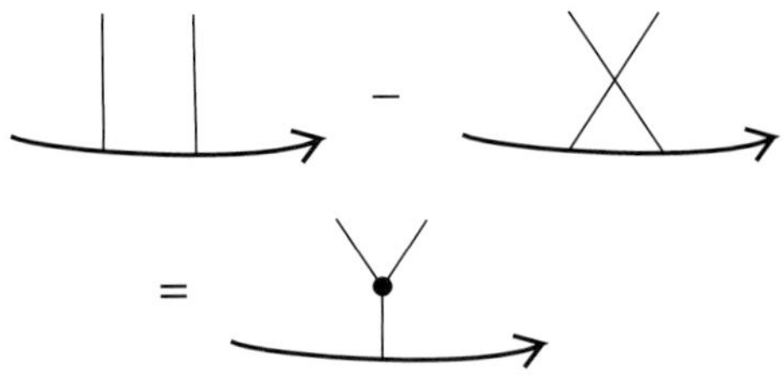

Fig. 21. The *STU* relation.

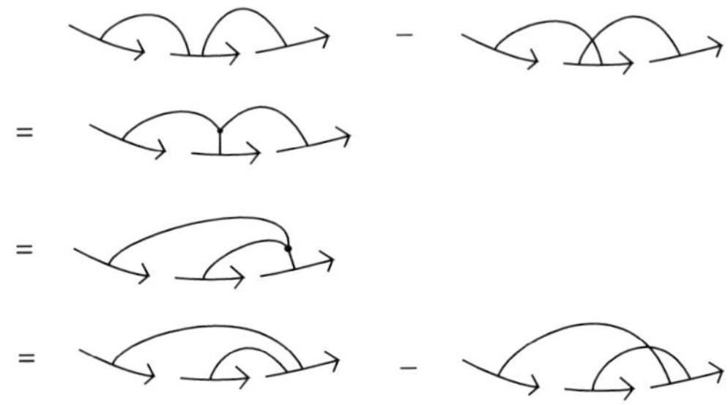

Fig. 22. A diagrammatic proof.

expresses the closure of the Lie algebra under commutators. Translate this equation into diagrams as shown in Figure 23, and see that this translation is STU with Lie algebraic clothing!

Here the structure tensor of the Lie algebra has been assumed (for simplicity) to be invariant under cyclic permutation of the indices. This invariance means that our previous lemma applies to this Lie algebraic interpretation of STU. The upshot is that we can manufacture weight systems for graphs that satisfy the four-term relation by replacing paired points on the chord diagram by an insertion of T^a in one point of the pair and a corresponding insertion of T^a at the other point in the pair and summing over all a. The result of all such insertions on a given chord diagram is a big sum of specific matrix products along the circle of the diagram, each of which (being a circular product) is interpreted as a trace.

Let us say this last matter more precisely: regard a graph with k nodes as obtained by identifying k *pairs* of points on a circle. Thus a code such as 1212 taken in cyclic order specifies such a graph by regarding the points 1, 2, 1, 2 as arrayed along a circle with the first and second 1s and 2s identified to form the graph. Define, for a code $a_1a_2...a_m$

$$\mathrm{wt}(a_1a_2...a_m) = \mathrm{trace}(T^{a_1}T^{a_2}T^{a_1}...T^{a_m}),$$

where the Einstein summation convention is in place for the double appearances of indices on the right-hand side. This gives the weight system.

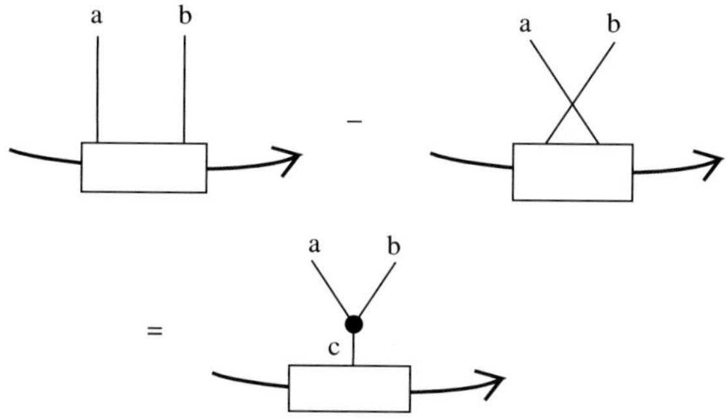

Fig. 23. Algebraic clothing.

The weight system described by the above procedure satisfies the four-term relation, but does not necessarily satisfy the vanishing condition for isolated pairings. This is because the framing compensation for converting an invariant of regular isotopy to ambient isotopy has not yet been introduced. We will show how to do this in the course of the discussion in the next paragraph. The main point to make here is that by starting with the idea of extending an invariant of knots to a Vassiliev invariant of embedded graphs and searching out the conditions on graph evaluation demanded by the topology, we have inevitably entered the domain of relations between Lie algebras and link invariants. Since the STU relation does not demand Lie algebras for its satisfaction we see that the landscape is wider than the Lie algebra context, but it is not yet understood how big is the class of link invariants derived from Lie algebras.

In fact, we can line up this weight system with the formalism related to the knot diagram by writing the Lie algebra insertions back on the four-valent graph. We then get a Casimir insertion at the node.

To get the framing compensation, note that an isolated pairing corresponds to the trace of the Casimir. Let γ denote this trace (see Figure 24).

$$\gamma = \operatorname{tr}\left(\sum_a T^a T^a\right).$$

Let D be the trace of the identity. Then it is easy to see that we must compensate the given weight system by subtracting (γ/D) multiplied by the result of dropping the identification of the two given points. We can diagram this by drawing two crossed arcs without a node drawn to bind them. Then the modified recursion formula becomes as shown in Figure 25.

For example, in the case of $SU(N)$ we have $D = N$, $\gamma = (N^2 - 1)/2$ so that we get the transformation shown in Figure 25, including the use of the Fierz identity.

For $N = 2$ the final formula of Figure 25 is, up to a multiple, exactly the top row formula that we deduced for the Jones polynomial from its combinatorial structure.

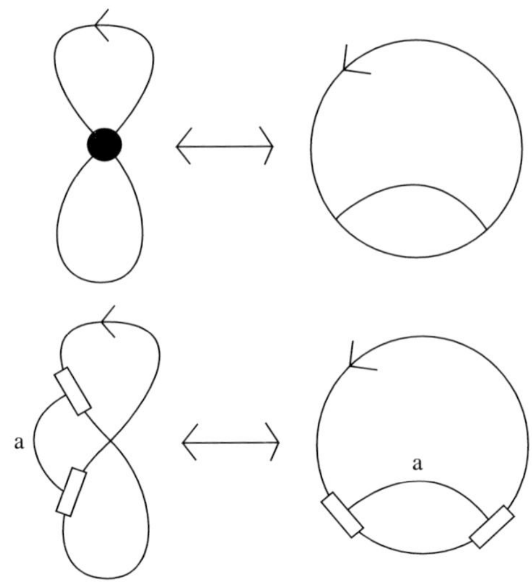

Fig. 24. Weight system and Casimir insertion.

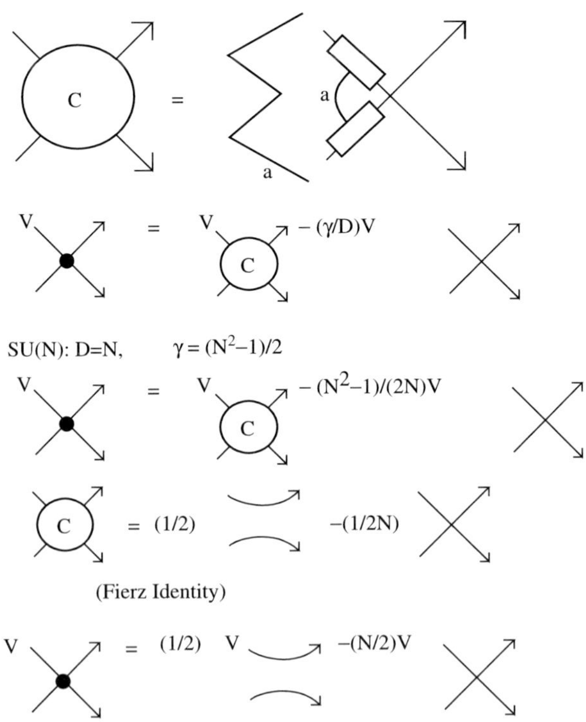

Fig. 25. Modified recursion formula.

3.2. *Vassiliev invariants and Witten's functional integral*

In [145] Edward Witten proposed a formulation of a class of three-manifold invariants as generalized Feynman integrals taking the form $Z(M)$ where

$$Z(M) = \int dA \exp[(ik/4\pi)S(M,A)].$$

Here M denotes a three-manifold without boundary and A is a gauge field (also called a gauge potential or gauge connection) defined on M. The gauge field is a one-form on a trivial G-bundle over M with values in a representation of the Lie algebra of G. The group G corresponding to this Lie algebra is said to be the gauge group. In this integral the "action" $S(M,A)$ is taken to be the integral over M of the trace of the Chern–Simons three-form $\mathrm{CS} = \mathrm{AdA} + (2/3)\mathrm{AAA}$. (The product is the wedge product of differential forms.)

$Z(M)$ integrates over all gauge fields modulo gauge equivalence (see [2] for a discussion of the definition and meaning of gauge equivalence.)

The formalism and internal logic of Witten's integral supports the existence of a large class of topological invariants of three-manifolds and associated invariants of knots and links in these manifolds.

The invariants associated with this integral have been given rigorous combinatorial descriptions [67,85,97,123,140] but questions and conjectures arising from the integral formulation are still outstanding [3,37,38,43,124]. Specific conjectures about this integral take the form of just how it involves invariants of links and three-manifolds, and how these invariants behave in certain limits of the coupling constant k in the integral. Many conjectures of this sort can be verified through the combinatorial models. On the other hand, the really outstanding conjecture about the integral is that it exists! At the present time there is no measure theory or generalization of measure theory that supports it. It is a fascinating exercise to take the speculation seriously, suppose that it does really work like an integral and explore the formal consequences. Here is a formal structure of great beauty. It is also a structure whose consequences can be verified by a remarkable variety of alternative means. Perhaps in the course of the exploration there will appear a hint of the true nature of this form of integration.

We now look at the formalism of the Witten integral in detail and see how it involves invariants of knots and links corresponding to each classical Lie algebra. In order to accomplish this task, we need to introduce the Wilson loop. The Wilson loop is an exponentiated version of integrating the gauge field along a loop K in three-space that we take to be an embedding (knot) or a curve with transversal self-intersections. For this discussion, the Wilson loop will be denoted by the notation $W_K(A) = \langle K|A\rangle$ to denote the dependence on the loop K and the field A. It is usually indicated by the symbolism $\mathrm{tr}(P\exp(\int_K A))$. Thus

$$W_K(A) = \langle K|A\rangle = \mathrm{tr}\left(P\exp\left(\int_K A\right)\right).$$

Here P denotes path ordered integration – we are integrating and exponentiating matrix-valued functions, and so must keep track of the order of the operations. The symbol "*tr*" denotes the trace of the resulting matrix.

With the help of the Wilson loop functional on knots and links, Witten writes down a functional integral for link invariants in a three-manifold M:

$$Z(M,K) = \int dA \exp[(ik/4\pi)S(M,A)]tr\left(P\exp\left(\int_K A\right)\right)$$

$$= \int dA \exp[(ik/4\pi)S]\langle K|A\rangle.$$

Here $S(M,A)$ is the Chern–Simons Lagrangian, as in the previous discussion.

We abbreviate $S(M,A)$ as S and write $\langle K|A\rangle$ for the Wilson loop. Unless otherwise mentioned, the manifold M will be the three-dimensional sphere S^3

An analysis of the formalism of this functional integral reveals quite a bit about its role in knot theory. This analysis depends upon key facts relating the curvature of the gauge field to both the Wilson loop and the Chern–Simons Lagrangian. The idea for using the curvature in this way is due to Lee Smolin [130,131] (see also [120]). To this end, let us recall the local coordinate structure of the gauge field $A(x)$, where x is a point in three-space. We can write $A(x) = A_a^k(x)T^a dx_k$ where the index a ranges from 1 to m with the Lie algebra basis $\{T^1, T^2, T^3, \ldots, T^m\}$. The index k goes from 1 to 3. For each choice of a and k, $A_a^k(x)$ is a smooth function defined on three-space. In $A(x)$ we sum over the values of repeated indices. The Lie algebra generators T^a are matrices corresponding to a given representation of the Lie algebra of the gauge group G. We assume some properties of these matrices as follows:

(1) $[T^a, T^b] = if_{abc}T^c$ where $[x,y] = xy - yx$, and f_{abc} (the matrix of structure constants) is totally antisymmetric. There is summation over repeated indices.

(2) $tr(T^aT^b) = \delta^{ab}/2$ where δ^{ab} is the Kronecker delta ($\delta^{ab} = 1$ if $a = b$) and zero otherwise).

We also assume some facts about curvature. (The reader may enjoy comparing with the exposition in [58]. But note the difference of conventions on the use of i in the Wilson loops and curvature definitions.) The first fact is the relation of Wilson loops and curvature for small loops:

FACT 1. *The result of evaluating a Wilson loop about a very small planar circle around a point x is proportional to the area enclosed by this circle times the corresponding value of the curvature tensor of the gauge field evaluated at x. The curvature tensor is written*

$$F_a^{rs}(x)T^a\, dx_r dy_s.$$

It is the local coordinate expression of $AdA + AA$.

APPLICATION OF FACT 1. Consider a given Wilson line $\langle K|S\rangle$. Ask how its value will change if it is deformed infinitesimally in the neighborhood of a point x on the line.

Approximate the change according to Fact 1, and regard the point x as the place of curvature evaluation. Let $\delta\langle K|A\rangle$ denote the change in the value of the line. $\delta\langle K|A\rangle$ is given by the formula

$$\delta\langle K|A\rangle = dx_r dx_s F_a^{rs}(x) T^a \langle K|A\rangle.$$

This is the first order approximation to the change in the Wilson line.

In this formula it is understood that the Lie algebra matrices T^a are to be inserted into the Wilson line at the point x, and that we are summing over repeated indices. This means that each $T^a\langle K|A\rangle$ is a new Wilson line obtained from the original line $\langle K|A\rangle$ by leaving the form of the loop unchanged, but inserting the matrix T^a into that loop at the point x. A Lie algebra generator is diagrammed by a little box with a single index line and two input/output lines which correspond to its role as a matrix (hence as mappings of a vector space to itself). See Figure 26.

REMARK. In thinking about the Wilson line $\langle K|A\rangle = tr(P\exp(\int_K A))$, it is helpful to recall Euler's formula for the exponential:

$$e^x = \lim_{n\to\infty}(1 + x/n)^n.$$

The Wilson line is the limit, over partitions of the loop K, of products of the matrices $(1 + A(x))$ where x runs over the partition. Thus we can write symbolically

$$\langle K|A\rangle = \prod_{x\in K}(1 + A(x)) = \prod_{x\in K}(1 + A_a^k(x)T^a dx_k).$$

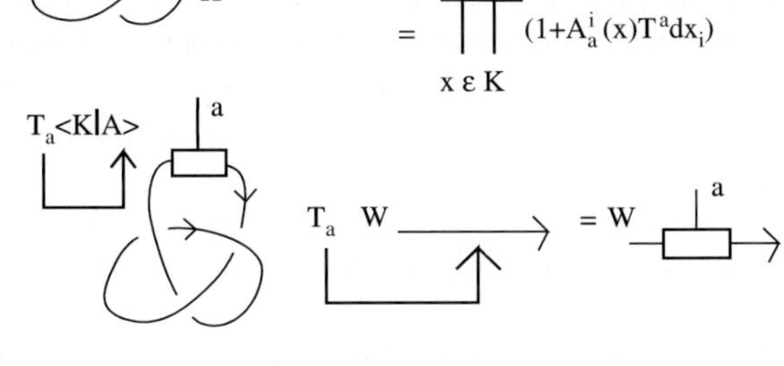

Fig. 26. Wilson loop insertion.

It is understood that a product of matrices around a closed loop connotes the trace of the product. The ordering is forced by the one-dimensional nature of the loop. Insertion of a given matrix into this product at a point on the loop is then a well-defined concept. If T is a given matrix then it is understood that $T\langle K|A\rangle$ denotes the insertion of T into some point of the loop. In the case above, it is understood from the context in the formula

$$dx_r dx_s F_a^{rs}(x)T^a\langle K|A\rangle$$

that the insertion is to be performed at the point x indicated in the argument of the curvature.

REMARK. The previous remark implies the following formula for the variation of the Wilson loop with respect to the gauge field:

$$\delta\langle K|A\rangle/\delta(A_a^k(x)) = dx_k T^a\langle K|A\rangle.$$

Varying the Wilson loop with respect to the gauge field results in the insertion of an infinitesimal Lie algebra element into the loop.

PROOF.

$$\begin{aligned}\delta\langle K|A\rangle/\delta(A_a^k(x)) &= \delta\prod_{y\in K}(1 + A_a^k(y)T^a dy_k)/\delta(A_a^k(x))\\ &= \prod_{y<x\in K}(1 + A_a^k(y)T^a dy_k)[T^a dx_k]\prod_{y>x\in K}(1 + A_a^k(y)T^a dy_k)\\ &= dx_k T^a\langle K|A\rangle.\end{aligned}$$

FACT 2. *The variation of the Chern–Simons Lagrangian S with respect to the gauge potential at a given point in three-space is related to the values of the curvature tensor at that point by the following formula:*

$$F_a^{rs}(x) = \varepsilon_{rst}\delta S/\delta(A_a^t(x)).$$

Here ε_{abc} is the epsilon symbol for three indices, i.e., it is $+1$ for positive permutations of 123 and -1 for negative permutations of 123 and zero if any two indices are repeated.

With these facts at hand we are prepared to determine how the Witten integral behaves under a small deformation of the loop K.

In accord with the theme of this paper, we shall use a system of abstract tensor diagrams to look at the differential algebra related to the functional integral. The translation to diagrams is accomplished with the aid of Figures 27 and 28. In Figure 27 we give diagrammatic equivalents for the component parts of our machinery. Tensors become labelled boxes. Indices become lines emanating from

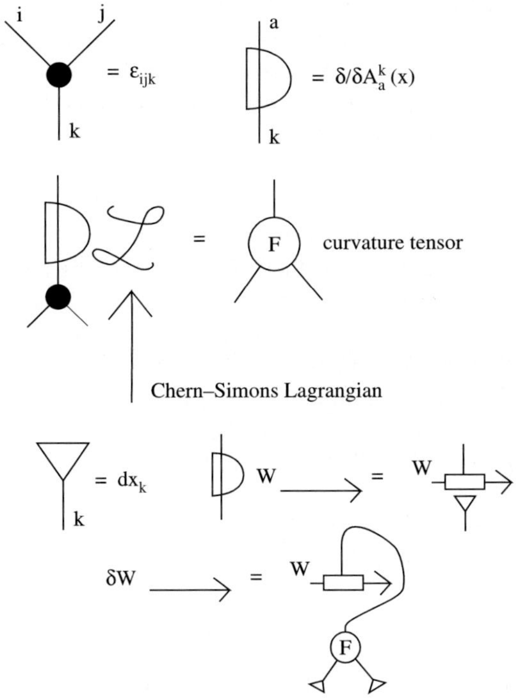

Fig. 27. Notation.

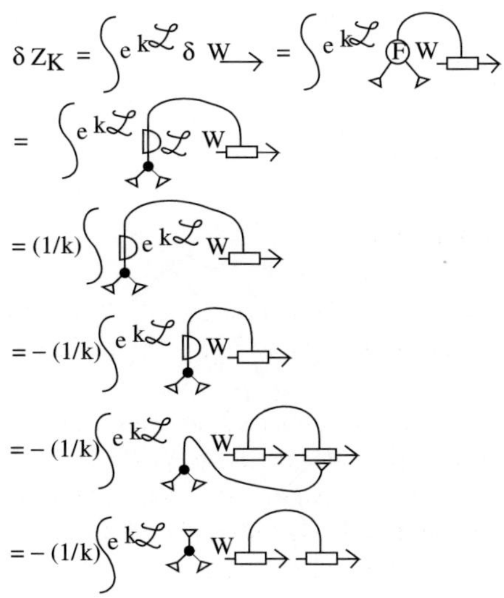

Fig. 28. Derivation.

the boxes. Repeated indices that we intend to sum over become lines from one box to another. (The eye can immediately apprehend the repeated indices and the tensors where they are repeated.) Note that we use a capital D with lines extending from the top and the bottom for the partial derivative with respect to the gauge field, a capital W with a link diagrammatic subscript for the Wilson loop, a cubic vertex for the three-index epsilon, little triangles with emanating arcs for the differentials of the space variables.

The Lie algebra generators are little boxes with single index lines and two input/output lines which correspond to their roles as matrices (hence as mappings of a vector space to itself). The Lie algebra generators are, in all cases of our calculation, inserted into the Wilson line either through the curvature tensor or through insertions related to differentiating the Wilson line.

In Figure 28 we give the diagrammatic calculation of the change in the functional integral corresponding to a tiny change in the Wilson loop. The result is a double insertion of Lie Algebra generators into the line, coupled with the presence of a volume form that will vanish if the deformation does not twist in three independent directions. This shows that the functional integral is formally invariant under regular isotopy since the regular isotopy moves are changes in the Wilson line that happen entirely in a plane. One does not expect the integral to be invariant under a Reidemeister move of type one, and it is not. This framing compensation can be determined by the methods that we are discussing, but we will not go into the details of those calculations here.

In Figure 29 we show the application of the calculation in Figure 28 to the case of switching a crossing. The same formula applies, with a different interpretation, to the case where x is a double point of transversal self-intersection of a loop K, and the deformation consists in shifting one of the crossing segments perpendicularly to the plane of

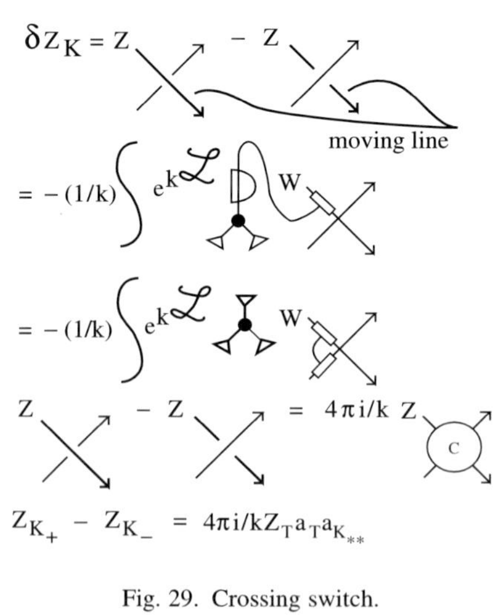

Fig. 29. Crossing switch.

intersection so that the self-intersection point disappears. In this case, one T^a is inserted into each of the transversal crossing segments so that $T^aT^a\langle K|A\rangle$ denotes a Wilson loop with a self-intersection at x and insertions of T^a at $x+\varepsilon_1$ and $x+\varepsilon_2$ where ε_1 and ε_2 denote small displacements along the two arcs of K that intersect at x. In this case, the volume form is non-zero, with two directions coming from the plane of movement of one arc, and the perpendicular direction is the direction of the other arc. The reason for the insertion into the two lines is a direct consequence of the calculational form of Figure 28: the first insertion is in the moving line, due to curvature. The second insertion is the consequence of differentiating the self-touching Wilson line. Since this line can be regarded as a product, the differentiation occurs twice at the point of intersection, and it is the second direction that produces the non-vanishing volume form.

Up to the choice of our conventions for constants, the switching formula is, as shown in Figure 29,

$$\begin{aligned} Z(K_+) - Z(K_-) &= (4\pi i/k)\int dA\, \exp[(ik/4\pi)S]T^aT^a\langle K_{**}|A\rangle \\ &= (4\pi i/k)Z(T^aT^aK_{**}). \end{aligned}$$

The key point is to notice that the Lie algebra insertion for this difference is exactly what we did to make the weight systems for Vassiliev invariants (without the framing compensation). Thus the formalism of the Witten functional integral takes us directly to these weight systems in the case of the classical Lie algebras. The functional integral is central to the structure of the Vassiliev invariants.

3.3. *Combinatorial constructions for Vassiliev invariants*

Perhaps the most remarkable thing about this story of the structure of the Vassiliev invariants is the way that Lie algebras are so naturally involved in the structure of the weight systems. This shows the remarkably close nature of the combinatorial structure of Lie algebras and the combinatorics of knots and links via the Reidemeister moves. A really complete story about the Vassiliev invariants at this combinatorial level would produce their existence on the basis of the weight systems with entirely elementary arguments.

As we have already mentioned, one can prove that a given set of weights for the top row, satisfying the abstract four-term relation does imply that there exists a Vassiliev invariant of finite type n realizing these weights for graphs with n nodes. Proofs of this result either use analysis [1,4] or non-trivial algebra [1,12]. There is no known elementary combinatorial proof of the existence of Vassiliev invariants for given top rows.

Of course quantum link invariants (see Section 4 of these lectures) do give combinatorial constructions for large classes of link invariants. These constructions rest on solutions to the Yang–Baxter equations, and it is not known how to describe the subset of finite type Vassiliev invariants that are so produced.

It is certainly helpful to look at the structure of Vassiliev invariants that arise from already-defined knot invariants. If $V(K)$ is an already defined invariant of knots (and possibly links), then its extension to a Vassiliev invariant is calculated on embedded graphs G by expanding each graphical vertex into a difference by resolving the vertex into a positive crossing and a negative crossing. If we know that $V(K)$ is of finite type n and G has n nodes then we can take any embedding of G that is convenient, and calculate $V(G)$ in terms of all the knots that arise in resolving the nodes of this chosen embedding. This is a finite collection of knots. Since there is a finite collection of four-valent graphs with n nodes, it follows that the top row evaluation for the invariant $V(K)$ is determined by the values of $V(K)$ on a finite collection of knots. Instead of asking for the values of the Vassiliev invariant on a top row, we can ask for this set of knots and the values of the invariant on this set of knots. A minimal set of knots that can be used to generate a given Vassiliev invariant will be called a *knots basis* for the invariant. Thus we have shown that the set consisting of the unknot, the right-handed trefoil and the left-handed trefoil is a knots basis for a Vassiliev invariant of type 3. See [113] for more information about this point of view.

A tantalizing combinatorial approach to Vassiliev invariants is due to Michael Polyak and Oleg Viro [118]. They give explicit formulas for the second, third and fourth Vassiliev invariants and conjecture that their method will work for Vassiliev invariants of all orders. The method is as follows.

First one makes a new representation for oriented knots by taking *Gauss diagrams*. A Gauss diagram is a diagrammatic representation of the classical *Gauss code* of the knot. The Gauss code is obtained from the oriented knot diagram by first labelling each crossing with a naming label (such as 1, 2,...) and also indicating the crossing type ($+1$ or -1). Then choose a basepoint on the knot diagram and begin walking along the diagram, recording the name of the crossings encountered, their sign and whether the walk takes you over or under that crossing. For example, if you go under crossing 1 whose sign is $+$ then you will record $o + 1$. Thus the Gauss code of the positive trefoil diagram is

$$(o1+)(u2+)(o3+)(u1+)(o2+)(u3+).$$

For prime knots the Gauss code is sufficient information to reconstruct the knot diagram. See [61] for a sketch of the proof of this result and for other references.

To form a Gauss diagram from a Gauss code, take an oriented circle with a basepoint chosen on the circle. Walk along the circle marking it with the labels for the crossings in the order of the Gauss code. Now draw chords between the points on the circle that have the same label. Orient each chord from overcrossing site to undercrossing site. Mark each chord with $+1$ or -1 according to the sign of the corresponding crossing in the Gauss code. The resulting labelled and basepointed graph is the Gauss diagram for the knot. See Figure 30 for examples.

The Gauss diagram is deliberately formulated to have the structure of a chord diagram (as we have discussed for the weight systems for Vassiliev invariants). If $G(K)$ is the Gauss diagram for a knot K, and D is an oriented (i.e., the chords as well as the circle in the diagram are oriented) chord diagram, let $|G(K)|$ denote the number of chords in $G(K)$ and $|D|$ denote the number of chords in D. If $|D| \leq |G(K)|$ then we may consider oriented

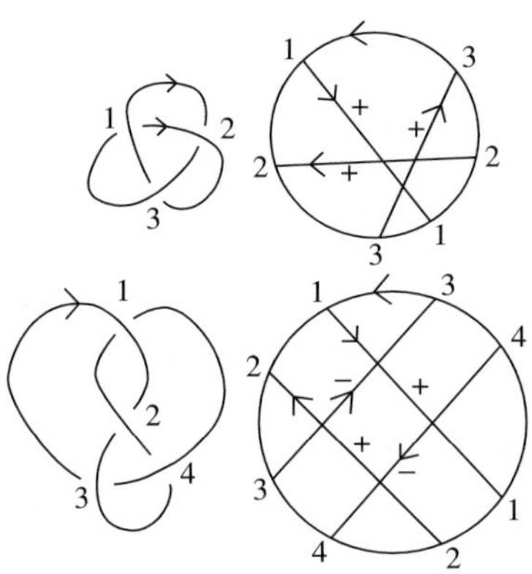

Fig. 30. Gauss diagrams.

embeddings of D in $G(K)$. For a given embedding $i : D \to G(K)$ define

$$\langle i(D) | G(K) \rangle = \mathrm{sign}(i),$$

where sign(i) denotes the product of the signs of the chords in $G(K) \cap i(D)$. Now suppose that C is a collection of *oriented* chord diagrams, each with n chords, and that

$$\mathrm{eval}: C \to R$$

is an evaluation mapping on these diagrams that satisfies the four-term relation at level n. Then we can define

$$\langle D | K \rangle = \sum_{i:D \to G(K)} \langle i(D) | G(K) \rangle$$

and

$$v(K) = \sum_{D \in C} \langle D | K \rangle \mathrm{eval}(D).$$

For appropriate oriented chord subsets this definition can produce Vassiliev invariants $v(K)$ of type n. For example, in the case of the Vassiliev invariant of type three taking value 0 on the unknot and value 1 on the right-handed trefoil, -1 on the left-handed trefoil, Polyak and Viro give the specific formula

$$v_3(K) = \langle A | K \rangle + (1/2) \langle B | K \rangle,$$

where A denotes the trefoil chord diagram as we described it in Section 3 and B denotes the three-chord diagram consisting of two parallel chords pierced by a third chord. In Figure 31 we show the specific orientations for the chord diagrams A and B. The key to this

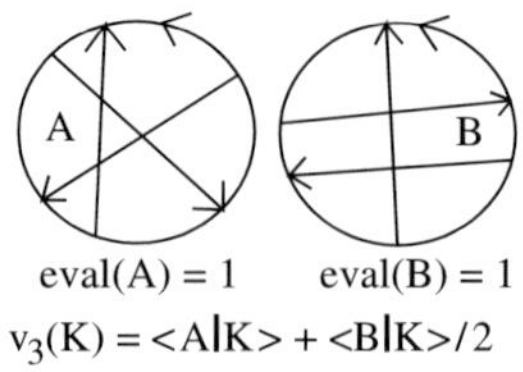

Fig. 31. Oriented chord diagrams for v_3.

construction is in the choice of orientations for the chord diagrams in $C = \{A, B\}$. It is a nice exercise in translation of the Reidemeister moves to Gauss diagrams to see that $v_3(K)$ is indeed a knot invariant.

It is possible that all Vassiliev invariants can be constructed by a method similar to the formula $v(K) = \sum_{D \in C} \langle D|K \rangle \mathrm{eval}(D)$. This remains to be seen.

3.4. 8_{17}

It is an open problem whether there are Vassiliev invariants that can detect the difference between a knot and its reverse (the reverse of an oriented knot is obtained by flipping the orientation). The smallest instance of a non-invertible knot is the knot 8_{17} depicted in Figure 32.

Thus, at the time of this writing there is no known Vassiliev invariant that can detect the non-invertibility of 8_{17}. On the other hand, the tangle decomposition shown in Figure 32

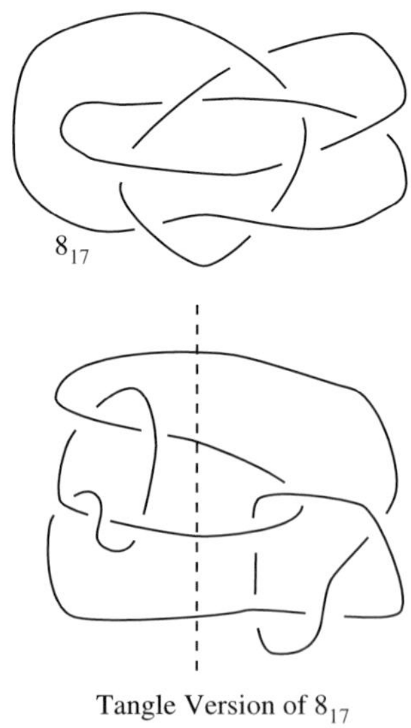

Fig. 32. Tangle decomposition of 8_{17}.

can be used in conjunction with the results of Siebenmann and Bonahon [128] and the formulations of John Conway [22] to show this non-invertibility. These tangle decomposition methods use higher level information about the diagrams than is easy to encode in Vassiliev invariants. The purpose of this section is to underline this discrepancy between different levels in the combinatorial topology.

4. Quantum link invariants

In this section we describe the construction of quantum link invariants from knot and link diagrams that are arranged with respect to a given direction in the plane. This special direction will be called "time". Arrangement with respect to the special direction means that perpendiculars to this direction meet the diagram transversely (at edges or at crossings) or tangentially (at maxima and minima). The designation of the special direction as time allows the interpretation of the consequent evaluation of the diagram as a generalized scattering amplitude.

In the course of this discussion we find the need to reformulate the Reidemeister moves for knot and link diagrams that are arranged to be transverse (except for a finite collection of standard critical points) to the specific special direction introduced in the previous paragraph. This brings us back to our theme of diagrams and related structures. This particular reformulation of the Reidemeister moves is quite far-reaching. It encompasses the relationship of link invariants with solutions to the Yang–Baxter equation and the relationship with Hopf algebras (to be dealt with in Section 5).

4.1. *Knot amplitudes*

Consider first a circle in a spacetime plane with time represented vertically and space horizontally as in Figure 33.

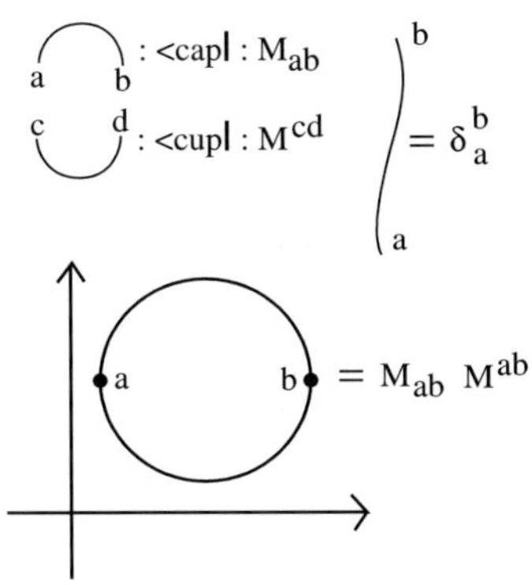

Fig. 33. Spacetime circle.

The circle represents a vacuum to vacuum process that includes the creation of two "particles" and their subsequent annihilation. We could divide the circle into these two parts (creation "cup" and annihilation "cap") and consider the amplitude $\langle \text{cap}|\text{cup}\rangle$. Since the diagram for the creation of the two particles ends in two separate points, it is natural to take a vector space of the form $V \otimes V$ as the target for the bra and as the domain of the ket. We imagine at least one particle property being catalogued by each factor of the tensor product. For example, a basis of V could enumerate the spins of the created particles.

Any non-self-intersecting differentiable curve can be rigidly rotated until it is in general position with respect to the vertical. It will then be seen to be decomposed into an interconnection of minima and maxima. We can evaluate an amplitude for any curve in general position with respect to a vertical direction. Any simple closed curve in the plane is isotopic to a circle, by the Jordan Curve Theorem. If these are topological amplitudes, then the value for any simple closed curve should be equal to the original amplitude for the circle. What condition on creation (cup) and annihilation (cap) will insure topological amplitudes? The answer derives from the fact that isotopies of the simple closed curves are generated by the cancellation of adjacent maxima and minima as illustrated in Figure 34.

This condition is articulated by taking a matrix representation for the corresponding operators. Specifically, let $\{e_1, e_2, \ldots, e_n\}$ be a basis for V. Let $e_{ab} = e_a \otimes e_b$ denote the elements of the tensor basis for $V \otimes V$. Then there are matrices M_{ab} and M^{ab} such that

$$|\text{cup}\rangle(1) = \sum M^{ab} e_{ab}$$

with the summation taken over all values of a and b from 1 to n. Similarly, $\langle \text{cap}|$ is described by

$$\langle \text{cap}|(e_{ab}) = M_{ab}.$$

Thus the amplitude for the circle is

$$\langle \text{cap}|\text{cup}\rangle(1) = \langle \text{cap}|\sum M^{ab} e_{ab} = \sum M^{ab} \langle \text{cap}|(e_{ab}) = \sum M^{ab} M_{ab}.$$

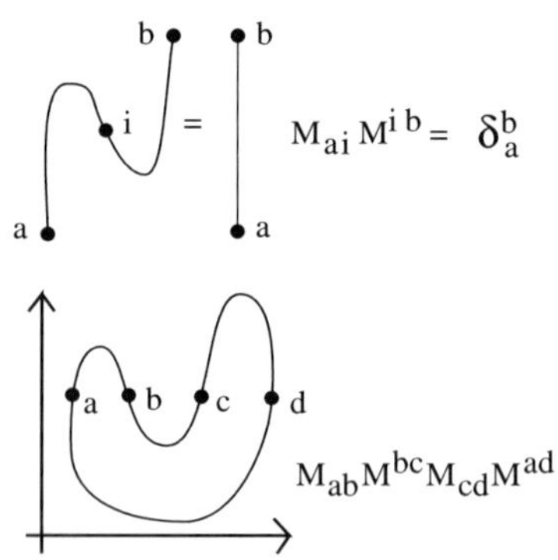

Fig. 34. Spacetime Jordan curve.

In general, the value of the amplitude on a simple closed curve is obtained by translating it into an "abstract tensor expression" in the M^{ab} and M_{ab}, and then summing over these products for all cases of repeated indices. Note that here the value "1" corresponds to the vacuum.

Returning to the topological conditions we see that they are just that the matrices M^{ab} and M_{cd} are inverses in the sense that

$$\sum_i M^{ai} M_{ib} = \delta^a_b,$$

where δ^a_b denotes the identity matrix (see Figure 34).

One of the simplest choices is to take a 2×2 matrix M such that $M^2 = I$ where I is the identity matrix. Then the entries of M can be used for both the cup and the cap. The value for a loop is then equal to the sum of the squares of the entries of M:

$$\langle \text{cap}|\text{cup}\rangle(1) = \sum M^{ab} M_{ab} = \sum M_{ab} M_{ab} = \sum M^2_{ab}.$$

In particular, consider the following choice for M. It has square equal to the identity matrix and yields a loop value of $d = -A^2 - A^{-2}$, just the right loop value for the bracket polynomial model for the Jones polynomial [52,53].

$$M = \begin{bmatrix} 0 & iA \\ -iA^{-1} & 0 \end{bmatrix}.$$

Any knot or link can be represented by a picture that is configured with respect to a vertical direction in the plane. The picture will decompose into minima (creations), maxima (annihilations) and crossings of the two types shown in Figure 35. Here the knots and links are unoriented. These models generalize easily to include orientation.

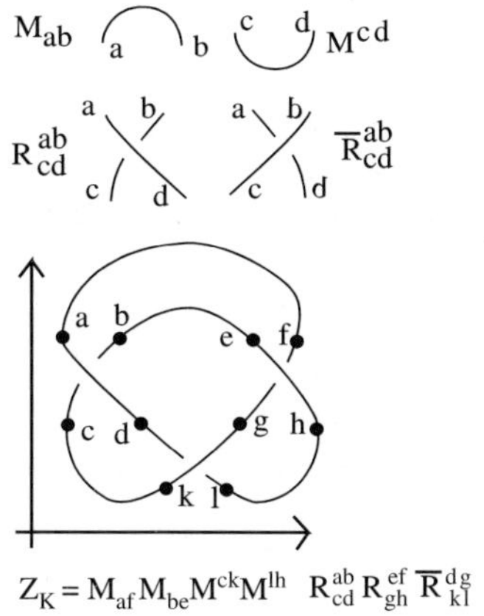

Fig. 35. Cups, caps and crossings.

Next to each of the crossings we have indicated mappings of $V \otimes V$ to itself, called R and R^{-1}, respectively. These mappings represent the transitions corresponding to elementary braiding. We now have the vocabulary of cup, cap, R and R^{-1}. Any knot or link can be written as a composition of these fragments, and consequently a choice of such mappings determines an amplitude for knots and links. In order for such an amplitude to be topological (i.e., an invariant of regular isotopy the equivalence relation generated by the second and third of the classical Reidemeister moves) we want it to be invariant under a list of local moves on the diagrams as shown in Figure 36. These moves are an augmented list of Reidemeister moves, adjusted to take care of the fact that the diagrams are arranged with respect to a given direction in the plane. The proof that these moves generate regular isotopy is composed in exact parallel to the proof that we gave for the classical Reidemeister moves in Section 2. In the piecewise linear setting, maxima and minima are replaced by upward and downward pointing angles. The fact that the triangle, in the Reidemeister piecewise linear triangle move, must be projected so that it is generically transverse to the vertical direction in the plane introduces the extra restriction that expands the move set.

In this context, the algebraic translation of Move III is the Yang–Baxter equation that occurred for the first time in problems of exactly solved models in statistical mechanics [9].

All the moves taken together are directly related to the axioms for a quasi-triangular Hopf algebra (aka quantum group). Many seeds of the structure of Hopf algebras are prefigured in the patterns of link diagrams and the structure of the category of tangles. The interested reader can consult [58,61,62,65,123,146] and Section 5 of this paper for more information on this point.

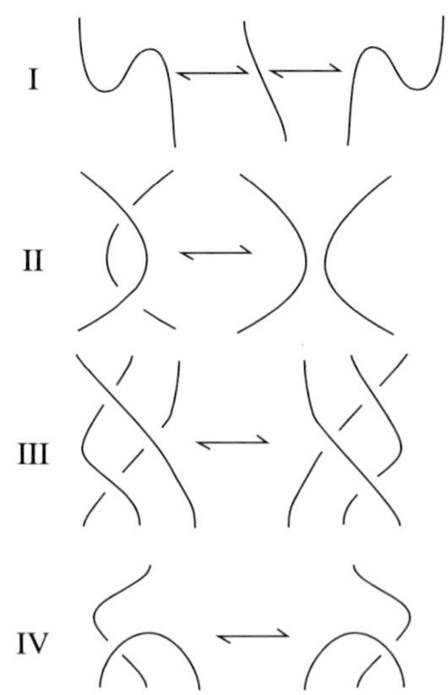

Fig. 36. Regular isotopy with respect to a vertical direction.

Here is the list of the algebraic versions of the topological moves. Move 0 is the cancellation of maxima and minima. Move II corresponds to the second Reidemeister move. Move III is the Yang–Baxter equation. Move IV expresses the relationship of switching a line across a maximum. (There is a corresponding version of IV where the line is switched across a minimum.)

$$\text{0.}\qquad M^{ai}M_{ib} = \delta^a_b,$$

$$\text{II.}\qquad R^{ab}_{ij}\overline{R^{ij}_{cd}} = \delta^a_c\delta^b_d,$$

$$\text{III.}\qquad R^{ab}_{ij}R^{jc}_{kf}R^{ik}_{de} = R^{bc}_{ij}R^{ai}_{dk}R^{kj}_{ef},$$

$$\text{IV.}\qquad R^{ai}_{bc}M_{id} = M_{bi}R^{ia}_{cd}.$$

In the case of the Jones polynomial we have all the algebra present to make the model. It is easiest to indicate the model for the bracket polynomial: let cup and cap be given by the 2×2 matrix M, described above so that $M_{ij} = M^{ij}$. Let R and R^{-1} be given by the equations

$$R^{ab}_{cd} = AM^{ab}M_{cd} + A^{-1}\delta^a_c\delta^b_d,$$

$$(R^{-1})^{ab}_{cd} = A^{-1}M^{ab}M_{cd} + A\delta^a_c\delta^b_d.$$

This definition of the R-matrices exactly parallels the diagrammatic expansion of the bracket, and it is not hard to see, either by algebra or diagrams, that all the conditions of the model are met.

4.2. *Oriented amplitudes*

Slight but significant modifications are needed to write the oriented version of the models we have discussed in the previous section (see [40,58,122,138]). In this section we sketch the construction of oriented topological amplitudes.

The generalization to oriented link diagrams naturally involves the introduction of right and left oriented caps and cups. These are drawn as shown in Figure 37.

A right cup cancels with a right cap to produce an upward pointing identity line. A left cup cancels with a left cap to produce a downward pointing identity line.

Just as we considered the simplifications that occur in the unoriented model by taking the cup and cap matrices to be identical, let us assume here that right caps are identical with left cups and that consequently left caps are identical with right cups. In fact, let us assume that the right cap and left cup are given by the matrix

$$M_{ab} = \lambda^{a/2}\delta_{ab},$$

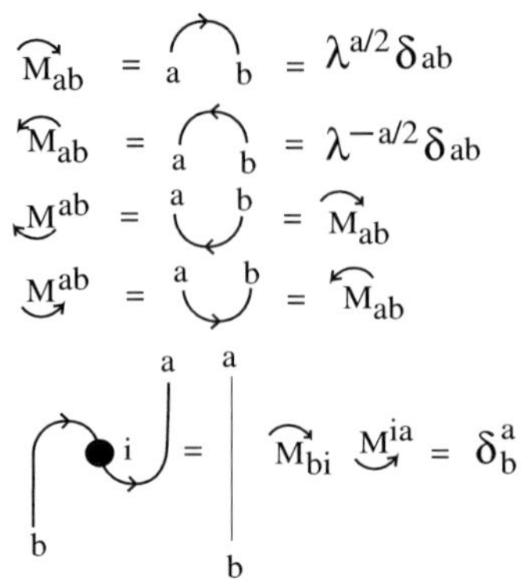

Fig. 37. Right and left cups and caps.

where λ is a constant to be determined by the situation, and δ_{ab} denotes the Kronecker delta. Then the left cap and right cup are given by the inverse of M:

$$M_{ab}^{-1} = \lambda^{-a/2}\delta_{ab}.$$

We assume that along with M we are given a solution R to the Yang–Baxter equation, and that in an oriented diagram the specific choice of R_{cd}^{ab} is governed by the local orientation of the crossing in the diagram. Thus a and b are the labels on the lines going into the crossing and c and d are the labels on the lines emanating from the crossing.

Note that with respect to the vertical direction for the amplitude, the crossings can assume the aspects: both lines pointing upward, both lines pointing downward, one line up and one line down (two cases; see Figure 38).

Call the cases of one line up and one line down the *mixed* cases and the upward and downward cases the *parallel* cases. A given mixed crossing can be converted, in two ways, into a combination of a parallel crossing of the same sign plus a cup and a cap (see Figure 39).

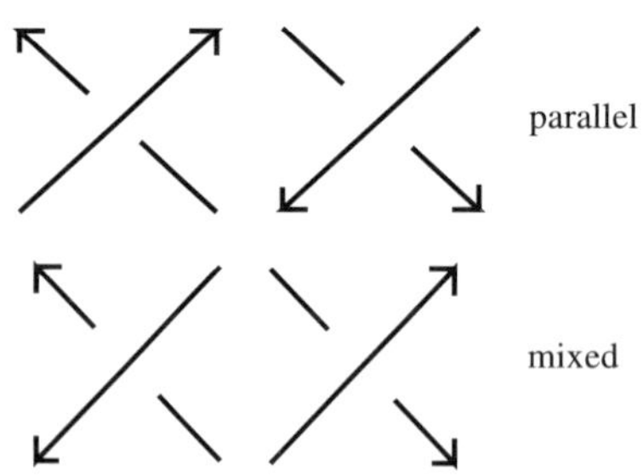

Fig. 38. Oriented crossings.

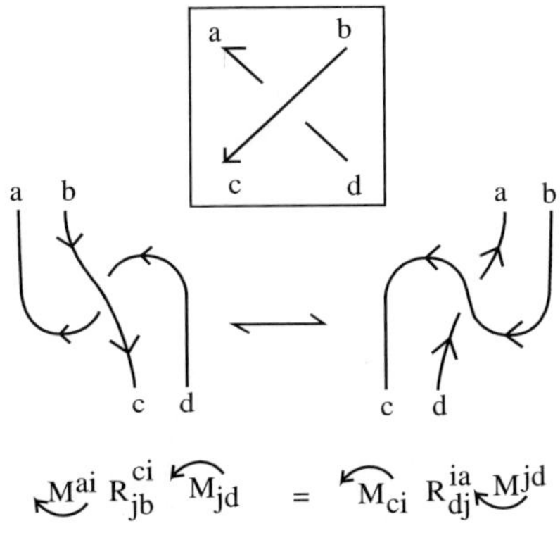

Fig. 39. Conversion.

This leads to an equation that must be satisfied by the R matrix in relation to powers of λ (again we use the Einstein summation convention):

$$\lambda^{a/2}\delta^{ai}R^{ci}_{jb}\lambda^{-d/2}\delta_{jd} = \lambda^{-c/2}\delta_{ic}R^{ia}_{dj}\lambda^{b/2}\delta^{jb}.$$

This simplifies to the equation

$$\lambda^{a/2}R^{ca}_{db}\lambda^{-d/2} = \lambda^{-c/2}R^{ca}_{db}\lambda^{b/2},$$

from which we see that R^{ca}_{db} is necessarily equal to zero unless $b + d = a + c$. We say that the R matrix is *spin preserving* when it satisfies this condition. Assuming that the R matrix is spin preserving, the model will be invariant under all orientations of the second and third Reidemeister moves just so long as it is invariant under the anti-parallel version of the second Reidemeister move as shown in Figure 40.

This antiparallel version of the second Reidemeister move places the following demand on the relation between λ and R:

$$\sum_{st}\lambda^{(s-b)/2}\lambda^{(t-c)/2}R^{bt}_{as}\overline{R^{cs}_{dt}} = \delta^a_c\delta^b_d.$$

Call this the $R - \lambda$ equation. The reader familiar with [44] or with the piecewise linear version as described in [58] will recognize this equation as the requirement for regular homotopy invariance in these models.

4.3. *Quantum link invariants and Vassiliev invariants*

Vassiliev invariants can be used as building blocks for all the presently known quantum link invariants.

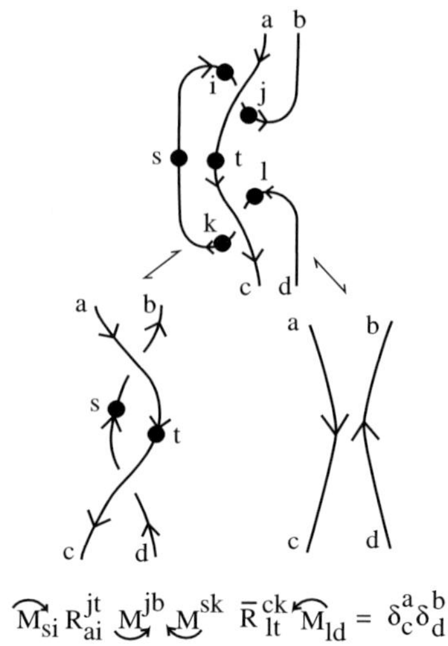

Fig. 40. Antiparallel second move.

It is this result that we can now make clear in the context of the models given in our section on quantum link invariants. Suppose that λ is written as a power series in a variable h, say $\lambda = \exp(h)$ to be specific. Suppose also, that the R-matrices can be written as power series in h with matrix coefficients so that $PR = I + r_+ h + O(h^2)$ and $PR^{-1} = I + r_- h + O(h^2)$ where P denotes the map of $V \otimes V$ that interchanges the tensor factors. Let $Z(K)$ denote the value of the oriented amplitude described by this choice of λ and R. Then we can write

$$Z(K) = Z_0(K) + Z_1(K)h + Z_2(K)h^2 + \cdots$$

where each $Z_n(K)$ is an invariant of regular isotopy of the link K. Furthermore, we see at once that h *divides the series for* $Z(K_+) - Z(K_-)$. By the definition of the Vassiliev invariants this implies that h^k divides $Z(G)$ if G is a graph with k nodes. Therefore, $Z_n(G)$ vanishes if n is less than the number of nodes of G. Therefore, Z_n is a Vassiliev invariant of finite type n. Hence the quantum link invariant is built from an infinite sequence of interlocked Vassiliev invariants.

It is an open problem whether the class of finite type Vassiliev invariants is greater than those generated from quantum link invariants. It is also possible that there are quantum link invariants that cannot be generated by Vassiliev invariants.

4.4. *Vassiliev invariants and infinitesimal braiding*

Kontsevich [4,90] proved that a weight assignment for a Vassiliev top row that satisfies the four-term relation and the framing condition (that the weights vanish for graphs with isolated double points) actually extends to a Vassiliev invariant defined on all knots. His method is motivated by the perturbative expansion of the Witten integral and by Witten's interpretation of the integral in terms of conformal field theory. This section will give a brief description of the Kontsevich approach and the questions that it raises about the functional integral itself.

The key to this approach is to see that the four-term relations are a kind of "infinitesimal braid relations". That is, we can re-write the four-term relations in the form of tangle operators as shown in Figure 41.

This shows that the commutator equation

$$[t^{12}, t^{13}] + [t^{13}, t^{23}] = 0$$

is an algebraic form of the four-term relation. The four-term relations translate exactly into these infinitesimal braid relations studied by Kohno [91]. Kohno showed that his version of infinitesimal braid relations corresponded to a flatness condition for a certain connection (the Knizhnik–Zamolodchikov connection) and that this meant that these relations constituted an integrability condition for making representations of the braid group via monodromy. Others have verified that the braid group representations related to the Chern–Simons form and the Witten integral arise in this same way from the Knizhnik–Zamolodchikov equations. In the case of Chern–Simons theory the weights in the K–Z equations come from the Casimir of a classical Lie algebra, just as we have discussed. Kontsevich observed that since the arbitrary four-term relations could also be regarded as an integrability condition it was possible to use them in a generalization of Kohno's ideas to produce braid group representations via iterated integration. He then generalized the process of producing these braid group representations to the production of

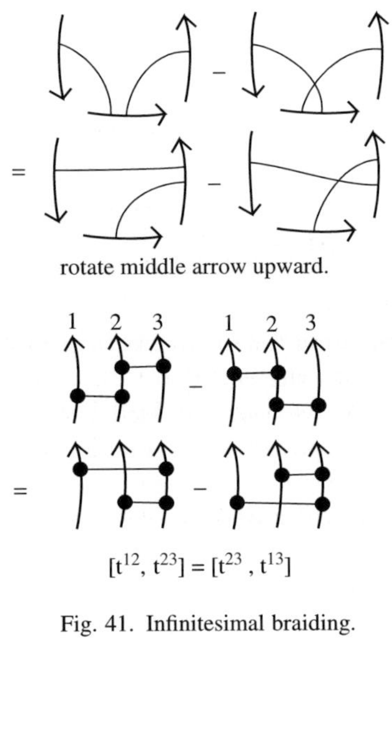

Fig. 41. Infinitesimal braiding.

knot invariants and these become realizations of Vassiliev invariants that have given admissible weight systems for their top rows.

The upshot of the Kontsevich work is a very specific integral formula for the Vassiliev invariants. See [4] for the specifics. It is clear from the nature of the construction that the Kontsevich formula captures the various orders of perturbative terms in the Witten integral. At this writing there is no complete published description of this correspondence.

4.5. *Weight systems and the classical Yang–Baxter equation*

Let us return momentarily to the series form of the solution to the Yang–Baxter equation, as we had indicated it in Section 4.4.

$$PR = I + r_+ h + \mathrm{O}(h^2),$$

$$PR^{-1} = I + r_- h + \mathrm{O}(h^2).$$

Since $RR^{-1} = P$, it follows that $r_- = -r'_+$ where a' denotes the transpose of a. In the case that R satisfies the $R - \lambda$ equation, it follows that

$$t = r_+ - r_- = r + r'$$

(letting r denote r_+) satisfies the infinitesimal braiding relations

$$[t^{12}, t^{13}] + [t^{13}, t^{23}] = 0.$$

It is interesting to contemplate this fact, since r, being the coefficient of h in the series for R, necessarily satisfies the *classical Yang–Baxter Equation*

$$[r^{13}, r^{23}] + [r^{12}, r^{23}] + [r^{12}, r^{13}] = 0.$$

(The classical Yang–Baxter equation for r is a direct consequence of the fact that R is a solution of the (quantum) Yang–Baxter equation.)

Via the quantum link invariants, we have provided a special condition (the assumption that r is the coefficient of h in a power series solution of the quantum Yang–Baxter equation R, and that R satisfies the $R - \lambda$ equation) ensuring that a solution r of the classical Yang–Baxter equation will produce a solution $t = r + r'$ of the infinitesimal braiding relation, whence a weight system for Vassiliev invariants. More work needs to be done to fully understand the relationship between solutions of the classical Yang–Baxter equation and the construction of Vassiliev invariants.

5. Hopf algebras and invariants of three-manifolds

This section is a rapid sketch of the relationship between the description of regular isotopy with respect to a vertical direction (as described in our discussion of quantum link invariants) and the way that this formulation of the Reidemeister moves is related to Hopf algebras and to the construction of link invariants and invariants of three-manifolds via Hopf algebras. More detailed presentations of this material can be found in [40,61,62,64,65].

Let us begin by recalling the Kirby calculus [84]. In the context of link diagrams the Kirby calculus has an elegant formulation in terms of (blackboard) framed links represented by link diagrams up to *ribbon equivalence*. Ribbon equivalence consists in diagrams up to regular isotopy coupled with the equivalence of a positive (negative) curl of Whitney degree 1 with a positive (negative) curl of Whitney degree -1 (see Figure 42).

Here we refer informally to the Whitney degree of a plane curve. The Whitney degree is the total turn of the tangent vector. If the curve is not closed, then it is assumed that the tangent direction of the initial point is the same as the tangent vector of the endpoint. In Figure 42 we illustrate how curls encode framings and how ribbon equivalent curls correspond to identical framings. A link is said to be *framed* if it is endowed with a smooth choice of normal vector field. Framing a link is equivalent to specifying an embedded band(s) of which it is the core. The *core* of a band is the center curve. Thus $S^1 \times \{0.5\}$ is the core of $S^1 \times [0, 1]$.

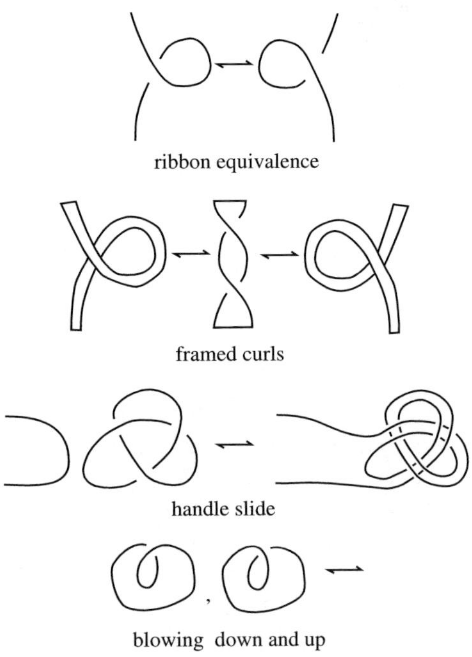

Fig. 42. Framing and Kirby calculus.

Now introduce two new moves on link diagrams called *handle sliding* and *blowing up and down*. These moves are illustrated in Figure 42. Handle sliding consists in duplicating a parallel copy of one link component and then band connect summing it with another component. Blowing up consists in adding an isolated unknotted component with a single curl. Blowing down consists in deleting such a component. These are the basic moves of the Kirby Calculus. Two link diagrams are said to be *KC-equivalent* if there is a combination of ribbon equivalence, handle-sliding and blowing up and blowing down that takes one diagram to the other.

The invariants of three-manifolds described herein are based on the representation of closed three-manifolds via surgery on framed links. Let $M^3(K)$ denote the three-manifold obtained by surgery on the blackboard framed link corresponding to the diagram K. In $M^3(K)$ the longitude associated with the diagram, as shown in Figure 43, bounds the meridian disk of the solid torus attached via the surgery.

The basic result about Kirby Calculus is that $M^3(K)$ is homeomorphic to $M^3(L)$ if and only if K and L are KC equivalent. Thus invariants of links that are also invariant under Kirby moves will produce invariants of three-manifolds. It is the purpose of this section to sketch one on the approaches to constructing such invariants.

The ideas behind this approach are quite simple. We are given a finite-dimensional quasitriangular Hopf algebra A. We associate to A a tensor category Cat(A). The objects in this category are the base field k of the Hopf algebra, and tensor powers of a formal object V. It is assumed that the tensor powers of V are canonically associative and that the tensor

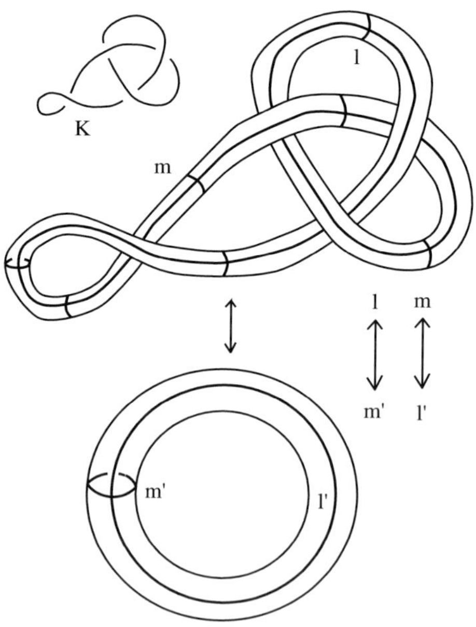

$M^3(K) = (S^3\text{-}N(K)) \cup (S^1 \times D^2)$

Fig. 43. Surgery on a blackboard framed link.

product of V with k on either side is canonically isomorphic to V. The morphisms in Cat(A) are represented by Hopf algebra decorated immersed curves arranged with respect to a vertical direction.

An *immersed curve diagram* is a link diagram where there is no distinction between undercrossings and overcrossings. Segments of the diagram can cross one another transversely as in a standard link diagram, and we can arrange such a diagram with respect to a vertical direction just as we did for link diagrams. A *vertical place* on such a diagram is a point that is not critical with respect to the vertical direction, and is not a crossing. A *decoration* of such an immersed curve diagram consists in a subset of vertical places labelled by elements of the Hopf algebra A. The diagrams can have endpoints and these are either at the bottom of the diagram or at the top (with respect to the vertical). The simplest decorated diagram is a vertical line segment with a label a (corresponding to a element a of the Hopf algebra) in its interior. In the category Cat(A) this segment is regarded as a morphism $[a]$: $V \longrightarrow V$ where V is the formal object alluded to above. Composition of these morphisms corresponds to multiplication in the algebra: $[a][b] = [ab]$. By convention, we take the order of multiplication from bottom to top with respect to the vertical direction.

A tensor product $a \otimes b$ in $A \otimes A$ is represented by two parallel segments, one decorated by a, the other decorated by b. It is our custom to place the decorations for a and for b at the same level in the diagram. In the Hopf algebra we have the coproduct Δ: $A \longrightarrow A \otimes A$. We shall write

$$\Delta(a) = \sum a_1 \otimes a_2,$$

where it is understood that this means that the coproduct of a is a sum over elements of the *form* $a_1 \otimes a_2$. It is also useful to use a version of the Einstein summation convention and just write

$$\Delta(a) = a_1 \otimes a_2,$$

where it is understood that the right-hand side is a summation. In diagrams, application of the antipode makes parallel lines with doubled decorations according to the two factors of the coproduct (see Figure 44).

A crossing of two undecorated segments is regarded as a morphism P: $V \otimes V \longrightarrow V \otimes V$. Since the lines interchange, we expect P to behave as the permutation of the two tensor factors. That is, we take the following formula to be axiomatic:

$$P \circ ([a] \otimes [b]) = ([b] \otimes [a]) \circ P.$$

A cap (see Figure 44) is regarded as a morphism from $V \otimes V$ to k, while a cup is regarded as a morphism form k to $V \otimes V$ As in the case of the crossing the relevance of these morphisms to the category is entirely encoded in their properties. The basic property of the cup and the cap is that *if you "slide" a decoration across the maximum or minimum in a counterclockwise turn, then the antipode S of the Hopf algebra is applied to*

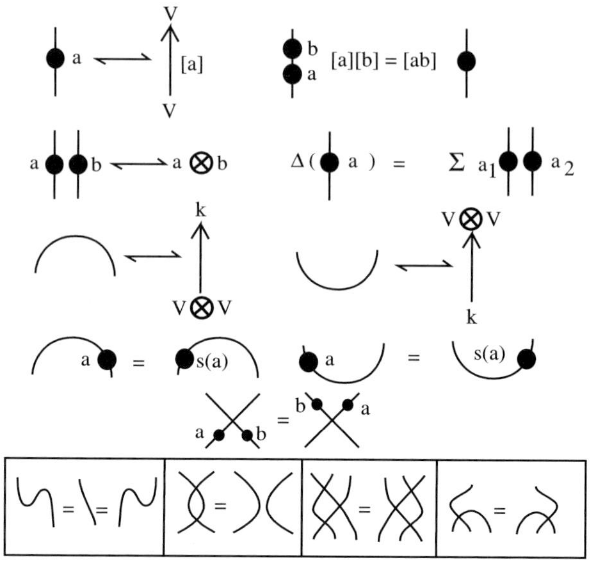

Fig. 44. Morphisms in Cat(A).

the decoration. In categorical terms this property says

$$\text{Cup}\circ([a]\otimes 1) = \text{Cup}\circ(1\otimes[Sa])$$

and

$$([Sa]\otimes 1)\circ\text{Cap} = (1\otimes[a])\circ\text{Cap}.$$

These properties and some other naturality properties of the cups and the caps are illustrated in Figure 44. These naturality properties of the flat diagrams include regular homotopy of immersions, as illustrated in Figure 44.

In Figure 45 we see how this property of the cups and the caps leads to a diagrammatic interpretation of the antipode. This, in turn, leads to the interpretation of the flat curl as a grouplike element G in A such that $S^2(a) = GaG^{-1}$ for all a in A. G is a flat curl diagram interpreted as a morphism in the category. We see that formally it is natural to interpret G as an element of A and that $\Delta(G) = G\otimes G$ is a direct consequence of the diagrams for Cat(A). In a so-called *ribbon Hopf algebra* there is such a grouplike already in the algebra. In the general case it is natural to extend the algebra to contain such an element.

We are now in a position to describe a functor F from the tangle category T to Cat(A). (The tangle category is defined for link diagrams without decorations. It has the same objects as Cat(A). The morphisms in the tangle category have relations corresponding to the augmented Reidemeister moves described in the section on quantum link invariants.) F simply decorates each positive (with respect to the vertical – see Figure 45) crossing of the tangle with the Yang–Baxter element (given by the quasi-triangular Hopf algebra A)

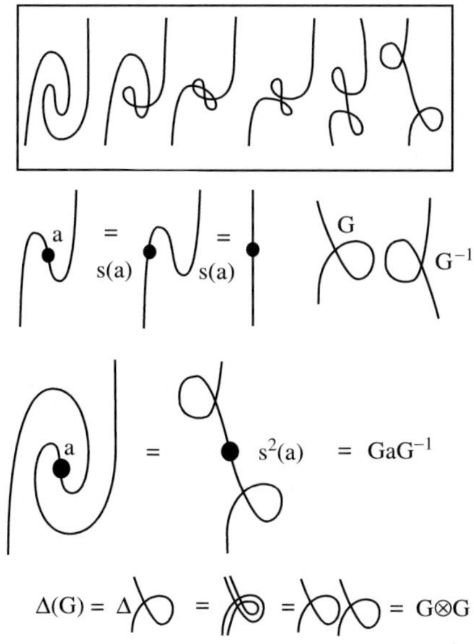

Fig. 45. Diagrammatics of the antipode.

$\rho = \sum e \otimes e'$ and each negative crossing (with respect to the vertical) with $\rho^{-1} = \sum S(e) \otimes e'$. The form of the decoration is indicated in Figure 46.

The key point about this category is that because Hopf algebra elements can be moved around the diagram, we can concentrate all the algebra in one place. Because the flat curls are identified with either G or G^{-1}, we can use regular homotopy of immersions to bring

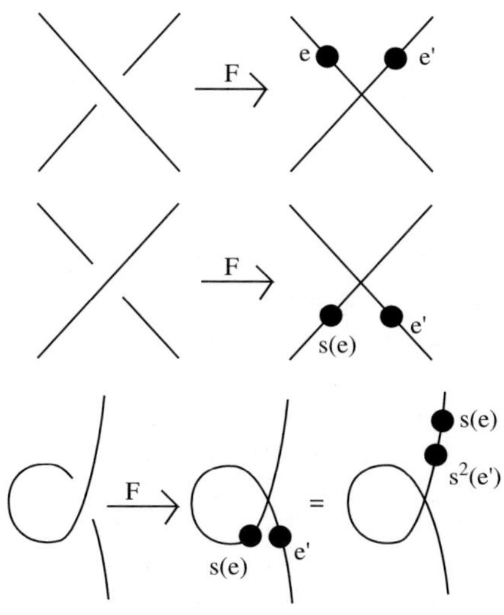

Fig. 46. The Functor $F : T \longrightarrow \mathrm{Cat}(A)$.

each component of a link diagram to the form of a circle with a single concentrated decoration (involving a sum over many products). An example is shown in Figure 46. Let us denote by $\lambda(a)$: $k \to k$ the morphism that corresponds to decorating the right-hand side of a standard circle with a. That is, $\lambda(a) = \mathrm{Cap} \circ (1 \otimes [a]) \circ \mathrm{Cup}$. We can regard λ as a linear functional defined on A as a vector space over k.

We wish to find out what properties of λ will be appropriate for constructing invariants of three-manifolds (view Figure 47). Handle sliding is accomplished by doubling a component and then band summing. The doubling corresponds to applying the antipode. As a result, we have that in order for λ to be invariant under handle-sliding it is sufficient that it have the property $\lambda(x)1 = \sum \lambda(x_1)x_2$. This is the formal defining property of a *right integral* on the Hopf algebra A. Finite-dimensional Hopf algebras have such functionals and suitable normalizations lead to well-defined three-manifold invariants. For more information see the references cited at the beginning of this section. This completes our capsule summary of Hopf algebras and invariants of three-manifolds.

There are a number of problems related to this formulation of invariants of three-manifolds. First of all, while it is the case that the invariants that come from integrals can be different from invariants defined through representations of Hopf algebras as in [123] it is quite difficult to compute them and consequently little is known. Another beautiful problem is related to the work of Greg Kuperberg [92,93]. Kuperberg defines invariants of three-manifolds associated via Hopf diagrams associated with a Heegard splitting of the three-manifold. Does our invariant on the Drinfeld double of a Hopf algebra H give the same result as Kuperberg's invariant for H? This conjecture is verified in the (easy) case where H is the group ring of a finite group. Finally, it should be mentioned that the way in which handle-sliding invariance is proven for the universal three-manifold invariant of finite type of Le and Murakami [95] is directly analogous to our method of relating handle sliding, coproduct and right integral. It remains to be seen what is the

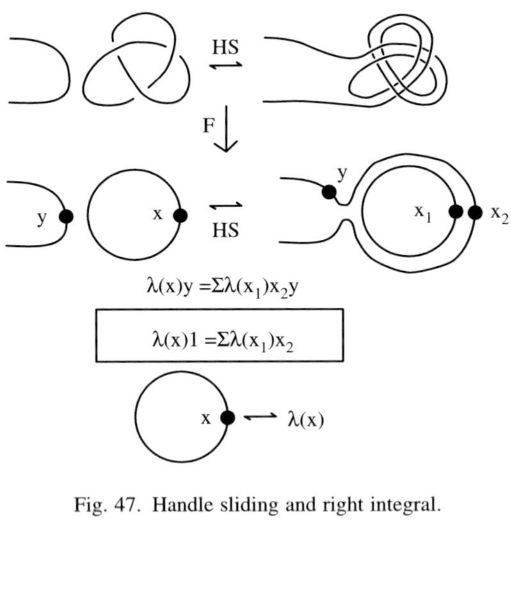

Fig. 47. Handle sliding and right integral.

relationship between three-manifold invariants of finite type and the formulations discussed here.

6. Temperley–Lieb algebra

This section is devoted to the structure of the Temperley–Lieb algebra as revealed by its diagrammatic interpretation. We begin with a combinatorial description of this algebra. It is customary, in referring to the Temperley–Lieb algebra to refer to a certain free algebra over an appropriate ring. This free algebra is the analog of the group ring of the symmetric group S_n on n letters. It is natural, therefore, to first describe that multiplicative structure that is analogous to S_n. We shall call this structure the *Temperley–Lieb Monoid* M_n. We shall describe the Temperley–Lieb algebra itself after first defining this monoid.

There is one Temperley–Lieb monoid, M_n, for each natural number n. The *connection elements* of M_n consist in diagrams in the plane that make connections involving two rows of n points. These rows will be referred to as the *top* and *bottom* rows. Each point in a row is paired with a unique point different from itself in either the top or the bottom row (it can be paired with a point in its own row). These pairings are made by arcs drawn in the space between the two rows. *No two arcs are allowed to intersect one another.* Such a connection element will be denoted by U, with subscripts to indicate specific elements. If the top row is the set $\mathrm{Top} = \{1, 2, 3, \ldots, n\}$ and the bottom row is $\mathrm{Bot} = \{1', 2', \ldots, n'\}$, then we can regard U as a function from Top $\cup$ Bot to itself such that $U(x)$ is never equal to x, $U(U(x)) = x$ for all x, and satisfying the planar non-intersection property described above. In topological terms U is an n-tangle with no crossings, taken up to regular isotopy of tangles in the plane.

If U and V are two elements of M_n as described above, then their product UV is the tangle product obtained by attaching the bottom row of U to the top row of V. Note that the result of taking such a product will produce a new connection structure plus some loops in the plane. Each loop is regarded as an instance of the *loop element* δ of the Temperley–Lieb monoid M_n. The loop element commutes with all other elements of the monoid and has no other relations with these elements. Thus $UV = \delta^k W$ for some non-negative integer k, and some connection element W of the monoid.

The Temperley–Lieb algebra T_n is the free additive module on M_n modulo the identification

$$\delta = -A^2 - A^{-2},$$

over the ring $Z[A, A^{-1}]$ of Laurent polynomials in the variable A. Products are defined on the connection elements and extended linearly to the algebra. The reason for this loop identification is the application of the Temperley–Lieb algebra for the bracket polynomial and for representations of the braid group [44,52,67].

The Temperley–Lieb monoid M_n is generated by the elements 1, U_1, $U_2, \ldots, U_{n-1}$ where the identity element 1 connects each i in the top row with its corresponding member i' in the bottom row. Here U_k connects i to i' for i not equal to k, $k+1$, k', $(k+1)'$.

U_k connects k to $k+1$ and k' to $(k+1)'$. It is easy to see that

$$U_k^2 = \delta U_k,$$

$$U_k U_{k\pm 1} U_k = U_k,$$

$$U_i U_j = U_j U_i, \quad |i-j| > 1.$$

See Figure 48.

We shall prove that the Temperley–Lieb Monoid is the universal monoid on $G_n = \{1, U_1, U_2, \ldots, U_{n-1}\}$ modulo these relations. In order to accomplish this end we give a direct diagrammatic method for writing any connection element of the monoid as a certain canonical product of elements of G_n. This method is illustrated in Figure 49.

As shown in Figure 49, we represent the connection diagram with vertical and horizontal straight arcs such that except for the height of the straight arcs, the form of the connection between any two points is unique – consisting in two vertical arcs and one horizontal arc. The horizontal arc has its endpoints on the vertical lines that go through the row points that are being connected. (Diagram is drawn so that each pair in the set $\{(i, i') : 1 \leq n\}$ determines such a vertical line.) The vertical arcs in the connection are chosen as segments from these vertical lines. All connections are chosen so that the connections do not intersect. It is from this diagram that we shall read out a factorization into a product of elements of G_n.

The factorization is achieved via a decoration of the straight arc diagram by dotted vertical arcs, as shown in Figure 49. Each dotted arc connects midpoints of the restrictions of

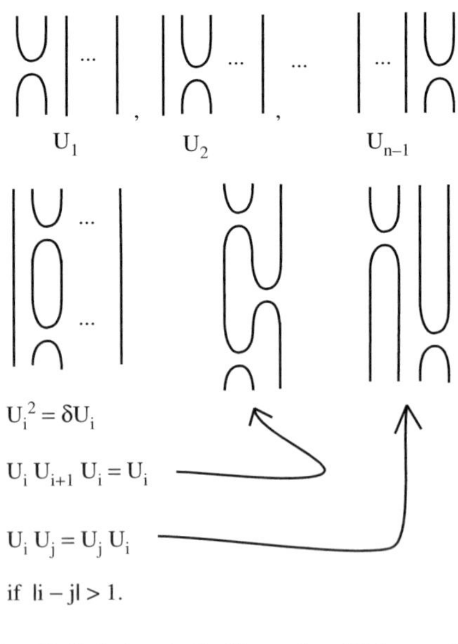

Fig. 48. Relations in the Temperley–Lieb monoid.

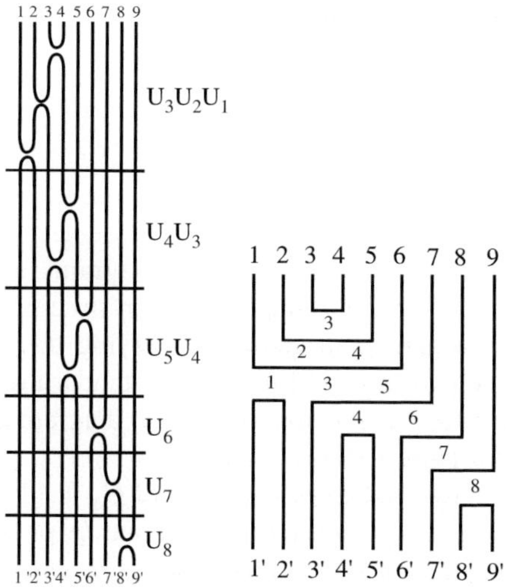

Fig. 49. Canonical factorization in the Temperley–Lieb monoid.

horizontal arcs to the *columns* of the diagram, where a column of the diagram is the space between two consecutive vertical lines (vertical lines described as in the previous paragraph). The *index* of a column is the row number associated to the left vertical boundary of the column. The dotted lines in a given column are uniquely determined by starting at the bottom of the column and pairing up the horizontal arcs in that column in vertical succession. Each dotted arc is labelled by the index of the column in which it stands.

In a given diagram a *sequence* of dotted arcs is a maximal set of dotted arcs (with consecutive indices) that are interconnected by horizontal segments such that one can begin at the top of the dotted arc (in that sequence) of highest index, go down the arc and left to the top of the next arc along a horizontal segment, continuing in this manner until the whole sequence is traversed. It is clear from the construction of the diagram that the dotted segments in the diagram collect into a disjoint union of sequences $\{s^1, s^2, \ldots, s^k\}$ where each s^i denotes the corresponding descending sequence of consecutive indices:

$$s^i = (m_i, m_i - 1, m_i - 2, \ldots, n_i + 1, n_i).$$

These indices satisfy the inequalities

$$m_1 < m_2 < m_3 < \cdots < m_k$$

and

$$n_1 < n_2 < n_3 < \cdots < n_k.$$

The sequences $\{s^1, s^2, \ldots, s^k\}$ occur in that order on the diagram read from left to right. Of course the descent of each sequence goes from right to left. If D is a diagram with sequence structure $s(D)$ as we have just described, let $U(s(D))$ be the following product of generators of the Temperley–Lieb monoid:

$$U(D) = U(s^1)U(s^2)U(s^3)\cdots U(s^k),$$

where

$$U(s^i) = U_{m_i}U_{m_i-1}\cdots U_{n_i+1}U_{n_i}.$$

By looking carefully at the combinatorics of these diagrams, as illustrated in Figure 49, one sees that D and $U(D)$ represent the same connection structure in the Temperley–Lieb monoid. Furthermore, any sequence structure s satisfying the inequalities given above (call these the canonical inequalities) will produce a standard diagram from the product $U(s)$ Thus the sequence structure of a Temperley–Lieb diagram completely classifies this diagram as a connection structure in the monoid.

We must now prove that any product of elements of $G_n = \{1, U_1, U_2, \ldots, U_{n-1}\}$ can be written, up to a loop factor, as $U(s(D))$ for some diagram D, or equivalently as $U(s)$ for a sequence structure satisfying the canonical inequalities. This is a simple exercise in using the relations we have already given for the products of elements of G_n. We leave the details to the reader. This completes the proof that the relations in Figure 48 are a complete set of relations for the Temperley–Lieb monoid.

6.1. *Parentheses*

Elements of the Temperley–Lieb monoid M_n are in one-to-one correspondence with well-formed parenthesis expressions using n pairs of parentheses. The proof of this statement is shown in Figure 50.

One result of this reformulation of the Temperley–Lieb monoid is that one can rewrite the product structure in terms of operations on parentheses, getting an interesting formal algebra that encodes properties of the topology of plane curves [54,66]. In this section we

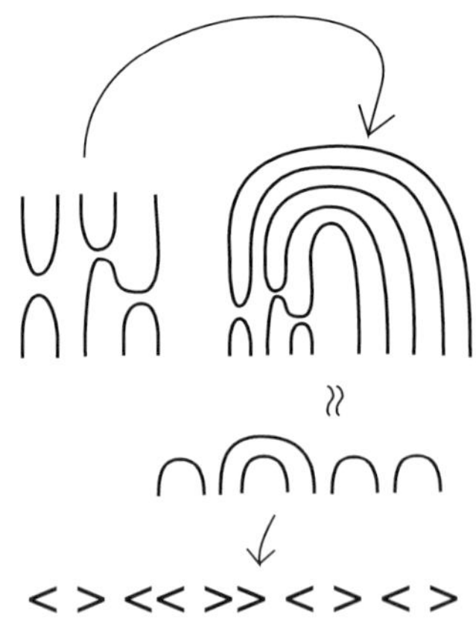

Fig. 50. Creating parentheses.

will take the relationship with parentheses as an excuse to indicate a further relationship with non-associative products.

First consider the abstract structure of non-associative products. They are usually written in the forms:

$$(a * b) * c,$$

$$a * (b * c),$$

$$((a * b) * c) * d,$$

$$(a * (b * c)) * d,$$

$$a * ((b * c) * d),$$

$$a * (b * (c * d)),$$

$$(a * b) * (c * d),$$

In writing products in this standard manner, one sees the same structure of parentheses occurring in different products, as in $((a * b) * c) * d$ and $(a * (b * c)) * d$. In each of these cases if we eliminated the algebraic literals and the operation symbol $*$, we would be left with the parenthetical: (( )).

There is another way to write the products so that different products correspond to different arrangements of parentheses. To do this, rewrite the products in the operator notation shown below:

$$a * b = a(b) \rightsquigarrow (\,),$$

$$(a * b) * c = a(b)(c) \rightsquigarrow (\,)(\,),$$

$$a * (b * c) = a(b(c)) \rightsquigarrow ((\,)),$$

$$((a * b) * c) * d = a(b)(c)(d) \rightsquigarrow (\,)(\,)(\,),$$

$$(a * (b * c)) * d = a(b(c))(d) \rightsquigarrow ((\,))(\,),$$

$$a * ((b * c) * d) = a(b(c)(d)) \rightsquigarrow ((\,)(\,)),$$

$$a * (b * (c * d)) = a(b(c(d))) \rightsquigarrow (((\,))),$$

$$(a * b) * (c * d) = a(b)(c(d)) \rightsquigarrow (\,)((\,)),$$

In the operator notation, each product of $n + 1$ terms is uniquely associated with an expression using n pairs of parentheses.

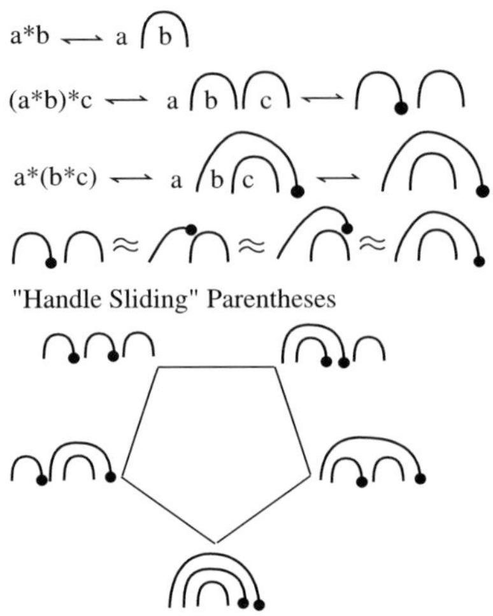

Fig. 51. Pentagon.

Of course we can replace the expressions in parentheses in terms of nested caps as we did in Figure 50. Once this is done, we notice the very interesting fact that *re-associations* can be visualized in terms of “handle-sliding” at the cap level. The meaning of this remark should be apparent to the reader from Figure 51.

In Figure 51 we illustrate the pentagon of re-associations of a product of four terms in terms of sliding caps. Notice that we can do this sliding without writing any algebraic literals by labelling the “active” right feet of the caps that do the sliding. In Figure 52 we show the structure of the Stasheff polyhedron with re-associations of five literals in the cap sliding formalism.

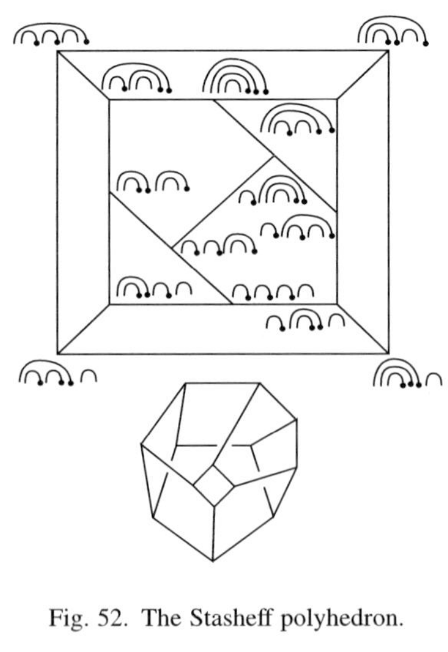

Fig. 52. The Stasheff polyhedron.

Cap sliding is a new formulation of a continuous background for these re-association moves. There is a related continuous background in terms of recoupling formalisms for trees. This formalism is intimately related to many topics in topological quantum field theory. The interested reader will find more about these points of view in [24,25,54] in the author's lecture notes with Scott Carter and Masahico Saito [14] and in his book with Sostenes Lins [67].

7. Virtual knot theory

Knot theory studies the embeddings of curves in three-dimensional space. Virtual knot theory studies the embeddings of curves in thickened surfaces of arbitrary genus, up to the addition and removal of empty handles from the surface. Virtual knots have a special diagrammatic theory that makes handling them very similar to the handling of classical knot diagrams. In fact, this diagrammatic theory simply involves adding a new type of crossing to the knot diagrams, a *virtual crossing* that is neither under nor over. From a combinatorial point of view, the virtual crossings are artifacts of the representation of the virtual knot or link in the plane. The extension of the Reidemeister moves that takes care of them respects this viewpoint. A virtual crossing (see Figure 53(a)) is represented by two crossing arcs with a small circle placed around the crossing point.

Moves on virtual diagrams generalize the Reidemeister moves for classical knot and link diagrams (see Figure 53(a)). One can summarize the moves on virtual diagrams by saying that the classical crossings interact with one another according to the usual Reidemeister moves. One adds the detour moves for consecutive sequences of virtual crossings and this completes the description of the moves on virtual diagrams. It is a consequence of moves (B) and (C) in Figure 53(a) that an arc going through any consecutive sequence of virtual crossings can be moved anywhere in the diagram keeping the endpoints fixed; the places where the moved arc crosses the diagram become new virtual crossings. This replacement is the *detour move* (see Figure 53(b)).

One can generalize many structures in classical knot theory to the virtual domain, and use the virtual knots to test the limits of classical problems such as the question whether the Jones polynomial detects knots and the classical Poincaré conjecture. Counterexamples to these conjectures exist in the virtual domain, and it is an open problem whether any of these counterexamples are equivalent (by addition and subtraction of empty handles) to classical knots and links. Virtual knot theory is a significant domain to be investigated for its own sake and for a deeper understanding of classical knot theory.

Another way to understand the meaning of virtual diagrams is to regard them as representatives for oriented Gauss codes (Gauss diagrams) [70,119]. Such codes do not always have planar realizations and an attempt to embed such a code in the plane leads to the production of the virtual crossings. The detour move makes the particular choice of virtual crossings irrelevant. Virtual equivalence is the same as the equivalence relation generated on the collection of oriented Gauss codes modulo an abstract set of Reidemeister moves on the codes.

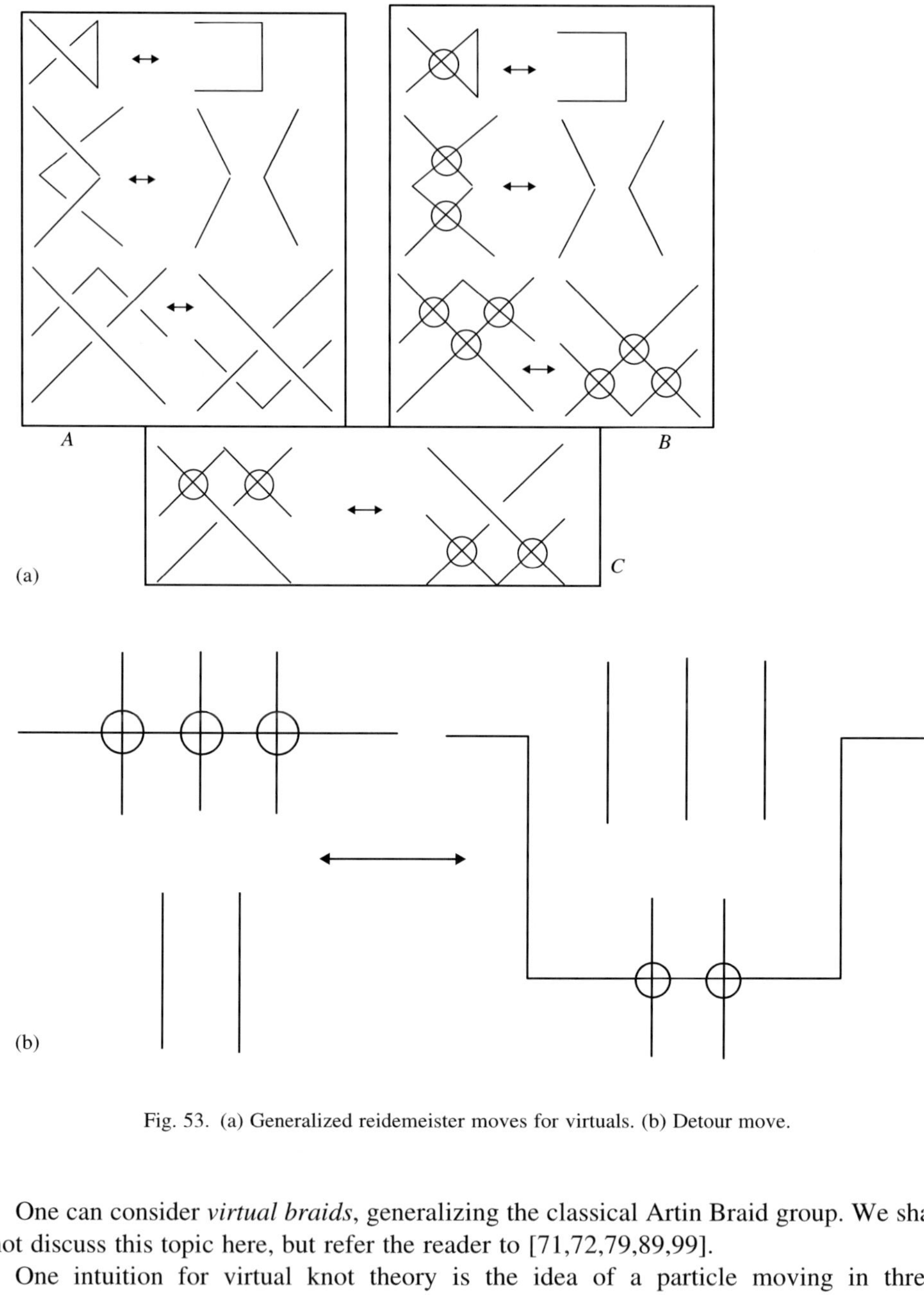

Fig. 53. (a) Generalized reidemeister moves for virtuals. (b) Detour move.

One can consider *virtual braids*, generalizing the classical Artin Braid group. We shall not discuss this topic here, but refer the reader to [71,72,79,89,99].

One intuition for virtual knot theory is the idea of a particle moving in three-dimensional space in a trajectory that occasionally disappears, and then reappears elsewhere. By connecting the disappearance points and the reappearance points with detour lines in the ambient space we get a picture of the motion, but the detours, being artificial, must be treated as subject to replacements.

7.1. *Flat virtual knots and links*

Every classical knot or link diagram can be regarded as a four-regular plane graph with extra structure at the nodes. This extra structure is usually indicated by the overcrossing and undercrossing conventions that give instructions for constructing an embedding of the link in three-dimensional space from the diagram. If we take the diagram without this extra structure, it is the shadow of some link in three-dimensional space, but the weaving of that link is not specified. It is well known that if one is allowed to apply the Reidemeister moves to such a shadow (without regard to the types of crossing since they are not specified) then the shadow can be reduced to a disjoint union of circles. This reduction is no longer true for virtual links. More precisely, let a *flat virtual diagram* be a diagram with virtual crossings as we have described them and *flat crossings* consisting in undecorated nodes of the four-regular plane graph. Virtual crossings are flat crossings that have been decorated by a small circle. Two flat virtual diagrams are *equivalent* if there is a sequence of generalized flat Reidemeister moves (as illustrated in Figure 53(a)) taking one to the other. A generalized flat Reidemeister move is any move as shown in Figure 53(a), but one can ignore the overcrossing or undercrossing structure. Note that in studying flat virtuals the rules for changing virtual crossings among themselves and the rules for changing flat crossings among themselves are identical. However, detour moves as in part C in Figure 53(a) are available for virtual crossings with respect to flat crossings and not the other way around.

We shall say that a virtual diagram *overlies* a flat diagram if the virtual diagram is obtained from the flat diagram by choosing a crossing type for each flat crossing in the virtual diagram. To each virtual diagram K there is an associated flat diagram $F(K)$ that is obtained by forgetting the extra structure at the classical crossings in K. Note that if K is equivalent to K' as virtual diagrams, then $F(K)$ is equivalent to $F(K')$ as flat virtual diagrams. Thus, if we can show that $F(K)$ is not reducible to a disjoint union of circles, then it will follow that K is a non-trivial virtual link.

Figure 54(a) illustrates an example of a flat virtual link H. This link cannot be undone in the flat category because it has an odd number of virtual crossings between its two components and each generalized Reidemeister move preserves the parity of the number of virtual crossings between components. Also illustrated in Figure 54(a) is a flat diagram D and a virtual knot K that overlies it. This example is given in [70]. The knot shown is undetectable by many invariants (fundamental group, Jones polynomial) but it is knotted and this can be seen either by using a generalization of the Alexander polynomial that we describe below, or by showing that the underlying diagram D is a non-trivial flat virtual knot using the filamentation invariant that is introduced in [41]. The filamentation invariant is a combinatorial method that is sometimes successful in indentifying irreducible flat virtuals. At this writing we know very few invariants of flat virtuals. The flat virtual diagrams present a strong challenge for the construction of new invariants. It is important to understand the structure of flat virtual knots and links. This structure lies at the heart of the comparison of classical and virtual links. We wish to be able to determine when a given virtual link is equivalent to a classical link. The reducibility or irreducibility of the underlying flat diagram is the first obstruction to such an equivalence.

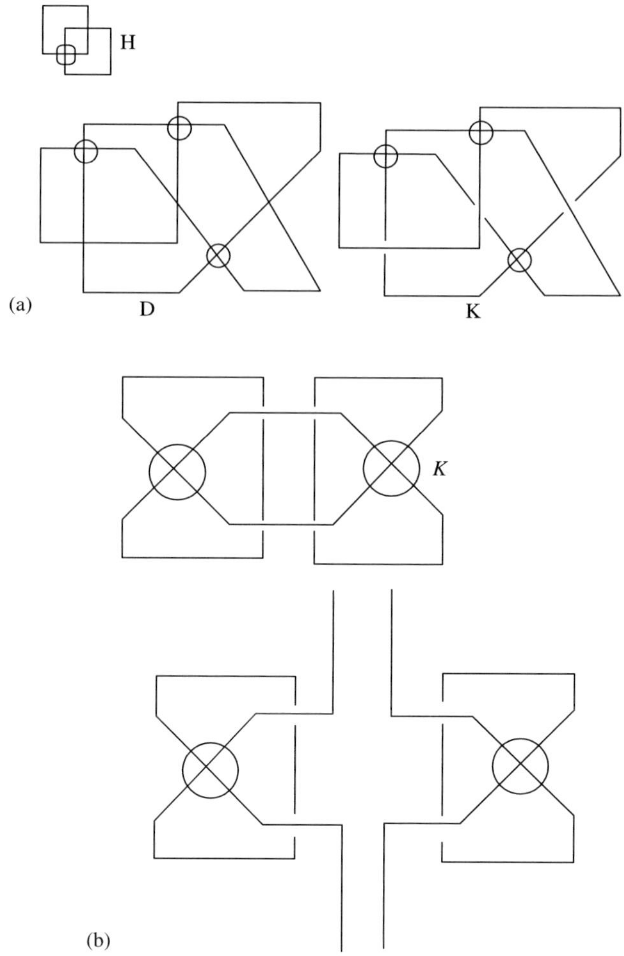

Fig. 54. (a) Flats H and D, and the knot K. (b) Kishino and parts.

7.2. *Interpretation of virtuals as stable classes of links in thickened surfaces*

There is a useful topological interpretation for this virtual theory in terms of embeddings of links in thickened surfaces (see [70,72,94]). Regard each virtual crossing as a shorthand for a detour of one of the arcs in the crossing through a 1-handle that has been attached to the 2-sphere of the original diagram. By interpreting each virtual crossing in this way, we obtain an embedding of a collection of circles into a thickened surface $S_g \times R$ where g is the number of virtual crossings in the original diagram L, S_g is a compact oriented surface of genus g and R denotes the real line. We say that two such surface embeddings are *stably equivalent* if one can be obtained from another by isotopy in the thickened surfaces, homeomorphisms of the surfaces and the addition or subtraction of empty handles. Then we have the following theorem.

THEOREM [70,81,94]. *Two virtual link diagrams are equivalent if and only if their correspondent surface embeddings are stably equivalent.*

Virtual knots and links give rise to a host of problems. As we saw in the previous section, there are non-trivial virtual knots with unit Jones polynomial. Moreover, there are non-trivial virtual knots with integer fundamental group and trivial Jones polynomial. (The fundamental group is defined combinatorially by generalizing the Wirtinger presentation.) These phenomena underline the question of how planarity is involved in the way the Jones polynomial appears to detect classical knots, and that the relationship of the fundamental group (and peripheral system) is a much deeper one than the surface combinatorics for classical knots. It is possible to take the connected sum of two trivial virtual diagrams and obtain a non-trivial virtual knot (the Kishino knot).

Here long knots (or, equivalently $1-1$ tangles) come into play. Having a knot, we can break it at some point and take its ends to infinity (say, in a way that they coincide with the horizontal axis line in the plane). One can study isotopy classes of such knots. A well-known theorem says that in the classical case, knot theory coincides with long knot theory. However, this is not the case for virtual knots. By breaking the same virtual knot at different points, one can obtain non-isotopic long knots [32]. Furthermore, even if the initial knot is trivial, the resulting long knot may not be trivial. The "connected sum" of two trivial virtual diagrams may not be trivial in the compact case. The phenomenon occurs because these two knot diagrams may be non-trivial in the long category. It is sometimes more convenient to consider long virtual knots rather than compact virtual knots, since connected sum is well-defined for long knots. It is important to construct long virtual knot invariants to see whether long knots are trivial and whether they are classical. One approach is to regard long knots as $1-1$ tangles and use extensions of standard invariants (fundamental group, quandle, biquandle, etc.). Another approach is to distinguish two types of crossings: those having early undercrossing and those having later undercrossing with respect to the orientation of the long knot. The latter technique is described in [117].

Unlike classical knots, the connected sum of long knots is not commutative. Thus, if we show that two long knots K_1 and K_2 do not commute, then we see that they are different and both non-classical.

A typical example of such knots is the two parts of the Kishino knot, see Figure 54(b).

We have a natural map

$$\langle \text{Long virtual knots} \rangle \longrightarrow \langle \text{Oriented compact virtual knots} \rangle,$$

obtained by taking two infinite ends of the long knots together to make a compact knot. This map is obviously well defined and allows one to construct (weak) long virtual knot invariants from classical invariants (by regarding compact knot invariants as long knot invariants). There is no well-defined inverse for this map. But, if we were able to construct the map from compact virtual knots to long virtual knots, we could apply the long techniques for the compact case. This map does have an inverse for *classical* knots. Thus, the long techniques are applicable to classical (long) knots. It would be interesting to obtain new classical invariants from it. The long category can also be applied in the case of flat virtuals, where all problems formulated above occur as well.

There are examples of virtual knots that are very difficult to prove knotted, and there are infinitely many flat virtual diagrams that appear to be irreducible, but we have no techniques to prove it. How can one tell whether a virtual knot is classical? One can ask: are there non-trivial virtual knots whose connected sum is trivial? The latter question cannot be shown by classical techniques, but it can be analyzed by using the surface interpretation for virtuals (see [112]).

In respect to virtual knots, we are in the same position as the compilers of the original knot tables. We are, in fact, developing tables. At Sussex, tables of virtual knots are being constructed, and tables will appear in a book being written by Kauffman and Manturov. The website "Knotilus" has tables as do websites develped by Zinn-Justin and Zuber. The theory of invariants of virtual knots, needs more development. Flat virtuals (whose study is a generalization of the classification of immersions) are a nearly unknown territory (but see [41,141]). The flat virtuals provide the deepest challenge since we have very few invariants to detect them.

7.3. *Jones polynomial of virtual knots*

We use a generalization of the bracket state summation model for the Jones polynomial to extend it to virtual knots and links. We call a diagram in the plane *purely virtual* if the only crossings in the diagram are virtual crossings. Each purely virtual diagram is equivalent by the virtual moves to a disjoint collection of circles in the plane.

Given a link diagram K, a state S of this diagram is obtained by choosing a smoothing for each crossing in the diagram and labelling that smoothing with either A or A^{-1} according to the convention that a counterclockwise rotation of the overcrossing line sweeps two regions labelled A, and that a smoothing that connects the A regions is labelled by the letter A. Then, given a state S, one has the evaluation $\langle K|S\rangle$ equal to the product of the labels at the smoothings, and one has the evaluation $\|S\|$ equal to the number of loops in the state (the smoothings produce purely virtual diagrams). One then has the formula

$$\langle K\rangle = \sum_{S} \langle K|S\rangle d^{\|S\|-1},$$

where the summation runs over the states S of the diagram K, and $d = -A^2 - A^{-2}$. This state summation is invariant under all classical and virtual moves except the first Reidemeister move. The bracket polynomial is normalized to an invariant $f_k(A)$ of all the moves by the formula $f_K(A) = (-A^3)^{-w(K)}\langle K\rangle$ where $w(K)$ is the writhe of the (now) oriented diagram K. The writhe is the sum of the orientation signs (± 1) of the crossings of the diagram. The Jones polynomial, $V_K(t)$, is given in terms of this model by the formula

$$V_K(t) = f_K(t^{-1/4}).$$

The reader should note that this definition is a direct generalization to the virtual category of the state sum model for the original Jones polynomial [52]. It is straightforward to verify the invariances stated above. In this way one has the Jones polynomial for virtual knots and links.

In terms of the interpretation of virtual knots as stabilized classes of embeddings of circles into thickened surfaces, our definition coincides with the simplest version of the Jones polynomial for links in thickened surfaces. In that version one counts all the loops in a state the same way, with no regard for their isotopy class in the surface. It is this equal treatment that makes the invariance under handle stabilization work. With this generalized version of the Jones polynomial, one has again the problem of finding a geometric/ topological interpretation of this invariant. There is no fully satisfactory topological interpretation of the original Jones polynomial and the problem is inherited by this generalization.

We have [72] the following theorem.

THEOREM. *To each non-trivial classical knot diagram of one component K there is a corresponding non-trivial virtual knot diagram* $\mathrm{Virt}(K)$ *with unit Jones polynomial.*

This theorem is a key ingredient in the problems involving virtual knots. Here is a sketch of its proof. The proof uses two invariants of classical knots and links that generalize to arbitrary virtual knots and links. These invariants are the *Jones polynomial* and the *involutory quandle* denoted by the notation $IQ(K)$ for a knot or link K.

Given a crossing i in a link diagram, we define $s(i)$ to be the result of *switching* that crossing so that the undercrossing arc becomes an overcrossing arc and vice versa. We also define the *virtualization* $v(i)$ of the crossing by the local replacement indicated in Figure 55. In this figure we illustrate how in the virtualization of the crossing the original crossing is replaced by a crossing that is flanked by two virtual crossings.

Suppose that K is a (virtual or classical) diagram with a classical crossing labeled i. Let $K^{v(i)}$ be the diagram obtained from K by virtualizing the crossing i while leaving the rest of the diagram just as before. Let $K^{s(i)}$ be the diagram obtained from K by switching the crossing i while leaving the rest of the diagram just as before. Then it follows directly

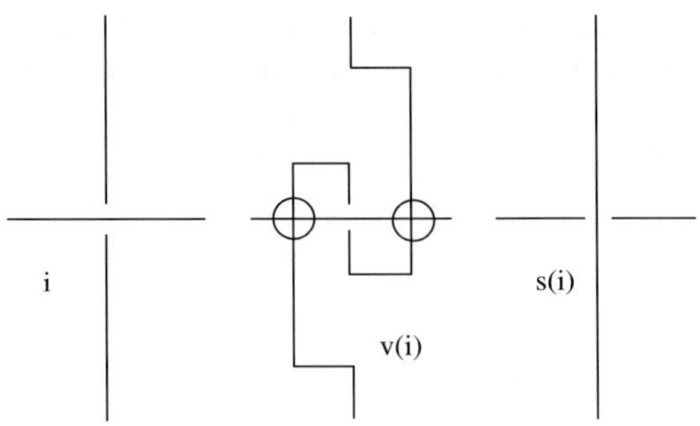

Fig. 55. Switching and virtualizing a crossing.

from the definition of the Jones polynomial that

$$V_{K^{s(i)}}(t) = V_{K^{v(i)}}(t).$$

As far as the Jones polynomial is concerned, switching a crossing and virtualizing a crossing look the same.

The involutory quandle [58] is an algebraic invariant equivalent to the fundamental group of the double branched cover of a knot or link in the classical case. In this algebraic system one associates a generator of the algebra $IQ(K)$ to each arc of the diagram K and there is a relation of the form $c = ab$ at each crossing, where ab denotes the (non-associative) algebra product of a and b in $IQ(K)$ (see Figure 56). In this figure we have illustrated through the local relations the fact that

$$IQ(K^{v(i)}) = IQ(K).$$

As far the involutory quandle is concerned, the original crossing and the virtualized crossing look the same.

If a classical knot is actually knotted, then its involutory quandle is non-trivial [144]. Hence if we start with a non-trivial classical knot, we can virtualize any subset of its crossings to obtain a virtual knot that is still non-trivial. There is a subset A of the crossings of a classical knot K such that the knot SK obtained by switching these crossings is an unknot. Let Virt(K) denote the virtual diagram obtained from A by virtualizing the crossings in the subset A. By the above discussion the Jones polynomial of Virt(K) is the same as the Jones polynomial of SK, and this is 1 since SK is unknotted. On the other hand, the IQ of Virt(K) is the same as the IQ of K, and hence if K is knotted, then so is Virt(K). We have shown that Virt(K) is a non-trivial virtual knot with unit Jones polynomial. This completes the proof of the theorem.

If there exists a classical knot with unit Jones polynomial, then one of the knots Virt(K) produced by this theorem may be equivalent to a classical knot. It is an intricate task to verify that specific examples of Virt(K) are not classical. This has led to an investigation of

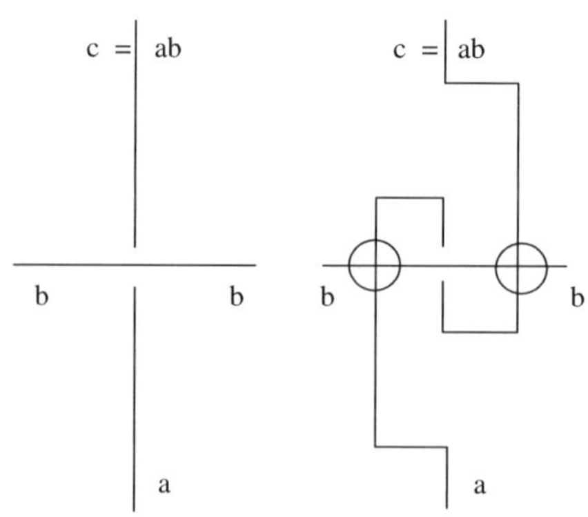

Fig. 56. IQ(Virt(K)) = IQ(K).

new invariants for virtual knots. In this investigation a number of issues appear. One can examine the combinatorial generalization of the fundamental group (or quandle) of the virtual knot and sometimes one can prove by pure algebra that the resulting group is not classical. This is related to observations by Silver and Williams [129], Manturov [104,105] and by Satoh [125] showing that the fundamental group of a virtual knot can be interpreted as the fundamental group of the complement of a torus embedded in four-dimensional Euclidean space. A very fruitful line of new invariants comes about by examining a generalization of the fundamental group or quandle that we call the *biquandle* of the virtual knot. The biquandle is discussed in Section 7.4. Invariants of flat knots (when one has them) are useful in this regard. If we can verify that the flat knot $F(\text{Virt}(K))$ is non-trivial, then $\text{Virt}(K)$ is non-classical. In this way the search for classical knots with unit Jones polynomial expands to the exploration of the structure of the infinite collection of virtual knots with unit Jones polynomial.

Another way of putting this theorem is as follows: in the arena of knots in thickened surfaces there are many examples of knots with unit Jones polynomial. Might one of these be equivalent via handle stabilization to a classical knot? In [94] Kuperberg shows the uniqueness of the embedding of minimal genus in the stable class for a given virtual link. The minimal embedding genus can be strictly less than the number of virtual crossings in a diagram for the link. There are many problems associated with this phenomenon.

There is a generalization of the Jones polynomial that involves surface representation of virtual knots (see [28,29,106,101]). These invariants essentially use the fact that the Jones polynomial can be extended to knots in thickened surfaces by keeping track of the isotopy classes of the loops in the state summation for this polynomial. In the approach of Dye and Kauffman, one uses this generalized polynomial directly. In the approach of Manturov, a polynomial invariant is defined using the stabilization description of the virtual knots.

7.4. *Biquandles*

In this section we give a sketch of some recent approaches to invariants of virtual knots and links.

A *biquandle* [7,8,16,32,72,76] is an algebra with 4 binary operations written a^b, a_b, $a^{\bar{b}}$, $a_{\bar{b}}$ together with some relations which we will indicate below. The *fundamental* biquandle is associated with a link diagram and is invariant under the generalized Reidemeister moves for virtual knots and links. The operations in this algebra are motivated by the formation of labels for the edges of the diagram (view Figure 57). In this figure we have shown the format for the operations in a biquandle. The overcrossing arc has two labels, one on each side of the crossing. There is an algebra element labeling each *edge* of the diagram. An edge of the diagram corresponds to an edge of the underlying plane graph of that diagram.

Let the edges oriented toward a crossing in a diagram be called the *input* edges for the crossing, and the edges oriented away from the crossing be called the *output* edges for the crossing. Let a and b be the input edges for a positive crossing, with a the label of the

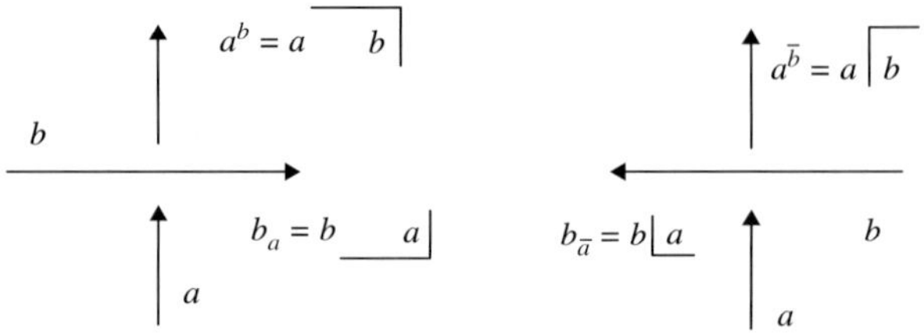

Fig. 57. Biquandle relations at a crossing.

undercrossing input and b the label on the overcrossing input. In the biquandle, we label the undercrossing output by

$$c = a^b,$$

while the overcrossing output is labeled

$$d = b_a.$$

The labelling for the negative crossing is similar using the other two operations.

To form the fundamental biquandle, $\mathrm{BQ}(K)$, we take one generator for each edge of the diagram and two relations at each crossing (as described above).

Another way to write this formalism for the biquandle is as follows:

$$a^b = a\ \ \overline{b}|,$$

$$a_b = a\ \ \underline{b}|,$$

$$a^{\bar b} = a\ \ |\overline{b},$$

$$a_{\bar b} = a\ \ |\underline{b}.$$

We call this the *operator formalism* for the biquandle.

These considerations lead to the following definition.

DEFINITION. A *biquandle* B is a set with four binary operations indicated above: a^b, $a^{\bar b}$, a_b, $a_{\bar b}$. We shall refer to the operations with barred variables as the *left* operations and the operations without barred variables as the *right* operations. The biquandle is closed under these operations and the following axioms are satisfied:

(1) Given an element a in B, then there exists an x in the biquandle such that $x = a_x$ and $a = x^a$. There also exists a y in the biquandle such that $y = a^{\bar y}$ and $a = y_{\bar a}$.

(2) For any elements a and b in B we have

$$a = a^{b\overline{b_a}} \quad \text{and} \quad b = b_{\overline{aa^b}} \quad \text{and} \quad a = a^{\bar b b_{\bar a}} \quad \text{and} \quad b = b_{\bar a a^{\bar b}}.$$

(3) Given elements a and b in B then there exist elements x, y, z, t such that $x_b = a$, $y^{\bar{a}} = b$, $b^x = y$, $a_{\bar{y}} = x$ and $t^a = b$, $a_t = z$, $z_{\bar{b}} = a$, $b^{\bar{z}} = t$. The biquandle is called *strong* if x, y, z, t are uniquely defined and we then write $x = a_{b^{-1}}$, $y = b^{\bar{a}^{-1}}$, $t = b^{a^{-1}}$, $z = a_{\bar{b}^{-1}}$ reflecting the inversive nature of the elements.

(4) For any a, b, c in B the following equations hold and the same equations hold when all right operations are replaced in these equations by left operations.

$$a^{bc} = a^{c_b b^c}, \quad c_{ba} = c_{a^b b_a}, \quad (b_a)^{c_{a^b}} = (b^c)_{a^{c_b}}.$$

These axioms are transcriptions of the Reidemeister moves. The first axiom transcribes the first Reidemeister move. The second axiom transcribes the directly oriented second Reidemeister move. The third axiom transcribes the reverse oriented Reidemeister move. The fourth axiom transcribes the third Reidemeister move. Much more work is needed in exploring these algebras and their applications to knot theory.

We may simplify the appearance of these conditions by defining

$$S(a,b) = (b_a, a^b), \quad \bar{S}(a,b) = (b^{\bar{a}}, a_{\bar{b}}),$$

and in the case of a strong biquandle,

$$S^{+}_{-}(a,b) = (b^{a_{b^{-1}}}, a_{b^{-1}}), \quad S^{-}_{+}(a,b) = (b^{a^{-1}}, a_{b^{a^{-1}}}),$$

and

$$\bar{S}^{+}_{-}(a,b) = (b_{\overline{a^{\bar{b}^{-1}}}}, a^{\bar{b}^{-1}}) = (b_{\overline{a^{b_{a^{-1}}}}}, a^{b_{a^{-1}}}),$$

and

$$\bar{S}^{-}_{+}(a,b) = (b_{\bar{a}^{-1}}, a^{\overline{b_{\bar{a}^{-1}}}}) = (b_{a^{b^{-1}}}, a^{\overline{b_{a^{b^{-1}}}}}),$$

which we call the *sideways* operators. The conditions then reduce to

$$S\bar{S} = \bar{S}S = 1,$$

$$(S \times 1)(1 \times S)(S \times 1) = (1 \times S)(S \times 1)(1 \times S),$$

$$\bar{S}^{-}_{+}S^{+}_{-} = S^{-}_{+}\bar{S}^{+}_{-} = 1,$$

and finally all the sideways operators leave the diagonal

$$\Delta = \{(a,a) | a \in X\}$$

invariant.

7.5. *The Alexander biquandle*

It is not hard to see that the following equations in a module over $Z[s, s^{-1}, t, t^{-1}]$ give a biquandle structure.

$$a^b = a \ \overline{b}| = ta + (1 - st)b, \qquad a_b = a \ \underline{b}| = sa$$

$$a^{\bar{b}} = a \ \overline{|b} = t^{-1}a + (1 - s^{-1}t^{-1})b, \qquad a_{\bar{b}} = a \ \underline{|b} = s^{-1}a.$$

We shall refer to this structure, with the equations given above, as the *Alexander Biquandle*.

Just as one can define the Alexander Module of a classical knot, we have the Alexander Biquandle of a virtual knot or link, obtained by taking one generator for each *edge* of the projected graph of the knot diagram and taking the module relations in the above linear form. Let ABQ(K) denote this module structure for an oriented link K. That is, ABQ(K) is the module generated by the edges of the diagram, factored by the submodule generated by the relations. This module then has a biquandle structure specified by the operations defined above for an Alexander Biquandle.

The determinant of the matrix of relations obtained from the crossings of a diagram gives a polynomial invariant (up to multiplication by $\pm s^i t^j$ for integers i and j) of knots and links that we denote by $G_K(s, t)$ and call the *generalized Alexander polynomial*. This polynomial vanishes on classical knots, but is remarkably successful at detecting virtual knots and links. In fact $G_K(s, t)$ is the same as the polynomial invariant of virtuals of Sawollek [126] and defined by an alternative method by Silver and Williams [129]. It is a reformulation of the invariant for knots in surfaces due to the principal investigator, Jaeger and Saleur [59,60].

We end this discussion of the Alexander Biquandle with two examples that show clearly its limitations (view Figure 58). In this figure we illustrate two diagrams labeled K and KI. It is not hard to calculate that both $G_K(s, t)$ and $G_{KI}(s, t)$ are equal to zero. However, The Alexander Biquandle of K is non-trivial – it is isomorphic to the free module over $Z[s, s^{-1}, t, t^{-1}]$ generated by elements a and b subject to the relation $(s^{-1} - t - 1) \times (a - b) = 0$. Thus K represents a non-trivial virtual knot. This shows that it is possible for a non-trivial virtual diagram to be a connected sum of two trivial virtual diagrams. However, the diagram KI has a trivial Alexander Biquandle. In fact the diagram KI, discovered by Kishino [19], is now known to be knotted and its general biquandle is non-trivial. The Kishino diagram has been shown non-trivial by a calculation of the three-strand Jones polynomial [86], by the surface bracket polynomial of Dye and Kauffman [28,29], by the Ξ-polynomial (the surface generalization of the Jones polynomial of Manturov [106], and its biquandle has been shown to be non-trivial by a quaternionic biquandle representation [7] of Fenn and Bartholomew which we will now briefly describe.

Referring back to the previous section define the linear biquandle by

$$S = \begin{pmatrix} 1+i & jt \\ -jt^{-1} & 1+i \end{pmatrix},$$

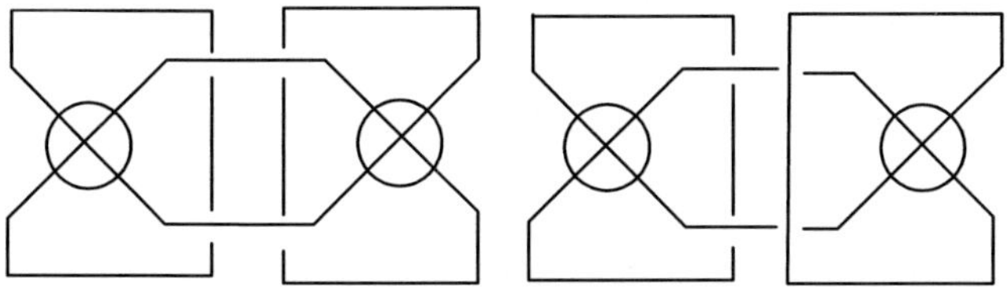

Fig. 58. The knot K and the Kishino diagram KI.

where i, j have their usual meanings as quaternions and t is a central variable. Let R denote the ring which they determine. Then as in the Alexander case considered above, for each diagram there is a square presentation of an R-module. We can take the (Study) determinant of the presentation matrix. In the case of the Kishino knot this is zero. However, the greatest common divisor of the codimension 1 determinants is $2 + 5t^2 + 2t^4$ showing that this knot is not classical.

7.5.1. *Virtual quandles.* There is another generalization of quandle [100] by means of which one can obtain the same polynomial as in [126,129] from the other point of view [104,105]. Namely, the formalism is the same as in the case of quandles at classical crossings but one adds a special structure at virtual crossings. The fact that these approaches give the same result in the linear case was proved recently by Roger Fenn and Andrew Bartholomew.

Virtual quandles (as well as biquandles) yield generalizations of the fundamental group and some other invariants. Also, virtual biquandles admit a generalization for multi-variable polynomials in the case of multicomponent links, see [104]. One can extend these definitions by bringing together the virtual quandle (at virtual crossings) and the biquandle (at classical crossings) to obtain what is called a *virtual biquandle*; this work is now in process [82].

7.6. *A Quantum model for $G_K(s,t)$, oriented and bi-oriented quantum algebras*

We can understand the structure of the invariant $G_K(s,t)$ by rewriting it as a quantum invariant and then analysing its state summation. The quantum model for this invariant is obtained in a fashion analogous to the construction of a quantum model of the Alexander polynomial in [59,60]. The strategy in those papers was to take the basic two-dimensional matrix of the Burau representation, view it as a linear transformation $T : V \longrightarrow V$ on a two-dimensional module V, and then take the induced linear transformation $\hat{T} : \Lambda^* V \longrightarrow \Lambda^* V$ on the exterior algebra of V. This gives a transformation on a four-dimensional module that is a solution to the Yang–Baxter equation. This solution of the Yang–Baxter equation then becomes the building block for the corresponding quantum invariant. In the present instance, we have a generalization of the Burau representation, and this same procedure can be applied to it.

The normalized state summation $Z(K)$ obtained by the above process satisfies a skein relation that is just like that of the Conway polynomial: $Z(K_+) - Z(K_-) = zZ(K_0)$. The basic result behind the correspondence of $G_K(s,t)$ and $Z(K)$ is the following theorem.

THEOREM [76]. *For a (virtual) link K, the invariants* $Z(K)(\sigma = \sqrt{s}, \tau = 1/\sqrt{t})$ *and* $G_K(s,t)$ *are equal up to a multiple of* $\pm s^n t^m$ *for integers n and m (this being the well-definedness criterion for G).*

It is the purpose of this section to place our work with the generalized Alexander polynomial in a context of bi-oriented quantum algebras and to introduce the concept of an oriented quantum algebra. In [74,75] Kauffman and Radford introduce the concept and show that *oriented quantum algebras encapsulate the notion of an oriented quantum link invariant.*

An *oriented quantum algebra* (A, ρ, D, U) is an abstract model for an oriented quantum invariant of classical links [74,75]. This model is based on a solution to the Yang–Baxter equation. The definition of an oriented quantum algebra is as follows: we are given an algebra A over a base ring k, an invertible solution ρ in $A \otimes A$ of the Yang–Baxter equation (in the algebraic formulation of this equation – differing from a braiding operator by a transposition), and commuting automorphisms $U, D : A \to A$ of the algebra, such that

$$(U \otimes U)\rho = \rho,$$

$$(D \otimes D)\rho = \rho,$$

$$[(1_A \otimes U)\rho)][(D \otimes 1_{A^{\mathrm{op}}})\rho^{-1}] = 1_{A \otimes A^{\mathrm{op}}},$$

and

$$[(D \otimes 1_{A^{\mathrm{op}}})\rho^{-1}][(1_A \otimes U)\rho)] = 1_{A \otimes A^{\mathrm{op}}}.$$

The last two equations say that $[(1_A \otimes U)\rho)]$ and $[(D \otimes 1_{A^{\mathrm{op}}})\rho^{-1}]$ are inverses in the algebra $A \otimes A^{\mathrm{op}}$ where A^{op} denotes the opposite algebra.

When $U = D = T$, then A is said to be *balanced.* In the case where D is the identity mapping, we call the oriented quantum algebra *standard.* In [75] we show that *the invariants defined by Reshetikhin and Turaev (associated with a quasi-triangular Hopf algebra) arise from standard oriented quantum algebras.* It is an interesting structural feature of algebras that we have elsewhere [61] called *quantum algebras* (generalizations of quasi-triangular Hopf algebras) that they give rise to standard oriented quantum algebras. We are continuing research on the relationships of quantum link invariants and oriented quantum algebras. In particular we are working on the reformulation of existing invariants such as the Links–Gould invariant [68] that are admittedly powerful but need a deeper understanding both topologically and algebraically.

We now extend the concept of oriented quantum algebra by adding a second solution to the Yang–Baxter equation γ that will take the role of the virtual crossing.

DEFINITION. A *bi-oriented quantum algebra* is a quintuple (A, ρ, γ, D, U) such that (A, ρ, D, U) and (A, γ, D, U) are oriented quantum algebras and γ satisfies the following properties:

(1) $\gamma_{12}\gamma_{21} = 1_{A\otimes A}$ (This is the equivalent to the statement that the braiding operator corresponding to γ is its own inverse.)
(2) Mixed identities involving ρ and γ are satisfied. These correspond to the braiding versions of the virtual detour move of type three that involves two virtual crossings and one real crossing. See [76] for the details.

By extending the methods of [75], it is not hard to see that a *bi-oriented quantum algebra will always give rise to invariants of virtual links up to the type one moves (framing and virtual framing).*

In the case of the generalized Alexander polynomial, the state model $Z(K)$ translates directly into a specific example of a bi-oriented balanced quantum algebra (A, ρ, γ, T). The main point about this bi-oriented quantum algebra is that the operator γ for the virtual crossing is *not* the identity operator; this non-triviality is crucial to the structure of the invariant. We will investigate bi-oriented quantum algebras and other examples of virtual invariants derived from them.

We have taken a path to explain not only the evolution of a theory of invariants of virtual knots and links, but also (in this section) a description of our oriented quantum algebra formulation of the whole class of quantum link invariants. Returning to the case of the original Jones polynomial, we want to understand its capabilities in terms of the oriented quantum algebra that generates the invariant.

7.7. *Invariants of three-manifolds*

As is well-known, invariants of three-manifolds can be formulated in terms of Hopf algebras and quantum algebras and spin recoupling networks. In formulating such invariants it is useful to represent the three-manifold via surgery on a framed link. Two framed links that are equivalent in the Kirby calculus of links represent the same three-manifold and conversely. To obtain invariants of three-manifolds one constructs invariants of framed links that are also invariant under the Kirby moves (handle sliding, blowing up and blowing down).

A classical three-manifold is mathematically the same as a Kirby equivalence class of a framed link. The fundamental group of the three-manifold associated with a link is equal to the fundamental group of the complement of the link modulo the subgroup generated by the framing longitudes for the link. We refer to the fundamental group of the three-manifold as the *three-manifold group*. If there is a counterexample to the classical Poincaré conjecture, then the counterexample would be represented by surgery on some link L whose three-manifold group is trivial, but L is not trivial in Kirby calculus (i.e., it cannot be reduced to nothing).

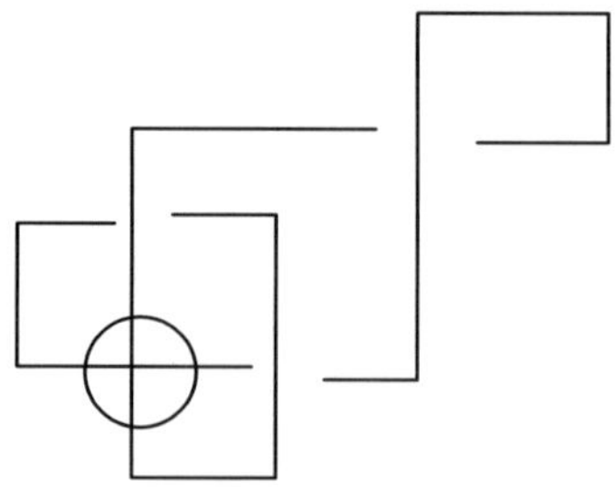

Fig. 59. A counterexample to the Poincaré conjecture for virtual three-manifolds.

Kirby calculus can be generalized to the class of virtual knots and links. We define a *virtual three-manifold* to be a Kirby equivalence class of framed virtual links. The three-manifold group generalizes via the combinatorial fundamental group associated to the virtual link (the framing longitudes still exist for virtual links). The *Virtual Poincaré Conjecture* to virtuals would say that a virtual three-manifold with trivial fundamental group is trivial in Kirby calculus. However, the virtual Poincaré conjecture is false [30]. There exist virtual links whose three-manifold group is trivial that are nevertheless not Kirby equivalent to nothing. The simplest example is the virtual knot in Figure 59. We detect the non-triviality of the Kirby class of this knot by computing that it has an SU(2) Witten invariant that is different from the standard three-sphere.

This counterexample to the Poincaré conjecture in the virtual domain shows how a classical counterexample might behave in the context of Kirby calculus. Virtual knot theory can be used to search for a counterexample to the classical Poincaré conjecture by searching for virtual counterexamples that are equivalent in Kirby calculus to classical knots and links. This is a new and exciting approach to the dark side of the classical Poincaré conjecture.

7.8. *Gauss diagrams and Vassiliev invariants*

The reader should recall the notion of a *Gauss diagram* for a knot. If K is a knot diagram, then $G(K)$ its Gauss diagram, is a circle comprising the Gauss code of the knot by arranging the traverse of the diagram from crossing to crossing along the circle and putting an arrow (in the form of a chord of the circle) between the two appearances of the crossing. The arrow points from the overcrossing segment to the undercrossing segment in the order of the traverse of the diagram. (Note: Turaev uses another convention [141].) Each chord is endowed with a sign that is equal to the sign of the corresponding crossing in the knot diagram. At the level of the Gauss diagrams, a virtual crossing is simply the absence of a chord. That is, if we wish to transcribe a virtual knot diagram to a Gauss diagram, we ignore the virtual crossings. Reidemeister moves on Gauss diagrams are defined by translation from the corresponding diagrams from planar representation. Virtual knot theory is precisely the theory of *arbitrary Gauss diagrams*, up to the Gauss diagram

Reidemeister moves. Note that an arbitrary Gauss diagram is any pattern of directed, signed chords on an oriented circle. When transcribed back into a planar knot diagram, such a Gauss diagram may require virtual crossings for its depiction.

In [119] Goussarov, Polyak and Viro initiate a very important program for producing Vassiliev invariants of finite type of virtual and classical knots. The gist of their program is as follows. They define the notion of a semi-virtual crossing, conceived as a dotted, oriented, signed chord in a Gauss diagram for a knot. An *arrow diagram* is a Gauss diagram all of whose chords are dotted. Let A denote the collection of all linear combinations of arrow diagrams with integer coefficients. Let G denote the collection of all arbitrary Gauss diagrams (hence all representatives of virtual knots). Define a mapping

$$i : \mathcal{G} \rightarrow \mathcal{A}$$

by expanding each chord of a Gauss diagram G into the sum of replacing the chord by a dotted chord and the removal of that chord. Thus

$$i(G) = \sum_{r \in R(G)} G^r,$$

where $R(G)$ denotes all ways of replacing each chord in G either by a dotted chord, or by nothing; and G^r denotes that particular replacement applied to G.

Now let $\mathcal{P}$ denote the quotient of $\mathcal{A}$ by the subalgebra generated by the relations in $\mathcal{A}$ corresponding to the Reidemeister moves. Each Reidemeister move is of the form $X = Y$ for certain diagrams, and this translates to the relation $i(X) - i(Y) = 0$ in $\mathcal{P}$, where $i(X)$ and $i(Y)$ are individually certain linear combinations in P. Let

$$I : \mathcal{G} \rightarrow \mathcal{P}$$

be the map induced by i. Then it is a formal fact that $I(G)$ is invariant under each of the Reidemeister moves, and hence that $I(G)$ is an invariant of the corresponding Gauss diagram (virtual knot) G. The algebra of relations that generate the image of the Reidemeister moves in $\mathcal{P}$ is called the *Polyak algebra.*

So far, we have only desribed a tautological and not a computable invariant. The key to obtaining computable invariants is to truncate. Let $\mathcal{P}_n$ denote $\mathcal{P}$ modulo all arrow diagrams with more than n dotted arrows. Now $\mathcal{P}_n$ is a finitely generated module over the integers, and the composed map

$$I_n : \mathcal{G} \rightarrow \mathcal{P}_n$$

is also an invariant of virtual knots. Since we can choose a specific basis for $\mathcal{P}_n$, the invariant I_n is in principle computable, and it yields a large collection of Vassiliev invariants of virtual knots that are of finite type. The paper by Goussarov, Polyak and Viro investigates specific methods for finding and representing these invariants. They show that every Vassiliev invariant of finite type for classical knots can be written as a combinatorial state sum for long knots. They use the virtual knots as an intermediate in the construction.

By directly constructing Vassiliev invariants of virtual knots from known invariants of virtuals, we can construct invariants that are not of finite type in the above sense (see [70]). These invariants also deserve further investigation.

8. Other invariants

The *Khovanov Categorification of the Jones polynomial* [6] is important for our concerns. This invariant is constructed by promoting the states in the bracket summation to tensor powers of a vector space V, where a single power of V corresponds to a single loop in the state. In this way a graded complex is constructed, whose graded Euler characteristic is equal to the original Jones polynomial, and the ranks of whose graded homology groups are themselves invariants of knots. It is now known that the information in the Khovanov construction exceeds that in the original Jones polynomial. It is an open problem whether a Khovanov type construction can generalize to virtual knots in the general case. The construction for the Khovanov polynomial for virtuals over $\mathbf{Z}_2$ was proposed in [102]. Recent work by other authors related to knots in thickened surfaces promises to shed light on this issue.

One of the more promising directions for relating Vassiliev invariants to our present concerns is the theory of gropes [20,21], where one considers surfaces spanning a given knot, and then recursively the surfaces spanning curves embedded in the given surface. This hierarchical structure of curves and surfaces is likely to be a key to understanding the geometric underpinning of the original Jones polynomial. The same techniques in a new guise could elucidate invariants of virtual knots and links.

9. The bracket polynomial and the Jones polynomial

It is an open problem whether there exist classical knots (single component loops) that are knotted and yet have unit Jones polynomial. In other words, it is an open problem whether the Jones polynomial can detect all knots. There do exist families of links whose linkedness is undectable by the Jones polynomial [136,137]. It is the purpose of this section of the paper to give a summary of some of the information that is known in this arena. We begin with a sketch of ways to calculate the bracket polynomial model of the Jones polynomial, and then discuss how to construct classical links that are undetectable by the Jones polynomial.

The formula for the bracket model of the Jones polynomial [52] can be indicated as follows: The letter chi, χ, denotes a crossing in a link diagram. The barred letter denotes the mirror image of this first crossing. A crossing in a diagram for the knot or link is expanded into two possible states by either smoothing (reconnecting) the crossing horizontally, $\asymp$, or vertically $><$. Any closed loop (without crossings) in the plane has value $\delta = -A^2 - A^{-2}$.

$$\chi = A \asymp + A^{-1} ><,$$

$$\bar{\chi} = A^{-1} \asymp + A ><.$$

One useful consequence of these formulas is the following *switching formula*

$$A\chi - A^{-1}\bar{\chi} = (A^2 - A^{-2}) \asymp.$$

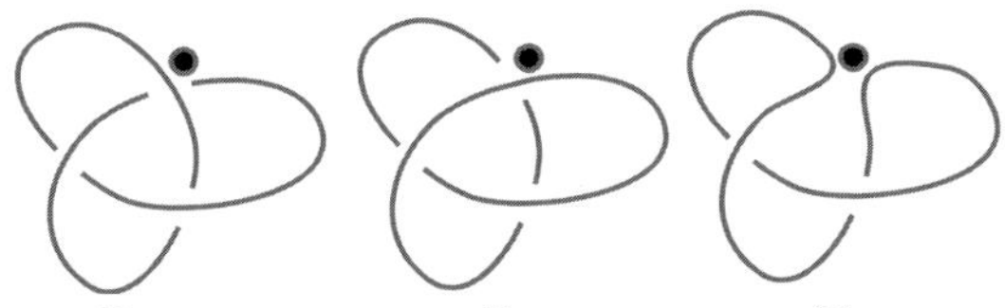

Fig. 60. Trefoil and two relatives.

Note that in these conventions the A-smoothing of χ is $\asymp$, while the A-smoothing of $\bar{\chi}$ is $><$. Properly interpreted, the switching formula above says that you can switch a crossing and smooth it either way and obtain a three-diagram relation. This is useful since some computations will simplify quite quickly with the proper choices of switching and smoothing. Remember that it is necessary to keep track of the diagrams up to regular isotopy (the equivalence relation generated by the second and third Reidemeister moves). Here is an example (view Figure 60).

You see in Figure 60, a trefoil diagram K, an unknot diagram U and another unknot diagram U' Applying the switching formula, we have

$$A^{-1}K - AU = (A^{-2} - A^2)U',$$

and $U = -A^3$ and $U' = (-A^{-3})^2 = A^{-6}$. Thus

$$A^{-1}K - A(-A^3) = (A^{-2} - A^2)A^{-6}.$$

Hence

$$A^{-1}K = -A^4 + A^{-8} - A^{-4}.$$

Thus

$$K = -A^5 - A^{-3} + A^{-7}.$$

This is the bracket polynomial of the trefoil diagram K. We have used to same symbol for the diagram and for its polynomial.

Since the trefoil diagram K has writhe $w(K) = 3$, we have the normalized polynomial

$$f_K(A) = (-A^3)^{-3}\langle K\rangle = -A^{-9}(-A^5 - A^{-3} + A^{-7}) = A^{-4} + A^{-12} - A^{-16}.$$

The asymmetry of this polynomial under the interchange of A and A^{-1} proves that the trefoil knot is not ambient isotopic to its mirror image.

In Figure 61 you see the knot $K = N_{42} = 9_{42}$ (the latter being its standard name in the knot tables) and a skein tree for it via switching and smoothing. In Figure 62 we show simplified (via regular isotopy) representatives for the end diagrams in the skein tree.

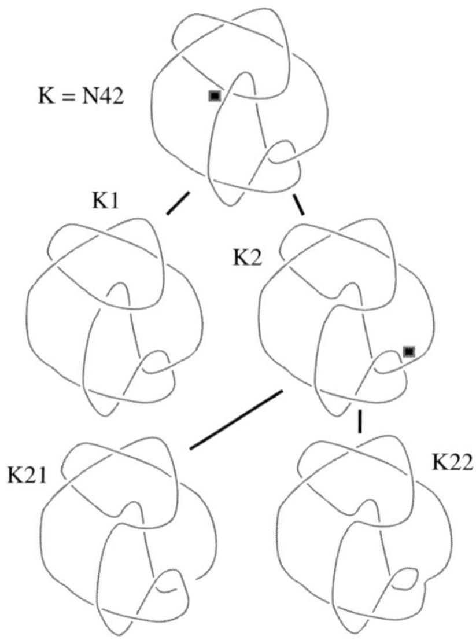

Fig. 61. Skein tree for 9_{42}.

It follows from the switching formula that for $K = 9_{42}$,

$$A^{-1}K - AK_1 = (A^{-2} - A^2)K_2,$$

$$AK_2 - A^{-1}K_{21} = (A^2 - A^{-2})K_{22},$$

and that K_1 is a connected sum of a right-handed trefoil diagram and a figure eight knot diagram, while K_{21} is a Hopf link (simple link of linking number one) with extra writhe of -2 while K_{22} is an unknot with writhe of 1. These formulas combine to give

$$\langle K_1 \rangle = -A^{-9} + A^{-5} - A^{-1} + A^3 - A^7 + A^{11} - A^{15}.$$

Since K has writhe one, we get

$$f_K = A^{-12} - A^{-8} + A^{-4} - 1 + A^4 - A^8 + A^{12}.$$

This shows that the normalized bracket polynomial does not distinguish 9_{42} from its mirror image. This knot is, in fact chiral (inequivalent to its mirror image), a fact that can be verified by other means. The knot 9_{42} is the first chiral knot whose chirality is undetected by the Jones polynomial.

REMARK. In writing computer programs for calculating the bracket polynomial it is useful to use the coding method illustrated in Figure 63. In this method each edge from one classical crossing to another has a label. Virtual crossings do not appear in the code.

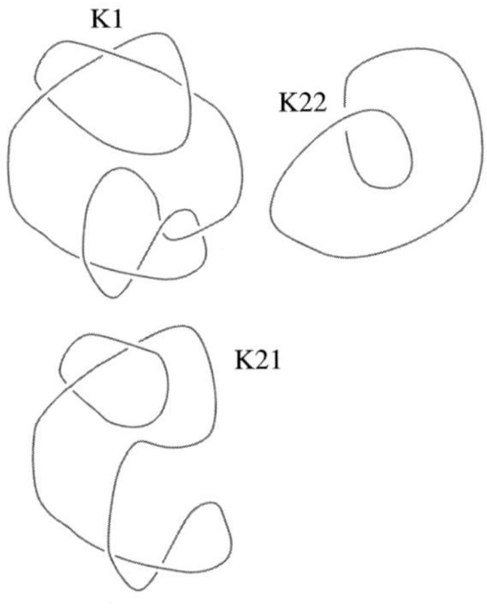

Fig. 62. Regular isotopy versions of bottom of skein tree for 9_{42}.

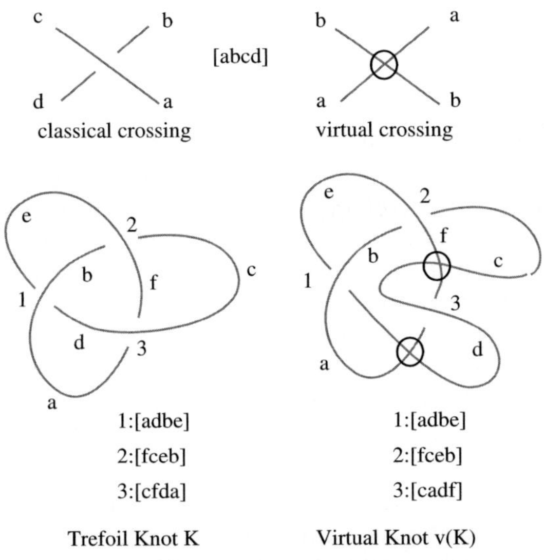

Fig. 63. Coding a link diagram.

Each classical crossing has a four-letter code in the form [*abcd*] connoting a clockwise encirclement of the crossing, starting at an overcrossing edge.

9.1. *Thistlethwaite's example*

View Figure 64. Here we have a version of a link L discovered by Morwen Thistlethwaite [136] in December 2000. We discuss some theory behind this link in the next section.

Fig. 64. Thistlethwaite's link.

It is a link that is linked but whose linking is not detectable by the Jones polynomial. One can verify such properties by using a computer program, or by the algebraic techniques described below.

9.2. *Present status of links not detectable by the Jones polynomial*

In this section we give a quick review of the status of our work [137] producing infinite families of distinct links all evaluating as unlinks by the Jones polynomial.

A tangle (2-tangle) consists in an embedding of two arcs in a three-ball (and possibly some circles embedded in the interior of the three-ball) such that the endpoints of the arcs are on the boundary of the three-ball. One usually depicts the arcs as crossing the boundary transversely so that the tangle is seen as the embedding in the three-ball augmented by four segments emanating from the ball, each from the intersection of the arcs with the boundary. These four segments are the *exterior edges* of the tangle, and are used for operations that form new tangles and new knots and links from given tangles. Two tangles in a given three-ball are said to be *topologically equivalent* if there is an ambient isotopy from one to the other in the given three-ball, fixing the intersections of the tangles with the boundary.

It is customary to illustrate tangles with a diagram that consists in a box (within which are the arcs of the tangle) and with the exterior edges emanating from the box in the NorthWest (NW), NorthEast (NE), SouthWest (SW) and SouthEast (SE) directions. Given tangles T and S, one defines the *sum*, denoted $T + S$ by placing the diagram for S to the right of the diagram for T and attaching the NE edge of T to the NW edge of S, and the SE edge of T to the SW edge of S. The resulting tangle $T + S$ has exterior edges correponding to the NW and SW edges of T and the NE and SE edges of S. There are two ways to create links associated to a tangle T. The *numerator* T^{N} is obtained, by attaching the (top) NW and NE edges of T together and attaching the (bottom) SW and SE edges together. The denominator T^{D} is obtained, by attaching the (left side) NW and SW edges together and attaching the (right side) NE and SE edges together. We denote by [0] the tangle with only unknotted arcs (no embedded circles) with one arc connecting, within the three-ball, the (top points) NW intersection point with the NE intersection point, and the other arc connnecting the (bottom points) SW intersection point with the SE intersection point. A 90° turn of the tangle [0] produces the tangle [∞]with connections between NW and SW and between NE and SE. One then can prove the basic formula for any tangle T

$$\langle T\rangle = \alpha_T\langle[0]\rangle + \beta_T\langle[\infty]\rangle,$$

where α_T and β_T are well-defined polynomial invariants (of regular isotopy) of the tangle T. From this formula one can deduce that

$$\langle T^N\rangle = \alpha_T d + \beta_T,$$

and

$$\langle T^D\rangle = \alpha_T + \beta_T d.$$

We define the *bracket vector* of T to be the ordered pair (α_T, β_T) and denote it by $\mathrm{br}(T)$, viewing it as a column vector so that $\mathrm{br}(T)^t = (\alpha_T, \beta_T)$ where v^t denotes the transpose of the vector v. With this notation the two formulas above for the evaluation for numerator and denominator of a tangle become the single matrix equation

$$\begin{bmatrix} \langle T^N\rangle \\ \langle T^D\rangle \end{bmatrix} = \begin{bmatrix} d & 1 \\ 1 & d \end{bmatrix} \mathrm{br}(T).$$

We then use this formalism to express the bracket polynomial for our examples. The class of examples that we considered are each denoted by $H(T, U)$ where T and U are each tangles and $H(T, U)$ is a satellite of the Hopf link that conforms to the pattern shown in Figure 65, formed by clasping together the numerators of the tangles T and U. Our method is based on a transformation $H(T, U) \longrightarrow H(T, U)^\omega$, whereby the tangles T and U are cut out and reglued by certain specific homeomorphisms of the tangle boundaries. This transformation can be specified by a modification described by a specific rational tangle and its mirror image. Like mutation, the transformation ω preserves the bracket polynomial. However, it is more effective than mutation in generating examples, as a trivial link can be transformed to a prime link, and repeated application yields an infinite sequence of inequivalent links.

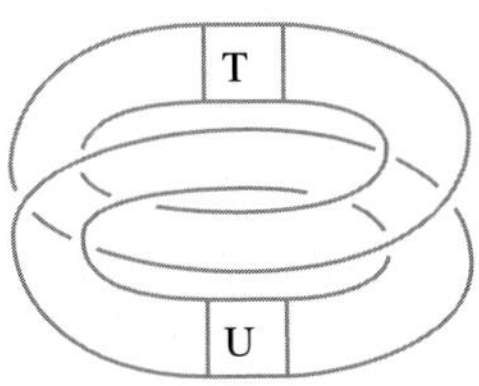

Fig. 65. Hopf link satellite $H(T, U)$.

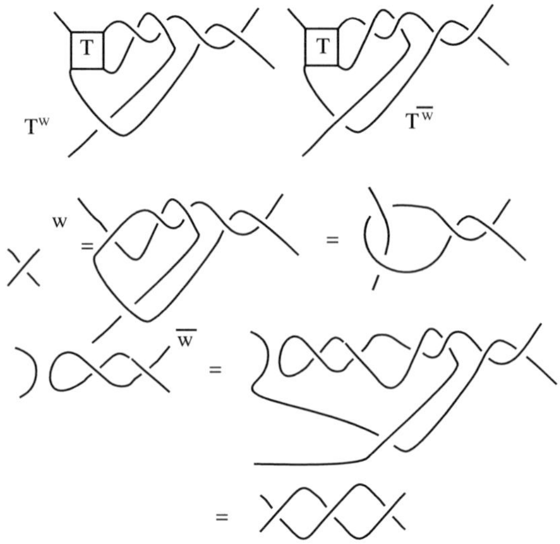

Fig. 66. The omega operations.

Specifically, the transformation $H(T,U)^{\omega}$ is given by the formula

$$H(T,U)^{\omega} = H(T^{\omega}, U^{\bar{\omega}}),$$

where the tangle operations T^{ω} and $U^{\bar{\omega}}$ are as shown in Figure 66. By direct calculation, there is a matrix M such that

$$\langle H(T,U)\rangle = \mathrm{br}(T)^{t}\mathrm{M}\mathrm{br}(U),$$

and there is a matrix Ω such that

$$\mathrm{br}(T^{\omega}) = \Omega\mathrm{br}(T),$$

and

$$\mathrm{br}(T^{\bar{\omega}}) = \Omega^{-1}\mathrm{br}(T).$$

One verifies the identity

$$\Omega^{t}M\Omega^{-1} = M,$$

from which it follows that $\langle H(T,U)\rangle^{\omega} = \langle H(T,U)\rangle$. This completes the sketch of our method for obtaining links that whose linking cannot be seen by the Jones polynomial. Note that the link constructed as $H(T^{\omega}, U^{\bar{\omega}})$ in Figure 67 has the same Jones polynomial as an unlink of two components. This shows how the first example found by Thistlethwaite fits into our construction.

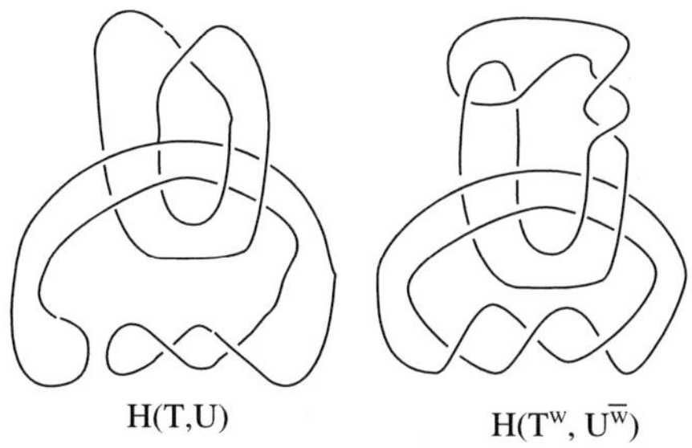

Fig. 67. Applying omega operations to an unlink.

9.3. *Switching a crossing*

If in Figure 67, we start with T replaced by Flip(T), switching the crossing, the resulting link $L = H(\mathrm{Flip}(T)^{\omega}, U^{\bar{\omega}})$ will still have Jones polynomial the same as the unlink, but the link L will be distinct from the link $H(T^{\omega}, U^{\bar{\omega}})$ of Figure 67. We illustrate this process in Figure 68.

The link L has the remarkable property that both it and the link obtained from it by flipping the crossing labeled A in Figure 68 have Jones polynomial equal to the Jones polynomial of the unlink of two components. (We thank Alexander Stoimenow for pointing out the possibility of this sort of construction.)

Now let us think about a link L with the property that it has the same Jones polynomial as a link L' obtained from L by switching a single crossing. We can isolate the rest of the link that is not this crossing into a tangle S so that (without any loss of generality) $L = N(S + [1])$ and $L' = N(S + [-1])$. Let us assume that orientation assignment to L and L' is as shown in Figure 69.

Since we are told that L and L' have the same Jones polynomial, it follows that $\langle L\rangle = -A^{-3}\kappa$ and $\langle L'\rangle = -A^{3}\kappa$ for some non-zero Laurent polynomial κ. Now suppose that

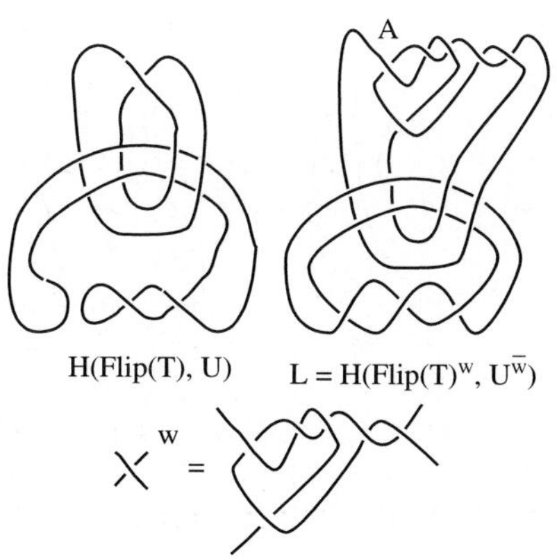

Fig. 68. Applying omega operations to an unlink with flipped crossing.

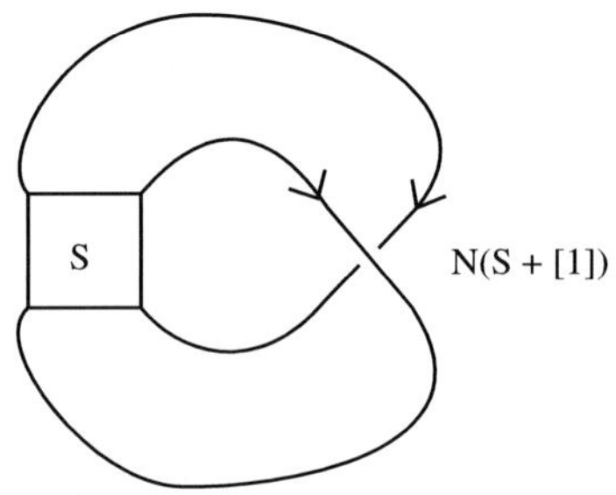

Fig. 69. $N(S+[1])$.

$\langle S\rangle = \alpha\langle[0]\rangle + \beta\langle[\infty]\rangle$. Then

$$\langle L\rangle = \alpha(-A^{3}) + \beta(-A^{-3}),$$

and

$$\langle L'\rangle = \alpha(-A^{-3}) + \beta(-A^{3}).$$

From this it follows that

$$\kappa = \alpha A^{6} + \beta,$$

and

$$\kappa = \alpha A^{-6} + \beta.$$

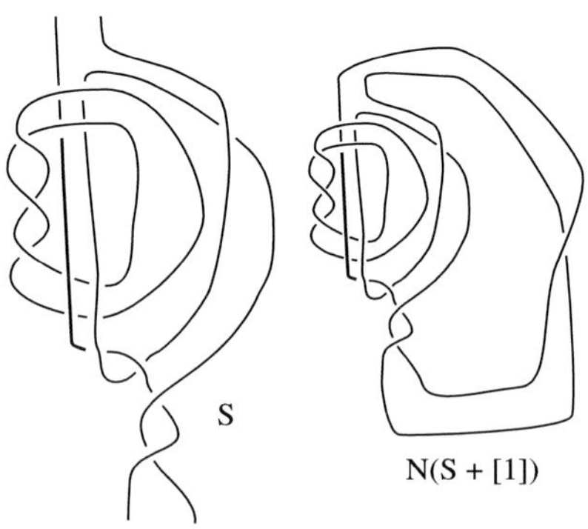

N(S + [1]) and N(S+[–1]) both have Jones poly same as the unlink of two components.

Fig. 70. The tangle S.

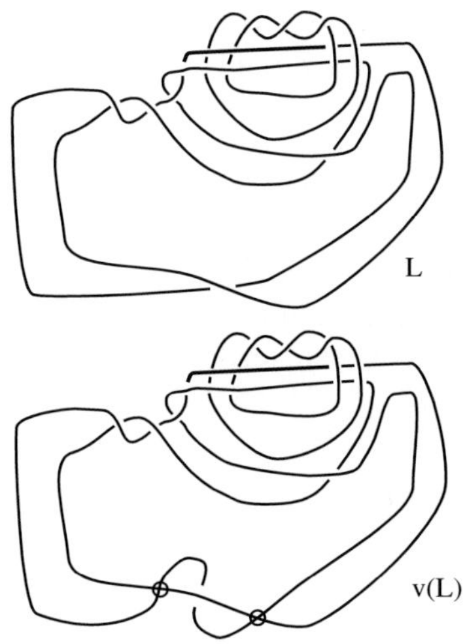

Fig. 71. The virtual link $v(L)$.

Thus

$$\alpha = 0,$$

and

$$\beta = \kappa.$$

We have shown that

$$\langle S \rangle = \kappa \langle [\infty] \rangle.$$

This means that we can, by using the example described above, produce a tangle S that is not splittable and yet has the above property of having one of its bracket coefficients equal to zero. The example is shown in Figure 70.

Finally, in Figure 71 we show L and the link $v(L)$ obtained by virtualizing the crossing corresponding to [1] in the decomposition $L = N(S + [1])$. The virtualized link $v(L)$ has the property that it also has Jones polynomial the same as an unlink of two components. We wish to prove that $v(L)$ is not isotopic to a classical link. The example has been designed so that surface bracket techniques will be difficult to apply.

Acknowledgements

Most of this effort was sponsored by the Defense Advanced Research Projects Agency (DARPA) and Air Force Research Laboratory, Air Force Materiel Command, USAF, under agreement F30602-01-2-05022. Some of this effort was also sponsored by the National

Institute for Standards and Technology (NIST). The US Government is authorized to reproduce and distribute reprints for Government purposes notwithstanding any copyright annotations thereon. The views and conclusions contained herein are those of the authors and should not be interpreted as necessarily representing the official policies or endorsements, either expressed or implied, of the Defense Advanced Research Projects Agency, the Air Force Research Laboratory, or the US Government. (Copyright 2004.) It gives the author great pleasure to acknowledge support from NSF Grant DMS-0245588. Along with the specified references within the text, we have also included a number of references that will be of further interest to readers of the paper. These are: 5, 11, 13, 15, 17, 18, 23, 26, 27, 33, 34–36, 39, 42, 45–51, 55, 69, 73, 77, 78, 80, 83, 87, 88, 96, 98, 103, 107–111, 114–117, 127, 133–135, 139, 143, 147.

References

[1] D. Altshuler and L. Friedel, *Vassiliev knot invariants and Chern–Simons perturbation theory to all orders*, Comm. Math. Phys. **187** (2) (1997), 261–287.

[2] M.F. Atiyah, *Geometry of Yang–Mills Fields*, Accademia Nazionale dei Lincei Scuola Superiore Lezioni Fermiare, Pisa (1979).

[3] M.F. Atiyah, *The Geometry and Physics of Knots*, Cambridge University Press, Cambridge (1990).

[4] D. Bar-Natan, *On the Vassiliev knot invariants*, Topology, **34** (2) (1995), 423–472.

[5] D. Bar-Natan, *Perturbative Aspects of the Chern–Simons Topological Quantum Field Theory*, PhD thesis, Princeton University (June 1991).

[6] D. Bar-Natan, *On Khovanov's categorification of the Jones polynomial*, Algebr. Geom. Topol. **2** (2002), 337–370.

[7] A. Bartholomew and R. Fenn, *Quaternionic invariants of virtual knots and links*, www.maths.sussex.ac.uk/Staff/RAF/Maths/Current/Andy/equivalent.ps, JLMS, to appear.

[8] S. Budden and R. Fenn, The equation $[B, (A-1)(A, B)] = 0$ and virtual knots and link, preprint available from R. Fenn's website.

[9] R.J. Baxter, *Exactly Solved Models in Statistical Mechanics*, Academic Press, London (1982).

[10] J. Birman and X.S. Lin, *Knot polynomials and Vassiliev's invariants*, Invent. Math. **111** (2) (1993), 225–270.

[11] R. Bott and C. Taubes, *On the self-linking of knots*, J. Math. Phys. **35** (1994), 5247–5287.

[12] P. Cartier, *Construction combinatoire des invariants de Vassiliev – Kontsevich des noeuds*, C.R. Acad. Sci. Paris Série I **316** (1993), 1205–1210.

[13] J.S. Carter, D.E. Flath and M. Saito, *The Classical and Quantum 6j-Symbols*, Mathematical Notes 43, Princeton University Press, Princeton, NJ (1995).

[14] J.S. Carter, L.H. Kauffman and M. Saito, *Diagrammatics, singularities and their algebraic interpretations*. 10th Brazilian Topology Meeting (So Carlos, 1996), Mat. Contemp. 13 (1997), 21–115.

[15] J.S. Carter, L.H. Kauffman and M. Saito, *Structures and diagrammatics of four-dimensional topological lattice field theories*, Adv. Math. **146** (1) (1999), 39–100.

[16] J.S. Carter and M. Saito, Diagrammatic invariants of knotted curves and surfaces (1992), unpublished manuscript.

[17] J.S. Carter and M. Saito, *Knotted Surfaces and Their Diagrams*, Mathematical Surveys and Monographs, 55, American Mathematical Society, Providence, RI (1998).

[18] J.S. Carter, S. Kamada and M. Saito, *Geometric interpretations of quandle homology*, J. Knot Theor. Ramif. **10** (3) (2001), 345–386.

[19] J.S. Carter and D. Silver, private conversation.

[20] J. Conant and P. Teichner, *Grope cobordism of classical knots*, Topology **43** (1) (2004), 119–156.

[21] J. Conant and P. Teichner, *Grope cobordism and Feynman diagrams*, Math. Ann. **328** (1/2) (2004), 135–171.

[22] J.H. Conway, private conversation.

[23] J.H. Conway, *An enumeration of knots and links and some of their algebraic properties*, Computational Problems in Abstract Algebra, Pergammon Press, New York (1970), 329–358.

[24] L. Crane and I. Frenkel, *Four-dimensional topological quantum field theory, Hopf categories and canonical bases*, J. Math. Phys. **35** (1994), 5136–5154.

[25] L. Crane, L.H. Kauffman and D. Yetter, *State sum invariants of 4-manifolds*, J. Knot Theor. Ramif. **6** (2) (1997), 177–234.

[26] J.W. Barrett and L. Crane, *A Lorentzian signature model for quantum general relativity*, Classical Quantum Gravity **17** (16) (2000), 3101–3118.

[27] P. Dehornoy, *Braids and Self-Distributivity*, Prog. Math. **192**, Birkhauser (2000).

[28] H. Dye, *Characterizing Virtual Knots*, PhD thesis (2002), UIC.

[29] H. Dye and L.H. Kauffman, Minimal surface representations of virtual knots and links. arXiv:math.AT/0401035 v1, January 2004.

[30] H. Dye and L.H. Kauffman, Virtual knot diagrams and the Witten–Reshetikhin–Turaev invariant, www.arxiv.org/math.GT/0407407, J. Knot Theor. Ramif., to appear.

[31] I.A. Dynnikov, Arc presentations of links – Monotonic simplification, www.arxiv.org/math.GT/0208153 v2 8 September 2003.

[32] R. Fenn, M. Jordan and L.H. Kauffman, *Biracks, biquandles and virtual knots*, Topol. Appl., **145** (1–3) (2004), 157–175.

[33] R. Fenn and C. Rourke, *Racks and links in codimension two*, J. Knot Theor. Ramif. **4** (1992), 343–406.

[34] R. Fenn, R. Rimanyi and C. Rourke, *The Braid permutation group*, Topology **36** (1) (1997), 123–135.

[35] R. Fenn and P. Taylor, Introducing doodles, Lecture Notes in Maths. LMS No. 722 (1977), 37–43.

[36] R. Fenn, *Techniques of Geometric Topology*, LMS Lecture Notes Ser. 57 (1983).

[37] S. Garoufalidis, Applications of TQFT to invariants in low dimensional topology (preprint 1993).

[38] D.S. Freed and R.E. Gompf, *Computer calculation of Witten's three-manifold invariant*, Commun. Math. Phys. **141** (1991), 79–117.

[39] B. Hasslacher and M.J. Perry, *Spin networks are simplicial quantum gravity*, Phys. Lett. **103B** (1981).

[40] M.A. Hennings, *Hopf algebras and regular isotopy invariants for link diagrams*, Math. Proc. Camb. Phil. Soc. **109** (1991), 59–77.

[41] D. Hrencecin, *On Filamentations and Virtual Knot Invariants*, PhD thesis, Unviversity of Illinois at Chicago (2001).

[42] D. Hrencecin and L.H. Kauffman, *On filamentations and virtual knots*, Topol. Appl. **134** (2003), 23–52.

[43] L.C. Jeffrey, *Chern–Simons–Witten invariants of lens spaces and torus bundles, and the semi-classical approximation*, Commun. Math. Phys. **147** (1992), 563–604.

[44] V.F.R. Jones, *A polynomial invariant of links via von Neumann algebras*, Bull. Am. Math. Soc. **29** (1985), 103–112.

[45] V.F.R. Jones, *Hecke algebra representations of braid groups and link polynomials*, Ann. Math. **126** (1987), 335–338.

[46] V.F.R. Jones, *On knot invariants related to some statistical mechanics models*, Pacific J. Math. **137** (2) (1989), 311–334.

[47] T. Kanenobu, *Forbidden moves unknot a virtual knot*, J. Knot Theor. Ramificat. **10** (1) (2001), 89–96.

[48] M. Khovanov, *Doodle groups*, Trans. Am. Math. Soc. **349** (6) (1997), 2297–2315.

[49] L.H. Kauffman, *The Conway polynomial*, Topology **20** (1980), 101–108.

[50] L.H. Kauffman, *Formal Knot Theory*, Lecture Notes Series 30, Princeton University Press, Princeton, NJ (1983).

[51] L.H. Kauffman, *On Knots*, Annals Study No. 115, Princeton University Press, Princeton, NJ (1987).

[52] L.H. Kauffman, *State models and the Jones polynomial*, Topology **26** (1987), 395–407.

[53] L.H. Kauffman, *Statistical mechanics and the Jones polynomial*, AMS Contemp. Math. Series **78** (1989), 263–297.

[54] L.H. Kauffman, *Map coloring, q-deformed spin networks, and Turaev–Viro invariants for 3-manifolds*, Int. J. Modern Phys. B **6** (11/12) (1992), 1765–1794.

[55] L.H. Kauffman, *An invariant of regular isotopy*, Trans. Am. Math. Soc. **318** (2) (1990), 417–471.

[56] L.H. Kauffman, *New invariants in the theory of knots*, Am. Math. Monthly **95** (3) (1988), 195–242.

[57] L.H. Kauffman and P. Vogel, *Link polynomials and a graphical calculus*, J. Knot Theor. Ramif. **1** (1) (1992).
[58] L.H. Kauffman, *Knots and Physics*, World Scientific, Singapore (1991, 1994, 2001).
[59] L.H. Kauffman and H. Saleur, *Free fermions and the Alexander–Conway polynomial*, Commun. Math. Phys. **141** (1991), 293–327.
[60] F. Jaeger, L.H. Kauffman and H. Saleur, *The Conway polynomial in R^3 and in thickened surfaces: a new determinant formulation*, J. Comb. Theory Ser. B **61** (1994), 237–259.
[61] L.H. Kauffman, *Gauss codes, quantum groups and ribbon Hopf algebras*, Rev. Math. Phys. **5** (1993), 735–773, Reprinted in [64], 551–596.
[62] L.H. Kauffman and D. Radford, *Invariants of 3-manifolds derived from finite dimensional Hopf algebras*, J. Knot Theor. Ramif. **4** (1) (1995), 131–162.
[63] L.H. Kauffman, *Knots and Diagrams*, Lectures at Knots '96, S. Suzuki, ed., World Scientific, Singapore (1997), 123–194.
[64] L.H. Kauffman, D. Radford and S. Sawin, *Centrality and the KRH invariant*, J. Knot Theor. Ramif. **7**(5) (1998), 571–624.
[65] L.H. Kauffman, *Hopf algebras and 3-manifold invariants*, J Pure Appl. Algebra **100** (1995), 73–92.
[66] L.H. Kauffman, *Spin Networks, Topology and Discrete Physics, Braid Group, Knot Theory and Statistical Mechanics*, 2nd edn., C.N. Yang and M.L. Ge, eds, World Scientific, Singapore (1994).
[67] L.H. Kauffman and S.L. Lins, *Temperley–Lieb Recoupling Theory and Invariants of 3-Manifolds*, Annals of Mathematics Study 114, Princeton University Press, Princeton, NJ (1994).
[68] L.H. Kauffman, D. De Wit and J. links, *On the Links–Gould invariant of links*, J. Knot Theor. Ramif. **8** (2) (1999), 165–199.
[69] L.H. Kauffman, *Knot theory and the heuristics of functional integration*, Physica A **281** (2000), 173–200.
[70] L.H. Kauffman, *Virtual knot theory*, Eur. J. Comb. **20** (1999), 663–690.
[71] L.H. Kauffman, *A Survey of Virtual Knot Theory*, Proceedings of Knots in Hellas '98, World Scientific, Singapore (2000), 143–202.
[72] L.H. Kauffman, *Detecting virtual knots*, Atti. Sem. Mat. Fis. Univ. Modena Supplemento al **IL** (2001), 241–282.
[73] L.H. Kauffman, *Right integrals and invariants of three-manifolds*, Proceedings of Conference in Honor of Robion Kirby's 60th Birthday, Geometry and Topology Monographs **2** (1999), 215–232.
[74] L.H. Kauffman and D.E. Radford, *Oriented quantum algebras and invariants of knots and links*, J. Algebra **246** (2001), 253–291.
[75] L.H. Kauffman and D.E. Radford, *Oriented quantum algebras, categories and invariants of knots and links*, J. Knot Theor. Ramif. **10** (7) (2001), 1047–1084.
[76] L.H. Kauffman and D.E. Radford, *Bi-oriented Quantum Algebras, and a Generalized Alexander Polynomial for Virtual Links*, Diagrammatic Morphisms and Applications, 113–140, Contemp. Math., 318, American Mathematical Society, Providence, RI (2003), San Francisco, CA, 2000.
[77] L.H. Kauffman and S. Lambropoulou, *On the classification of rational tangles*, Adv. Appl. Math. **33** (2) (2004), 199–237.
[78] L.H. Kauffman and S. Lambropoulou, *On the classification of rational knots*, L'Enseign. Math. 2 **49** (3/4) (2003), 357–410.
[79] L.H. Kauffman and S. Lambrpoulou, *Virtual Braids*, www.arxiv.org/math.GT/0407349, Fundamenta Mathematicae, to appear.
[80] L.H. Kauffman, M. Saito and M. Sullivan, *Quantum invariants of templates*, J. Knot Theor. Ramif. **12** (5) (2003), 653–681.
[81] L.H. Kauffman, Diagrammatic knot theory, in preparation.
[82] L.H. Kauffman and V.O. Manturov, *Virtual biquandles*, Fundamenta Mathematicae, to appear.
[83] L.H. Kauffman and S.J. Lomonaco, *Quantum entanglement and topological entanglement*, New J. Phys. **4** (2002), 73.1–73.18.
[84] R. Kirby, *A calculus for framed links in S^3*, Invent. Math. **45** (1978), 35–56.
[85] R. Kirby and P. Melvin, *On the 3-manifold invariants of Reshetikhin–Turaev for sl(2,C)*, Invent. Math. **105** (1991), 473–545.

[86] T. Kishino and S. Satoh, *A note on non-classical virtual knots*, J. Knot Theor. Ramif. **13** (7) (2004), 845–856.

[87] T. Kishino, 6 kouten ika no kasou musubime no bunrui ni tsiuti (On classification of virtual links whose crossing number is less than or equal to 6), Master Thesis, Osaka City University (2000).

[88] T. Kadokami, Detecting non-triviality of virtual links, J. Knot Theor. Ramif. **12** (6) (2003), 781–803.

[89] S. Kamada, Braid presentation of virtual knots and welded knots (preprint, March 2000).

[90] M. Kontsevich, Graphs, homotopical algebra and low dimensional topology (preprint 1992).

[91] T. Kohno, *Linear representations of braid groups and classical Yang–Baxter equations*, Contemporary Math. **78** (1988), 339–364.

[92] G. Kuperberg, *Involutory Hopf algebras and 3-manifold invariants*, Int. J. Math. (1991), 41–66.

[93] G. Kuperberg, *Non-involutory Hopf algebras and 3-manifold invariants*, Duke Math. J. **84** (1) (1996), 83–129.

[94] G. Kuperberg, *What is a virtual link? arXiv:math.GT / 0208039 v1 5 Aug 2002*, Algebr. Geom. Topol **3** (2003), 587–591.

[95] T.Q.T. Le, J. Murakami and T. Ohtsuki, *On a universal perturbative invariant of 3-manifolds*, Topology **37** (3) (1998), 539–574.

[96] R. Lawrence, *Asymptotic expansions of Witten–Reshetikhin–Turaev invariants for some simple 3-manifolds*, J. Math. Phys. **36** (11) (1995), 6106–6129.

[97] W.B.R. Lickorish, *The Temperley–Lieb Algebra and 3-manifold invariants*, J. Knot Theor. Ramif. **2** (1993), 171–194.

[98] W.B.R. Lickorish and K.C. Millett, *Some evaluations of link polynomials*, Comment Math. Helvetici **61** (1986), 349–359.

[99] V.O. Manturov, *Knot Theory*, CRC Press, Boca Raton, FL (2004).

[100] V.O. Manturov, *Invariants of virtual links*, Doklady Mathematics **384** (1) (2002), 11–13.

[101] V.O. Manturov, *Curves on surfaces, virtual knots, and the Jones–Kauffman polynomial*, Doklady Mathematics **390** (2) (2003), 155–157.

[102] V.O. Manturov, *The Khovanov polynomial for virtual knots*, Doklady Mathematics, to appear.

[103] V.O. Manturov, *Atoms and minimal diagrams of virtual links*, Doklady Mathematics **391** (2) (2003), 136–138.

[104] V.O. Manturov, *Two-variable invariant polynomial for virtual links*, Russ. Math. Surveys **5** (2002), 141–142.

[105] V.O. Manturov, *Multivariable polynomial invariants for virtual knots and links*, J. Knot Theor. Ramif. **12** (8) (2003), 1131–1144.

[106] V.O. Manturov, *Kauffman-like polynomial and curves in 2-surfaces*, J. Knot Theor. Ramif. **12** (8) (2003), 1145–1153.

[107] V.O. Manturov, *Vassiliev invariants for virtual links, curves on surfaces, and the Jones–Kauffman polynomial*, J. Knot Theor. Ramif., to appear.

[108] V.O. Manturov, *On invariants of virtual links*, Acta Applicandae Mathematicae **73** (2) (2002), 295–309.

[109] V.O. Manturov, *Long virtual knots and their invariants*, J. Knot Theor. Ramif. **13** (3) (2004), 221–233.

[110] V.O. Manturov, O raspoznavanii virtual'nyh kos (On the recognition of virtual braids), POMI Scientific Seminars, Geometry and Topology 8, St. Petersburg (2003), 267–286.

[111] V.O. Manturov, *Atoms and minimal diagrams of virtual links*, Doklady Mathematics **68** (1) (2003), 37–39.

[112] V.O. Manturov, in preparation.

[113] J. Mathias, PhD thesis, University of Illinois at Chicago (1996).

[114] K. Murasugi, *The Jones polynomial and classical conjectures in knot theory*, Topology **26** (1987), 187–194.

[115] S. Nelson, *Unknotting virtual knots with Gauss diagram forbidden moves*, J. Knot Theor. Ramif. **10** (6) (2001), 931–935.

[116] R. Penrose, Angular momentum: an approach to combinatorial spacetime, *Quantum Theory and Beyond*, T. Bastin, ed., Cambridge University Press, Cambridge (1969).

[117] R. Penrose, Applications of negative dimensional tensors, *Combinatorial Mathematics and Its Applications*, D.J.A. Welsh, ed., Academic Press, London (1971).

[118] M. Polyak and O. Viro, *Gauss diagram formulas for Vassiliev invariants*, Intl Math. Res. Notices **11** (1994), 445–453.

[119] M. Goussarov, M. Polyak and O. Viro, *Finite type invariants of classical and virtual knots*, Topology **39** (2000), 1045–1068.
[120] P. Cotta-Ramusino, E. Guadagnini, M. Martellini and M. Mintchev, Quantum field theory and link invariants, (preprint 1990).
[121] K. Reidemeister, *Knotentheorie*, Chelsea, New York (1948), Copyright 1932, Springer, Berlin.
[122] N.Y. Reshetikhin, Quantized universal enveloping algebras, the Yang–Baxter equation and invariants of links, I and II, *LOMI reprints E-4-87 and E-17-87*, Steklov Institute, Leningrad, USSR (1988).
[123] N.Y. Reshetikhin and V. Turaev, *Invariants of three-manifolds via link polynomials and quantum groups*, Invent. Math. **103** (1991), 547–597.
[124] L. Rozansky, Witten's invariant of 3-dimensional manifolds: loop expansion and surgery calculus, *Knots and Applications*, L. Kauffman, ed., World Scientific, Singapore (1995).
[125] S. Satoh, *Virtual knot presentation of ribbon torus-knots*, J. Knot Theor. Ramif. **9** (4) (2000), 531–542.
[126] J. Sawollek, On Alexander–Conway polynomials for virtual knots and links, arXiv: math.GT/9912173, 21 December 1999.
[127] J. Sawollek, An orientation-sensitive, Vassiliev invariant for virtual knots, arXiv: math.GT/0203123v3, 2002.
[128] L. Siebenmann and F. Bonahon, Unpublished Manuscript.
[129] D.S. Silver and S.G. Williams, *Alexander Groups and Virtual Links*, J. Knot Theor. Ramif. **10** (2001), 151–160.
[130] L. Smolin, *Link polynomials and critical points of the Chern–Simons path integrals*, Mod. Phys. Lett. A **4** (12) (1989), 1091–1112.
[131] L. Smolin, The physics of spin networks, *The Geometric Universe*, (Oxford, 1996), Oxford University Press, Oxford (1998), 291–304.
[132] T. Stanford, Finite-type invariants of knots, links and graphs (preprint 1992).
[133] H. Trotter, *Non-invertible knots exist*, Topology **2** (1964), 275–280.
[134] W. Menasco and M.B. Thistlethwaite, *The classification of alternating links*, Ann. Math. **138** (1993), 113–171.
[135] M.B. Thistlethwaite, *A spanning tree expansion of the Jones polynomial*, Topology **26** (3) (1987), 297–309.
[136] M.B. Thistlethwaite, *Links with trivial Jones polynomial*, J. Knot Theor. Ramif. **10** (4) (2001), 641–643.
[137] S. Eliahou, L. Kauffman and M.B. Thistlethwaite, *Infinite families of links with trivial Jones polynomial*, Topology, **42** (2003), 155–169.
[138] V.G. Turaev, The Yang–Baxter equations and invariants of links, LOMI preprint E-3-87, Steklov Institute, Leningrad, USSR., Inventiones Math., **92**, Fasc.3, 527–553.
[139] V.G. Turaev and O.Y. Viro, *State sum invariants of 3-manifolds and quantum 6j symbols*, Topology **31** (1992), 865–902.
[140] V.G. Turaev and H. Wenzl, *Quantum invariants of 3-manifolds associated with classical simple Lie algebras*, Int. J. Math. **4** (2) (1993), 323–358.
[141] V. Turaev, Virtual strings and their cobordisms, math.GT/0311185.
[142] V. Vassiliev, Cohomology of knot spaces, *Theory of Singularities and Its Applications*, V.I. Arnold, ed., American Mathematical Society, (1990), 23–69.
[143] J.H. White, *Self-linking and the Gauss integral in higher dimensions*, Am. J. Math. **91** (1969), 693–728.
[144] S. Winker, PhD thesis, University of Illinois at Chicago (1984).
[145] E. Witten, *Quantum field theory and the Jones polynomial*, Commun. Math. Phys. **121** (1989), 351–399.
[146] D. Yetter, *Quantum groups and representations on monoidal categories*, Math. Proc. Camb. Philos. Soc. **108** (1990), 197–229.
[147] P. Zinn-Justin and J.B. Zuber, *Matrix integrals and the generation and counting of virtual tangles and links*, J. Knot Theor. Ramif. **13** (3) (2004), 325–355.

CHAPTER 7

A Survey of Classical Knot Concordance

Charles Livingston

Department of Mathematics, Indiana University, Bloomington, IN 47405, USA

E-mail: livingst@indiana.edu

Contents

HANDBOOK OF KNOT THEORY
Edited by William Menasco and Morwen Thistlethwaite

Abstract

This survey provides an overview of the concordance group of knots in three-dimensional space. It begins with a review of the definitions of knots and concordance and then presents aspects of the algebraic theory of concordance. Following this, Casson–Gordon invariants are examined in detail. Recent results from the topological locally flat category are presented, as are new applications from smooth geometry. A discussion of the interplay between 3-dimensional knot properties and concordance is presented. The survey concludes with a brief list of open problems.

In 1926, Artin [3] described the construction of certain knotted 2-spheres in $\mathbf{R}^4$. The intersection of each of these knots with the standard $\mathbf{R}^3 \subset \mathbf{R}^4$ is a nontrivial knot in $\mathbf{R}^3$. Thus, a natural problem is to identify which knots can occur as such slices of knotted 2-spheres. Initially it seemed possible that every knot is such a *slice knot* and it was not until the early 1960s that Murasugi [86] and Fox and Milnor [24,25] succeeded at proving that some knots are not slice.

Slice knots can be used to define an equivalence relation on the set of knots in S^3: knots K and J are equivalent if $K\# -J$ is slice. With this equivalence the set of knots becomes a group, the *concordance group* of knots. Much progress has been made in studying slice knots and the concordance group, yet some of the most easily asked questions remain untouched.

There are two related theories of concordance, one in the smooth category and the other topological. Our focus will be on the smooth setting, though the distinctions and main results in the topological setting will be included. Related topics must be excluded, in particular the study of link concordance. Our focus lies entirely in the classical setting; higher dimensional concordance theory is only mentioned when needed to understand the classical setting.

1. Introduction

Two smooth knots, K_0 and K_1, in S^3 are called *concordant* if there is a smooth embedding of $S^1 \times [0,1]$ into $S^3 \times [0,1]$ having boundary the knots K_0 and $-K_1$ in $S^3 \times \{0\}$ and $S^3 \times \{1\}$, respectively. Concordance is an equivalence relation, and the set of equivalence classes forms a countable abelian group, $\mathcal{C}$, under the operation induced by connected sum. A knot represents the trivial element in this group if it is *slice*; that is, if it bounds an embedded disk in the 4-ball.

The concordance group was introduced in 1966 by Fox and Milnor [25], though earlier work on slice knots was already revealing aspects of its structure. Fox [24] described the use of the Alexander polynomial to prove that the figure eight knot is of order two in C and Murasugi [86] used the signature of a knot to obstruct the slicing of a knot, thus showing that the trefoil is of infinite order in $\mathcal{C}$. (These results, along with much of the introductory material, are presented in greater detail in the body of this article.) The application of abelian knot invariants (those determined by the cohomology of abelian covers or, equivalently, by the Seifert form) to concordance culminated in 1969 with Levine's classification of higher dimensional knot concordance, [62,63], which applied in the classical dimension to give a surjective homomorphism, $\phi : \mathcal{C} \rightarrow \mathbf{Z}^{\infty} \oplus \mathbf{Z}_2^{\infty} \oplus \mathbf{Z}_4^{\infty}$.

In 1975, Casson and Gordon [8,9] proved that Levine's homomorphism is not an isomorphism, constructing nontrivial elements in the kernel, and Jiang [42] expanded on this to show that the kernel contains a subgroup isomorphic to $\mathbf{Z}^{\infty}$. Along these lines it was shown in [72] that the kernel also contains a subgroup isomorphic to $\mathbf{Z}_2^{\infty}$. The 1980s saw two significant developments in the study of concordance. The first was based on Freedman's work [26,27] studying the structure of topological 4-manifolds. One consequence was that methods of Levine and those of Casson–Gordon apply in the topological locally flat category, rather than only in the smooth setting. More significant, Freedman proved that all knots with trivial Alexander polynomial are in fact slice in the topological locally flat category.

The other important development concerns the application of differential geometric techniques to the study of smooth 4-manifolds, beginning with the work of Donaldson [20,21] and including the introduction of Seiberg–Witten invariants and their application to symplectic manifolds, the use of the Thurston–Bennequin invariant [2,101], and recent work of Ozsváth and Szabó [96]. This work quickly led to the construction of smooth knots of Alexander polynomial one that are not smoothly slice, along with a much deeper understanding of related issues, such as the 4-ball genus of knots. Using these methods it has recently been shown that the results of Ozsváth and Szabó [96] imply that the kernel of Levine's homomorphism contains a summand isomorphic to $\mathbf{Z}$ and thus contains elements that are not divisible [78]. References are too numerous to enumerate here; a few will be included as applications are mentioned.

Recent work of Cochran et al. [14,15] has revealed a deeper structure to the knot concordance group. In that work a filtration of $\mathcal{C}$ is defined:

$$\cdots \mathcal{F}_{2.0} \subset \mathcal{F}_{1.5} \subset \mathcal{F}_{1} \subset \mathcal{F}_{.5} \subset \mathcal{F}_{0} \subset \mathcal{C}.$$

It is shown that $\mathcal{F}_0$ corresponds to knots with trivial Arf invariant, $\mathcal{F}_5$ corresponds to knots in the kernel of ϕ and all knots in $\mathcal{F}_{1.5}$ have vanishing Casson–Gordon invariants. Using von Neumann η-invariants, it has been proved in [16] that each quotient is infinite. This work places Levine's obstructions and those of Casson–Gordon in the context of an infinite sequence of obstructions, all of which reveal a finer structure to $\mathcal{C}$.

Outline: Section 2 is devoted to the basic definitions related to concordance and algebraic concordance. In Section 3, algebraic concordance invariants are presented, including the description of Levine's homomorphism. Sections 4 and 5 present Casson–Gordon invariants and their application. In Section 6, the consequences of Freedman's work on topological surgery in dimension four are described. Section 7 concerns the application of the results of Donaldson and more recent differential geometric techniques to concordance. In Section 8, the recent work of Cochran, Orr and Teichner on the structure of the topological concordance group is outlined. Section 9 relates 3-dimensional knot properties and concordance. Finally, Section 10 presents a few outstanding problems in the study of knot concordance.

2. Definitions

We will work in the smooth setting. In Section 6, there will be a discussion of the necessary modifications and main results that apply in the topological locally flat category. Knots are usually thought of as isotopy classes of embeddings of S^1 into S^3. However, to simplify the discussion of orientation and symmetry issues, it is worthwhile to begin with the following precise definitions of knots, slice knots, concordance and Seifert surfaces.

2.1. *Knot theory and concordance*

DEFINITION 2.1.

(1) A knot is an oriented diffeomorphism class of a pair of oriented manifolds, $K = (\Sigma^3, \Sigma^1)$, where Σ^n is diffeomorphic to the n-sphere.
(2) A knot is called slice if there is a pair (B^4, D^2) with $\partial(B^4, D^2) = K$, where B^4 is the 4-ball and D^2 is a smoothly embedded 2-disk.
(3) Knots K_1 and K_2 are called concordant if $K_1\# - K_2$ is slice. (Here $-K$ denotes the knot obtained by reversing the orientation of each element of the pair and connected sum is defined in the standard way for oriented pairs.) The set of concordance classes is denoted $\mathcal{C}$.
(4) A Seifert surface for a knot K is an oriented surface F embedded in S^3 such that $K = (S^3, \partial F)$.

The basic theorem in the subject is the following.

THEOREM 2.2. *The set of concordance classes of knots forms a countable abelian group, also denoted $\mathcal{C}$, with its operation induced by connected sum and with the unknot representing the identity.*

Related to the notion of slice knots there is the stronger condition of being a ribbon knot.

DEFINITION 2.3. A knot K is called ribbon if it bounds an embedded disk D in B^4 for which the radial function on the ball restricts to be a smooth Morse function with no local maxima in the interior of D.

There is no corresponding group of ribbon concordance. Casson observed that for every slice knot K there is a ribbon knot J such that $K\#J$ is ribbon. Hence, if any equivalence relation identifies ribbon knots, it also identifies all slice knots. There is however a notion of ribbon concordance, first studied in [39].

2.2. *Algebraic concordance*

An initial understanding of $\mathcal{C}$ is obtained via the algebraic concordance group, defined by Levine in terms of Seifert pairings.

DEFINITION 2.4. A Seifert pairing for a knot K with Seifert surface F is the bilinear mapping

$$V : H_1(F) \times H_1(F) \to \mathbf{Z}$$

given $V(x, y) = \text{lk}(x, i_* y)$, where lk denotes the linking number and i_* is the map induced by the positive pushoff, $i : F \to S^3 - F$.

(Here and throughout, homology groups will be taken with integer coefficients unless indicated otherwise.) A Seifert matrix is the matrix representation of the Seifert pairing with respect to some free generating set for $H_1(F)$.

If the transpose pairing V^τ is defined by $V^\tau(x, y) = V(y, x)$ then $V - V^\tau$ represents the unimodular intersection form on $H_1(F)$. Hence, in general we define an abstract Seifert pairing on a finitely generated free **Z**-module M to be a bilinear form $V : M \times M \rightarrow \mathbf{Z}$ satisfying $V - V^\tau$ is unimodular. (In order for this to make sense for the trivial knot with Seifert surface B^2, the Seifert form on the 0-dimensional **Z**-module is defined to be unimodular.)

DEFINITION 2.5. An abstract Seifert form V on M is metabolic if $M = M_1 \oplus M_2$ with $M_1 \cong M_2$ and $V(x, y) = 0$ for all x and $y \in M_1$. Such an M_1 is called a metabolizer for V.

THEOREM 2.6. *If K is slice and F is a Seifert surface for K, then the associated Seifert form is metabolic.*

PROOF. Let D be a slice disk for K. The union $F \cup D$ bounds a 3-manifold R embedded in B^4. Such an R can be constructed explicitly, or an obstruction theory argument can be used to construct a smooth mapping $B^4 - D \rightarrow S^1$ which has $F \cup D$ as the boundary of the pull-back of a regular value. (Note that this construction depends on the triviality of the normal bundle to D.)

A duality argument implies that $\text{rank}(\ker(H_1(F) \rightarrow H_1(R))) = (1/2)\text{rank}(H_1(F))$. For any x and y in that kernel, $V(x, y) = 0$: since x bounds a 2-chain in R, $i_*(x)$ bounds a 2-chain in $B^4 - R$ which is disjoint from the chain bounded in R by y.

Since V vanishes on this kernel, it vanishes on the summand M generated by the kernel, and hence V is metabolic. □

COROLLARY 2.7. *If K_1 is concordant to K_2 and these knots have Seifert forms V_1 and V_2 (with respect to arbitrary Seifert surfaces), then $V_1 \oplus - V_2$ is metabolic.*

In general, abstract Seifert forms V_1 and V_2 are called *algebraically concordant* if $V_1 \oplus - V_2$ is metabolic. This is an equivalence relation. (The proof is based on cancellation: if V and $V \oplus W$ are metabolic, then so is W. See [49].)

THEOREM 2.8. *The set of algebraic concordance classes forms a group, denoted $\mathcal{G}$, with its operation induced by direct sum. The trivial 0-dimensional* **Z**-module serves as the identity.

In the following theorem, defining Levine's homomorphism, we temporarily use the notation $[K]$ to denote concordance class of a knot and $[V_F]$ to represent the algebraic concordance class of a Seifert form associated to an arbitrarily chosen Seifert surface F for K.

THEOREM 2.9. *The map $\phi : \mathcal{C} \rightarrow \mathcal{G}$ defined by $\phi([K]) = [V_F]$ is a surjective homomorphism.*

PROOF. That this map is well-defined follows from the previous discussion and, in particular, Corollary 2.7. Surjectivity follows from an explicit construction of a surface with desired Seifert form [6,102]. □

3. Algebraic concordance invariants

Levine [62] defined a collection of homomorphisms from $\mathcal{G}$ to the groups $\mathbf{Z}$, $\mathbf{Z}_2$ and $\mathbf{Z}_4$. These can be properly combined to give an isomorphism Φ from $\mathcal{G}$ to the infinite direct sum $\mathbf{Z}^\infty \oplus \mathbf{Z}_2^\infty \oplus \mathbf{Z}_4^\infty$. The proof of this will be left to [62].

We should remark that what Levine actually did was to classify the rational algebraic concordance group, based on rational matrices. He also showed that the integral group injects into the rational group, with image sufficiently large to contain $\mathbf{Z}^\infty \oplus \mathbf{Z}_2^\infty \oplus \mathbf{Z}_4^\infty$. Stolzfus [104] completed the classification of the integral concordance group.

In this section, we will describe a collection of invariants that are sufficient to show that $\mathcal{G}$ contains a summand isomorphic to $\mathbf{Z}^\infty \oplus \mathbf{Z}_2^\infty \oplus \mathbf{Z}_4^\infty$. The invariants will be applied to a particular family of knots, which we now describe.

Figure 1 illustrates a basic knot that we denote $K(a, b, c)$; the curves J_1 and J_2 can be ignored for now. The integers a and b indicate the number of full twists in each band. The integer c is odd and represents the number of half twists between the bands; those twists between the bands are so placed as to not add twisting to the individual bands. Figure 2 illustrates a particular example, $K(2, 0, 3)$, along with a basis for the first homology of the Seifert surface, indicated with oriented dashed curves on the surface.

The knot $K(a, b, c)$ bounds a genus one Seifert surface with Seifert form represented by the following matrix with respect to the indicated basis of $H_1(F)$.

$$\begin{pmatrix} a & (c+1)/2 \\ (c-1)/2 & b \end{pmatrix}$$

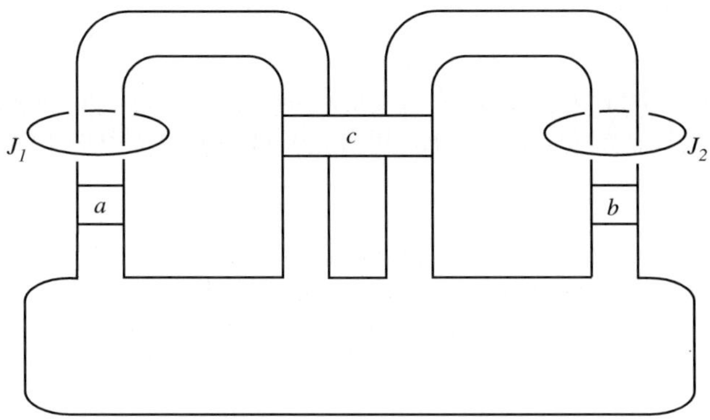

Fig. 1. The knot $K(a, b, c)$.

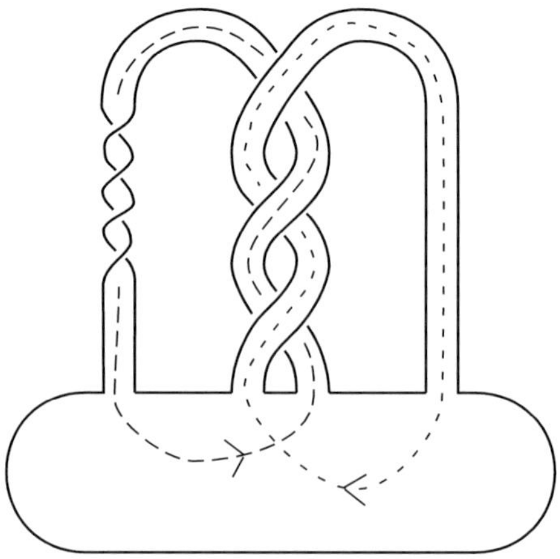

Fig. 2. The knot $K(2, 0, 3)$.

3.1. *Integral invariants, signatures*

Let V be a Seifert matrix and V^{τ} its transpose. If ω is a unit complex number that is not a root of the Alexander polynomial of V, $\Delta_V(t) = \det(V - tV^{\tau})$, then the form $V_\omega = \frac{1}{2}(1 - \omega)V + \frac{1}{2}(1 - \bar{\omega})V^{\tau}$ is nonsingular. In this case, if V is metabolic, the signature of V_ω is 0. To adjust for the possibility of the V_ω being singular, for general ω on the unit circle the signature $\sigma_\omega(V)$ is defined to be the limiting average of the signatures of $V_{\omega+}$ and $V_{\omega-}$, where ω_+ and ω_- are unit complex numbers approaching ω from different sides. For all ω, σ_ω defines a homomorphism from $\mathcal{G}$ to $\mathbf{Z}$. It is onto $2\mathbf{Z}$ if $\omega \neq 1$. For the set of ω given by roots of unity $e^{2\pi i/p}$ where p is a prime, the functions σ_ω are independent on $\mathcal{G}$ (this can be seen using the b-twisted doubles of the unknot, $K(1, b, 1)$, $b > 0$), and hence together these give a map of $\mathcal{G}$ onto $\mathbf{Z}^\infty$. In the case of $\omega = -1$, this signature, defined by Trotter [111], was shown to be a concordance invariant by Murasugi [86]. The more general formulation is credited to Levine and Tristram [110] and is referred to as the Levine–Tristram signature.

In [43,44] the identification of these signatures with signatures of the branched covers of B^4 branched over a pushed in Seifert surface of a knot was made. In [12] it was shown that the set of σ_ω over all ω with positive imaginary part are independent.

3.2. *The Arf invariant:* $\mathbf{Z}_2$

Given a $(2g) \times (2g)$ Seifert matrix V one defines a $\mathbf{Z}_2$-valued quadratic form on $\mathbf{Z}_2^{2g}$ by $q(x) = xVx^{\tau}$. This is a nonsingular quadratic form in the sense that $q(x + y) - q(x) - q(y) = x \cdot y$ where the nonsingular bilinear pairing $x \cdot y$ is given by the matrix $V + V^{\tau}$. (Recall that the determinant of $V + V^{\tau}$ is odd.)

The simplest definition of the Arf invariant of a nonsingular quadratic form on a $\mathbf{Z}_2$-vector space W is that $\mathrm{Arf}(q) = 0$ or $\mathrm{Arf}(q) = 1$ depending on whether q takes value 0 or 1, respectively, on a majority of elements in W. See, for instance [5,41]. The Arf invariant defines a homomorphism on the Witt group of $\mathbf{Z}_2$ quadratic forms and in particular vanishes on metabolic forms. Hence, the Arf invariant gives a well-defined $\mathbf{Z}_2$-valued homomorphism from $\mathcal{G}$ to $\mathbf{Z}_2$.

This invariant was first defined by Robertello [97]. Murasugi [87] observed that $\mathrm{Arf}(V) = 0$ if and only if $\Delta_V(-1) = \pm 1 \bmod 8$.

3.3. *Polynomial invariants:* $\mathbf{Z}_2$

The Alexander polynomial of a Seifert matrix is defined to be $\Delta_V(t) = \det(V - tV^\tau) \in \mathbf{Z}[t, t^{-1}]$. If different Seifert matrices associated to the same knot are used to compute an Alexander polynomial, the resulting polynomials will differ by multiplication by a unit in $\mathbf{Z}[t, t^{-1}]$, that is by $\pm t^n$ for some n. Hence, two Alexander polynomials are considered equivalent if they differ by multiplication by $\pm t^n$ for some n.

If V is metabolic, then $\Delta_V(t) = \pm t^n f(t) f(t^{-1})$ for some integral polynomial f. For concordance considerations, if $p(t)$ is an irreducible symmetric polynomial ($p(t^{-1}) = \pm t^n p(t)$) then the exponent of $p(t)$ in the irreducible factorization of $\Delta_V(t)$ taken modulo 2 yields a $\mathbf{Z}_2$ invariant of $\mathcal{G}$. Fox and Milnor [25] used this to define a surjective homomorphism of $\mathcal{G}$ to $\mathbf{Z}_2^\infty$. The knots $K(a, -a, 1)$ (see Figure 1) are of order at most 2 in $\mathcal{C}$ since for each, $K(a, -a, 1) = -K(a, -a, 1)$. On the other hand, these have distinct irreducible Alexander polynomials if $a > 0$. The existence of an infinite summand of $\mathcal{G}$ isomorphic to $\mathbf{Z}_2^\infty$ follows. Note that the knot $K(1, -1, 1)$ is the figure eight knot.

3.4. *W*($\mathbf{Q}$): $\mathbf{Z}_2$ *and* $\mathbf{Z}_4$ *invariants*

The matrix $V + V^\tau$ defines an element in the Witt group of $\mathbf{Q}$, $W(\mathbf{Q})$. We will now summarize the theory of this Witt group and associated Witt groups of finite fields. Details can be found in [41]. Notice that the determinant of $V + V^\tau$ is odd; hence, in the following discussion we restrict attention to odd primes p.

Recall that the Witt group of an arbitrary field $\mathbf{F}$ consists of finite dimensional $\mathbf{F}$-vector spaces with nonsingular symmetric forms and forms W_1 and W_2 are equivalent if $W_1 \oplus - W_2$ is metabolic. Addition is via direct sums.

There is a surjective homomorphism $\oplus \partial_p : W(\mathbf{Q}) \to \oplus W(\mathbf{F}_p)$. Here $\mathbf{F}_p$ is the field with p elements, and the direct sums are over the set of all primes. For p odd, the group $W(\mathbf{F}_p)$ is isomorphic to either $\mathbf{Z}_2$ or $\mathbf{Z}_4$, depending on whether p is 1 or 3 modulo 4. We next define ∂_p and then discuss the invariants of $W(\mathbf{F}_p)$. (For completeness, we note here that the kernel of $\oplus \partial_p$ is $W(\mathbf{Z})$ which is isomorphic to $\mathbf{Z}$ via the signature [41].)

3.4.1. *Reducing to finite fields.* There is a simple algorithm giving the map ∂_p. A symmetric rational matrix A can be diagonalized using simultaneous row and column operations, and this form decomposes as the direct sum of forms: $\oplus_{i=1}^{n}(a_i p^{\epsilon_i})$ where $\gcd(a_i, p) = 1$, $\epsilon_i = 1$ for $i \le m$ and $\epsilon_i = 0$ for $m + 1 \le i \le n$. The map ∂_p takes A to the $\mathbf{F}_p$ form represented by the direct sum $\oplus_{i=1}^{m}(a_i)$.

3.4.2. *$W(\mathbf{F}_p)$: $\mathbf{Z}_2$ and $\mathbf{Z}_4$ invariants.* For p odd, any form on a finite dimensional $\mathbf{F}_p$-vector space can be diagonalized with ± 1 as the diagonal entries. In the Witt group the form represented by the matrix $(1)\oplus(-1)$ is trivial. A little more work shows that the form $4(1)$ is Witt trivial: find elements a and b such that $a^2 + b^2 = -1$ and consider the subspace spanned by $(1, 0, a, b)$ and $(0, 1, b, -a)$. Hence, $W(\mathbf{F}_p)$ is generated by (1), an element of order 2 or 4.

In the case that $p \equiv 1$ modulo 4, -1 is a square. It follows quickly that $W(\mathbf{F}_p) \cong \mathbf{Z}_2$. On the other hand, in the case that $p \cong 3$ modulo 4, -1 is not a square, and $W(\mathbf{F}_p) \cong \mathbf{Z}_4$.

As a simple example, if one starts with the Seifert form for the knot $K(1, -5, 1)$,

$$V = \begin{pmatrix} 1 & 1 \\ 0 & -5 \end{pmatrix}, \qquad V + V^{\tau} = \begin{pmatrix} 2 & 1 \\ 1 & -10 \end{pmatrix}.$$

Diagonalizing over the rationals yields

$$V = \begin{pmatrix} 2 & 0 \\ 0 & -(2)(3)(7) \end{pmatrix}.$$

With $p = 3$ this form maps to the element (-14) of $W(\mathbf{F}_3)$, which is equivalent to the form (1), a generator of order 4. The same is true working with $p = 7$.

As a consequence of the next theorem we will see that this particular form V is actually of order four in $\mathcal{G}$.

3.5. *Quadratic polynomials*

A special case of a theorem of Levine (Section 23 of [63]) gives the following result, which implies in particular that the form just described is of order 4 in $\mathcal{G}$.

THEOREM 3.1. *Suppose that $\Delta_V(t)$ is an irreducible quadratic. Then V is of finite order in the algebraic concordance group if and only if $\Delta_V(1)\Delta_V(-1) < 0$. In this case V is of order 4 if $|\Delta_V(-1)| = p^a q$ for some prime p congruent to 3 modulo 4, a odd, and p and q relatively prime; otherwise it is of order 2.*

3.6. *Other approaches to algebraic invariants*

There are alternative approaches to algebraic obstructions to a knot being slice that do not depend on Seifert forms. For instance, Milnor [83] described signature invariants based on his duality theorem for the infinite cyclic cover of a knot complement. The equivalence of these signatures and those of Tristram and Levine is proved in [82]. There is also an interpretation of the algebraic concordance group in terms of the Blanchfield pairing of the knot.

4. Casson–Gordan invariants

In the case that K is algebraically slice, Casson–Gordon invariants offer a further obstruction to a knot being slice. We follow the basic description of [8].

4.1. *Definitions*

We begin by reviewing the *linking form* on torsion $(H_1(M))$ for an oriented 3-manifold M. If x and y are curves representing torsion in the first homology, then $\mathrm{lk}(x, y)$ is defined to be $(d \cap y)/n \in \mathbf{Q}/\mathbf{Z}$, where d is a 2-chain with boundary nx. Intersections are defined via transverse intersections of chains, and of course one must check that the value of the linking form is independent of the many choices in its definition. For a closed oriented 3-manifold the linking form is nonsingular in the sense that it induces an isomorphism from torsion $(H_1(M))$ to hom(torsion$(H_1(M), \mathbf{Q}/\mathbf{Z})$.

Such a symmetric pairing on a finite abelian group, $l : H \times H \to \mathbf{Q}/\mathbf{Z}$, is called *metabolic* with *metabolizer* L if the linking form vanishes on $L \times L$ for some subgroup L with $|L|^2 = |H|$.

Let M_q denote the q-fold branched cover of S^3 branched over a given knot K, and let $\bar{M}_q$ denote 0-surgery on M_q along $\tilde{K}$, where $\tilde{K}$ is the lift of K to M_q. Here q will be a prime power.

Let x be an element of self-linking 0 in $H_1(M_q)$ and suppose that x is of prime power order, say p. Linking with x defines a homomorphism $\chi_x : H_1(M_q) \to \mathbf{Z}_p$. Furthermore, χ_x extends to give a $\mathbf{Z}_p$-valued character on $H_1(\bar{M}_q)$ which vanishes on the meridian of $\tilde{K}$. In turn, this character extends to give $\bar{\chi}_x : H_1(\bar{M}_q) \to \mathbf{Z}_p \oplus \mathbf{Z}$. Since x has self-linking 0, bordism theory implies that the pair $(\bar{M}_q, \bar{\chi}_x)$ bounds a 4-manifold, character, pair, (W, η).

More generally, for any character $\chi : H_1(M_q) \to \mathbf{Z}_p$, there is a corresponding character $\bar{\chi} : H_1(\bar{M}_q) \to \mathbf{Z}_p \oplus \mathbf{Z}$. This character might not extend to a 4-manifold, but since the relevant bordism groups are finite, for some multiple $r\bar{M}_q$ the character given by $\bar{\chi}$ on each component does extend to a 4-manifold, character pair, (W, η).

Let Y denote the $\mathbf{Z}_p \times \mathbf{Z}$ cover of W corresponding to η. Using the action of $\mathbf{Z}_p \times \mathbf{Z}$ on $H_2(Y, \mathbf{C})$ one can form the twisted homology group $H_2^t(W, \mathbf{C}) = H_2(Y, \mathbf{C}) \otimes_{\mathbf{C}[\mathbf{Z}_p \times \mathbf{Z}]} \mathbf{C}(t)$. (The action of $\mathbf{Z}_p$ on $\mathbf{C}(t)$ is given by multiplication by $e^{2\pi i/p}$.) There is a nonsingular Hermitian form on $H_2^t(W, \mathbf{C})$ taking values in $\mathbf{C}(t)$. The Casson–Gordon invariant is defined to be the difference of this form and the intersection form of $H_2(W, \mathbf{C})$, both tensored with $1/r$, in $W(\mathbf{C}[t, t^{-1}]) \otimes \mathbf{Q}$. (In showing that this Witt class yields a well-defined obstruction to slicing a knot, the fact that $\Omega_4(\mathbf{Z}_p \oplus \mathbf{Z})$ is nonzero appears, and as a consequence one must tensor with $\mathbf{Q}$ to arrive at a well-defined invariant, even in the case of χ_x in which it is possible to take $r = 1$.)

DEFINITION 4.1. The Casson–Gordon invariant $\tau(M_q, \chi)$ is the class $(H_2^t(W, \mathbf{C}) - H_2(W), \mathbf{C}) \otimes 1/r \in W(\mathbf{C}(t)) \otimes \mathbf{Q}$.

4.2. *Main theorem*

The main theorem of [8] states:

THEOREM 4.2. *If K is slice, there is a metabolizer L for the linking form on $H_1(M_q)$ such that, for each prime power p and each element $x \in L$ of order p, $\tau(M_q, \chi_x) = 0$.*

The proof shows that if K is slice with slice disk D, then covers of $B^4 - D$ can be used as the manifold W, and for this W the invariant vanishes.

COMMENT. There are a number of extensions of this theorem. With care the definition of the Casson–Gordon invariant can be refined and τ can be viewed as taking values in $W(\mathbf{Q}[\zeta_p](t)) \otimes \mathbf{Z}[1/p]$. This yields finer invariants (see, for instance [34]). The observation that L can be assumed to be equivariant with respect to the deck transformation of M_q can give stronger constraints (see, for example [57]). In [50] it is demonstrated that a factorization of the Alexander polynomial of a knot yields further constraints on the metabolizer L.

4.3. *Invariants of* $W(\mathbf{C}(t)) \otimes \mathbf{Q}$

In the next section, we will describe examples of algebraically slice knots which can be proved to be nonslice using Casson–Gordon invariants. We conclude this section with a description of the types of algebraic invariants associated to the Witt group $W(\mathbf{C}(t)) \otimes \mathbf{Q}$.

4.3.1. *Signatures.* Let ξ be a unit complex number. Let $A \otimes a/b \in W(\mathbf{C}(t)) \otimes \mathbf{Q}$. Then A can be represented by a matrix of rational functions, $A(t)$. The signature $\sigma_\xi(A \otimes a/b)$ is defined, roughly, to be $(a/b)\sigma(A(\xi))$ where σ denotes the standard Hermitian signature. There is the technical point arising that $A(\xi)$ might be singular, so the precise definition of $\sigma_\xi(A \otimes a/b)$ takes the two-sided average over unit complex numbers close to ξ. This limit is defined to be the *Casson–Gordon signature invariant*, $\sigma_\xi(K, \chi)$. For $\xi = 1$ this is abbreviated as $\sigma(K, \chi)$.

4.3.2. *Discriminants.* If the matrix $A(t)$ represents $0 \in W(\mathbf{C}(t))$, the discriminant, $\mathrm{dis}(A(t)) = (-1)^k \det(A(t))$ (where k is half the dimension of A) will be of the form $f(t)\bar{f}(t)$ for some rational function f. Let $g(t) = t^2 + \lambda t + 1$, $|\lambda| > 2$ be an irreducible real symmetric polynomial. It follows that for a matrix $A(t)$, the exponent of $g(t)$ in the factorization of $\mathrm{dis}(A(t))$ gives a $\mathbf{Z}_2$-valued invariant of the Witt class of $A(t)$.

More generally, in the case that p is odd, the exponent of g in the determinant of $A(t)^a$ gives a $\mathbf{Z}_2$ invariant of the class represented by $A(t) \otimes a/b$ in $W(\mathbf{C}(t)) \otimes \mathbf{Z}[1/p]$.

These discriminants were first discussed in unpublished work of Litherland [68]. Later developments and applications are included in [34].

In [56,57] a 3-dimensional approach to the definition of Casson–Gordon discriminant invariants is presented. In short, the representation $\bar{\chi} : H_1(\bar{M}_q) \to \mathbf{Z}_p \times \mathbf{Z}$ determines the twisted homology group: $H_1^t(\bar{M}_q, \mathbf{Q}(\zeta_p)[t])$. This is a $\mathbf{Q}[\zeta_p][t]$ module, and the discriminant of the Casson–Gordon invariant is given by the order of this module. Although this 3-dimensional approach does not give the signature invariant, it has the advantage of being completely algorithmic in computation via a procedure first developed in [66,112] and applied in [56–58]. A computer implementation of that algorithm facilitated the classification of the order of low-crossing number knots in concordance [108] and the proof that most low-crossing number knots which are not reversible are not concordant to their reverses, in [107].

In a different direction, we note that some effort has been made in removing the restriction on prime power covers and characters. In the case of ribbon knots, it was known that stronger results could be attained. Recent work of Kim [52] has developed examples of nonslice algebraically slice knots for which all prime power branched covers are homology spheres. Other work in this realm includes that of Letsche [61] and recent work of Friedl [28,29].

5. Companionship and Casson–Gordon invariants

In Casson and Gordon's original work the computation of Casson–Gordon invariants was quite difficult, largely limited to restricted classes of knots. Litherland [69] studied the behavior of these invariants under companionship and, independently, Gilmer [31] found interpretations of particular Casson–Gordon invariants in terms of signatures of simple closed curves on a Seifert surface for a knot. Further work addressing companionship and Casson–Gordon invariants includes [1]. In this section, we describe the general theory and its application to genus one knots.

5.1. *Construction of companions*

Let U be an unknotted circle in the complement of a knot K. If S^3 is modified by removing a neighborhood of U and replacing it with the complement of a knot J in S^3 (via a homeomorphism of boundaries that identifies the meridian of J with the longitude of U and vice versa) then the resulting manifold is again diffeomorphic to S^3. The image of K in this manifold will be denoted $K(J)$ (the choice of U will be suppressed in the notation). In the language of classical knot theory, $K(J)$ is a satellite knot with companion J and satellite K.

If M_q is the q-fold branched cover of S^3 branched over K, then U has q' lifts, denoted U_i, $i = 1, \ldots, q'$, where. $q' = \gcd(q, \mathrm{lk}(U, K))$ It follows that M'_q, the q-fold branched cover of S^3 branched over $K(J)$, is formed from M_q by removing neighborhoods of the

U_i and replacing each with the q/q'-cyclic cover of the complement of J. If χ is a $\mathbf{Z}_p$-valued homomorphism on $H_1(M_q)$, there is a naturally associated homomorphism χ' on $H_1(M'_q)$.

5.2. *Casson–Gordon invariants and companions*

In the case that $\text{lk}(U, K) = 0$, we have the following theorem of Litherland [69] (see also [35]).

THEOREM 5.1. *In the situation just described, with* $\text{lk}(U, K) = 0$

$$\sigma(K(J), \chi') = \sigma(K, \chi) + \sum_{i=1}^{q} \sigma_{\chi(U_i)/p}(J).$$

The main idea of the proof is fairly simple. If (W, η) is the chosen pair bounding $(\bar{M}_q, \bar{\chi})$ in the definition of the Casson–Gordon invariant, then for the new knot K' a 4-manifold W' can be built from W by attaching copies of a 4-manifold with character (Y, η) bounding 0-surgery on J with its canonical representation to $\mathbf{Z}$. Signatures of cyclic covers of Y are related to the signatures of J. A similar analysis can be done for the discriminant of the Casson–Gordon invariant. This was detailed in [34], and further explored in [58] where it was no longer assumed that J was null homologous.

EXAMPLE. Consider the knot illustrated in Figure 1 with $a = 0$, $b = 0$ and $c = 3$. The bands have knots J_1 and J_2 tied in them. (We will also refer to the pair of unknotted circles as J_1 and J_2 in this situation, as the meaning is unambiguous.) Call the resulting knot $K(J_1, J_2)$. The homology of the 2-fold cover is isomorphic to $\mathbf{Z}_3 \oplus \mathbf{Z}_3$ with the linking form vanishing on the two summands. Call generators of the summands x_1 and x_2. An analysis of the cover shows that χ_{x_1} is a $\mathbf{Z}_3$-valued character that vanishes on the lifts of J_1 and takes value ± 1 on the two lifts of J_2. Similarly for χ_{x_2}.

Since K is slice, by the Casson–Gordon theorem, either $\sigma(K, \chi_{x_1})$ or $\sigma(K, \chi_{x_1})$ must vanish. Hence, using Theorem 5.1, if $K(J_1, J_2)$ is slice, either $2\sigma_{1/3}(J_1)$ or $2\sigma_{1/3}(J_2)$ must vanish. By choosing J_1 and J_2 so that this is not the case, one constructs basic examples of algebraically slice knots which are not slice.

5.3. *Genus one knots and the Seifert form*

Gilmer [31,32] observed that for genus one knots the computation of Casson–Gordon invariants is greatly simplified. Roughly, he interpreted the Casson–Gordon signature invariants of an algebraically slice genus one knot in terms of the signatures of knots tied in the bands of the Seifert surface. The previous example offers an illustration of the appearance of these signatures. This work is now most easily understood via the use of companionship just described.

In short, if an algebraically slice knot K bounds a genus one Seifert surface F, then some nontrivial primitive class in $H_1(F)$ has trivial self-linking with respect to the Seifert form. If that class is represented by a curve α, the surface can be deformed to be a disk with two bands attached, one of which is tied into the knot α. If a new knot is formed by adding the knot $-\alpha$ to the band, the knot becomes slice and certain of its Casson–Gordon invariants will vanish. However, the previous results on companionship determine how the modification of the knot changes the Casson–Gordon invariant. The situation is made somewhat more delicate in that α is not unique: for genus one algebraically slice knots there are two metabolizers. The following represents the sort of result that can be proved.

THEOREM 5.2. *Let K be a genus one slice knot. The Alexander polynomial of K is given* $(at-(a+1))((a+1)t-a)$ *for some a. For some simple closed curve* α *representing a generator of a metabolizer of the Seifert form and for some infinite set of primes powers q, one has*

$$\sum_{i=1}^{q}\sigma_{bm^i/p}(\alpha)=0$$

for all prime power divisors p of $(a-1)^q-a^q$, *and for all integers b.*

(The appearance of the term $(a-1)^q-a^q$ represents the square root of the order of the homology of the q-fold branched cover.) Since the sum is taken over a coset of the multiplicative subgroup of $\mathbf{Z}_p$, by combining these cosets one has the following.

COROLLARY 5.3. *If K is a genus one slice knot with nontrivial Alexander polynomial, then for some simple closed curve* α *representing a generator of a metabolizer of the Seifert form, there is an infinite set of prime powers p for which*

$$\sum_{i=1}^{p-1}\sigma_{i/p}(\alpha)=0.$$

A theorem of Cooper [17] follows quickly:

COROLLARY 5.4. *If K is a genus one slice knot with nontrivial Alexander polynomial, then for some simple closed curve x representing a generator of a metabolizer of the Seifert form*

$$\int_0^{1/2}\sigma_t(x)\mathrm{d}t=0.$$

(This theorem reappears in [15] where the integral is reinterpreted as a metabelian von Neumann signature of the original knot K, giving a direct reason why it is a concordance invariant. For more on this, see Section 8.)

EXAMPLE. Consider the knot $K(0,0,3)$, as in Figure 1. Replacing the curves labeled J_1 and J_2 with the complements of knots J_1 and J_2 yields a knot for which the metabolizers of the Seifert form are represented by the knots J_1 and J_2. The knot is algebraically slice, but by the previous corollary, if both of the knots have signature functions with nontrivial integral, the knot is not slice.

6. The topological category

Freedman [26] developed surgery theory in the category of topological 4-manifolds, proving roughly that for manifolds with fundamental groups that are not too complicated (in particular, finitely generated abelian groups) the general theory of higher dimensional surgery descends to dimension 4. The most notable consequence of this work was the proof the 4-dimensional Poincaré conjecture: a closed topological 4-manifold that is homotopy equivalent to the 4-sphere is homeomorphic to the 4-sphere.

Two significant contributions to the study of concordance quickly followed from Freedman's original paper. The first of these, proved in [27], is that a locally flat surface in a topological 4-manifold has an embedded normal bundle. The use of such a normal bundle was implicit in the proof that slice knots are algebraically slice. It is also used in a key step in the proof of the Casson–Gordon theorem, as follows. Casson–Gordon invariants of slice knots are shown to vanish via the observation that for a slice knot K, if 0-surgery is performed on K, the resulting 3-manifold $M(K,0)$ bounds a homology $S^1 \times B^3$, W. This W is constructed by removing a tubular neighborhood of a slice disk for K in the 4-ball. The existence of the tubular neighborhood is equivalent to the existence of the normal bundle.

In a different direction, Freedman's theorem implied that in the topological locally flat category all knots of Alexander polynomial one are slice. To understand why this is a consequence, note first the following.

THEOREM 6.1. *For a knot K, if $M(K,0)$ bounds a homology $S^1 \times B^3$, W, with $\pi_1(W) = \mathbf{Z}$ then K is slice.*

PROOF. We have that $M(K,0)$ is formed from S^3 by removing a solid torus and replacing it with another solid torus. Performing 0-surgery on the core, C, of that solid torus returns S^3. Attach a 2-handle to W with framing 0 to C. The resulting manifold is a homotopy ball with boundary S^3, and hence, by the Poincaré conjecture, is homeomorphic to B^4. The cocore of that added 2-handle is a slice disk for the boundary of the cocore, which can seen to be the original K. □

Freedman observed that a surgery obstruction to finding such a manifold W is determined by the Seifert form, and for a knot of Alexander polynomial one that is the only obstruction, and it vanishes.

6.1. *Extensions*

Is it possible that more delicate arguments using 4-dimensional surgery might yield stronger results, showing that other easily identified classes of algebraically slice knots are slice, based only on the Seifert form of the knot? The following result indicates that the answer is no.

THEOREM 6.2. *If $\Delta_K(t)$ is nontrivial then there are two nonconcordant knots having that Alexander polynomial.*

This result was first proved in [76] where there was the added constraint that the Alexander polynomial is not the product of cyclotomic polynomials $\phi_n(t)$ with n divisible by three distinct primes. The condition on Alexander polynomials is technical, assuring that some prime power branched cover is not a homology sphere. Kim [52] has shown this condition is not essential in particular cases, and in unpublished work he has shown that the result applies for all nontrivial Alexander polynomials.

7. Smooth knot concordance

In 1983, Donaldson [20] discovered new constraints on the intersection forms of smooth 4-manifolds. This and subsequent work soon yielded the following theorem.

THEOREM 7.1. *Suppose that X is a smooth closed 4-manifold and $H_1(X,\mathbf{Z}_2) = 0$. If the intersection form on $H_2(X)$ is positive definite then the form is diagonalizable. If the intersection form is even and definite, and hence of the type $nE_8 \oplus mH$, where H is the standard 2-dimensional hyperbolic form, then if $n > 0$, it follows that $m > 2$.*

This result is sufficient to prove that many knots of Alexander polynomial one are not slice. The details of any particular example cannot be presented here, but the connections with Theorem 7.1 are easily explained.

Let $M(K,1)$ denote the 3-manifold constructed as 1-surgery on K. Then $M(K,1)$ bounds the 4-manifold W constructed by adding a 2-handle to the 4-ball along K with framing 1. If K is slice, the generator of $H_2(W)$ is represented by a 2-sphere with self-intersection number 1. A tubular neighborhood of that sphere can be removed and replaced with a 4-ball, showing that $M(K,1)$ bounds a homology ball, X. If $M(K,1)$ also bounds a 4-manifold Y (say simply connected) with intersection form of the type obstructed by Theorem 1, then a contradiction is achieved using the union of X and Y.

As an alternative approach, notice that if K is slice, the 2-fold branched cover of S^3 branched over K, M_2,bounds the $\mathbf{Z}_2$-homology ball formed as the 2-fold branched cover of B^4 branched over the slice disk. Hence, if M_2 is known to bound a simply connected 4-manifold with one of the forbidden forms of Theorem 7.1, then again a contradiction is achieved.

It seems that prior to Donaldson's work it was known that either of these approaches would be applicable to proving that particular polynomial one knots are not slice, but these arguments were not published. In particular, following the announcement of Donaldson's theorem it immediately was known that the pretzel knot $K(-3,5,7)$ and the untwisted double of the trefoil (Akbulut) are not slice. Early papers presenting details of such arguments include [36] where it was shown that there are topologically slice knots of infinite order in smooth concordance. See [13] for further examples.

7.1. *Further advances*

Continued advances in smooth 4-manifold theory have led to further understanding of the knot slicing problem. In particular, proving that large classes of Alexander polynomial one knots are not slice has fallen to algorithmic procedures. Notable among this work is that of Rudolph [99–101]. Here, we outline briefly the approach using Thurston–Bennequin numbers, as described by Akbulut and Matveyev in the paper [2].

The 4-ball has a natural complex structure. If a 2-handle is added to the 4-ball along a knot K with appropriate framing, which we call f for now, the resulting manifold W will itself be complex. According to Lisca and Matić [67], W will then embed in a closed Kahler manifold X. Further restrictions on the structure of X are known to hold, and with these constraints the adjunction formula of Kronheimer and Mrowka [59,60] applies to show that no essential 2-sphere in X can have self-intersection greater than or equal to -1.

On the other hand, if K were slice and the framing f of K were greater than -2, such a sphere would exist. The appropriate framing f mentioned above depends on the choice of representative of K, not just its isotopy class. If the representative is $\mathbf{K}$, then $f = tb(\mathbf{K}) - 1$, where $tb(\mathbf{K})$ is the Thurston–Bennequin number, easily computed from a diagram for $\mathbf{K}$.

Applying this, both Akbulut–Matveyev [2] and Rudolph [101] have given simple proofs that, for instance, all iterated positive twisted doubles of the right handed trefoil are not slice.

Although these powerful techniques have revealed a far greater complexity to the concordance group than had been expected, as of yet they seem incapable of addressing some of the basic questions: for instance, the slice implies ribbon conjecture and problems related to torsion in the concordance group.

8. Higher order obstructions and the filtration of $\mathcal{C}$

Recent work of Cochran, Orr and Teichner has demonstrated a deep structure to the topological concordance group. This is revealed in a filtration of the concordance group by an infinite sequence of subgroups:

$$\cdots \mathcal{F}_{2.0} \subset \mathcal{F}_{1.5} \subset \mathcal{F}_1 \subset \mathcal{F}_{.5} \subset \mathcal{F}_0 \subset \mathcal{C}.$$

This approach has successfully placed known obstructions to the slicing problem – the Arf invariant, algebraic sliceness, and Casson–Gordon invariants – as the first in an infinite sequence of invariants. Of special significance is that each level of the induced filtration of the concordance group has both an algebraic interpretation and a geometric one. Here, we can offer a simplified view of the motivations and consequences of their work, and in that interest will focus on the $\mathcal{F}_n$ with n a nonnegative integer.

To begin, suppose that $M(K,0)$, 0-surgery on a knot K, bounds a 4-manifold W with the homology type and intersection form of $S^1 \times B^3 \#_n S^2 \times S^2$. Such a W will exist if and only if the Arf invariant of K is trivial. Constructing one such W is fairly simple in this case. Push a Seifert surface F for K into B^4 and perform surgery on B^4 along a set of curves on F representing a basis of a metabolizer for its intersection form, with the additional condition that it represents a metabolizer for the $\mathbf{Z}_2$-Seifert form. (Finding such a basis is where the Arf invariant condition appears.) When performing the surgery, the surface F can be ambiently surgered to become a disk, and the complement of that disk is the desired W.

If a generating set of a metabolizer for the intersection form on $H_2(W)$ could be represented by disjoint embedded 2-spheres, then surgery could be performed on W to convert it into a homology $S^1 \times B^3$. It would quickly follow that K would be slice in a homology 4-ball bounded by S^3.

In the higher dimensional analog (of the concordance group of knotted $(2k-1)$-spheres in S^{2k+1}, $k > 1$), there is an obstruction (to finding this family of spheres) related to the twisted intersection form on $H_{k+1}(W, \mathbf{Z}[\pi_1(W)])$, or, equivalently, related to the intersection form on the universal cover of W. In short, the intersection form of W should have a metabolizer that lifts to a metabolizer in the universal cover of W. In this higher dimensional setting, if the obstruction vanishes then, via the Whitney trick, the metabolizer for W can be realized by embedded spheres and W can be surgered as desired. This viewpoint on knot concordance has its roots in the work of Cappell and Shaneson [7].

Whether in high dimensions or in the classical setting, the explicit construction of a W described earlier in this section yields a W with cyclic fundamental group. This obstruction is thus determined solely by the infinite cyclic cover and vanishes for algebraically slice knots. Of course, in higher dimensions algebraically slice knots are slice. Clearly something more is needed in the classical case.

In light of the Casson–Freedman approach to 4-dimensional surgery theory, in addition to finding immersed spheres representing a metabolizer for W, one needs to find appropriate dual spheres in order to convert the immersed spheres into embeddings. The Cochran–Orr–Teichner filtration can be interpreted as a sequence of obstructions to finding a family of spheres and dual spheres. To describe the filtration, we denote $\pi^{(0)} = \pi = \pi_1(W)$ and let $\pi^{(n)}$ be the derived subgroup: $\pi^{(n+1)} = [\pi^{(n)}, \pi^{(n)}]$.

DEFINITION 8.1. A knot K is called n-solvable if there exists a (spin) 4-manifold W with boundary $M(K,0)$ such that: (a) the inclusion map $H_1(M(K,0)) \to H_1(W)$ is an isomorphism; (b) the intersection form on $H_2(W, \mathbf{Z}[\pi/\pi^{(n)}])$ has a dual pair of self-annihilating submodules (with respect to intersections and self-intersections), L_1 and L_2; and (c) the images of L_1 and L_2 in $H_2(W)$ generate $H_2(W)$.

(Here and in what follows we leave the description of $n.5$-solvability to [14].)

There are the following basic corollaries of the work in [14].

THEOREM 8.2. *If the Arf invariant of a knot K is 0, then K is 0-solvable. If K is 1-solvable, K is algebraically slice. If K is 2-solvable, Casson–Gordon type obstructions to K being slice vanish. If K is slice, K is n-solvable for all n.*

One of the beautiful aspects of [14] is that this very algebraic formulation is closely related to the underlying topology. For those familiar with the language of Whitney towers and gropes, we have the following theorem from [14].

THEOREM 8.3. *If K bounds either a Whitney tower or a grope of height $n+2$ in B^4, then K is n-solvable.*

Define $\mathcal{F}_n$ to be the subgroup of the concordance group consisting of n-solvable knots. One has the filtration (where we have dropped the $n.5$-subgroups).

$$\cdots\mathcal{F}_3 \subset \mathcal{F}_2 \subset \mathcal{F}_1 \subset \mathcal{F}_0 \subset \mathcal{C}.$$

Beginning with [14] and culminating in [16], there is the following result.

THEOREM 8.4. *For all n, the quotient group $\mathcal{F}_n/\mathcal{F}_{n+1}$ is infinite and $\mathcal{F}_2/\mathcal{F}_3$ is infinitely generated.*

Describing the invariants that provide obstructions to a knot being in $\mathcal{F}_n$ is beyond the scope of this survey. However, two important aspects should be mentioned. First, Cochran et al. [14] identifies a connection between n-solvability and the structure and existence of metabolizers for linking forms on

$$H_1(M(K,0), \mathbf{Z}[\pi_1(M(K,0))/\pi_1(M(K,0))^{(k)}]), \qquad k \leq n,$$

generalizing the fact that for algebraically slice knots the Blanchfield pairing of the knot vanishes.

The second aspect of proving the nontriviality of $\mathcal{F}_n/\mathcal{F}_{n+1}$ is the appearance of von Neumann signatures for solvable quotients of the knot group. Though difficult to compute in general, Cochran et al. [14] demonstrates that if K is built as a satellite knot, then in special cases, as with the Casson–Gordon invariant, the value of this complicated invariant is related to the Tristram–Levine signature function of the companion knot. More precisely, if a knot K is built from another knot by removing an unknot U that lies in $\pi^{(n)}$ of the complement and replacing it with the complement of a knot J, then the change in a particular von Neumann η-invariant of the $\pi^{(n)}$-cover is related to the integral of the Tristram–Levine signature function of J, taken over the entire circle. The Cheeger–Gromov estimate for these η-invariants can then be applied to show the nonvanishing of the invariant by choosing J in a way that the latter integral exceeds the estimate. This construction generalizes in a number of ways the one used in applications of the Casson–Gordon invariant described earlier, which applied only in the case that $U \in \pi^{(1)}$

and $U \notin \pi^{(2)}$. Furthermore, the Casson–Gordon invariant is based on a finite dimensional representation where here the representation becomes infinite dimensional. In the construction of [16] it is also required that one work with a family of unknots; a single curve U will not suffice.

9. Three-dimensional knot properties and concordance

9.1. *Primeness*

The first result of the sort to be discussed here is the theorem of Kirby and Lickorish [55]:

THEOREM 9.1. *Every knot is concordant to a prime knot.*

Shorter proofs of this were given in [70,94]. In these constructions it was shown that the concordance can be chosen so that the Seifert form, and hence the algebraic invariants, of the knot are unchanged. Myers [89] proved that every knot is concordant to a knot with hyperbolic complement, and hence to one with no incompressible tori in its complement. Later, Soma [103] extended Myers's result by showing that fibered knots are (fibered) concordant to fibered hyperbolic knots.

In the reverse direction, one might ask if every knot is concordant to a composite knot, but the answer here is obviously yes: K is concordant to $K\#J$, for any slice knot J. However, when the Seifert form is taken into consideration the question becomes more interesting. Here we have the following example, the proof of which is contained in [74 version 1].

THEOREM 9.2. *There exists a knot K with Seifert form $V_K = V_{J_1} \oplus V_{J_2}$, but K is not concordant to a connected sum of knots with Seifert forms V_{J_1} and V_{J_2}.*

Notice that by Levine's classification of higher dimensional concordance, such examples cannot exist in dimensions greater than 3.

9.2. *Knot symmetry: amphicheirality*

For the moment, view a knot K formally as a smooth oriented pair (S, K) where S is diffeomorphic to S^3 and K is diffeomorphic to S^1. Equivalence is up to orientation preserving diffeomorphism. (In dimension three it does not matter whether the smooth or locally flat topological category is used.)

DEFINITION 9.3. A knot (S, K) is called reversible (or invertible), negative amphicheiral, or positive amphicheiral, if it is equivalent to $K^r = (S, -K)$, $-K = (-S, -K)$, or $-K^r = (-S, K)$ respectively. It is called strongly reversible, strongly positive amphicheiral, or strongly negative amphicheiral if there is an equivalence that is an involution.

Each of these properties constrains the algebraic invariants of a knot, and hence can constrain the concordance class of a knot. For instance, according to Hartley [40], if a knot K is negative amphicheiral, then its Alexander polynomial satisfies $\Delta_K(t^2) = F(t)F(t^{-1})$ for some symmetric polynomial F. It follows quickly from the condition that slice knots have polynomials that factor as $g(t)g(t^{-1})$ that if a knot K is concordant to a negative amphicheiral knot, $\Delta_K(t^2)$ must factor as $F(t)F(t^{-1})$. Further discussion of amphicheirality and knot concordance is included in [18], where the focus is on higher dimensions, but some results apply in dimension three.

EXAMPLE. Let K be a knot with Seifert form

$$V_{\mathrm{a}} = \begin{pmatrix} 1 & 1 \\ 0 & -a \end{pmatrix}.$$

If a is positive, it follows from Levine's characterization of knots with quadratic Alexander polynomial (Theorem 3.1) that K is of order two in the algebraic concordance group if every prime of odd exponent in $4a + 1$ is congruent to 1 modulo 4. It follows as one example that any knot with Seifert form V_3, for instance the 3-twisted double of the unknot, is of order 2 in algebraic concordance but is not concordant to a negative amphicheiral knot.

This gives insight into the following conjecture based on a long standing question of Gordon [38]:

CONJECTURE 9.4. *If K is of order two in $\mathcal{C}$, then K is concordant to a negative amphicheiral knot.*

(Gordon's original question did not have the "negative" constraint in its statement.)

In a different direction, it was noted by Long [81] that the example of a knot K for which $K\# - K^r$ is not slice (described in the next subsection) yields an example of a nonslice strongly positive amphicheiral knot. Flapan [23] subsequently found a prime example of this type. It has since been shown that the concordance group contains infinitely many linearly independent such knots [73].

9.3. *Reversibility and mutation*

Every knot is algebraically concordant to its reverse. A stronger result, but the only proof in print, follows from Long [81]: if K is strongly positive amphicheiral then it is algebraically slice. For any knot, $K\# - K^r$ is strongly positive amphicheiral, so K and K^r are algebraically concordant. It is proved in [71] that there are knots that are not concordant to their reverses. Further examples have been developed in [56,90,107].

Kearton [48] observed that since $K\# - K^r$ is a (negative) mutant of the slice knot $K\# - K$, an example of a knot which is not concordant to its reverse yield an example of mutation changing the concordance class of a knot. Similar examples for positive mutants proved harder to find and were developed in [56,58].

9.4. *Periodicity*

A knot K is called periodic if it is invariant under a periodic transformation T of S^3 with the fixed point set of T a circle disjoint from K. Some of the strongest results concerning periodicity are those of Murasugi [88] constraining the Alexander polynomials of such knots. Naik [91] used Casson–Gordon invariants to obstruct periodicity for knots for which all algebraic invariants coincided with those of a periodic knot.

A theory of periodic concordance has been developed. Basic results in the subject include those of Cha and Ko [11] and Naik [92] obstructing knots from being periodically slice and those of Davis and Naik [19] giving a characterization of the Alexander polynomials of periodically ribbon knots.

9.5. *Genus*

The 4-ball genus of a knot K, $g_4(K)$, is the minimal genus of an embedded surface bounded by K in the 4-ball. It is a concordance invariant of a knot which is clearly bounded by its 3-sphere genus.

This invariant has been studied extensively. It is known to be bounded below by half the classical signature and the Tristram–Levine signature [43,44,86,110]. In the case that a knot is algebraically slice, Gilmer developed bounds on the 4-ball genus using Casson–Gordon invariants [30]. In [51] it is shown that for any pair of nonnegative integers m and n there is a knot K with a mutant K^* such that $g_4(K) = m$ and $g_4(K^*) = n$; a knot and its mutant are algebraically concordant. Beyond that, there are many results giving bounds on the 4-ball genus in the smooth setting based on differential geometric results. See, for instance [101,109].

Nakanishi [93] and Casson observed that there are knots that bound surfaces of genus one in the 4-ball but which are not concordant to knots of 3-sphere genus 1. In [77] this observation was the starting point of the definition of the concordance genus of a knot K: the minimum genus among all knots concordant to K. It is shown that this invariant can be arbitrarily large, even for knots of 4-ball genus 1, and even among algebraically slice knots.

9.6. *Fibering*

A knot is called fibered if its complement is a surface bundle over S^1. It is relatively easy to see that not all knots are concordant to fibered knots, as follows. The Alexander polynomial of a fibered knot is monic. Consider a knot K with $\Delta_K(t) = 2t^2 - 3t + 2$. If K were concordant to a fibered knot, then $\Delta_K(t)g(t) = f(t)f(t^{-1})$ for some monic polynomial g and integral f. However, since $\Delta_K(t)$ is irreducible and symmetric, it would have to be a factor of $f(t)$ and of $f(t^{-1})$, giving it even exponent in $\Delta_K(t)g(t)$, implying it is a factor of $g(t)$, contradicting monotonicity.

As mentioned above, Soma [103] proved that fibered knots are concordant to hyperbolic fibered knots.

The most significant result associating fibering and concordance is the theorem of Casson and Gordon [10].

THEOREM 9.5. *If K is a fibered ribbon knot, then the monodromy of the fibration extends over some solid handlebody.*

9.7. *Unknotting number*

The unknotting number of a knot K is the least number of crossing changes that must be made in any diagram of K to convert it to an unknot. This is closely related to the 4-ball genus of a knot (see the discussion above) and questions regarding the slicing of a knot in manifolds bounded by S^3 other than B^4, for instance, a once punctured connected sums of copies of $S^2 \times S^2$. A related invariant that is more closely tied to concordance was introduced by Askitas [4,95], which we call the slicing number of a knot: $u_s(K)$ is the minimum number of crossing changes required to convert a knot into a slice knot. It is relatively easy to see that the 4-ball genus of a knot provides a lower bound on the slicing number; it was shown in [85] and later in [75] that these two need not be equal.

10. Problems

Past problem sets that include questions related to the knot concordance group include [38,54].

(1) *Is every slice knot a ribbon knot?* A knot is ribbon if it bounds an embedded disk in B^4 having no local maxima (with respect to the radial function) in its interior. In the topological category this is not defined, so one asks the following instead: *is every slice knot homotopically ribbon?* (That is, does K bound a disk D in B^4 such that $\pi_1(S^3 - K) \to \pi_1(B^4 - D)$ is surjective?) In the smooth setting one then has the additional question: *is every homotopically ribbon knot a ribbon knot?*

    One has little basis to conjecture here. Perhaps obstructions will arise (in either category) but the lack of potential examples is discouraging. On the other hand, topological surgery might provide a proof in that category, but would give little indication concerning the smooth setting.

(2) *Describe all torsion in $\mathcal{C}$.* Beginning with [25] the question of whether there is any odd torsion has been open. More generally, the only known torsion in $\mathcal{C}$ is two torsion that arises from knots that are concordant to negative amphicheiral knots, and Conjecture 9.4 (first suggested in [38]) states that negative amphicheirality is the source of all (two) torsion in $\mathcal{C}$.

As described in Section 9, the Seifert form

$$V_3 = \begin{pmatrix} 1 & 1 \\ 0 & -3 \end{pmatrix}$$

represents 2-torsion in $\mathcal{G}$ but cannot be represented by a negative amphicheiral knot.

The prospects for understanding 4-torsion look better. A start has been made in [79,80] where it is shown, for instance, that no knot with Seifert form

$$V_5 = \begin{pmatrix} 1 & 1 \\ 0 & -5 \end{pmatrix}$$

can be of order 4 in $\mathcal{C}$, although every such knot is of order 4 in $\mathcal{G}$.

Closely related to questions of torsion is the question: *Does Levine's homomorphism split?* That is, is there a homomorphism $\psi : \mathcal{G} \to \mathcal{C}$ such that $\phi \circ \psi$: is the identity? An affirmative answer would yield elements of order 4 in $\mathcal{C}$ as well as elements of order 2 that do not arise from negative amphicheiral knots.

See [45,84] for computations of the algebraic orders of small crossing number knots.

(3) *If the knots K and K#J are doubly slice, that is cross-sections of unknotted 2-spheres in R^4, is J doubly slice?* The study of double knot concordance has a long history, with some of the initial work appearing in [106]. Other references include [33,46,47,64,65,105]. The property of double sliceness can be used to define a double concordance group which maps onto $\mathcal{C}$ and there is a corresponding algebraic double concordance group formed using quotienting by the set of hyperbolic Seifert forms rather than metabolic forms. Algebraic invariants show that the kernel is infinitely generated, and Casson–Gordon invariants and Cochran–Orr–Teichner methods apply in the case that algebraic invariants do not [33,53]. Although a variety of questions regarding double null concordance can be asked, this problem points to the underlying geometric difficulty of the topic.

(4) *Describe the structure of the kernel of Levine's homomorphism,* $\mathcal{A} = \ker(\phi : \mathcal{C} \to \mathcal{G})$. It is known [42,72] that $\mathcal{A}$ contains a subgroup isomorphic to $\mathbf{Z}^\infty \oplus \mathbf{Z}_2^\infty$. A reasonable conjecture is that $\mathcal{A} \cong \mathbf{Z}^\infty \oplus \mathbf{Z}_2^\infty$. It has recently been shown by the author [78] that results of Ozsváth and Szabó [96] imply that $\mathcal{A}$ has a summand isomorphic to $\mathbf{Z}$. This implies that $\mathcal{A}$ contains elements that are not divisible and that $\mathcal{A}$ is not a divisible group. There remains the unlikely possibility that $\mathcal{A}$ does contain infinitely divisible elements, perhaps including summands isomorphic to $\mathbf{Q}$ and $\mathbf{Q}/\mathbf{Z}$.

(5) *Describe the kernel of the map from $\mathcal{C}$ to the topological concordance group,* $\mathcal{C}_{\text{top}}$. It is known that the kernel is nontrivial, containing for instance nonsmoothly slice Alexander polynomial one knots. (See [13,36] for early references.) In fact it contains an infinitely generated such subgroup [22]. What more can be said about this kernel?

(6) *Identify new relationships between the various unknotting numbers and genera of a knot*. Here is a problem that seems to test the limits of presently known techniques. If K can be converted into a slice knot by making m positive crossing changes and n negative crossing changes, then a geometric construction yields a surface bounded by K in the 4-ball of genus $\max\{m,n\}$. Conversely: *If the 4-ball genus of K is g_4, can K be converted into a slice knot by making g_4 positive and g_4 negative crossing changes?* A simpler question ask the same thing except for the 3-sphere genus g_3 instead of g_4. (It is interesting to note that at this time it seems unknown if the classical unknotting number satisfies $u(K) \leq 2g_3(K)$.)

Acknowledgements

This survey benefited from the suggestions of many readers. In particular, conversations with Pat Gilmer were very helpful. Also, the careful reading of Section 8 by Tim Cochran, Kent Orr and Peter Teichner improved the exposition there. General references concerning concordance which were of great benefit to me as I learned the subject include [37,98].

References

[1] H. Abchir, *Note on the Casson–Gordon invariant of a satellite knot*, Manuscr. Math. **90** (1996), 511–519.

[2] S. Akbulut and R. Matveyev, *Exotic structures and adjunction inequality*, Turk. J. Math. **21** (1997), 47–53.

[3] E. Artin, *Zur Isotopie zweidimensionalen Flachen im R_4*, Abh. Math. Sem. Univ. Hamburg **4** (1926), 174–177.

[4] N. Askitas, *Multi-# unknotting operations: a new family of local moves on a knot diagram and related invariants of knots*, J. Knot Theory Ramificat. **7** (1998), 857–871.

[5] W. Browder, *Surgery on simply-connected manifolds*, Ergebnisse der Mathematik und ihrer Grenzgebiete, Band 65, Springer, New York (1972).

[6] G. Burde and H. Zieschang, *Knots*, de Gruyter Studies in Mathematics, 5, Walter de Gruyter & Co., Berlin (1985).

[7] S. Cappell and J. Shaneson, *The codimension two placement problem and homology equivalent manifolds*, Ann. Math. **99** (2) (1974), 277–348.

[8] A. Casson and C.McA. Gordon, Cobordism of classical knots. *A la recherche de la Topologie perdue*, Progress in Mathematics, Vol. 62, L. Guillou, and A. Marin, eds (1986). (Originally published as Orsay Preprint, 1975.).

[9] A. Casson and C.McA. Gordon, *On Slice Knots Dimension Three*, Algebraic and Geometric Topology (Proc. Symp. Pure Math., Stanford Univ., Stanford, CA, 1976), Part 2, Proc. Symp. Pure Math., XXXII, American Mathematical Society, Providence, RI (1978), pp. 39–53.

[10] A. Casson and C.McA. Gordon, *A loop theorem for duality spaces and fibred ribbon knots*, Invent. Math. **74** (1983), 119–137.

[11] J.C. Cha and K.H. Ko, *On equivariant slice knots*, Proc. Am. Math. Soc. **127** (1999), 2175–2182.

[12] J. Cha and C. Livingston, Knot signatures are independent. Preprint 2002. arxiv.org/math.GT/0208225.

[13] T. Cochran and R. Gompf, *Applications of Donaldson's theorems to classical knot concordance, homology 3-spheres and property P*, Topology **27** (1988), 495–512.

[14] T. Cochran, K. Orr and P. Teichner, *Knot concordance, Whitney towers and L^2 signatures*, Ann. Math. **157** (2003), 433–519.

[15] T. Cochran, K. Orr and P. Teichner, Structure in the classical knot concordance group. *Comment. Math. Helv.*, to appear, arxiv.org/math.GT/0206059.

[16] T. Cochran and P. Teichner, Knot concordance and von Neumann η-invariants. Preprint 2004. arxiv.org/math.GT/0411057.

[17] D. Cooper, *Detecting knots that are not slice*, Thesis, Warwick University (1982).
[18] D. Coray and F. Michel, *Knot cobordism and amphicheirality*, Comment. Math. Helv. **58** (1983), 601–616.
[19] J. F. Davis and S. Naik, Alexander polynomials of equivariant slice and ribbon knots in S^3. Preprint 2002. arxiv.org/math.GT/0201179.
[20] S. Donaldson, *An application of gauge theory to four-dimensional topology*, J. Differ. Geom. **18** (1983), 279–315.
[21] S. Donaldson and P. Kronheimer, *The Geometry of Four-Manifolds*, Oxford Mathematical Monographs, Oxford Science Publications. The Clarendon Press, Oxford University Press, New York (1990).
[22] H. Endo, *Linear independence of topologically slice knots in the smooth cobordism group*, Topol. Appl. **63** (1995), 257–262.
[23] E. Flapan, *A prime strongly positive amphicheiral knot which is not slice*, Math. Proc. Cambridge Philos. Soc. **100** (1986), 533–537.
[24] R. Fox, *A Quick Trip Through Knot Theory*, Topology of 3-Manifolds, M.K. Fort, ed., Prentice-Hall, Englewood Cliffs, NJ (1962), 120–167.
[25] R. Fox and J. Milnor, *Singularities of 2-spheres in 4-space and cobordism of knots*, Osaka J. Math. **3** (1966), 257–267.
[26] M. Freedman, *The topology of four-dimensional manifolds*, J. Differ. Geom. **17** (1982), 357–453.
[27] M. Freedman and F. Quinn, *Topology of 4-Manifolds*, Princeton Mathematical Series, Vol. 39., Princeton University Press, Princeton, NJ (1990).
[28] S. Friedl, L^2-eta-invariants and their approximation by unitary eta-invariants. Preprint 2003. arxiv.org/math.GT/0305401.
[29] S. Friedl, Eta invariants as sliceness obstructions and their relation to Casson–Gordon invariants. Preprint, 2003 arxiv.org/mathGT/0305402.
[30] P. Gilmer, *On the slice genus of knots*, Invent. Math. **66** (1982), 191–197.
[31] P. Gilmer, *Slice knots in S^3*, Quart. J. Math. Oxford Ser. **34** (2) (1983), 305–322.
[32] P. Gilmer, *Classical knot and link concordance*, Comment. Math. Helv. **68** (1993), 1–19.
[33] P. Gilmer and C. Livingston, *On embedding 3-manifolds in 4-space*, Topology **22** (1983), 241–252.
[34] P. Gilmer and C. Livingston, *Discriminants of Casson–Gordon invariants*, Math. Proc. Cambridge Philos. Soc. **112** (1992), 127–139.
[35] P. Gilmer and C. Livingston, *The Casson–Gordon invariant and link concordance*, Topology **31** (1992), 475–492.
[36] R. Gompf, *Smooth concordance of topologically slice knots*, Topology **25** (1986), 353–373.
[37] C.McA. Gordon, *Some Aspects of Classical Knot Theory*, Knot Theory (Proc. Sem., Plans-sur-Bex, 1977), Lecture Notes in Mathematics., Vol. 685, Springer, Berlin (1978), pp. 1–60.
[38] C.McA. Gordon, *Problems in Knot Theory*, Knot Theory (Proc. Sem., Plans-sur-Bex, 1977), Lecture Notes in Mathematics., Vol. 685, Springer, Berlin (1978).
[39] C.McA. Gordon, *Ribbon concordance of knots in the 3-sphere*, Math. Ann. **257** (2) (1981), 157–170.
[40] R. Hartley, *Invertible amphicheiral knots*, Math. Ann. **252** (1979/1980), 103–109.
[41] D. Husemoller and J. Milnor, *Symmetric Bilinear Forms*, Ergebnisse der Mathematik und ihrer Grenzgebiete, Band 73, Springer, New York (1973).
[42] B. Jiang, *A simple proof that the concordance group of algebraically slice knots is infinitely generated*, Proc. Am. Math. Soc. **83** (1981), 189–192.
[43] L. Kauffman, *Branched coverings, open books and knot periodicity*, Topology **13** (1974), 143–160.
[44] L. Kauffman and L. Taylor, *Signature of links*, Trans. Am. Math. Soc. **216** (1976), 351–365.
[45] A. Kawauchi, *A Survey of Knot Theory*, Birkhauser, Basel (1996).
[46] C. Kearton, *Simple knots which are doubly-null-cobordant*, Proc. Am. Math. Soc. **52** (1975), 471–472.
[47] C. Kearton, *Hermitian signatures and double-null-cobordism of knots*, J. London Math. Soc. **23** (2) (1981), 563–576.
[48] C. Kearton, *Mutation of knots*, Proc. Am. Math. Soc. **105** (1989), 206–208.
[49] M. Kervaire, *Knot Cobordism in Codimension Two*, Manifolds-Amsterdam 1970 (Proc. Nuffic Summer School), Lecture Notes in Mathematics, Vol. 197, Springer, Berlin (1971), pp. 83–105.

[50] S. Kim, Polynomial splittings of Casson–Gordon invariants. Preprint 2003. arxiv.org/math.math.GT/0102112.

[51] S. Kim and C. Livingston, Knot mutation: 4-genus of knots and algebraic concordance. Preprint 2003. arxiv.org/math.math.GT/0310100.

[52] T. Kim, Filtration of the classical knot concordance group and Casson–Gordon invariants. Preprint 2002. arxiv.org/math.GT/0207221.

[53] T. Kim, Obstructions to slicing and doubly slicing knots it, Thesis, Indiana University (2003).

[54] R. Kirby, Problems in low-dimensional topology, *Geometric Topology (Athens, GA, 1993)*, AMS/IP Stud. Adv. Math., 2.2, R. Kirby, ed., American Mathematical Society, Providence, RI (1997), 35–473.

[55] R. Kirby and W.B.R. Lickorish, *Prime knots and concordance*, Math. Proc. Cambridge Philos. Soc. **86** (1979), 437–441.

[56] P. Kirk and C. Livingston, *Twisted Alexander invariants, Reidemeister torsion, and Casson–Gordon invariants*, Topology **38** (1999), 635–661.

[57] P. Kirk and C. Livingston, *Twisted knot polynomials: inversion, mutation and concordance*, Topology **38** (1999), 663–671.

[58] P. Kirk and C. Livingston, *Concordance and mutation*, Geom. Topol. **5** (2001), 831–883.

[59] P. Kronheimer and T. Mrowka, *Gauge theory for embedded surfaces. I*, Topology **32** (1993), 773–826.

[60] P. Kronheimer and T. Mrowka, *Gauge theory for embedded surfaces. II*, Topology **34** (1995), 37–97.

[61] C. Letsche, *An obstruction to slicing knots using the eta invariant*, Math. Proc. Cambridge Philos. Soc. **128** (2000), 301–319.

[62] J. Levine, *Knot cobordism groups in codimension two*, Comment. Math. Helv. **44** (1969), 229–244.

[63] J. Levine, *Invariants of knot cobordism*, Invent. Math. **8** (1969), 98–110.

[64] J. Levine, *Doubly sliced knots and doubled disk knots*, Mich. Math. J. **30** (1983), 249–256.

[65] J. Levine, *Metabolic and hyperbolic forms from knot theory*, J. Pure Appl. Algebra **58** (1989), 251–260.

[66] X.-S. Lin, *Representations of knot groups and twisted Alexander polynomials*, Acta Math. Sin. **17** (2001), 361–380.

[67] P. Lisca and G. Matić, *Tight contact structures and Seiberg Witten invariants*, Invent. Math. **129** (1997), 509–525.

[68] R. Litherland, A formula for the Casson–Gordon invariant of a knot. Preprint.

[69] R. Litherland, Cobordism of satellite knots, *Four-Manifold Theory*, Contemporary Mathematics, C. Gordon and R. Kirby, eds, American Mathematical Society, Providence, RI (1984), 327–362.

[70] C. Livingston, *Homology cobordisms of 3-manifolds, knot concordances, and prime knots*, Pac. J. Math. **94** (1981), 193–206.

[71] C. Livingston, *Knots which are not concordant to their reverses*, Quart. J. Math. Oxford Ser. **34** (2) (1983), 323–328.

[72] C. Livingston, *Order 2 algebraically slice knots. Proceedings of the Kirbyfest (Berkeley, CA, 1998)*, Geom. Topol. Monogr., 2, Geom. Topol. Publ., Coventry (1999), 335–342.

[73] C. Livingston, *Infinite order amphicheiral knots*, Algebr. Geom. Topol. **1** (2001), 231–241.

[74] C. Livingston, Examples in Concordance. Preprint 2001. arxiv.org/math.GT/0101035v1.

[75] C. Livingston, *The slicing number of a knot*, Algebr. Geom. Topol. **2** (2002), 1051–1060.

[76] C. Livingston, *Seifert forms and concordance*, Geom. Topol. **6** (2002), 403–408.

[77] C. Livingston, The concordance genus of knots. Preprint 2001. arxiv.org/math.GT/0107141.

[78] C. Livingston, Splitting the concordance group of algebraically slice knots. Preprint 2003. arxiv.org/math.GT/0305363.

[79] C. Livingston and S. Naik, *Obstructing four-torsion in the classical knot concordance group*, J. Differ. Geom. **51** (1999), 1–12.

[80] C. Livingston and S. Naik, *Knot concordance and torsion*, Asian J. Math. **5** (2001), 161–167.

[81] D.D. Long, *Strongly plus-amphicheiral knots are algebraically slice*, Math. Proc. Cambridge Philos. Soc. **95** (1984), 309–312.

[82] T. Matumoto, *On the signature invariants of a non-singular complex sesquilinear form*, J. Math. Soc. Jpn **29** (1977), 67–71.

[83] J. Milnor, *Infinite cyclic coverings, Topology of Manifolds*, Complementary Series in Mathematics, Vol. 13, J.G. Hocking, ed., Prindle, Weber & Schmidt, Boston (1968).

[84] T. Morita, *Orders of knots in the algebraic knot cobordism group*, Osaka J. Math. **25** (1988), 859–864.

[85] H. Murakami and A. Yasuhara, *Four-genus and four-dimensional clasp number of a knot*, Proc. Am. Math. Soc. **128** (2000), 3693–3699.

[86] K. Murasugi, *On a certain numerical invariant of link types*, Trans. Am. Math. Soc. **117** (1965), 387–422.

[87] K. Murasugi, *The Arf invariant for knot types*, Proc. Am. Math. Soc. **21** (1969), 69–72.

[88] K. Murasugi, *On periodic knots*, Comment. Math. Helv. **46** (1971), 162–174.

[89] R. Myers, *Homology cobordisms, link concordances, and hyperbolic 3-manifolds*, Trans. Am. Math. Soc. **278** (1983), 271–288.

[90] S. Naik, *Casson–Gordon invariants of genus one knots and concordance to reverses*, J. Knot Theory Ramificat. **5** (1996), 661–677.

[91] S. Naik, *New invariants of periodic knots*, Math. Proc. Cambridge Philos. Soc. **122** (1997), 281–290.

[92] S. Naik, *Equivariant concordance of knots in S^3*, KNOTS '96 (Tokyo), World Scientific Publishing, River Edge, NJ (1997), pp. 81–89.

[93] Y. Nakanishi, *A note on unknotting number*, Math. Sem. Notes Kobe Univ. **9** (1981), 99–108.

[94] Y. Nakanishi, *Prime links, concordance and Alexander invariants*, Math. Sem. Notes Kobe Univ. **8** (1980), 561–568.

[95] T. Ohtsucki, Problems on invariants of knots and 3-manifolds, *Invariants of Knots and 3-Manifolds, Kyoto University 2001*, Geometry and Topology Monographs, Vol. 4 (2002).

[96] P. Ozsváth and Z. Szabó, Knot Floer homology and the four-ball genus. Preprint 2003. arxiv.org/math.GT/0301149.

[97] R. Robertello, *An invariant of knot cobordism*, Commun. Pure Appl. Math. **18** (1965), 543–555.

[98] D. Rolfsen, *Knots and Links*, Publish or Perish, Berkeley, CA (1976).

[99] L. Rudolph, *An obstruction to sliceness via contact geometry and "classical" gauge theory*, Invent. Math. **119** (1995), 155–163.

[100] L. Rudolph, *Quasipositivity as an obstruction to sliceness*, Bull. Am. Math. Soc. **29** (1993), 51–59.

[101] L. Rudolph, *The slice genus and the Thurston–Bennequin invariant of a knot*, Proc. Am. Math. Soc. **125** (1997), 3049–3050.

[102] H. Seifert, *Über das Geschlecht von Knoten*, Math. Ann. **110** (1934), 571–592.

[103] T. Soma, *Hyperbolic, fibred links and fibre-concordances*, Math. Proc. Cambridge Philos. Soc. **96** (1984), 283–294.

[104] N. Stoltzfus, *Unraveling the integral knot concordance group*, AMS Mem. (1977) no. 192.

[105] N. Stoltzfus, *Equivariant concordance of invariant knots*, Trans. Am. Math. Soc. **254** (1979), 1–45.

[106] D.W. Sumners, *Invertible knot cobordisms*, Comment. Math. Helv. **46** (1971), 240–256.

[107] A. Tamulis, *Concordance of classical knots*, Thesis, Indiana University, Bloomington (1999).

[108] A. Tamulis, *Knots of ten or fewer crossings of algebraic order 2*, J. Knot Theory Ramificat. **11** (2002), 211–222.

[109] T. Tanaka, *Four-genera of quasipositive knots*, Topol. Appl. **83** (1998), 187–192.

[110] A. Tristram, *Some cobordism invariants for links*, Proc. Cambridge Philos. Soc. **66** (1969), 251–264.

[111] H.F. Trotter, *Homology of group systems with applications to knot theory*, Ann. Math. **76** (2) (1962), 464–498.

[112] M. Wada, *Twisted Alexander polynomial for finitely presentable groups*, Topology **33** (1994), 241–256.

CHAPTER 8

Knot Theory of Complex Plane Curves

Lee Rudolph*

Department of Mathematics and Computer Science and Department of Psychology, Clark University, Worcester MA 01610 USA
E-mail: lrudolph@black.clarku.edu

Contents

*During the preparation and completion of this survey, the author was partially supported by the Fonds National Suisse and by a National Science Foundation Interdisciplinary Grant in the Mathematical Sciences (DMS-0308894).

HANDBOOK OF KNOT THEORY
Edited by William Menasco and Morwen Thistlethwaite

Abstract

The primary objects of study in the "knot theory of complex plane curves" are *ℂ-links*: links (or knots) cut out of a 3-sphere in $\mathbb{C}^2$ by complex plane curves. There are two very different classes of ℂ-links, *transverse* and *totally tangential*. Transverse ℂ-links are naturally oriented. There are many natural classes of examples: links of singularities; links at infinity; links of divides, free divides, tree divides, and graph divides; and – most generally – quasipositive links. Totally tangential ℂ-links are unoriented but naturally framed; they turn out to be precisely the real-analytic Legendrian links, and can profitably be investigated in terms of certain closely associated transverse ℂ-links.

The knot theory of complex plane curves is attractive not only for its own internal results but also for its intriguing relationships and interesting contributions elsewhere in mathematics. Within low-dimensional topology, related subjects include braids, concordance, polynomial invariants, contact geometry, fibered links and open books, and Lefschetz pencils. Within low-dimensional algebraic and analytic geometry, related subjects include embeddings and injections of the complex line in the complex plane, line arrangements, Stein surfaces, and Hilbert's 16th problem. There is even some experimental evidence that nature favors quasipositive knots.

1. Foreword

In the past two decades, knot theory in general has seen much progress and many changes. "Classical knot theory" – the study of knots as objects in their own right – has taken great strides, documented throughout this Handbook (see the chapters by Birman and Brendle, Hoste, Kauffman, Livingston, and Scharlemann). Simultaneously, there have been extraordinarily wide and deep developments in what might be called "modern knot theory": the study of knots and links in the presence of extra structure,[1] for instance, a hyperbolic metric on the knot complement (as in the chapters by Adams and Weeks) or a contact structure on the knot's ambient 3-sphere (as in the chapter by Etnyre). In these terms, the knot theory of complex plane curves is solidly part of modern knot theory – the knots and links in question are $\mathbb{C}$-links, and the extra structures variously algebraic, analytic, and geometric.

"Some knot theory of complex plane curves" (Rudolph, 1983d) was a broad view of the state of the art in 1982. Here, I propose to look at the subject through a narrower lens, that of quasipositivity. Section 2 is devoted to general notations and definitions. Section 3 is a treatment of braids and braided surfaces tailored to quasipositivity.[2] General transverse $\mathbb{C}$-links are constructed and described in Section 4, while Section 5 is a brief look at the special transverse $\mathbb{C}$-links that arise from complex algebraic geometry "in the small" and "in the large" – to wit, links of singularities and links at infinity.[3] Totally tangential $\mathbb{C}$-links are constructed and described in Section 6. The material in Sections 4–6 is related to other research areas in Section 7. In Section 8 I give some fairly explicit, somewhat programmatic suggestions of directions for future research in the knot theory of complex plane curves.

Original texts of some motivating problems in the knot theory of complex plane curves are collected in an Appendix. This survey concludes with an index of definitions and notations and a bibliography. Open questions are distributed throughout.

2. Preliminaries

Terms being defined are set in *this* typeface; mere emphasis is indicated *thus*. A definition labeled as such is either of greater (local) significance, or non-standard to an extent which might lead to confusion; labeled or not, potentially startling definitions and notations are flagged with the symbol ☡ in the margins of both the text and the index. The end or omission of a proof is signaled by □. Both $A := B$ and $B =: A$ mean "A is defined as B". The symbol $\cong$ is reserved for a *natural isomorphism* (in an appropriate category). ☡

[1] Some observers have also detected "postmodern knot theory": the study of extra structure in the absence of knots.

[2] Consult the chapter by Birman and Brendle for a deeper and broader account of braid theory.

[3] One consequence of this survey's bias towards quasipositivity is a de-emphasis of other aspects of the knot theory of links of singularities and links at infinity; the reader is referred to Boileau and Fourrier, 1998 (who include sections on both these topics), to the discussions of singularities and their higher-dimensional analogues by Durfee (1999, Section 2) and Neumann (2001, Section 1), and of course to the extensive literature on both subjects – particularly links of singularities – referenced in those articles.

2.1. *Sets and groups*

The set of real (resp., complex) numbers is $\mathbb{R}$ (resp., $\mathbb{C}$); write $z \mapsto \bar{z}$ for *complex conjugation* $\mathbb{C} \to \mathbb{C}$, and Re (resp., Im) for *real part* (resp., *imaginary part*) $\mathbb{C} \to \mathbb{R}$. For $x \in \mathbb{R}$, let $\mathbb{R}_{\geq x} := \{t \in \mathbb{R}: t \geq x\}$, $\mathbb{R}_{>x} := \{t \in \mathbb{R}: t > x\}$, $\mathbb{R}_{\leq x} := \{t \in \mathbb{R}: t \leq x\}$, $\mathbb{R}_{<x} := \{t \in \mathbb{R}: t < x\}$. Let $\mathbb{C}_{\pm} := \{w \in \mathbb{C}: \pm \operatorname{Im} w \geq 0\}$. Let $\mathbb{N} := \mathbb{Z} \cap \mathbb{R}_{\geq 0}$ and $\mathbb{N}_{>0} := \mathbb{N} \cap \mathbb{R}_{>0}$. For $n \in \mathbb{N}$, let $\mathbf{n} := \{k \in \mathbb{N}_{>0}: k \leq n\}$. Denote projection on the ith factor of a Cartesian product by pr_i. The Euclidian norm on $\mathbb{R}^n$ or $\mathbb{C}^n$ is $\|\cdot\|$. For $\mathbf{u}$, $\mathbf{v}$ in a real vectorspace, $[\mathbf{u}, \mathbf{v}]$ is $\{(1 - t)\mathbf{u} + t\mathbf{v}: 0 \leq t \leq 1\}$.

Let X be a set. Denote the identity map $X \to X$ by id_X, and the cardinality of X by $\mathrm{card}(X)$. A *characteristic function* on X is an element of $\{0,1\}^X$; for $Y \subset X$, let $c_{Y \hookrightarrow X}: X \to \{0,1\}$ denote *the characteristic function of* Y in X, so $c_{Y \hookrightarrow X}(x) = 1$ for $x \in Y$, $c_{Y \hookrightarrow X}(x) = 0$ for $x \in X \setminus Y$. A *multicharacteristic function* on X is an element of $\mathbb{N}^X$. Let $c \in \mathbb{N}^X$. The *total multiplicity* of c is $\Sigma_{x \in X} c(x) \in \mathbb{N} \cup \{\infty\}$. For $n \in \mathbb{N}$, an *n-multisubset* of X is a pair $(\mathrm{supp}(c), c|\mathrm{supp}(c))$ where c is a multicharacteristic function on X of total multiplicity n, Call the set of all n-multisubsets of X the nth *multipower set* of X, denoted $\mathcal{MP}^{[n]}(X)$. (Note that if $X \neq X'$ then no multicharacteristic function on X is a multicharacteristic function on X', whereas $\mathcal{MP}^{[n]}(X) \cap \mathcal{MP}^{[n]}(X') \neq \varnothing$ if and only if $X \cap X' \neq \varnothing$.) For any A, any f: $A \to \mathcal{MP}^{[n]}(X)$ can be construed as a *multivalued* (specifically, an *n-valued*) *function from A to X*; typically $\mathrm{gr}(f) := \{(a,x) \in A \times X: x \in f(a)\}$, the *multigraph of f*, determines neither n nor f (unless n is – *and is known to be* – equal to 1), but the notation is still useful. The type of $c \in \mathbb{N}^X$ is $\tau(c) := \mathrm{card} \circ (c^{-1}|\mathbb{N}_{>0}) \in \mathbb{N}^{\mathbb{N}_{>0}}$ (a multicharacteristic function on $\mathbb{N}$). Identify an *n-subset* of X (i.e., $Y \subset X$ with $\mathrm{card}(Y) = n$) with the n-multisubset $(Y, 1)$ of type $nc_{\{1\} \hookrightarrow \mathbb{N}}$, and the set of n-subsets of X with the nth *configuration set*

$$\binom{X}{n} := \{(\mathrm{supp}(c), c|\mathrm{supp}(c)) \in \mathcal{MP}^{[n]}(X) : \tau(c) = nc_{\{1\} \hookrightarrow \mathrm{N}}\}$$

of X. Call $\Delta_n(X) := \mathcal{MP}^{[n]}(X) \setminus \binom{X}{n}$ the nth *discriminant set* of X.

Let G be a group. For g, $h \in G$, let ${}^g h$ (resp., $[g,h]$;$\lll g,h \ggg$) denote the *conjugate* (resp., *commutator; yangbaxter*) ghg^{-1} (resp., $ghg^{-1}h^{-1} = {}^g h h^{-1}$;$ghgh^{-1}g^{-1}h^{-1} = {}^{gh}gh^{-1}$). For $A \subset G$, let $\langle A \rangle_G$ be the subgroup generated by A, i.e., $\cap \{H: A \subset H$ and H is a subgroup of $G\}$, and let $\langle a \rangle_G := \langle \{a\} \rangle_G$ (when G is understood, it may be dropped from these notations). The *normal closure* of A in G is $\langle \{{}^g a$: $g \in G$, $a \in A\} \rangle_G$. A *presentation* of G, denoted

$$G = \mathrm{gp}\big(\mathbf{g}_i (i \in I) : \mathbf{r}_j (j \in J)\big), \tag{2.1}$$

consists of: (1) a short exact sequence $R \subset F \overset{\pi}{\twoheadrightarrow} G$ in which F is a free group and R is a subgroup of F; (2) a family $\{\mathbf{g}_i : i \in I\} \subset F$ of free generators of F, the *generators of* (2.1); and (3) a family $\{\mathbf{r}_j : j \in J\} \subset R$ with normal closure R, the *relators of* (2.1). (Sometimes the elements $\pi(\mathbf{g}_i)$ of G are also, abusively, called *the generators of G* with respect to (2.1).) The presentation (2.1) is *Wirtinger* in case every relator $\mathbf{r}_j$ is of the form ${}^{w(j)}\mathbf{g}_{s(j)} \mathbf{g}_{t(j)^{-1}}$ for some $s,t: J \to I$ and $w: J \to F$. Denote the *free product* of groups G_0 and G_1 by $G_0 * G_1$.

A *partition* of a set X is the quotient set $X/\equiv$ of X by an equivalence relation $\equiv$ on X. Call $X/\approx$ a *refinement* of $X/\equiv$ in case each $\approx$-class is a union of $\equiv$-classes. Given $f: X \to Y$, write X/f for $X/\equiv_f$, where $x_0 \equiv_f x_1$ iff $f(x_0) = f(x_1)$. Given a (right) action of a group G on X, write X/G for $X/\equiv_G$, where $x_0 \equiv_G x_1$ iff $x_1 = x_0 g$ for some $g \in G$; as usual, xG stands for the $\equiv_G$-class (i.e., the G-orbit) of x. The group $\mathfrak{S}_n$ of bijections $\mathbf{n} \to \mathbf{n}$ acts in a standard way on $X^n = X^{\mathbf{n}} = \{f : f : \mathbf{n} \to X\}$. An *unordered n-tuple* in X is an element $U(x_1,\ldots,x_n) := (x_1,\ldots,x_n)\mathfrak{S}_n$ of the nth *symmetric power* $X^n/\mathfrak{S}_n$ of X, where $U = U_{X,n}: X^n \to X^n/\mathfrak{S}_n$ is the *unordering map*. The map

$$\begin{aligned} X^n/\mathfrak{S}_n \to \mathcal{MP}^{[n]}(X) : U(\overbrace{x_1,\ldots,x_1}^{n_1}, \overbrace{x_2,\ldots,x_2}^{n_2}, \ldots, \overbrace{x_k,\ldots,x_k}^{n_k}) \\ \mapsto (\{x_1,x_2,\ldots,x_k\}, (\{x_1,x_2,\ldots,x_k\} \to \mathbb{N}_{>0} : x_i \mapsto n_i)) \end{aligned} \tag{2.2}$$

(where $x_i \neq x_j$ for $i \neq j$, $n_i > 0$, and $n = \sum_{i=1}^k n_i$) is a bijection. *Except* in (2.2), notation is abused in the standard way, so that $\{x_1,\ldots, x_n\}$ denotes $U(x_1,\ldots, x_n)$. In case X is ordered by $\preceq$, notations like $\{x_1 \preceq \ldots \preceq x_k\} \in \binom{X}{k}$, $\{i \preceq j\} : C \to \binom{X}{2}$, and so on, mean "$\{x_1,\ldots x_k\} \in \binom{X}{k}$ and $x_1 \preceq \ldots \preceq x_k$", "$\{i(c), j(c)\} \in \binom{X}{2}$ and $i(c) \preceq j(c)$ for all $c \in C$", and so on. In case X is totally ordered by $<$, call $\{s < t\}, \{s' < t'\} \in \binom{X}{2}$ *linked* (resp., *unlinked; in touch at u*) iff either $s < s' < t < t'$ or $s' < s < t' < t$ (resp., either $s < s' < t' < t$ or $s' < s < t < t'$; $\{u\} = \{s,t\} \cap \{s',t'\}$).

The bijection (2.2) induces a definition of $\tau(\{x_1,\ldots,x_n\})$ as itself an unordered n-tuple, such that, e.g., $\tau(\{1,1,1,1\}) = \{4\}$, $\tau(\{1,1,2,3\}) = \{1,1,2\}$, etc.

2.2. *Spaces*

A simplicial complex is not necessarily finite. A geometric realization of a simplicial complex $\mathcal{K}$ is denoted $|||\mathcal{K}|||$. A *triangulation* of a topological space X is a homeomorphism $|||\mathcal{K}||| \to X$ for some $\mathcal{K}$. A *polyhedron* is a topological space X which is the target of some triangulation. Let X be a polyhedron. The set of components of X is denoted $\pi_0(X)$. The Euler characteristic of X is denoted $\chi(X)$. The fundamental group of X with *base point* $*$ is denoted by $\pi_1(X; *)$, or simply $\pi_1(X)$ in contexts where $*$ can be safely suppressed. Call $\pi_1(X \setminus K)$ the *knotgroup* of K in case X is connected and $K \subset X$.

Manifolds are smooth (C^∞) unless otherwise stated. A manifold M may have boundary, but corners (possibly cuspidal) only when so noted. Denote by ∂M (resp., Int M; $T(M)$) the interior (resp., boundary; tangent bundle) of M. Call M *closed* (resp., *open*) in case M is compact (resp., M has no compact component) and $\partial M = \varnothing$. Manifolds are (not only orientable, but) oriented, unless otherwise stated: in particular, a complex manifold (e.g., $\mathbb{C}^n$ or $\mathbb{P}_n(\mathbb{C})$) has a *natural orientation*, and $\mathbb{R}$, $D^{2n} := \{(z_1,\ldots,z_n) \in \mathbb{C}^n : \|(z_1,\ldots, z_n)\| \leq 1\}$, and $S^{2n-1} := \partial D^{2n}$ have *standard orientations*, as do S^2 (identified with $\mathbb{P}_1(\mathbb{C}) := \mathbb{C} \cup \{\infty\}$), $\mathbb{R}^3$ (identified with $S^3 \setminus (0,1)$) and the *bidisk* $D^2 \times D^2$ (with corners $S^1 \times S^2$). The tangent space $T_x(M)$ to M at $x \in M$ is an oriented vectorspace. Let $-M$ (resp., $+M$; $|M|$) denote M with orientation reversed (resp., preserved; forgotten); in case $\mathsf{M}: M \to M$ is a diffeomorphism reversing orientation, Mir $Q := \mathsf{M}(Q)$ is a *mirror image* of $Q \subset M$.

Call $Q \subset M$ *interior* (resp., *boundary*) in case $Q \subset \operatorname{Int} M$ (resp., $Q \subset \partial M$). In case Q has a closed regular neighborhood $\operatorname{Nb}(Q \subset M)$ in $(M, \partial M)$, the *exterior* of Q in $(M, \partial M)$ is $\operatorname{Ext}(Q \subset M) := M \setminus \operatorname{Int} \operatorname{Nb}(Q \subset M)$.

A *connected sum* (resp., *boundary-connected sum*) of n-manifolds M_1 and M_2 is denoted $M_1 \# M_2$ (resp., $M_1 \natural M_2$); notations like $M_1 \#_{D^n} M_2$, $M_1 \natural_{D^{n-1}} M_2$, and so on, can be used for greater precision.

DEFINITIONS 2.1. A *stratification* of a topological space X is a locally finite partition $X/\equiv$ such that (1) each $\equiv$-equivalence class, equipped with the topology induced from X, is a connected, not necessarily oriented, manifold (called a $\equiv$*-stratum* of X, or simply a *stratum* of X when $\equiv$ is understood), and (2) for every stratum M, the closure of M in X is a union of strata. A *vertex* of $X/\equiv$ is a point x such that $\{x\}$ is a stratum; let $\mathsf{V}(X/\equiv)$ denote the set of vertices of $X/\equiv$. An *edge* of $X/\equiv$ is the closure of a stratum of dimension 1; let $\mathsf{E}(X/\equiv)$ denote the set of edges of $X/\equiv$. A *cellulation* is a stratification such that each stratum is diffeomorphic to some $\mathbb{R}^k$. The cellulation *associated to a triangulation* $h{:}|||\mathcal{K}||| \to P$ is that with strata the h-images of open simplices of $|||\mathcal{K}|||$; a fixed or understood triangulation h of a polyhedron P determines edges and vertices of P and gives sense to the notations $\mathsf{V}(P)$ and $\mathsf{E}(P)$.

An *arc* is a manifold diffeomorphic to $[0,1]$; in particular, in a real vectorspace if $[\mathbf{u},\mathbf{v}] \neq \{\mathbf{u}\}$ then $[\mathbf{u},\mathbf{v}]$ is an arc oriented from $\mathbf{u}$ to $\mathbf{v}$. An *edge* is an unoriented arc (this is mildly inconsistent with the definition above). A *simple* (*collection of*) *closed curve*(*s*) is a manifold diffeomorphic to (a disjoint union of copies of) S^1. A *graph* is a polyhedron G of dimension ≤ 1 equipped with a cellulation $G/\equiv$ (which need not be associated to a triangulation of G). For every graph G, there exists val_G: $\mathsf{V}(G) \to \mathbb{N}$ such that, for every triangulation of G, $\operatorname{val}_G(x) = \operatorname{card}(\{\mathbf{e} \in \mathsf{E}(G){:}\ x \in \mathbf{e}\})$; $\operatorname{val}_G(x)$ is the *valence of* $x \in \mathsf{V}(G)$ *in* G. Call $x \in \mathsf{V}(G)$ an *endpoint* (resp., *isolated point*; *intrinsic vertex*) of G in case $\operatorname{val}_G(x) = 1$ (resp., $\operatorname{val}_G(x) = 0$; $\operatorname{val}_G(x) > 2$). Call $x \in G$ an *ordinary point* in case either $x \in \mathsf{V}(G)$ and $\operatorname{val}_G(x) = 2$ or $x \notin \mathsf{V}(G)$. A graph embedded in $\mathbb{C}$ is *planar*. A *tree* is a finite, connected, acyclic graph. Let $n \in \mathbb{N}_{>0}$. An n-star is a tree with $n + 1$ vertices of which at least n are endpoints; an n*-gon* is a 2-disk P equipped with a cellulation having exactly n edges, all in ∂P.

DEFINITIONS 2.2. A *surface* is a compact 2-manifold no component of which has empty boundary. The *genus* of a connected surface S is denoted $g(S)$. A surface is *annular* in case each component is an annulus. A subset X of a surface S is *full* provided that no component of $S \setminus X$ is contractible.

The *standard* (2-dimensional) 0*-handle* is $h^{(0)} := D^2$. Fix some continuous function H: $[-2,2] \to [1,2]$ such that: (1) H is even; (2) $H(x) = 1$ for $|x| \leq 1$, $H(2) = 2$, and $H|[1,2]$ is strictly increasing; (3) $H|\,]-2,2[$ and $(H|]1,2])^{-1}$ are smooth; (4) for $n \in \mathbb{N}_{>0}$, $D^n((H|]1,2])^{-1})(2) = 0$. The *standard* (2-dimensional) 1*-handle* is $h^{(1)} := \{z \in \mathbb{C}{:}\ |\operatorname{Im} z| \leq H(\operatorname{Re} z),\ |\operatorname{Re} z| \leq 2\}$, a 2-manifold with cuspidal corners. The *attaching* (resp., *free*) *arcs* of $h^{(1)}$ are the intervals $[\pm(2 - 2i), \pm(2 + 2i)] \subset \mathbb{C}$ (resp., the arcs $\mp\{z \in \mathbb{C}{:}\ \operatorname{Im} z = \pm H(\operatorname{Re} z),\ |\operatorname{Re} z| \leq 2\}$). The union of the attaching

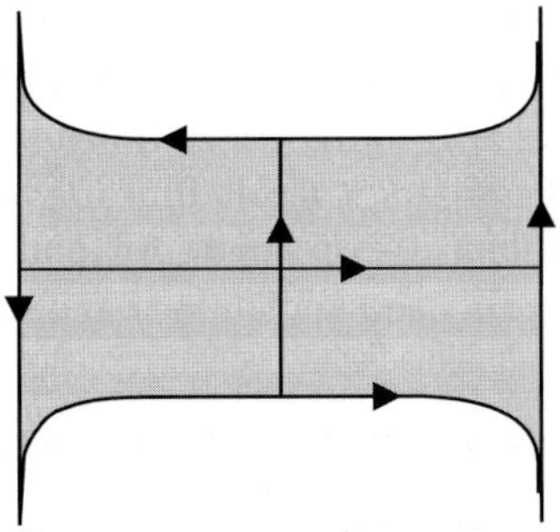

Fig. 1. The standard 1-handle and associated arcs.

arcs of $h^{(1)}$ is the *attaching region* $A(h^{(1)})$ of $h^{(1)}$. The *standard core* (resp., *transverse*) *arc* of $h^{(1)}$ is $\kappa(h^{(1)}) := [-2,2]$ (resp., $\tau(h^{(1)}) := [-i,i]$); a *core arc* (resp., *transverse arc*) of $h^{(1)}$ is any arc isotopic in $h^{(1)}$ to $\kappa(h^{(1)})$ (resp., $\tau(h^{(1)})$). (See Figure 1.)

DEFINITIONS 2.3. Fix some smooth function $g\colon \mathbb{R} \to [0,\pi]$ such that: (1) g is odd, and periodic with period 8; (2) $g'(x) > 0$ for $x \in]0,1[$, $g(x) = \pi$ for $x \in [1,3]$, $g'(x) < 0$ for $x \in]3,4[$, and $g(4) = 0$. The *standard bowtie* is $\bowtie\ := \Theta(h^{(1)})$, where $\Theta\colon \mathbb{C} \to \mathbb{C}$: $z \mapsto \operatorname{Re} z + i\cos(\pi + g(\operatorname{Re} z))\operatorname{Im} x$. Give each of the two halves $\bowtie \cap\, i\mathbb{C}_\pm$ of $\bowtie$ the orientation induced on it from $h_0^{(1)}$ by Θ. The *attaching region* of $\bowtie$ is $A(\bowtie) := \Theta(A(h^{(1)}))$; the *attaching arcs* of $\bowtie$ are the components of $A(\bowtie)$. The *standard core arc* of $\bowtie$ is $\Theta(\kappa(h^{(1)}))$. The *crossed arcs* of $\bowtie$ are the Θ-images of the free arcs of $h_0^{(1)}$, and the *crossing* of $\bowtie$ is their point of intersection. (see Figure 2.)

Let S be a 2-manifold. Call

$$S = \bigcup_{x\in X} h_x^{(0)} \cup \bigcup_{t\in T} h_t^{(1)} \tag{2.3}$$

a (0, 1)-*handle decomposition* of S provided that: (1) X and T are finite; (2) each 0-*handle* $h_x^{(0)}$ is diffeomorphic to the standard 0-handle D^2; (3) each 1-*handle* $h_t^{(1)}$ is diffeomorphic,

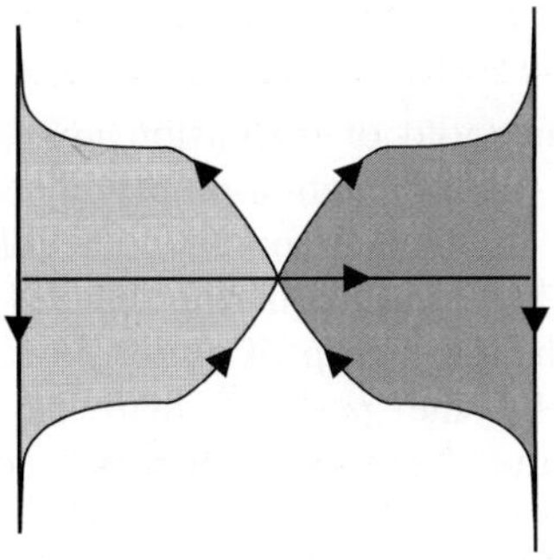

Fig. 2. The standard bowtie and associated arcs.

as a manifold with corners, to the standard 1-handle $h^{(1)}$ and thereby equipped with an attaching region $A(h_t^{(1)})$, attaching arcs, free arcs, a standard core arc $\kappa(h_t^{(1)})$ and other core arcs, a standard transverse arc $\tau(h_t^{(1)})$ and other transverse arcs; (4) the 0-handles are pairwise disjoint, as are the 1-handles; (5) $h_t^{(1)} \cap \bigcup_{x\in X} h_x^{(0)} = h_t^{(1)} \cap \partial \bigcup_{x\in X} h_x^{(0)} = A(h_t^{(1)})$ for each $t \in T$; and (6) the orientations of S and all the 0- and 1-handles are compatible. A 2-manifold S has a (0, 1)-handle decomposition if and only if S is a surface.

2.3. *Smooth maps*

Maps between manifolds are smooth (C^∞), and isotopies are ambient, unless otherwise stated. Given manifolds M and N, let Diff (M, N) denote the set of maps $M \to N$, with a suitable topology. Let $f \in$ Diff (M, N). The *f-multiplicity* $x \in M$ is card$(f^{-1}(x))$. A point of f-multiplicity 1 (resp., 2; at least 2) is a *simple point* (resp., *double point*; *multiple point*) of f; the image by f of a simple (resp., double; multiple) point of f is a *simple* (resp., *double*; *multiple*) *value* of f. Let simp$(f) := \{x \in M$: x is a simple point of $f\}$, doub$(f) := \{x \in M$: x is a double point of $f\}$, mult$(f) := \{x \in M$: x is a multiple point of $f\}$. Denote by $Df(x)$: $T_x(M) \to T_{f(x)}(N)$ the derivative of f at $x \in M$, and by Df: $T(M) \to T(N)$ the map induced by f on tangent bundles. A *critical point* of f is any $x \in M$ with rank$(Df(x)) <$ dim$T_{f(x)}(N)$. The set of critical points of f is denoted crit(f), so f (crit(f)) is the set of *critical values* of f. As usual, $y \in N\setminus f$ (crit(f)) is called a *regular value* of f (even if $y \notin f(M)$). For dim$(N) = 1$, the index of f at $x \in$ crit(f) is denoted ind$(f; x)$. Call f a *Morse function* (resp., *Morse map*) in case $N = \mathbb{R}$ (resp., $N = S^1$), f is constant on ∂M, and every $x \in$ crit(f) is non-degenerate; in case also ind$(f; x) <$ dim M for all $x \in$ crit(f), call f *topless*. Call $f \in$ Diff(M, N) an *immersion* in case $Df(x)$: $T_x(M) \to T_{f(x)}(N)$ is injective for every $x \in M$. An *embedding* is an immersion that is a homeomorphism onto its image. Write f: $M \looparrowright N$ (resp., f: $M \hookrightarrow N$) to indicate that f is an immersion (resp., embedding). The *normal bundle* of f: $M \looparrowright \mathbb{N}$ is denoted $\nu(f)$; given a submanifold $M \subset N$, the *normal bundle* of M in N is $\nu(\mathrm{i}_{M\hookrightarrow N})$, where $\mathrm{i}_{M\hookrightarrow N}$ denotes inclusion.

Here are various constructions with normal bundles, in the course of which assorted notations and definitions are established.

DEFINITIONS 2.4. Let M be a manifold of dimension m.

2.4.1. Let $Q \subset M$ be a submanifold of dimension $m - 1$ with trivial normal bundle. A *collaring* of Q in M is an orientation-preserving embedding col$_{Q\subset M}$:$Q \times [0,1] \hookrightarrow$ Nb$(Q \subset M)$ with col$_{Q\subset M}$ $(q, 0) = q$ for all q; a *collar* of Q in M is the image Col$(Q \subset M) :=$ col$_{Q\subset M}$ $(Q \times [0,1])$ of a collaring. Let $M \wr Y := M\setminus$col$_{Q\subset M}(Q \times]0,1[)$ be called M *cut* along Y. The *push-off map* of Q is $Q \hookrightarrow M\setminus Q$: $q \mapsto$ col$_{Q\subset M}(q,1)$. Call the image Y^+ of $Y \subset Q$ by the push-off map the *push-off* of Y. (Note that $Q^+ \subset \partial$ Col$(Q \subset M)$ has the "outward normal" orientation, whereas $Q \hookrightarrow \partial$ Col$(Q \subset M)$ reverses orientation.) It is convenient to define various *standard* collarings, thus. (a) Given $Q \subset S^n = \partial D^{n+1}$, let col$_{Q\subset D^{n+1}}$: $(x, t) \mapsto \varepsilon(1 - t)x$ for a suitable $\varepsilon \in]0,1[$. (b) Given a manifold W, $u \in \mathbb{R}_{>0} \cup \{\infty\}$, and $Q \subset$ Int

$W \times \{0\} \subset \partial W \times [0, \pm u[$, let $\mathrm{col}_{Q \subset \partial W \times [0,\pm u[}$: $(x, t) \mapsto (x, \pm \varepsilon t)$ for a suitable $\varepsilon \in \mathbb{R}_{>0}$. (c) For a suitable $\varepsilon :]-2,2[\to]0,1[$, $(\pm 2 + iy,\ t) \mapsto \pm 2 + i\varepsilon(t)y$ is a collaring $\mathrm{col}_{\mathrm{Int}\, A(h^{(1)}) \subset h^{(1)}}$. (The cusps of $h^{(1)}$ prevent the existence of a collaring of $A(h^{(1)})$ in $h^{(1)}$.)

2.4.2. Let Q be a manifold of dimension $q < m$, $j{:}Q \looparrowright M$ an immersion, $B := j(Q)$. For $x \in B \setminus j(\partial Q \cup \mathrm{mult}(j))$, a *meridional* $(m - q)$*-disk of* B *at* x is the image $\pitchfork(B;x)$ of an embedding f: $D^{m-q} \hookrightarrow M$ such that $f^{-1}(B) = f^{-1}(x) = 0$, f is transverse to j, and $\pitchfork(B;x)$ intersects B positively (with respect to given orientations of M and Q). In case M is connected and $q = m - 2$, any element of $\pi_1(M \setminus B)$ represented by a loop freely homotopic to $\partial\pitchfork(B;x)$ in $M \setminus B$ is called a *meridian* in that knotgroup.

2.4.3. Let $B \subset M$ be a submanifold of dimension $m - 2$ with trivial normal bundle. Let n: $B \times \mathbb{C} \to \nu(B)$ be a fixed trivialization of $\nu(B)$. In the standard way, using (inexplicit) metrics, etc., identify $\mathrm{Nb}(B \subset M)$ to $n(B \times D^2)$ in such a way that if $x \in B$, then the image of $D^2 \to M$: $z \mapsto n(x,z)$ is a meridional 2-disk $\pitchfork(B;x)$. Call f: $M \setminus B \to S^1$ *weakly adapted* to n in case for every $Q \in \pi_0(B)$ there is an integer $d(Q)$ such that, if $d(Q) \neq 0$, then $f(n(x, z)) = (z/|z|)^{d(Q)}$ for all $x \in Q$, $z \in D^2 \setminus \{0\}$. Call f: $M \setminus B \to S^1$ *adapted* in case f is weakly adapted and, in addition, if $d(Q) = 0$, then f extends to $f_Q \in \mathrm{Diff}(M \setminus B \setminus Q, S^1)$, $f_Q | Q{:}Q \to S^1$ is an immersion, and $f_Q | \pitchfork(B;x)$ is constant for each $x \in Q$.

DEFINITIONS 2.5. Let $f : M \to N$ be smooth; let Q be a codimension-0 submanifold of ∂M. Call f *proper along* Q, *relative to* $\mathrm{col}_{\partial N \subset N}$ with $\mathrm{Col}(\partial N \subset N) \subset \mathrm{Nb}(\partial N \subset N)$ and $\mathrm{col}_{Q \subset M}$, provided that: (1) $f(Q) \subset \partial N$ and $f(M \setminus Q) \subset \mathrm{Int}\, N$; (2) $f^{-1}(\mathrm{Nb}(\partial N \subset N)) \subset M$ is a submanifold, and $f | f^{-1}(\mathrm{Nb}(\partial N \subset N))$ is an embedding; (3) $f \circ \mathrm{col}_{Q \subset M}{:}\mathrm{Col}(Q \subset M) \hookrightarrow N$ is an embedding into $\mathrm{Nb}(\partial N \subset N)$ and $\mathrm{pr}_2 \circ (\mathrm{col}_{\partial N \subset N})^{-1} \circ f \circ \mathrm{col}_{Q \subset M} = \mathrm{pr}_2$: $\mathrm{Col}(Q \subset M) \to [0,1]$. Call *f proper along* Q in case there exist collarings $\mathrm{col}_{Q \subset M}$ and $\mathrm{col}_{\partial N \subset N}$ such that f is *proper along* Q relative to $\mathrm{col}_{Q \subset M}$ and $\mathrm{col}_{\partial N \subset N}$ ("along Q" is dropped when $Q = \partial M$). A properly embedded submanifold is simply *proper*.

Some special cases of low-dimensional immersions and embeddings of particular interest, and associated ancillary constructions, need extra terminology.

DEFINITIONS 2.6. Let M be a compact m-manifold, N an n-manifold. Let f: $M \looparrowright N$ be an immersion with $\mathrm{mult}(f) = \mathrm{doub}(f)$.

2.6.1. Let $m = 1$, and suppose M is an arc or an edge. Call *f half-proper* provided that it is proper along a single component of ∂M. A *half-proper* arc or edge is the image of a half-proper embedding. An n-star embedded in N is *proper* provided each of its edges is half-proper.

2.6.2. Let $m = 1$, $n = 2$. Call *f normal* provided that f is proper (whence $\mathrm{doub}(f) \subset \mathrm{Int}\, M$), $\mathrm{doub}(f)$ is finite, and if $f(x_1) = f(x_2)$ with $x_1 \neq x_2$, then the lines $Df(x_1)(T_{x_1}(M))$

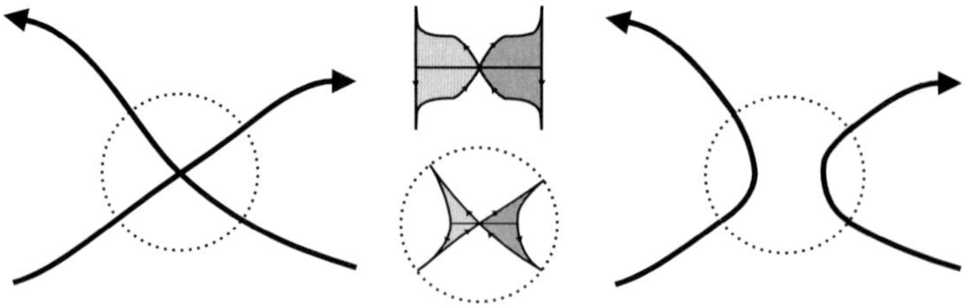

Fig. 3. Local smoothing, using a bowtie.

and $Df(x_2)(T_{x_2}(M))$ in $T_{f(x_1)}(N)$ are transverse. A *crossing* of a normal immersion $f\colon M \looparrowright N$ is a double value of f; the *crossing number* of f is $\operatorname{cross}(f) := \operatorname{card}(f(\operatorname{doub}(f)))$. A *branch* of f at a crossing y is the (germ at y of the) f-image of either component of $\operatorname{Nb}(f^{-1}(y) \subset M \setminus (\operatorname{doub}(f) \setminus f^{-1}(y)))$. The image of a normal immersion of S^1 (resp., an arc; a finite disjoint union of copies of S^1; a finite disjoint union of arcs) is a *normal closed curve* (resp., a *normal arc*; a *normal collection of closed curves*; a *normal collection of arcs*); the *crossing number* of the normal collection of closed curves or arcs $f(M)$ is $\operatorname{cross}(f(M)) := \operatorname{cross}(f)$. A *bowtie* in N is the image of $\bowtie \subset h^{(1)}$ by an *embedding* of $\bowtie$ in N, that is, a map $\delta\colon \bowtie \to N$ that extends to an embedding $h^{(1)} \hookrightarrow N$. Let $C := f(M)$ be a normal collection of closed curves. For each $y \in f(\operatorname{doub}(f))$, let $\bowtie_y = \delta_y(\bowtie)$ be a bowtie with $\bowtie_y \subset \operatorname{Nb}(\{y\} \subset N \setminus f(\operatorname{doub}(f)) \setminus \{y\})$, such that the *crossed arcs* of $\bowtie_y$ (i.e., the δ_y-images of the crossed arcs of $\bowtie$) are the branches of f at y, correctly oriented. The *local smoothing* of C at y (see Figure 3) is the normal collection of closed curves $\operatorname{sm}(C; y) =: \operatorname{sm}(f;y)(\operatorname{sm}(M; y))$ created by replacing the crossed arcs of $\bowtie_y$ with $-\delta_y(A(\bowtie))$. Here, $\operatorname{sm}(M; y)$ is unique up to diffeomorphism, and $\operatorname{sm}(C;y)$ up to (arbitrarily small) isotopy; and $\operatorname{cross}(\operatorname{sm}(C;y)) = \operatorname{cross}(C) - 1$. The *smoothing* of C (see Figure 4) is the simple collection of closed curves $\operatorname{sm}(C) := \operatorname{sm}(\cdots \operatorname{sm}(\operatorname{sm}(C; y_1); y_2) \cdots; y_{\operatorname{cross}(C)}) \subset N$, independent of the enumeration $\{y_1, \ldots, y_{\operatorname{cross}(C)}\}$ of $f(\operatorname{doub}(f))$.

2.6.3. Let $m = 2$, $n = 3$. Call f *clasp* provided that $\operatorname{doub}(f)$ is the union of finitely many pairwise disjoint edges $A'_1, A''_1, \ldots, A'_s, A''_s$ with $f(A'_i) = f(A''_i) \subset \operatorname{Int} N$, both A'_i and A''_i half-proper ($i = 1, \ldots, s$); call s the *clasp number* of the clasp immersion f, and denote it by

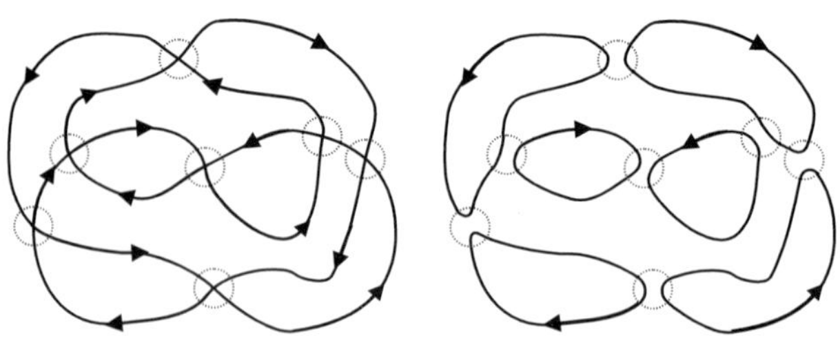

Fig. 4. A normal closed curve in $\mathbb{R}^2$ and its smoothing.

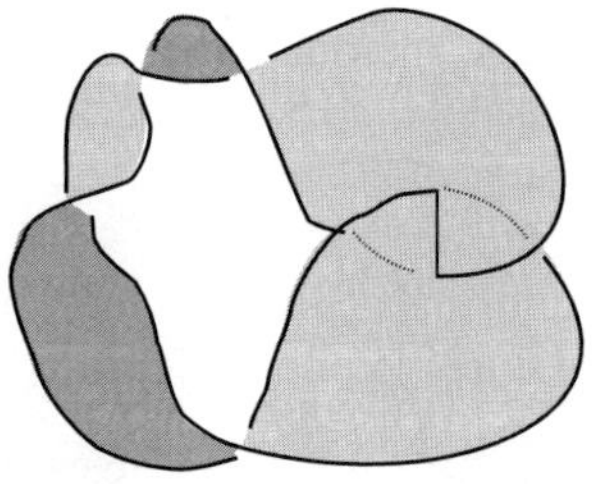

Fig. 5. A clasp surface in $\mathbb{R}^3$.

clasp(f). The image $f(S)$ is a *clasp surface* (see Figure 5); the *clasp number* of $f(S)$ is $\mathrm{clasp}(f(S)) := \mathrm{clasp}(f)$.

2.6.4. Let $m = 2$, $n = 3$. Call f *ribbon* provided that doub(f) is the union of finitely many pairwise disjoint edges $A'_1, A''_1, \ldots, A'_s, A''_s$ with A'_i proper, A''_i interior, and $f(A'_i) = f(A''_i) \subset \mathrm{Int}\, N$. The image of a ribbon immersion of a surface is a *ribbon surface* (see Figure 6). For connected S, the *genus* of the ribbon surface $f(S)$ is $g(f(S)) := g(S)$. The *smoothing* of a ribbon surface $R = f(S)$ is an embedded surface $\mathrm{sm}(R) \subset N$, unique up to isotopy, constructed as follows. Let $S_0 := S \wr (\cup_{i=1}^s A'_i \cup \bigcup_{i=1}^s \mathrm{Int}\, A''_i)$. Let $\equiv$ be the equivalence relation on S_0 having as its non-trivial equivalence classes $\{x, y\}$ and $\{x^+, y^+\}$ ($x \in \mathrm{Int}\, A'_i$, $f(x) = f(y)$) and $\{x, x^+, y\}$ ($x \in \partial A'_i$, $f(x) = f(y)$), $i = 1, \ldots, s$. There is a natural way to smooth the quotient manifold $S_1 := S_0/\equiv$, and a natural map $f_1 : S_1 \to N$ with $f_1(S_1) = f(S)$, which after an arbitrarily small perturbation yields an embedding $S_1 \hookrightarrow N$ with image $\mathrm{sm}(f(S))$.[4]

2.6.5. Let $m = 2$, $n = 4$. Call f *nodal* provided that f is proper, doub(f) is finite, and if $f(x_1) = f(x_2)$ and $x_1 \neq x_2$, then the planes $Df(x_1)(T_{x_1}(M))$ and $Df(x_2)(T_{x_2}(M))$ in $T_{f(x_1)}(N)$ are transverse. A *node* of a nodal immersion f is a double value of f; the *node number* of f is $\mathrm{node}(f) := \mathrm{card}(f(\mathrm{doub}(f)))$. A *branch* f at a node y is the (germ at y of the) f-image of either component of $\mathrm{Nb}(f^{-1}(y) \subset M \setminus \mathrm{doub}(f) \setminus f^{-1}(y))$. The *sign* $\varepsilon(y)$ of the node y is the sign ($+$ or $-$) of the given orientation of $T_y(N)$ with respect to its orientation as the direct sum of the oriented 2-planes $Df(x_1)(T_{x_1}(M))$ and $Df(x_2)(T_{x_2}(M))$ (in either order), where $f(x_1) = f(x_2) = y$, $x_1 \neq x_2$. The image of a nodal immersion of a surface is a *nodal surface*. The *node number* of a nodal surface $f(M)$ is $\mathrm{node}(f(M)) := \mathrm{node}(f)$, and its *smoothing* is an embedded surface $\mathrm{sm}(f(M)) \subset N$, unique up to isotopy, constructed by replacing $\mathrm{Nb}(f(\mathrm{doub}(f)) \subset f(M))$ with $\mathrm{card}(f(\mathrm{doub}(f)))$ annuli embedded in $\mathrm{Nb}(f(\mathrm{doub}(f)) \subset N)$ in a standard way.

[4] Fox (1962) introduced the word "ribbon" into knot theory (specifically, in the context of ribbon-immersed 2-disks). His usage was soon generalized (Tristram, 1969) and widely adopted. In a distinct chain of development, the biologist Crick (1976), followed by physicists (Grundberg, Hansson, Karlhede, and Lindström, 1989) and other scientists applying mathematics, gave the word quite a different meaning (perhaps closer to its everyday use), essentially to refer to twisted 2-dimensional bands. More recently, this conflicting usage has been adopted by some knot-theorists (see, e.g., Reshetikhin and Turaev, 1990), particularly of a categorical bent. *Caveat lector.*

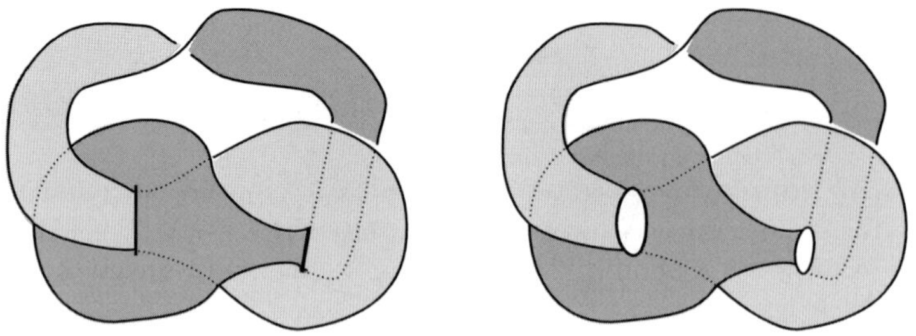

Fig. 6. A ribbon surface in $\mathbb{R}^3$ and its smoothing.

(In appropriate local coordinates on N, the neighborhood on $f(M)$ of a positive node is $\{(z, w) \in D^4 \subset \mathbb{C}^2 : zw = 0\}$, and is replaced by $\{(z, w) \in D^4 : zw = \epsilon(1 - \|(z, w)\|^4)\}$, where $\epsilon: [0,1] \to \mathbb{R}_{\geq 0}$ is smooth, 0 near 0, positive near 1, and sufficiently small.)

2.6.6. Let $m = 2$, $n = 4$. Call f *slice* provided that f is proper. A *slice surface* is the image of a slice embedding of a surface (i.e., a proper surface).[5]

2.6.7. Let $m = 2$, $n = 4$, and suppose that N is compact and $\rho: N \to \mathbb{R}$ is a topless Morse function. Call f *ρ-ribbon* provided that f is slice and $\rho \circ f$ is a topless Morse function on M. In case $N = D^4$ and card(crit(ρ)) = 1 (so that, up to diffeomorphisms of N and $\mathbb{R}$, $\rho = \|\cdot\|^2$), a ρ-ribbon embedding is called simply a *ribbon embedding*, and its image is called a *ribbon surface* in D^4.

There is a close relation between ribbon surfaces in D^4 and in S^3.

PROPOSITION 2.7. *Let M be a surface. If $f : M \looparrowright S^3 = \partial D^4$ is a ribbon immersion, then there is a non-ambient isotopy $\{f_t : M \looparrowright D^4\}_{t\in[0,1]}$ such that $f_0 = f$, $f_t|\partial M = f_0|\partial M$ for $t \in [0, 1]$, and $f_t : M \looparrowright D^4$ is a ribbon embedding for $t \in [0, 1]$; conversely, if $g : M \hookrightarrow \partial D^4$ is ribbon, then $g = f_1$ for some such non-ambient isotopy $\{f_t : M \looparrowright D^4\}_{t\in[0,1]}$ with $f_0 : M \looparrowright S^3$ ribbon. (Although the first of these non-ambient isotopies is unique up to ambient isotopy, the second enjoys no such uniqueness.)* □

See Tristram (1969), Hass (1983), or Rudolph (1985b) for more detailed statements and proofs.

A (smooth) *covering map* is an orientation-preserving immersion $f : M \looparrowright N$ that is also a topological covering map (it is not required that the domain of a topological covering map be connected). The usual theory of covering maps is assumed – in particular, the well-behaved, albeit many–many, correspondence for connected N between permutation representations $\rho : \pi_1(N; *) \to \mathfrak{S}_s$ and covering maps f with target N and *base fiber* $f^{-1}(*) = \mathrm{s}$. Given $s \in \mathbb{N}_{>0}$ and $\mathrm{g} \in \mathfrak{S}_s$, let $\rho_{\mathrm{g}} : \pi_1(S^1; 1) \to \mathfrak{S}_s$ be the permutation

[5] The redundant term "slice surface" has been retained for the sake of tradition. See Rudolph (1993, Section 1) for a history of the use of the word "slice" in knot theory.

representation $\circlearrowleft^k \mapsto \mathbf{g}^k$, where $\circlearrowleft$ denotes the *positive generator* of $\pi_1(S^1;1) \cong \pi_1(D^2\setminus\{0\};1)$, that is, the homotopy class of id_{S^1}. Call $\Xi_{\rho_\mathbf{g}} := (D^2\setminus\{0\}) \times \mathbf{s}/\langle\mathbf{g}\rangle$ the *standard covering space of* D^2 *of type* $\mathbf{g}$ and $\xi_{\rho_\mathbf{g}}:\Xi_{\rho_\mathbf{g}} \to D^2 : (z, t\langle \mathrm{g}\rangle) \mapsto z^{\mathrm{card}(t\langle\mathrm{g}\rangle)}$ the *standard covering map of* D^2 *of type* $\mathbf{g}$.

The theory of *branched covering maps* originated in complex analysis and algebraic geometry. Following earlier work of Heegard (1898) and Tietze (1908), the theory was adapted to combinatorial manifolds by Alexander (1920) and Reidemeister (1926), then to more general spaces by Fox (1957). Durfee and Kauffman (1975) made the construction more precise in the smooth category. The *ad hoc* approach of 2.8 and 2.9, below, suffices to handle the cases that are most important for the knot theory of complex plane curves.

DEFINITIONS 2.8. Let s, $\mathbf{g}$, etc., be as above. The *standard branched covering space* (resp., *map*) of D^2 *of type* $\mathbf{g}$ is $\tilde{\Xi}_{\rho_\mathbf{g}} := D^2 \times \mathrm{s}/\langle\mathbf{g}\rangle$ (resp., $\tilde{\xi}_{\rho_\mathbf{g}} : \tilde{\Xi}_{\rho_\mathbf{g}} \to D^2 : (z, t\langle\mathbf{g}\rangle) \mapsto z^{\mathrm{card}(t\langle\mathbf{g}\rangle)}$). Let N be a connected n-manifold equipped with a stratification $N/\equiv$ such that: (1) every stratum of $N/\equiv$ is smoothly immersed in N; (2) no stratum of $N/\equiv$ has dimension $n-1$ (whence there is a unique stratum N_0 of dimension n) or $n-3$; and (3) $N\setminus N_0$ is the closure in N of the union B of all $(n-2)$-dimensional strata. A smooth map $f\colon M \to N$ is a *branched covering map of* N *branched over* B, and M is a *branched covering space of* N *branched over* B, provided that: (4) $f(\mathrm{crit}(f)) \subset B$; (5) $f|(f^{-1}(N\setminus B)$ is a covering map of degree s; and (6) for every $x \in B$, there exist $\mathbf{g}(x) \in \mathfrak{S}_s$ (the *type of* f *at* x), an embedding $\varphi : D^2 \hookrightarrow N$ onto a meridional disk $⫛(B;x)$, and an embedding $\Phi : \tilde{\Xi}_{\rho_{\mathbf{g}(x)}} \hookrightarrow M$ with $(f|(f^{-1}(⫛(B;x)))) \circ \Phi = \varphi \circ \tilde{\xi}_{\rho_{\mathbf{g}(x)}}$. The conjugacy class of $\mathbf{g}(x)$ in $\mathfrak{S}_s$ is constant on each component of B, and $B = \{x \in f(\mathrm{crit}(f))\colon \mathbf{g}(x) \neq \mathrm{id}_\mathbf{s}\}$. The branched covering f is called *simple* in case $\mathbf{g}(x)$ is a transposition for each $x \in B$.

CONSTRUCTION 2.9. Let N be a connected 2-manifold, $B \subset \mathrm{Int}\ N$ a non-empty finite subset. Fix an enumeration $B =: \{x_1,\ldots,x_q\}$. Let $*_0 \in \mathrm{Ext}(B \subset \mathrm{Int}\ N)$. The regular neighborhood $\mathrm{Nb}(B \subset N)$ is a union of pairwise disjoint meridional 2-disks $D_i^2 := ⫛(B;x_i)$, $i \in \mathbf{q}$. Let $*_i \in \partial D_i$. Fix a proper q-star $\psi \subset \mathrm{Ext}(B \subset N)$ with $\mathsf{V}(\psi) = \{*_0, *_1,\ldots, *q\}$ and $\mathsf{E}(\psi) =: \{\mathbf{e}_1,\ldots,\mathbf{e}_q\}$ such that $\partial \mathrm{e}_k = \{*_0, *_k\}$ and the cyclic order of $\mathsf{E}(\psi)$ at $*_0$ (with respect to the orientation of N) is the cyclic order of their indices $1,\ldots,q$. Let $\Psi := \psi \cup \bigcup_{i=1}^q (D_i\setminus x_i)$. Let $\mathbf{g}_i \in \pi_1(\Psi; *_0)$ be the element represented by a loop that traverses the edge $\mathbf{e}_i$ of ψ from $*_0$ to $*_i$, represents $\circlearrowleft$ in $\pi_1(\partial D_i^2; *_i) \cong \pi_1(S^1;1)$, and returns to $*_0$ along $\mathbf{e}_i$. Evidently $\pi_1(\Psi; *_0)$ is the free group $\mathrm{gp}\,(\mathbf{g}_i, i \in \mathbf{q} : \varnothing) \cong \pi_1(\partial D_1^2) * \cdots * \pi_1(\partial D_q^2)$. Call $\mathbf{g} =: (\mathbf{g}_1, \ldots, \mathbf{g}_q) \in \mathfrak{S}_s^q$ *compatible with* ψ (or Ψ) in case there exists a permutation representation $\rho_\mathbf{g} : \pi_1(\mathrm{Ext}(B \subset N); *_0) \to \mathfrak{S}_s$ such that $\rho_\mathbf{g}(\mathbf{g}_i') = \mathrm{g}_i (i \in \mathbf{q})$, where $\mathbf{g} \mapsto \mathbf{g}'$ is the inclusion-induced homomorphism $\pi_1(\Psi;\ *_0) \to \pi_1\ (N\setminus B; *_0) \cong \pi_1(\mathrm{Ext}(B \subset N);\ *_0)$. (If N is closed then compatibility is a genuine restriction; if N is not closed, then every $\mathbf{g}$ is compatible with ψ, but $\rho_\mathbf{g}$ may not be unique if N is not contractible.) Finally, given Ψ, a compatible q-tuple $\mathbf{g}$, and $\rho_\mathbf{g}$, construct $\tilde{\Xi}_{\rho_\mathbf{g}}$ as the identification space (with an appropriate, easily defined smooth structure) of the disjoint union of copies of $\tilde{\Xi}_{\rho_{\mathbf{g}_i}}$ $(i \in \mathbf{q})$ and $\Xi_{\rho_\mathbf{g}}$ along their tautologously diffeomorphic boundaries $\tilde{\xi}_{\rho_{\mathbf{g}_i}}^{-1}(\partial D_i^2) = \partial\tilde{\Xi}_{\rho_{\mathbf{g}_i}}$ $(i \in \mathbf{q})$ and $\xi_{\rho_\mathbf{g}}^{-1}(\partial D_i^2) \subset \partial\Xi_{\rho_\mathbf{g}}$; let $\tilde{\xi}_{\rho_\mathbf{g}}|\tilde{\Xi}_{\rho_{\mathbf{g}_i}} = \tilde{\xi}_{\rho_{\mathbf{g}_i}}$ and $\tilde{\xi}_{\rho_\mathbf{g}}|\Xi_{\rho_\mathbf{g}} = \xi_{\rho_\mathbf{g}}$.

2.4. *Knots, links, and Seifert surfaces*

A *link* is a simple collection of closed curves embedded in S^3; except where otherwise stated, isotopic links are treated as identical. A 1-component link is a *knot*. For $n \in \mathbb{N}$, $O^{(n)}$ denotes the *n-component unlink*, that is, the boundary of n pairwise disjoint copies of D^2 embedded in S^3; the *unknot* is $O := O^{(1)}$.

DEFINITIONS 2.10. A *link diagram* is a pair $\mathbf{D}(L) =: (\mathbf{P}(\mathbf{D}(L)), \mathbf{I}(\mathbf{D}(L)))$, where (1) $L \subset \mathbb{R}^3 \subset S^3$ is a link (called *the link of* the diagram); (2) $\mathbf{P}(\mathbf{D}(L))$, the $\mathbf{D}(L)$-*picture* of L, is the image of L by an affine projection $\mathbf{p}_{\mathbf{D}(L)}\colon \mathbb{R}^3 \to \mathbb{C}$; and (3) $\mathbf{I}(\mathbf{D}(L))$, the $\mathbf{D}(L)$-*information* about L, is information sufficient to reconstruct from $\mathbf{P}(\mathbf{D}(L))$ the embedding of L into $\mathbb{R}^3$, up to isotopy respecting the fibers of $\mathbf{p}_{\mathbf{D}(L)}$. The *mirror image* of $\mathbf{D}(L)$ is the link diagram $\mathbf{D}(\text{Mir } L)$, where $\mathbf{P}(\mathbf{D}(\text{Mir } L)) := \overline{\mathbf{P}(\mathbf{D}(L))}$, and $\mathbf{p}_{\mathbf{D}(L)}$ and $\mathbf{I}(\mathbf{D}(L))$ are modified accordingly to produce $\mathbf{p}_{\mathbf{D}(\text{Mir } L)}$ and $\mathbf{I}(\mathbf{D}(\text{Mir } L))$; of course the link of $\mathbf{D}(\text{Mir } L)$ is the mirror image Mir L of L as already defined.

The nature of the information $\mathbf{I}(\mathbf{D}(L))$ can be of various sorts, depending on context. For instance, $\mathbf{D}(L)$ is a *standard link diagram* for L provided that (1) $\mathbf{P}(\mathbf{D}(L))$ is a normal collection of closed curves, and (2) $\mathbf{I}(\mathbf{D}(L))$ consists of (a) the global information that $\mathbf{p}_{\mathbf{D}(L)}|L\colon L \to \mathbb{C}$ is a normal immersion with image $\mathbf{P}(\mathbf{D}(L))$, supplemented by (b) local information at each crossing specifying which branch is "under" and which is "over" – equivalently, which of the two points of $L \cap \mathbf{p}_{\mathbf{D}(L)}^{-1}(z)$ is the *undercrossing* $z_{\frown}$ (initial endpoint) and which is the *overcrossing* $z^{\smile}$ (terminal endpoint) of the interval between them on $\mathbf{p}_{\mathbf{D}(L)}^{-1}(z)$, when the standard orientations of $\mathbb{R}^3$ and $\mathbb{C}$ are used to orient the fibers of $\mathbf{p}_{\mathbf{D}(L)}$. (It is usual to depict crossings in the style of the left half of Figure 7.) Every link has many different standard link diagrams (some satisfying further conditions), as well as non-standard link diagrams of various types, some of which will be introduced later as needed.

DEFINITIONS 2.11. Let $\mathbf{D}(L)$ be a standard link diagram.

2.11.1. A *Seifert cycle*[6] of $\mathbf{D}(L)$ is any $\mathsf{o} \in \mathsf{O}_{\mathbf{D}(L)} := \pi_0(\text{sm}(\mathbf{P}(\mathbf{D}(L))))$. The *inside* of a Seifert cycle o is the 2-disk $\iota(\mathsf{o}) \subset \mathbb{C}$ oriented so that $\partial\iota(\mathsf{o}) = \mathsf{o}$. The *sign* $\varepsilon(\mathsf{o})$ of o is the sign, positive $(+)$ or negative $(-)$, of the orientation of $\iota(\mathsf{o})$ with respect to the standard orientation of $\mathbb{C}$. Let $\mathsf{O}^{+}_{\mathbf{D}(L)}$(resp., $\mathsf{O}^{-}_{\mathbf{D}(L)}$) be the set of positive (resp., negative) Seifert cycles of $\mathbf{D}(L)$. Given $\mathsf{o}, \mathsf{o}' \in \mathsf{O}_{\mathbf{D}(L)}$, say that o *encloses* o', and write $\mathsf{o} \Supset \mathsf{o}'$, in case $\text{Int}\, \iota(\mathsf{o}) \supset \mathsf{o}'$. Call $\mathbf{D}(L)$ *nested* (resp., *scattered*) in case $\mathsf{O}_{\mathbf{D}(L)}$ is an $\Supset$-chain (resp., $\Supset$-antichain).

[6] The more commonly used term "Seifert circle" seems to have been popularized, if not coined, by Fox (1962; see also a 1963 review in which Fox glosses Murasugi's "standard loops" as "Seifert circles"). Certainly, Seifert's term "Kreis" does mean "circle", but it can also be translated as "cycle", and in the expositon of Seifert's construction the latter translation has two apparent advantages over the former: it does not connote geometric rigidity, and does connote intrinsic orientation.

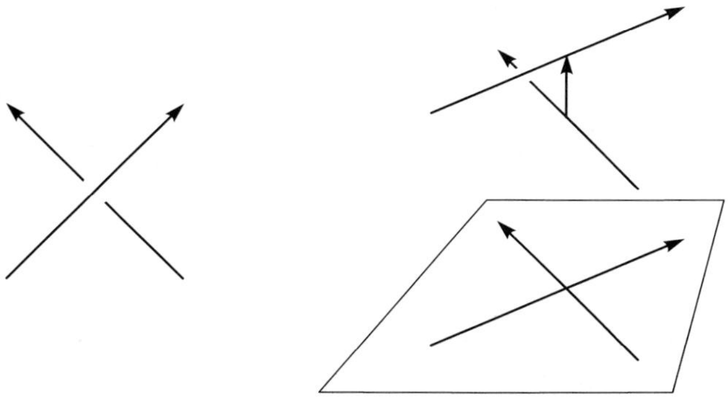

Fig. 7. A positive crossing in a standard link diagram.

2.11.2. A *crossing* of $\mathbf{D}(L)$ is any $\mathsf{x} \in \mathsf{X}_{\mathbf{D}(L)} := \mathbf{p}_{\mathbf{D}(L)}(\mathrm{doub}(\mathbf{p}_{\mathbf{D}(L)}))$. Let $\mathsf{x} \in \mathsf{X}_{\mathbf{D}(L)}$. An *over-arc of* $\mathbf{D}(L)$ *on* L *at* x is any arc $\mathrm{over}_{\mathsf{x}}(\mathbf{D}(L)) \subset L$ such that (a) $\mathsf{x}^{\smile} \in \mathrm{over}_{\mathsf{x}}(\mathbf{D}(L))$, (b) $\mathrm{over}_{\mathsf{x}}(\mathbf{D}(L)) \cap \{\mathsf{y}_{\frown} : \mathsf{y} \in \mathsf{X}_{\mathbf{D}(L)}\} = \varnothing$, and (c) among arcs $\alpha \subset L$ satisfying (a) and (b), $\mathrm{over}_{\mathsf{x}}(\mathbf{D}(L))$ maximizes $\mathrm{card}(\alpha \cap \{\mathsf{y}^{\smile} : \mathsf{y} \in \mathsf{X}_{\mathbf{D}(L)}\})$. An *over-arc of* $\mathbf{D}(L)$ *in* $\mathbf{P}(\mathbf{D}(L))$ *at* x is any arc $\mathbf{p}_{\mathbf{D}(L)}(\mathrm{over}_{\mathsf{x}}(\mathbf{D}(L))) \subset \mathbf{P}(\mathbf{D}(L))$. Let $\mathbf{u}^{\smile}$ (resp., $\mathbf{u}_{\frown}$) be a positively oriented basis vector for $T_{\mathsf{x}^{\smile}}(L)$ (resp., $T_{\mathsf{x}_{\frown}}(L)$). The *sign* $\varepsilon(\mathsf{x})$ of x is the sign, positive $(+)$ or negative $(-)$, of the frame $(\mathbf{u}_{\frown}, \mathsf{x}^{\smile} - \mathsf{x}_{\frown}, \mathbf{u}^{\smile})$ with respect to the standard orientation of $\mathbb{R}^3$. (The crossing in Figure 7 is positive.) Let $\mathsf{X}^{+}_{\mathbf{D}(L)}$ (resp., $\mathsf{X}^{-}_{\mathbf{D}(L)}$) be the set of positive (resp., negative) crossings of $\mathbf{D}(L)$.

2.11.3. Call $\mathsf{o} \in \mathsf{O}_{\mathbf{D}(L)}$ *adjacent* to $C \subset \mathsf{X}_{\mathbf{D}(L)}$ in case, for some $\mathsf{x} \in C$, o contains an attaching arc of the bowtie $\bowtie_{\mathsf{x}}$ used to construct $\mathrm{sm}(\mathbf{P}(\mathbf{D}(L)))$. Let $\mathsf{O}^{\geqq}_{\mathbf{D}(L)}$ (resp., $\mathsf{O}^{<}_{\mathbf{D}(L)}$; $\mathsf{O}^{\varnothing}_{\mathbf{D}(L)}$) denote the set of Seifert cycles adjacent to $\mathsf{X}^{+}_{\mathbf{D}(L)}$ (resp., not adjacent to $\mathsf{X}^{+}_{\mathbf{D}(L)}$; not adjacent to $\mathsf{X}_{\mathbf{D}(L)}$).

THEOREM 2.12. *Let* $\mathbf{D}(L)$ *be a standard link diagram. If* $C \subset \mathsf{X}_L$ *is any set of minimal cardinality such that* $\mathsf{X}_L \subset \bigcup_{\mathsf{x} \in C} \mathbf{p}_{\mathbf{D}(L)}(\mathrm{over}_{\mathsf{x}}(\mathbf{D}(L)))$, *then the knotgroup of* L *has a Wirtinger presentation*

$$\pi_1(S^3 \setminus L) = \mathrm{gp}\Big(\mathbf{g}_{\mathsf{x}}\ (\mathsf{x} \in C), \mathbf{g}_{\mathsf{o}}\ (\mathsf{o} \in \mathsf{O}^{\varnothing}_{\mathbf{D}(L)}) : \mathbf{r}_{\mathsf{x}}\ (\mathsf{x} \in \mathsf{X}_L)\Big) \tag{2.4}$$

where $\mathbf{r}_{\mathsf{x}} := {}^{\mathbf{g}_{\mathsf{x}}}\mathbf{g}_{\mathsf{y}}\mathbf{g}_{\mathsf{z}}^{-1}$ *in case* $\mathbf{P}(\mathbf{D}(L))$ *looks locally like Figure 8 near* x. □

As Epple (1995) points out, Wirtinger (1905) discovered (2.4) in the course of a study of the topology of holomorphic curves. A proof of 2.12 is given by Crowell and Fox (1977).

A *Seifert surface* is a surface $S \subset S^3$. The boundary of a Seifert surface S is a link L, and S is called a Seifert surface for L. Similarly, a ribbon (resp., clasp) surface S with $L = \partial S$ is called a *ribbon* (resp., *clasp*) *surface for* L. It is a well known fact (apparently first stated

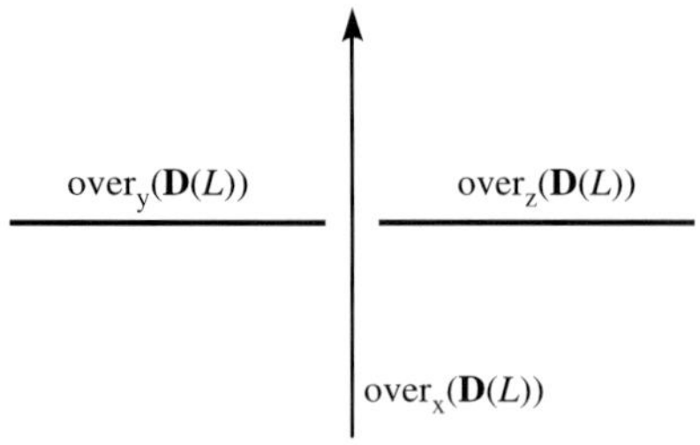

Fig. 8. Labeled over-arcs (the unlabelled crossing need not be x).

by Frankl and Pontrjagin, 1930) that, if L is a link, then there exist Seifert surfaces (and, *a fortiori*, ribbon surfaces and clasp surfaces) for L. For the purposes of this survey the construction sketched by Seifert (1934, 1935) is especially convenient.

CONSTRUCTION 2.13. Let $\mathbf{D}(L)$ be a standard link diagram for a link $L \subset \mathbb{R}^3 \subset S^3$, equipped with a fixed smoothing $\mathrm{sm}(\mathbf{P}(\mathbf{D}(L)))$ of $\mathbf{P}(\mathbf{D}(L))$ given by a fixed family $\{\delta_{\mathsf{x}} : \bowtie \to \mathbb{C} : \mathsf{x} \in \mathsf{X}_{\mathbf{D}(L)}\}$ of embeddings of $\bowtie$. To implement Seifert's construction, choose embeddings $\eta_{\mathsf{o}} : \iota(\mathsf{o}) \hookrightarrow \mathbb{C} \times \mathbb{R}_{\leq 0}$ for $\mathsf{o} \in \mathsf{O}_{\mathbf{D}(L)}$, and $\eta_{\mathsf{x}} : h^{(1)} \hookrightarrow \mathbb{C} \times \mathbb{R}_{\geq 0}$ for $\mathsf{x} \in \mathsf{X}_{\mathbf{D}(L)}$, subject to the following conditions. As suggested in Figure 9, for each Seifert cycle o,

(1) η_{o} is proper relative to the standard collarings of o in $\iota(\mathsf{o})$ and $\mathbb{C} \times \{0\}$ in $\mathbb{C} \times \mathbb{R}_{\leq 0}$,
(2) $\mathrm{pr}_1(\eta_{\mathsf{o}}(\iota(\mathsf{o}) \setminus \mathrm{Int}\,\mathrm{Col}(\mathsf{o} \subset \iota(\mathsf{o})))) = \iota(\mathsf{o}) \subset \mathbb{C} \times \{0\} \subset \mathbb{C} \times \mathbb{R}_{\leq 0}$, and
(3) if $\mathsf{o}' \neq \mathsf{o}$, then $\eta_{\mathsf{o}}(\iota(\mathsf{o}))$ and $\eta_{\mathsf{o}}(\iota(\mathsf{o}'))$ are disjoint.

As suggested in Figure 10, for each crossing x,

(4) $\eta_{\mathsf{x}} : h^{(1)} \hookrightarrow \mathbb{C} \times \mathbb{R}_{\geq 0}$ is proper relative to the standard collarings of $\mathrm{Int}\, A(h^{(1)})$ in $h^{(1)}$ and $\mathbb{C} \times \{0\}$ in $\mathbb{C} \times \mathbb{R}_{\geq 0}$,
(5) $\mathrm{pr}_1(\eta_{\mathsf{x}}(h^{(1)} \setminus \mathrm{Int}\,\mathrm{Col}(A(h^{(1)}) \subset (h^{(1)})) = \delta_{\mathsf{x}}(\bowtie)$,
(6) $\eta_{\mathsf{x}}(h^{(1)}) \cap \mathbf{p}_{\mathbf{D}(L)}^{-1}(\mathsf{x}) = \kappa(h^{(1)})$,

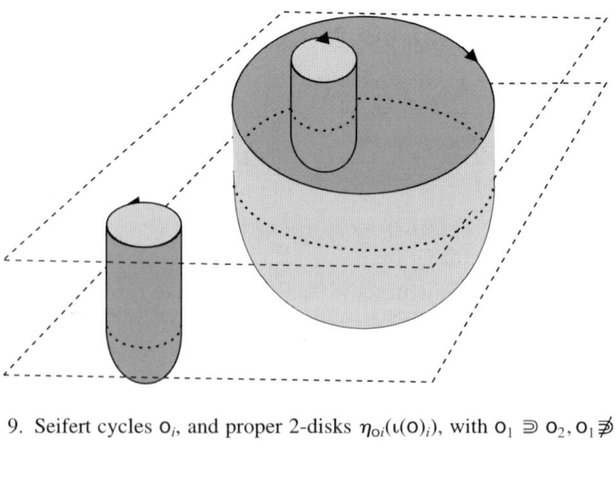

Fig. 9. Seifert cycles o_i, and proper 2-disks $\eta_{\mathsf{o}_i}(\iota(\mathsf{o})_i)$, with $\mathsf{o}_1 \Supset \mathsf{o}_2, \mathsf{o}_1 \not\Supset \mathsf{o}_3$.

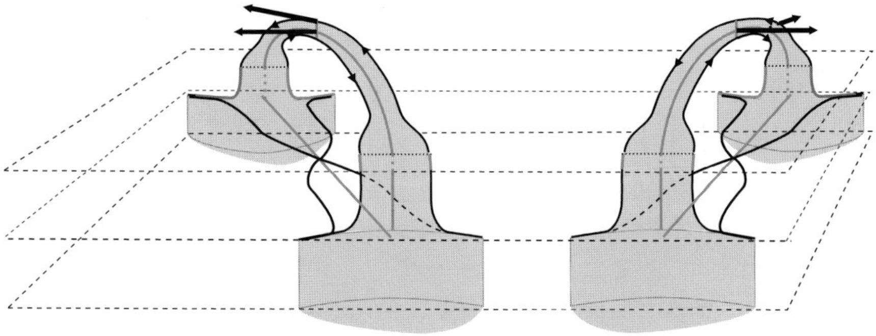

Fig. 10. Attaching $\eta_{\mathsf{x}_i}(h^{(1)})$ to $\cup_{\mathsf{o}\in\mathsf{o}_{\mathbf{D}(L)}}\mathsf{o}$ (left, positive; right, negative).

(7) $\eta_{\mathsf{x}}|A(h^{(1)}) : \eta_{\mathsf{x}}(A(h^{(1)})) \hookrightarrow \cup_{\mathsf{o}\in\mathsf{o}_{\mathbf{D}(L)}}\mathsf{o}$ is orientation reversing, and
(8) the sign of the crossing of $\eta_{\mathsf{x}}(\partial h^{(1)}\setminus \mathrm{A}(h^{(1)}))$ is equal to $\varepsilon(\mathsf{x})$.
It follows that

$$\Sigma := \bigcup_{\mathsf{o}\in\mathsf{O}_{\mathrm{D}(L)}} \eta_{\mathsf{o}}(\iota(\mathsf{o})) \cup \bigcup_{\mathsf{x}\in\mathsf{X}_{\mathrm{D}(L)}} \eta_{\mathsf{x}}(h^{(1)}) \subset \mathbb{C}\times\mathbb{R} \tag{2.5}$$

is a (0,1)-handle decomposition (3.13) of a surface.

PROPOSITION 2.14. (1) *There exists a diffeomorphism* δ: $\mathbb{C}\times\mathbb{R}\to\mathbb{R}^3$ *such that* $\mathbf{p}_{\mathrm{D}(L)}\circ\delta = \mathrm{pr}_1 : \mathbb{C}\times\mathbb{R}\to\mathbb{C}$ *and* $L=\delta(\partial\Sigma)$. (2) *The Seifert surface* $\delta(\Sigma)\subset\mathbb{R}^3\subset S^3$ *for L is independent of* δ, *up to isotopy fixing L pointwise.* □

Any Seifert surface for L of the form $\delta(\Sigma)$ in 1.14(2) is denoted $S(\mathbf{D}(L))$, and called a *diagrammatic Seifert surface* for L. (A Seifert surface need not be isotopic to any diagrammatic Seifert surface.) Figures 11 and 12 depict two diagrammatic Seifert surfaces.

Many operations on links are (most conveniently, and sometimes necessarily) defined using Seifert surfaces. Here are two examples: a *connected sum* of links L_1, L_2 bounding Seifert surfaces S_1, S_2 is $L_1 \# L_2 := \partial(S_1 \natural S_2) \subset S^3 \# S^3 \cong S^3$; the *split sum* of L_1, L_2 is $L_1 \sqcup L_2 := \partial(S_1 \# S_2)$. It is well known that if $L_1 = K_1$ and $L_2 = K_2$ are knots, then $K_1 \# K_2$ is well defined up to isotopy, and independent of S_1 and S_2; in any case, $L_1 \sqcup L_2$ is well defined. In particular, for $n\in\mathbb{N}$ the n-component unlink is the split sum of n unknots. For any link L, let $L^{(n)}$ denote the (well defined) link $L\# O^{(n)}$.

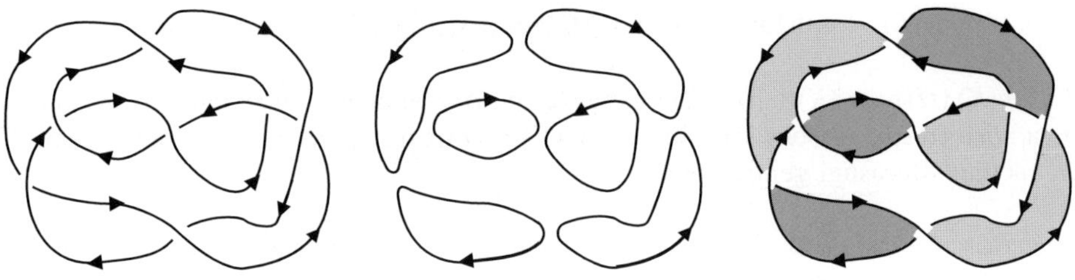

Fig. 11. Seifert's construction applied to a scattered diagram.

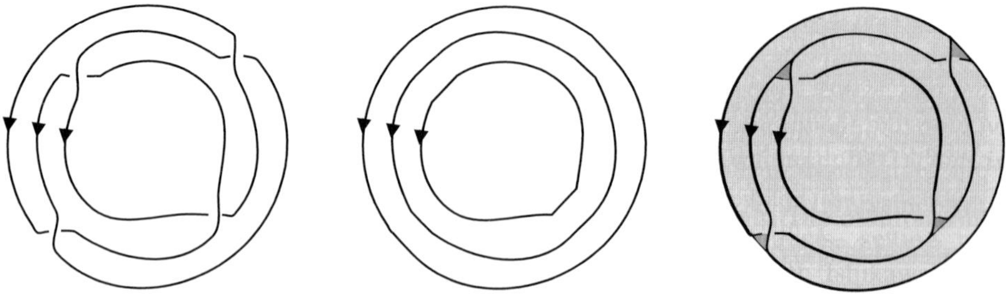

Fig. 12. Seifert's construction applied to a nested diagram.

Similarly, many knot and link invariants are defined using Seifert surfaces.

PROPOSITION 2.15. *If L and L' are disjoint links, then the algebraic number of intersections of L' with a Seifert surface S for L is independent of S (provided only that L' intersects S transversely).* □

This integer invariant of the pair (L, L'), denoted link (L, L') and called the *linking number* of L and L', satisfies $\operatorname{link}(L, L') = \operatorname{link}(L, L') = -\operatorname{link}(-L, L') = -\operatorname{link}(\operatorname{Mir} L, \operatorname{Mir} L')$.

DEFINITIONS 2.16. Let K be a knot, L a link. Define invariants

$$g(K) := \min\{g(S) : S \text{ is a Seifert surface for } K\},$$

$$g_r(K) := \min\{g(S) : S \text{ is a ribbon surface for } K\},$$

$$g_s(K) := \min\{g(S) : S \text{ is a slice surface for } K\},$$

$$X(L) := \max\{\chi(S) : S \text{ is a Seifert surface for } L\},$$

$$X_r(L) := \max\{\chi(S) : S \text{ is a ribbon surface for } L\},$$

$$X_s(L) := \max\{\chi(S) : S \text{ is a slice surface for } L\},$$

$$\operatorname{clasp}(L) := \min\{\operatorname{clasp}(S) : S = f(D^2 \times \pi_0(L)) \text{ is a clasp surface for } L\},$$

$$\operatorname{node}(L) := \min\{\operatorname{node}(S) : S = f(D^2 \times \pi_0(L)) \text{ is a nodal surface for } L\}.$$

Call $g(K)$ (resp., $g_r(K)$; $g_s(K)$) the *genus* (resp., *ribbon genus*; *slice genus*) of K, and say K is a *slice* (resp., *ribbon*) knot in case $g_s(K) = 0$ (resp., $g_s(K) = 0$). Another name for $g_s(K)$ is the "Murasugi genus" of K.

DEFINITION 2.17. Let $\mathbf{D}(L)$ be a standard link diagram, $n := \operatorname{card}(\pi_0(L))$. It is easy to prove and well known (see Hoste or Kauffman, this Handbook) that there is a standard link diagram $\mathbf{D}(O^{(n)})$ for an unlink $O^{(n)}$ such that $\mathbf{P}(\mathbf{D}(O^{(n)})) = \mathbf{P}(\mathbf{D}(L))$ and $\mathbf{I}(\mathbf{D}(O^{(n)}))$ differs from $\mathbf{I}(\mathbf{D}(L))$ precisely to the extent that some number $u \geq 0^{(n)}$ of crossings of $\mathbf{P}(\mathbf{D}(O^{(n)})) = \mathbf{P}(\mathbf{D}(L))$ have opposite signs in $\mathbf{I}(\mathbf{D}(O^{(n)}))$ and $\mathbf{I}(\mathbf{D}(L))$. The *unknotting number* of $\mathbf{D}(L)$ is the least such u. The *unknotting number* $\ddot{u}(L)$ of L is the least unknotting number of all standard link diagrams for L. (The unknotting number is also called the *Überschneidungszahl* Wendt (1937); Milnor (1968) and the *Gordian number* Boileau and Weber (1983); Bennequin (1993); A'Campo (1998).) More generally, the *Gordian distance* $d_G(L, L')$ between two links L, L' with $\text{card}(\pi_0(L)) = \text{card}(\pi_0(L'))$ is the minimum number of sign changes at crossings needed to transform some standard link diagram $\mathbf{D}(L)$ to some standard link diagram $\mathbf{D}(L')$ (Murakami, 1985); thus $\ddot{u}(L) = d_G(L, O^{(n)})$.

Various more or less obvious inequalities relate $\ddot{u}$ and the several invariants named in 2.16 (see Shibuya, 1974; Rudolph, 1983b).

2.5. *Framed links; Seifert forms*

Let L be a link A *framing* of L is a function $f\colon (\pi_0(L) \to \mathbb{Z}$; the pair (L, f) is a *framed link*. A framing of a knot is identified with the integer, which is its value. For framings f, g of L, write $f \preccurlyeq g$, and say f is *less twisted* than g provided that $f(K) \leq g(K)$ for every $K \in \pi_0(L)$.

PROPOSITION 2.18. *The normal bundle of L is trivial. Given a Seifert surface S for L, there exists a trivialization* $n\colon L \times \mathbb{C} \to \nu(L)$ *such that, under the identification of* $Nb(L \subset S^3)$ *with* $n(L \times D^2)$ *(as in* 1.4.3), $S \cap \text{Nb}(L \subset S^3) = \text{Nb}(L \subset S)$ *is identified with* $n(L \times [0, 1])$. *The homotopy class of n is well-defined, independent of S.* □

Let (K, k) be a framed knot. A *k-twisted annulus of type K* is any annulus $A(K, k) \subset S^3$ such that $K \subset \partial A(K, f)$ and $\text{link}(K, \partial A(K, k)\backslash K)$ is $-k$; note that, since $\partial A(K, k)\backslash K$ is clearly isotopic to $-K$, all four of $A(K, k)$, $A(-K, k)$, $-A(K, k)$, and $-A(-K, k)$ are isotopic. For a framed link (L, f), $A(L, f)$ is defined componentwise. Given a 2-submanifold $S \subset S^3$ and a link $L \subset S$, the *S-framing of L* is the framing $f_{L\subset S}$ such that $\text{Col}(L \subset S) = A(-L, f_{L\subset S})$. A framed link (L, f) is *embedded* on a Seifert surface S in case $L \subset S$ and $f = f_{L\subset S}$.

Let S be a Seifert surface with collaring $\text{col}_{S\subset S^3}$. The *Seifert pairing* (on S) of an ordered pair of links (L_0, L_1) with $L_0, L_1 \subset S$ is $(L_0, L_1)_S := \text{link}(L_0, L_1^+)$; if $K \subset S$ is a knot, then $(K, K)_S = f_{K\subset S}$. Given an ordered μ-tuple $(L_1,\ldots,L_\mu)$ of links on S the homology classes of which form a basis for $H_1(S;\mathbb{Z})$, the *Seifert matrix* of S with respect to that basis is the $\mu \times \mu$ matrix $[(L_i, L_j)_S]$, and the *Seifert form* is the (typically non-symmetric) bilinear form on $H_1(S; \mathbb{Z})$ represented by $[(L_i, L_j)_S]$.

2.6. *Fibered links, fiber surfaces, and open books*

Let L be a link. Let $n\colon L \times \mathbb{C} \to \nu(L)$ be a trivialization, as in 2.18, in the homotopy class corresponding to any Seifert surface S for L. Call L *fibered* in case there exists a map φ: $S^3\backslash L \to S^1$ (called a *fiber map* for L) which is adapted to n, has $d(K) = 1$ for all $K \in \pi_0(L)$, and is a fibration (in particular, a Morse map). If L is a fibered link with fiber map φ, then for each $e^{i\theta} \in S^1$, $L \cap \varphi^{-1}(e^{i\theta})$ is a Seifert surface for L. A *fiber surface* is any Seifert surface F_L isotopic to $L \cap \varphi^{-1}(e^{i\theta})$ for any fibered link L with fiber map φ. The *Milnor number* of L is $\mu(L) := \dim_{\mathbb{R}} H_1(F;\mathbb{R})$.

Let S be a Seifert surface. The top of S is $\mathrm{top}(S) := \mathrm{Col}(S \subset S^3)$. A 2-disk $D \subset \mathrm{top}(S)$ is a *top-compression disk* in case $\partial D = D \cap S$ and ∂D bounds no disk on S. Call S *compressible* in case there exists a top-compression disk for at least one of S and $-S$, and *incompressible* in case it is not compressible. Call S *least-genus* provided that $\chi(S) = X(L)$. The following facts are well known (see Stallings, 1978; Gabai, 1983a,b, 1986).

PROPOSITION 2.19. (1) *S is a fiber surface if and only if S is connected and a push-off map induces an isomorphism $\pi_1(\mathrm{Int}\, S; *) \to \pi_1(S^3 \backslash S; *^+)$.* (2) *A least-genus surface S is incompressible.* (3) *A fiber surface is least-genus, and up to isotopy it is the unique incompressible surface with its boundary.* (4) *$A(K, n)$ is least-genus if and only if $(K, n) \neq (O, 0)$.* (5) *$A(K, n)$ is a fiber surface if and only if $(K, n) = (O, \mp 1)$.* □

The fiber surface $A(O, -1)$ (resp., $A(O, 1)$) is called a *positive* (resp., *negative*) *Hopf annulus* (sometimes "Hopf band"); the choice of the adjectives "positive" and "negative" reflects the linking number of the components of $\partial A(O, \mp 1)$.

Fibered links are slightly flaccid. They may be rigidified as follows. An *open book* is a map $\mathbf{f}\colon S^3 \to \mathbb{C}$ such that 0 is a regular value and $\arg(\mathbf{f}) := \mathbf{f}/|\mathbf{f}|\colon S^3 \backslash \mathbf{f}^{-1}(0) \to S^1$ is a fibration. The *binding* $\mathbf{f}^{-1}(0)$ of $\mathbf{f}$ clearly is a fibered link, and for each $e^{i\theta} \in S^1$, the *θth page* $F_\theta := \mathbf{f}^{-1}(\{re^{i\theta}\colon r \geq 0\})$ of $\mathbf{f}$ is a fiber surface. Every fibered link is the binding of various open books; any two fibered books with the same binding are equivalent in a straightforward sense (cf. Kauffman and Neumann, 1977).

Milnor (1968) discovered a rich source (now called *Milnor fibrations*; see 5.114) of open books as part of his investigation of the topology of singular points of complex hypersurfaces. The simplest special cases are fundamental to the knot theory of complex plane curves and easy to write down. Let $m, n \in \mathbb{N}$, $(m, n) \neq (0,0)$.

THEOREM 2.20. $\mathbf{o}\{m, n\}\colon S^3 \to \mathbb{C}^2\colon (z, w) \mapsto z^m + w^n$ *is an open book.* □

The binding $\mathbf{o}\{m, n\}^{-1}(0)$ is a *torus link of type* $\{m, n\}$, sometimes (as in Rudolph, 1982a, 1988, cf. Litherland, 1979) denoted $O\{m, n\}$. Call $\mathbf{o} := \mathbf{o}\{1,0\}$ (resp., $\mathbf{o}' := \mathbf{o}\{0,1\}$) the *vertical* (resp., *horizontal*) *unbook* and its binding $\mathbb{O} := O\{1,0\}$ (resp., $\mathbb{O}' := O\{0,1\}$) the *vertical* (resp., *horizontal*) *unknot* (Rudolph, 1988).

2.7. *Polynomial invariants of knots and links*

The intent of this section is to establish notations and conventions, and to state without proof two useful theorems. For a thorough treatment of polynomial link invariants, see Kauffman (this Handbook).

For any ring $\mathcal{R}$, for any Laurent polynomial $H(s) \in \mathcal{R}[s^{\pm 1}]$, write $\mathrm{ord}_s\, H(s) := \sup\{n \in \mathbb{Z}\colon s^{-n}H(s) \in \mathcal{R}[s] \subset \mathcal{R}[s^{\pm 1}]\}$, $\deg_s H(s) := -\mathrm{ord}_s\, H(s^{-1})$.

DEFINITION 2.21. Let K be a knot. Let S be a Seifert surface for K. Let $A = [(L_i, L_j)_S]$ be a Seifert matrix of S, with transpose A^{T}. The *unnormalized Alexander polynomial* of K is

$\det(A^{\mathrm{T}} - tA) \in \mathbb{Z}[t]$. It is easily shown that $\mu := \deg_t \det(A^{\mathrm{T}} - tA)$ is even. The *Alexander polynomial* of K is $\Delta_K(t) := t^{-\mu/2}\det(A^{\mathrm{T}} - tA) \in \mathbb{Z}[t, t^{-1}]$.

PROPOSITION 2.22. *Δ_K depends only on K, not on the choice of S* or A. □

Let $L \subset \mathbb{R}^3 \subset S^3$ be a link. Let $\mathbf{D}(L)$ be a standard link diagram for L. Let $\mathsf{x} \in \mathsf{X}_{\mathbf{D}(L)}$. Call x *homogeneous* (resp., *heterogeneous*) in case $\mathrm{card}(\pi_0(\mathrm{sm}(L; \mathsf{x})))$ equals $\mathrm{card}(\pi_0(L)) + 1$ (resp., $\mathrm{card}(\pi_0(L)) - 1$). In case $\mathsf{x} \in \mathsf{X}^+_{\mathbf{D}(L)}$, let $L_+ := L$, and define standard link diagrams $\mathbf{D}(L_-)$ and $\mathbf{D}(L_0)$ (and thereby links L_- and L_0) as follows: (1) $\mathbf{P}(\mathbf{D}(L_-)) = \mathbf{P}(\mathbf{D}(L_+))$, and $\mathbf{I}(\mathbf{D}(L_-))$ differs from $\mathbf{I}(\mathbf{D}(L_+))$ precisely to the extent that $\mathsf{x} \in \mathsf{X}^-_{\mathbf{D}(L_-)}$; (2) $\mathbf{P}(\mathbf{D}(L_0))$ is the local smoothing $\mathrm{sm}(\mathbf{P}(\mathbf{D}(L_\pm)); \mathsf{x})$ of $\mathbf{P}(\mathbf{D}(L_\pm))$ at x, and $\mathbf{I}(\mathbf{D}(L_0))$ differs from $\mathbf{I}(\mathbf{D}(L_\pm))$ precisely to the extent that $\mathsf{x} \notin \mathsf{X}_{\mathbf{D}(L_0)}$. In case $\mathsf{x} \in \mathsf{X}^-_{\mathbf{D}(L)}$, modify these definitions accordingly; the two sets of definitions are consistent.

Up to isotopy, the local situation at x is as in Figure 13(A). In case x is homogeneous (resp., heterogeneous), let $\mathbf{D}(L_\infty)$ be the standard link diagram differing from $\mathbf{D}(L_\pm)$ and $\mathbf{D}(L_0)$ only as required by case (1) (resp., case (2)) of Figure 13(B), let p (resp., q) be the linking number of the right-hand visible component of L_0 with the rest of L_0 (resp., of the lower visible component of L_+ with the rest of L_+), and define $r := 4p + 1$ (resp., $r := 4q - 1$).

DEFINITION 2.23. The *oriented polynomial* $P_L(v, z) \in \mathbb{Z}[v^{\pm 1}, z^{\pm 1}]$ and *semi-oriented polynomial* $F_L(a, x) \in \mathbb{Z}[a^{\pm 1}, x^{\pm 1}]$ of L are defined recursively as follows, with the initial conditions $P_O(v, z) = 1 = F_O(a, x)$.

$$P_{L_+}(v,z) = vzP_{L_0}(v,z) + v^2P_{L_-}(v,z) \tag{2.6}$$

$$F_{L_+}(a,x) = a^{-1}xF_{L_0}(a,x) - a^{-2}F_{L_-}(a,x) + a^{-r}xF_{L_\infty}(a,x) \tag{2.7}$$

The nomenclature is that of Lickorish (1986); the choice of variables v, z in (2.6) follows Morton (1986). The oriented (resp., semi-oriented) polynomial is often known, eponymously, as the *FLYPMOTH* (Freyd et al., 1985; Przytycki and Traczyk, 1988) (resp., *Kauffman* (Kauffman, 1987)) polynomial.

DEFINITIONS 2.24. Several other polynomials, though mere adaptations of the oriented or semi-oriented polynomials, nonetheless have their uses.

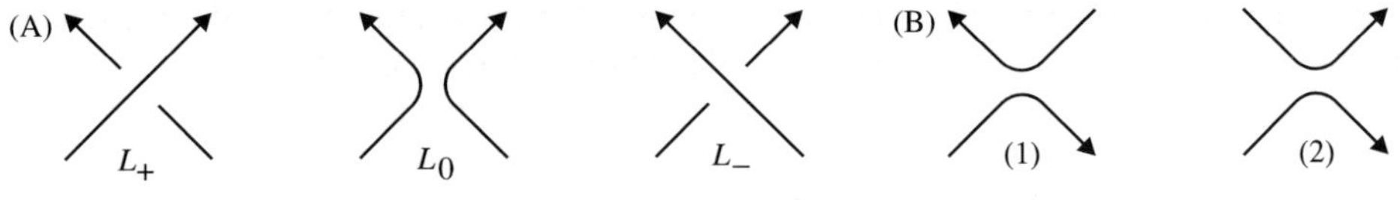

Fig. 13. (A) L_+, L_0, and L_-. (B) L_∞ (homogeneous and heterogeneous cases).

2.24.1. Let (L, f) be a framed link. The *framed polynomial* $\{L, f\}(v, z) \in \mathbb{Z}[v^{\pm 1}, z^{\pm 1}]$ is

$$(-1)^{\operatorname{card}(\pi_0(L))}\left(1 + (v^{-1} - v)z^{-1} \sum_{\emptyset \neq C \subset \pi_0(L)} (-1)^{\operatorname{card}(C)} P_{\partial A(\cup C, f | \cup C)}\right) \tag{2.8}$$

(see Rudolph, 1990). For any (L, f),$\{L, f\} = v^{-2\sum_{K\in\pi_0(L)}^{f(K)}}\{L, 0\}$.

2.24.2. Let $R_L(v) := (z^{\operatorname{card}(\pi_0(L))-1} P_L(v, z))|_{z=0}$. R_L can be calculated from $R_O(v) = 1$ and $R_{L_+}(v) = hR_{L_0}(v) + v^2 R_{L_-}(v)$, where h *is* 0 (resp., 1) in case x is heterogeneous (resp., homogeneous).

2.24.3. Let $F_L^*(a, x) := (F_L \pmod 2) \in (\mathbb{Z}/2\mathbb{Z})[a^{\pm 1}, x^{\pm 1}]$. For $k = 0$, 1, let $G_L^k(a) := (x^{1-c(L)} F_L^*(a, x))|_{x=k} \in (\mathbb{Z}/2\mathbb{Z})[a^{\pm 1}]$, so $G_L^0(a) = R_L(a^{-1}) \pmod 2$ and can be calculated using 2.24.2, while G_L^1 can be calculated from $G_O^1(a) = 1$ and $G_{L+}^1(a) = a^{-2}G_{L-}^1(a) + a^{-1}G_{L_0}^1(a) + a^{-r}G_{L\infty}^1(a)$.

The result underlying most applications of polynomial invariants to the knot theory of complex plane curves, due to Morton (1986) and Franks and Williams (1987), is rephrased here to fit the expository order of this survey; the usual statement, in terms of braids, is given in 3.58.

THEOREM 2.25. *If* $\mathbf{D}(L)$ *is a standard link diagram such that* (1) $\mathbf{D}(L)$ *is nested and* (2) $\mathsf{O}_{\mathbf{D}(L)} = \mathsf{O}^+_{\mathbf{D}(L)}$, *then*

$$\operatorname{ord}_v P_L \geq \operatorname{card}(\mathsf{X}^+_{\mathbf{D}(L)}) - \operatorname{card}(\mathsf{X}_{\mathbf{D}(L)^-}) - \operatorname{card}(\mathsf{O}_{\mathbf{D}(L)}) + 1. \qquad \square$$

The framed polynomial provides a bridge between the oriented and semi-oriented polynomials, as the following result (Rudolph, 1990) makes plain.

THEOREM 2.26. $(1 + (v^{-2} + v^2)z^{-2})F_L(v^{-2}, z^2) \equiv v^{4\tau(L)}\{L, 0\}(v, z) \pmod 2$. $\square$

2.8. *Polynomial and analytic maps; algebraic and analytic sets*

This section recalls needed definitions from real and complex algebraic and analytic geometry, and establishes notations. General background, and proofs of stated results, can be found in Whitney (1957, 1972); Milnor (1968); Narasimhan (1960); Gunning and Rossi (1965), and, for 2.27.1, Appendix B (Abraham and Robbin, 1967).

Let $\mathbb{F}$ be one of the fields $\mathbb{R}$ or $\mathbb{C}$, with its metric topology. The algebra of polynomials (resp., somewhere-convergent power series) in n variables with ground field $\mathbb{F}$ is denoted $\mathbb{F}[\varphi_1,\ldots,\varphi_n]$ (resp., $\mathbb{F}\{\varphi_1,\ldots,\varphi_n\}$), where φ stands for x (resp., z) in case $\mathbb{F}$ is $\mathbb{R}$ (resp., $\mathbb{C}$). As usual, $f \in \mathbb{F}[\varphi_1,\ldots,\varphi_n]$ is conflated with the *polynomial function* $f\colon \mathbb{F}^n \to \mathbb{F}$ that it defines, and $f \in \mathbb{F}\{\varphi_1,\ldots,\varphi_n\}$ with both the $\mathbb{F}$-analytic function that it defines in a neighborhood of $0 \in \mathbb{F}^n$ and the germ of that function at $0 \in \mathbb{F}^n$. (In particular φ_s is conflated with the coordinate projection $\mathrm{pr}_s\colon \mathbb{F}^n \to \mathbb{F}$ for every $s \in \mathbf{n}$, no notational distinction being made between the functions φ_s on F^m and F^n so long as $s \in \mathbf{m} \cap \mathbf{n}$.) Given a non-empty open set $\Omega \subset \mathbb{F}^n$, a function $f{:}\,\Omega \longrightarrow \mathbb{F}$ is called $\mathbb{F}$-*analytic* (simply *analytic* when $\mathbb{F}$ is clear from context; also *holomorphic* for $\mathbb{F} = \mathbb{C}$, and *entire* when $\Omega = \mathbb{C}^n$) in case, for every $(\varphi_1^{(0)},\ldots,\varphi_n^{(0)}) \in \Omega$, the germ of $f(\varphi_1 - \varphi_1^{(0)},\ldots,\varphi_n - \varphi_n^{(0)})$ at $(0,\ldots,0)$ belongs to $\mathbb{F}\{\varphi_1,\ldots,\varphi_n\}$. The set $\mathcal{O}(\Omega)$ of all $\mathbb{F}$-analytic functions on Ω is an algebra containing (a natural isomorphic image of) $\mathbb{F}[\varphi_1,\ldots,\varphi_n]$.

A *polynomial* (resp., $\mathbb{F}$-*analytic*) *map* $F = (f_1,\ldots, f_m)$ from $\mathbb{F}^n$ (resp., Ω) to $\mathbb{F}^m$ is one with components f_s that are polynomial (resp., $\mathbb{F}$-analytic) functions. An *algebraic set* (resp., *global analytic set*) is a subset $\mathcal{V}_F := F^{-1}(0,\ldots,0)$ of $\mathbb{F}^m$ (resp., Ω), where F is a polynomial (resp., analytic) map. An *analytic set* is a subset X of Ω for which every point of Ω has an open neighborhood U such that $X \cap U$ is a global analytic set $\mathcal{V}_F$ for some $F \in \mathcal{O}(U)$. Every global analytic set (in particular, every algebraic set) is an analytic set; the converse fails for many Ω when $n > 1$.

Let F: $\Omega \longrightarrow \mathbb{F}^m$ be an analytic map (allowing the possibility that $\Omega = \mathbb{F}^n$ and F is polynomial). It may happen that

$$\mathrm{Reg}(F) := \{(y_1,\ldots,y_m) \in \mathcal{V}_F\colon \mathrm{rank}_{\mathbb{F}}\, DF(y_1,\ldots,y_m) = n - m\}$$

is not dense in $\mathcal{V}_F$. However, theorems of commutative algebra show that there is another analytic map F_0: $\Omega \longrightarrow \mathbb{F}^m$ such that $\mathcal{V}_{F_0}$ and $\mathcal{V}_F$ are equal *as sets* (that is, ignoring multiplicities), and $\mathrm{Reg}(F_0)$ is dense in $\mathcal{V}_{F_0} = \mathcal{V}_F$. Call $(y_1,\ldots,y_m) \in \mathcal{V}_F$ a *regular* (*resp., singular point*) of the global analytic set $\mathcal{V}_F$ in case $\mathrm{rank}_{\mathbb{F}}\, DF_0(y_1,\ldots,y_m)$ equals (resp., is less than) $n - m$; these definitions are independent of the particular choice of F_0. Regularity in $\mathcal{V}_F$ is clearly a local property, and is therefore well-defined in any analytic set X. The set $\mathrm{Reg}(X)$ of regular points of X is called the *regular locus* of X; it is an $\mathbb{F}$-analytic manifold. The *singular locus* $X\backslash\mathrm{Reg}(X) =: \mathrm{Sing}(X)$ of X is an algebraic, global analytic, or analytic set according as X is. Let $\mathrm{Sing}^0(X)$ and $\mathrm{Reg}^1(X)$ both mean $\mathrm{Reg}(X)$; for $s \in \mathbb{N}_{>0}$, let $\mathrm{Sing}^{s+1}(X) := \mathrm{Sing}(\mathrm{Sing}^s(X))$, $\mathrm{Reg}^{s+1}(X) := \mathrm{Reg}(\mathrm{Sing}^s(X))$. An $\mathbb{F}$-analytic set X is partitioned by the (finitely many) non-empty sets in the sequence $\{\mathrm{Reg}^s(X)\}_{s\in\mathbb{N}_{>0}}$. In case X is algebraic, the refinement of this partition obtained by separating each $\mathrm{Reg}^s(X)$ into its connected components is a finite stratification (Whitney, 1957; see Milnor, 1968, Theorems 2.3 and 2.4); call it the *naïve stratification* of X.

A *basic semi-algebraic set* in $\mathbb{R}^n$ is the intersection of an $\mathbb{R}$-algebraic set $\mathcal{V}_F$ and finitely many sets of the form $G^{-1}(\mathbb{R}_{>0})$, with $G \in \mathbb{R}[x_1,\ldots,x_n]$; a *semi-algebraic set* is the union of finitely many basic semi-algebraic sets. An algebraic set is semi-algebraic.

PROPOSITION 2.27. (1) *The image of a semi-algebraic set by a polynomial map is semi-algebraic.* (2) *A semi-algebraic set has a finite naïve stratification.* □

2.27(1) is due to Tarski and to Seidenberg. 2.27(2) is a result of Whitney (1957).

Let $U \subset \mathbb{C}^2$ be an open set. A *holomorphic curve* in U is a $\mathbb{C}$-analytic set Γ such that the complex manifold $\mathrm{Reg}(\Gamma)$ is non-empty, everywhere of real dimension 2, and dense in Γ.

PROPOSITION 2.28. *Let* $\Gamma \subset U$ *be a holomorphic curve in an open set in* $\mathbb{C}^2$. *There exists a complex manifold G of real dimension* 2, *and a holomorphic map* $R: G \to U$, *such that:* (1) $R(G) = \Gamma$; (2) $\operatorname{crit}(R) \subset R^{-1}(\operatorname{Sing}(\Gamma))$, *and* $R|R^{-1}(\operatorname{Sing}(\Gamma))$ *has finite fibers;* (3) $R|R^{-1}$ $(\operatorname{Reg}(\Gamma))$ *is a holomorphic diffeomorphism.* □

The map R is essentially unique, and is called the *resolution* of Γ. A *branch* of Γ at $P \in \Gamma$ is the image by R of a component of Int $\operatorname{Nb}(\{Q\} \subset G \backslash \operatorname{crit}(R))$ for some $Q \in R^{-1}(P)$ (or the germ of the image of such a component). If $P \in \operatorname{Reg}(\Gamma)$ then there is only one branch of Γ at P, but there can also be singular points of Γ at which there is only one branch of Γ.

EXAMPLES 2.29. Classical algebraic geometers gave names to quite a few special cases of branches (and resolutions). Two examples are of particular importance in the knot theory of complex plane curves.

2.29.1. Define $f \in \mathcal{O}(\operatorname{Int} D^4)$ by $f(z, w) = z^2 + w^2$. The holomorphic curve $\mathcal{V}_f$ has two branches at (0,0). Its resolution is R: Int $D^2 \times \{+,-\} \to \mathcal{V}_f$: $(\zeta, \pm) \mapsto 2^{-1/2}(\zeta, \pm i\zeta)$. A point P of a holomorphic curve $\Gamma \subset \Omega$ such that there exist an open neighborhood U of P in Ω and a diffeomorphism (which may in fact be required to be holomorphic) $h: (U, U \cap \Gamma, P) \to (\operatorname{Int} D^4, \mathcal{V}_f, (0, 0))$ is called a *node* of Γ.

2.29.2. Define $f \in \mathcal{O}(\operatorname{Int} D^4)$ by $f(z, w) = z^2 + w^3$. The holomorphic curve $\mathcal{V}_f$ has one branch at (0,0). Its resolution is R: Int $D^2 \to \mathcal{V}_f$: $(\zeta, \pm) \mapsto 2^{-1/2}(\zeta^3, -\zeta^2)$. A point P of a holomorphic curve $\Gamma \subset \Omega$ such that there exist an open neighborhood U of P in Ω and a diffeomorphism (which may in fact be required to be holomorphic) $h: (U, U \cap \Gamma, P) \to (\operatorname{Int} D^4, \mathcal{V}_f, (0,0))$ is called a *cusp* of Γ.

A holomorphic curve Γ such that every point of $\operatorname{Sing}(\Gamma)$ is a node (resp., either a node or a cusp) is called a *node* (resp., *cusp*) *curve*.

2.9. *Configuration spaces and spaces of monic polynomials*

Let X be a topological space. For $n \in \mathbb{N}$, the sets $\mathcal{MP}^{[n]}(X)$, $\binom{X}{n}$ and $\Delta_n(X)$ are endowed with topologies by the application of the bijection (2.2) to the quotient topology induced on $X^n/\mathfrak{S}_n$ from the product topology on X^n; with these topologies, they are called the nth *multipower space*, the nth *configuration space*, and the nth *discriminant space* of X, respectively.

If M is a manifold, then clearly each equivalence class of the partition by type $\mathcal{MP}^{[n]}(M)/\tau$ of $\mathcal{MP}^{[n]}(M)$ is a manifold. However, even for connected M it often happens that not every fiber of τ is connected. The stratification $\mathcal{MP}^{[n]}(M)/\tau^c$ of $\mathcal{MP}^{[n]}(M)$ by the connected components of the fibers of τ will be called the *standard stratification* of $\mathcal{MP}^{[n]}(M)$; the standard stratification of $\mathcal{MP}^{[n]}(M)$ induces a *standard stratification* on each of $\mathcal{MP}^{[n]}(M)/\tau^c$ and $\Delta_n(M)$, since they are evidently unions of strata of $\mathcal{MP}^{[n]}(M)/\tau^c$.

If $\mathbb{F}$ is a field, then of course $\mathbb{F}^n$ is an algebraic set over $\mathbb{F}$, and the standard action of $\mathfrak{S}_n$ on $\mathbb{F}^n$ is algebraic. In general this is not enough to ensure that the set $\mathbb{F}^n/\mathfrak{S}_n$ can be endowed with as much structure as might be desirable to full-fledged algebraic geometers. However, for $\mathbb{F} = \mathbb{C}$ (or any algebraically closed field), on general principles $\mathbb{C}^n/\mathfrak{S}_n$ does have a natural structure as an algebraic set (more or less naturally embedded in an affine space $\mathbb{C}^N$; see, e.g., Cartan, 1957) with respect to which the unordering map $U_{\mathbb{C},n}$: $\mathbb{C}^n \to \mathbb{C}^n/\mathfrak{S}_n$ is a polynomial map. By contrast, if $n > 1$, then $\mathbb{R}^n/\mathfrak{S}_n$ not an algebraic set (in any natural way), although it is semi-algebraic.

Denote by $MP_n := \{p(w) \in \mathbb{C}[w]:p(w) = w^n + c_1w^{n-1} + \cdots + c_{n-1}w + c_n\}$ the n-dimensional complex affine space of *monic polynomials* of degree $n \in \mathbb{N}$. Define the *roots map* r: $MP_n \to \mathcal{MP}^{[n]}(\mathbb{C})$ by $r(p) := (p^{-1}(0), \mathbf{m}_p|P^{-1}(0))$, where $\mathbf{m}_p(z)$ is the usual *multiplicity of* $p(w) \in MP_n$ *at* $z \in \mathbb{C}$. Let V be the polynomial map $\mathbb{C}^n \to MP_n: (z_1, \ldots, z_n) \mapsto (w - z_1)\cdots(w - z_n)$ The diagram

$$\begin{array}{ccccc}
\mathbb{R}^n & \xleftarrow{\text{Re}} & \mathbb{C}^n & \xrightarrow{V} & MP_n \\
\downarrow U_{\mathbb{R},n} & & \downarrow U_{\mathbb{C},n} & & \downarrow r \\
\mathbb{R}^n/\mathfrak{S}_n & \xleftarrow{\text{Re}} & \mathbb{C}^n/\mathfrak{S}_n & \underset{(2.2)}{\xrightarrow{\cong}} & \mathcal{MP}^{[n]}(\mathbb{C}) \\
\downarrow \tau^c & & \downarrow \tau & & \\
(\mathbb{R}^n/\mathfrak{S}_n)/\tau^c & & (\mathbb{C}^n/\mathfrak{S}_n)/\tau & &
\end{array} \tag{2.9}$$

is commutative, and defines $R: MP_n \to \mathbb{C}^n/\mathfrak{S}_n$. The following results are standard; some go back, in essence, to Viète (1593)[7] and Descartes (1637).

PROPOSITION 2.30. *The naïve stratification of* $\mathbb{C}^n/\mathfrak{S}_n$ *as an algebraic set coincides with its stratification by type* $(\mathbb{C}^n/\mathfrak{S}_n)/\tau$. *In particular*:

(1) *there are no strata of odd (real) codimension*;

(2) *the only codimension-*0 *stratum is* $\text{Reg}(\mathbb{C}^n/\mathfrak{S}_n) = \binom{\mathbb{C}}{n} = \tau^{-1}(\{1, \ldots, 1\})$, *so* $\text{Sing}(\mathbb{C}^n/\mathfrak{S}_n) = \Delta_n(\mathbb{C})$ *is the union of the strata of codimension* > 0;

(3) *the only codimension-*2 *stratum is* $\text{Reg}^2(\mathbb{C}^n/\mathfrak{S}_n) = \tau^{-1}(\{1, \ldots, 1, 2\})$. □

Call the stratification in 2.30 the *complex stratification* of $\mathbb{C}^n/\mathfrak{S}_n$, and denote it by $(\mathbb{C}^n/\mathfrak{S}_n)/{\equiv_{\mathbb{C}}}$.

[7] V stands for *Viète map*, a coinage due (apparently) to Arnol'd, now widely used.

PROPOSITION 2.31. *R is a homeomorphism and $R|r^{-1}(\Delta_n(\mathbb{C}))$ is a diffeomorphism; in fact, R is a normalization and minimal resolution of the algebraic set $\mathbb{C}^n/\mathfrak{S}_n$.* □

The resolution R and the complex stratification of $\mathbb{C}^n/\mathfrak{S}_n$ together impose a *complex stratification* $MP_n/\equiv_{\mathbb{C}}$ on MP_n.

PROPOSITION 2.32. *$(\mathbb{R}^n/\mathfrak{S}_n)/\tau^c$ is a cellulation; in particular, $\binom{\mathbb{R}}{n}$, its unique codimension-0 stratum, is an n-cell.* □

PROPOSITION 2.33. (1) *If S is a stratum of $(\mathbb{C}^n/\mathfrak{S}_n)/\equiv_{\mathbb{C}}$, then $S/(\tau^c\circ\mathrm{Re})$ is a stratification of S, and in fact a cellulation of S; thus, the partition $(\mathbb{C}^n/\mathfrak{S}_n)/\equiv_{\mathbb{R}}$, such that each $\equiv_{\mathbb{R}}$-class is a $\tau^c\circ\mathrm{Re}$-class of some stratum S of $(\mathbb{C}^n/\mathfrak{S}_n)/\equiv_{\mathbb{C}}$, is a cellulation.* (2) *Each cell of $(\mathbb{C}^n/\mathfrak{S}_n)/\equiv_{\mathbb{R}}$ is a real semi-algebraic set.*

PROOF. (1) is apparently originally due to Fox and Neuwirth (1962) (see also Fuchs, 1970; Vaĭnsteĭn, 1978; Napolitano, 1998). (2) follows from 2.27(1). □

Call the cellulation in 2.33 the *real cellulation of* $\mathbb{C}^n/\mathfrak{S}_n$ and denote it by $(\mathbb{C}^n/\mathfrak{S}_n)/\equiv_{\mathbb{R}}$. The resolution R and the real cellulation of $\mathbb{C}^n/\mathfrak{S}_n$ together impose a *real cellulation* $MP_n/\equiv_{\mathbb{R}}$ on MP_n. The real cellulations of $\mathbb{C}^n/\mathfrak{S}_n$ and MP_n in turn define real cellulations of $\mathcal{MP}^{[n]}(\mathbb{C})$, $\Delta_n(\mathbb{C})$, $R^{-1}(\mathcal{MP}^{[n]}(\mathbb{C}))$, and $R^{-1}(\Delta_n(\mathbb{C}))$.

EXAMPLES 2.34. For small n, very explicit descriptions of the complex stratifications are easily given.

2.34.1. $MP_1/\equiv_{\mathbb{C}}$ consists of a single stratum, necessarily of codimension 0.

2.34.2. $MP_2/\equiv_{\mathbb{C}}$ consists of two incident strata: $R^{-1}\left(\binom{\mathbb{C}}{2}\right)$, of codimension 0, is diffeomorphic to $\mathbb{C}\times(\mathbb{C}\setminus\{0\})$; $R^{-1}(\Delta_2(\mathbb{C}))$, of codimension 2, is diffeomorphic to $\mathbb{C}$. Explicitly, $\mathbb{C}^2/\mathfrak{S}_2\to\mathbb{C}^2:\{w_1,w_2\}\mapsto(w_1+w_2,(w_1-w_2)^2)$ is a homeomorphism that maps $\binom{\mathbb{C}}{2}$ (resp., $\Delta_2(\mathbb{C})$) diffeomorphically onto $\mathbb{C}\setminus\{0\}$ (resp., $\mathbb{C}\times 0$).

2.34.3. $MP_3/\equiv_{\mathbb{C}}$ consists of three mutually incident strata: $\binom{\mathbb{C}}{3}$, of codimension 0, is diffeomorphic to $\mathbb{C}\times(\mathbb{C}^2\setminus\mathcal{V}_f)$, where $f(z_1,z_2)=4z_1^3+9z_2^2$ and so $\mathcal{V}_f$ is a cuspidal cubic curve, homeomorphic to $\mathbb{C}$ and having a single singular point; $\mathrm{Reg}(\Delta_3(\mathbb{C}))$, of codimension 2, is diffeomorphic to $\mathbb{C}\times\mathrm{Reg}(\mathcal{V}_f)$ and therefore to $\mathbb{C}\times\mathbb{C}\setminus\{0\}$; and $\mathrm{Sing}(\Delta_3(\mathbb{C}))$, of codimension 4, is diffeomorphic to $\mathbb{C}\times\mathrm{Sing}(\mathcal{V}_f)$ and therefore to $\mathbb{C}$. It is easy to write down an explicit polynomial homeomorphism $\mathbb{C}^3/\mathfrak{S}_3\to\mathbb{C}^3$ giving an isomorphic stratification.

The real cellulations of MP_n, and thus of $\binom{\mathbb{C}}{n}$ and $\Delta_n(\mathbb{C})$, can be described very explicitly, in all dimensions (see Fox and Neuwirth, 1962; Napolitano, 2000). For the purposes of this survey, it is sufficient to describe the cells of dimension $2n$, $2n - 1$, and $2n - 2$ only, along with their incidence relations; this can be done in a uniform manner for all n.

EXAMPLE 2.35. In $\binom{\mathbb{C}}{n}/\equiv_{\mathbb{R}}$ there is exactly one cell of dimension $2n$, exactly $n - 1$ cells of dimension $2n - 1$, and exactly $(n - 1)(n - 2)/2$ cells of dimension $2n - 2$.

(1) The cell C_0 of dimension $2n$ consists of all $\{z_1, \ldots, z_n\}$ with $\operatorname{Re} z_1 < \cdots < \operatorname{Re} z_n$.

(2) For $k = 1,\ldots,n - 1$, there is a cell C_k of dimension $2n - 1$ consisting of all $\{z_1, \ldots, z_n\}$ with $\operatorname{Re} z_1 < \cdots < \operatorname{Re} z_k = \operatorname{Re} z_{k+1} < \cdots < \operatorname{Re} z_n$ and $\operatorname{Im} z_k \neq \operatorname{Im} z_{k+1}$; C_k is transversely oriented by the complex orientation of MP_n, and C_0 is incident on C_k from both sides – more precisely, there is a simple closed curve in $\binom{\mathbb{C}}{n}$ that intersects C_k in a single point, transversely, and is otherwise contained in C_0.

(3) For $1 \leq k \leq n - 2$, there is a cell $C_{k,k+1}$ of dimension $2n - 2$ consisting of all $\{z_1, \ldots, z_n\}$ with $\operatorname{Re} z_1 < \cdots < \operatorname{Re} z_k = \operatorname{Re} z_{k+1} = \operatorname{Re} z_{k+2} < \cdots < \operatorname{Re} z_n$ and $\operatorname{card}(\{\operatorname{Im} z_k, \operatorname{Im} z_{k+1}, \operatorname{Im} z_{k+2}\}) = 3$. The two cells C_k and C_{k+1} of dimension $2n - 1$ are each triply incident on $C_{k,k+1}$ – more precisely, the stratification induced on a small 2-disk in MP_n intersecting $C_{k,k+1}$ in a single point, transversely, is as pictured to the left of Figure 14.

(4) For $1 \leq i < j - 1 \leq n - 2$, there is a cell $C_{i,j}$ of dimension $2n - 2$ consisting of all $\{z_1, \ldots, z_n\}$ with $\operatorname{Re} z_1 < \cdots < \operatorname{Re} z_i = \operatorname{Re} z_{i+1} < \cdots < \operatorname{Re} z_j = \operatorname{Re} z_{j+1} < \cdots < \operatorname{Re} z_n$, $\operatorname{Im} z_i \neq \operatorname{Im} z_{i+1}$, and $\operatorname{Im} z_j \neq \operatorname{Im} z_{j+1}$. The two cells C_i and C_j of dimension $2n - 1$ are each doubly incident on $C_{i,j}$ – more precisely, the stratification induced on a small 2-disk in MP_n intersecting $C_{i,j}$ in a single point, transversely, is as pictured in the middle of Figure 14.

In $\Delta_n(\mathbb{C})/\equiv_{\mathbb{R}}$, there are no cells of dimension $2n$ or $2n - 1$. For $1 \leq k \leq n - 2$, there is one cell D_k of dimension $2n - 2$ consisting of all $\{z_1,\ldots, z_n\}$ with $\operatorname{Re} z_1 < \cdots < \operatorname{Re} z_k = \operatorname{Re} z_{k+1} < \cdots < \operatorname{Re} z_n$ and $\operatorname{Im} z_k = \operatorname{Im} z_{k+1}$. The stratification induced on a small 2-disk in MP_n intersecting D_k transversely at a single point is as pictured at the right of Figure 14.

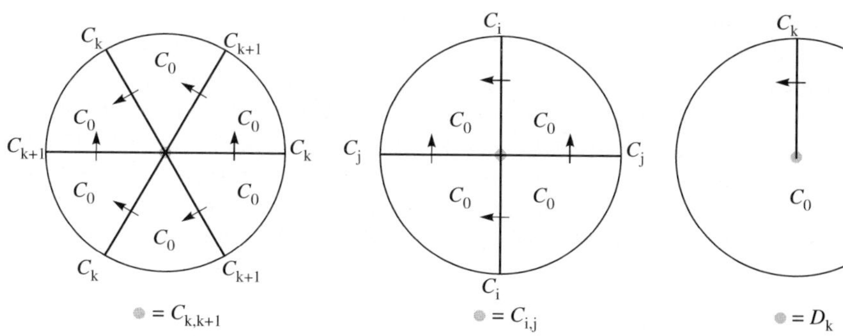

Fig. 14. The transverse structure of $MP_n/\equiv_{\mathbb{R}}$ along its codimension-2 cells.

2.10. *Contact 3-manifolds, Stein domains, and Stein surfaces*

This section simply resumes basic definitions and needed results. For details on contact structures and contact 3-manifolds, see Etnyre (this Handbook). For the topology of Stein domains and Stein surfaces, see Gompf (1998). For the complex function theory of Stein domains, Stein manifolds, and Stein spaces (in general dimensions), see Gunning and Rossi (1965).

DEFINITIONS 2.36. Let M be a 3-manifold without boundary. A *contact structure* on M is a completely non-integrable field ξ of tangent 2-planes on M; for instance, on any round 3-sphere $S^3 \subset \mathbb{C}^2$, the field ξ of tangent 2-planes that are actually complex lines is a contact structure, called the *standard* contact structure on that sphere. A 3-manifold with a contact structure is called a *contact manifold.* A (closed) 1-submanifold L in a contact manifold M with contact structure ξ is *Legendrian* in case $T_x(L) \subset \xi_x \subset T_x(M)$ for all $x \in L$. Of course, L is Legendrian if and only if $-L$ is.[8] In any contact manifold M, a Legendrian 1-submanifold L is naturally endowed with a normal line field $\xi_L^\perp$ (unique up to isotopy), which determines an annular surface $A \subset M$ (also unique up to isotopy) containing L as a retract and such that $\nu(\mathrm{i}_{L \hookrightarrow A}) = \xi_L^\perp$. In particular, a Legendrian link L in S^3 with its standard contact structure has a *natural framing* $f_L^{\mathcal{L}}$ for which $A(L, f_L^{\mathcal{L}})$ is such an annular surface. The *Thurston–Bennequin number* of a Legendrian knot $K \subset S^3$ is $\mathrm{tb}(K) := f_K^{\mathcal{L}}(K)$. For an arbitrary knot $K \subset S^3$, denote by $\mathrm{TB}(K)$ the *maximal Thurston–Bennequin number* $\max\{\mathrm{tb}(K') : K'$ is a Legendrian knot isotopic to $K\}$ of K; $\mathrm{TB}(K)$ is an integer (i.e., neither $-\infty$ nor ∞; see Bennequin, 1983).

DEFINITIONS 2.37. An *open Stein manifold* is a complex manifold that is holomorphically diffeomorphic to a topologically closed complex submanifold of some complex affine space $\mathbb{C}^N$ (equivalently, to a non-singular global analytic set $\mathcal{V}_f$ with $f \in \mathcal{O}(\mathbb{C}^N)$; see Gunning and Rossi, 1965). For instance, $\mathbb{C}^N$ itself is an open Stein manifold, as is (non-obviously) any open subset of $\mathbb{C}$. An *open Stein surface* is an open Stein manifold of real dimension 4. Let M be an open Stein surface. An *exhausting strictly plurisubharmonic function* on a non-empty open set U of M is a smooth function $\rho : U \to \mathbb{R}$ that is bounded below, *proper* (in the sense that $\rho^{-1}([a, b])$ is compact for all $a, b \in \mathbb{R}$), and such that for each $c \in \mathbb{R}$, the field of tangent complex lines on the 3-manifold $\rho^{-1}(c) \setminus \mathrm{crit}(\rho)$ is a contact structure, called the *natural* contact structure on that 3-manifold. (Thus, the standard structure on S^3 is the natural structure for the standard embedding $S^3 \hookrightarrow \mathbb{C}^2$.) A *Stein domain* in M is a compact codimension-0 submanifold $X \subset M$ such that X is a sublevel set $\rho^{-1}(\mathbb{R}_{\le c})$ ($c \in \mathbb{R} \setminus \rho(\mathrm{crit}(\rho))$) of an exhausting strictly plurisubharmonic Morse function $\rho : U \to \mathbb{R}$ on an open set $U \subset M$. A *Stein surface with boundary* is a compact 4-manifold

[8] In particular, in the context of links in S^3 (equipped with its standard contact structure) it is common practice to refer to either L or $|L|$ as a *Legendrian link* (or knot, as the case may be), and – contrary to the conventions established earlier, which require that a link (or knot) be oriented – this practice will be followed here.

X with a complex structure on Int Y such that X is diffeomorphic to a Stein domain in some open Stein surface by a diffeomorphism that is holomorphic on Int X. A *Stein disk* is a Stein domain D in $\mathbb{C}^2$ diffeomorphic to D^4 (for instance, D^4 itself, where $\rho(z,w)$ can be taken to be $\|z,w\|^2$). Any non-singular level 3-manifold of an exhausting strictly plurisubharmonic function on a Stein surface is called a *strictly pseudoconvex* 3-manifold. A closed contact manifold is *Stein-fillable* in case it is diffeomorphic to a strictly pseudoconvex 3-manifold N by a diffeomorphism carrying its contact structure to the natural contact structure on N.

Let $X \subset \mathbb{C}^2$ be a Stein domain. It is convenient to establish notation for several subsets of $C(X) := \{f : X \to \mathbb{C} : f \text{ is continuous}\}$; although $C(X)$, equipped with its sup norm, is a well-known Banach algebra, here it (and its subsets) will not be endowed with any topology.

$$\begin{aligned}
\mathcal{A}(X) &:= \{f \in C(X) : f|\text{Int } X \text{ is holomorphic}\} \\
\mathcal{O}(X) &:= \{f \in C(X) : f = F|X \text{ for some open neighborhood } U \text{ of} \\
&\qquad\qquad X \text{ in } \mathbb{C}^2 \text{ and some holomorphic } F : U \to \mathbb{C}\} \\
\mathcal{U}(X) &:= \{f \in \mathcal{O}(X) : 0 \notin f(X)\} \\
\mathcal{W}(X) &:= \{f \in \mathcal{O}(X) : 0 \notin f(\text{Int } X)\}
\end{aligned} \tag{2.10}$$

An element of $\mathcal{A}(X)$ is called a *germ* of a holomorphic function on X. Both $\mathcal{A}(X)$ and $\mathcal{O}(X)$ are algebras. By a standard argument, $\mathcal{U}(X)$ is the group of units of $\mathcal{O}(X)$. Clearly $\mathcal{W}(X)$ is a multiplicative subsemigroup of $\mathcal{O}(X)$ properly containing $\mathcal{U}(X)$.

Major reasons for complex analysts' interest in Stein manifolds include general theorems of which the following are special cases.

THEOREM 2.38. *If $X \subset \mathbb{C}^2$ is a Stein disk and $\Gamma \subset U$ is a holomorphic curve in an open neighborhood U of X, then $\Gamma \cap X = f^{-1}(0)$ for some $f \in \mathcal{O}(X)$.* □

THEOREM 2.39. *Any holomorphic function on a Stein disk X can be arbitrarily closely uniformly approximated, along with any finite number of its derivatives, by the restriction to X of a polynomial function.* □

The following results are especially useful for topological applications.

THEOREM 2.40. *If $X \subset \mathbb{C}^2$ is a Stein disk with exhausting plurisubharmonic Morse function ρ, and $f \in \mathcal{O}(X)$ is such that $\text{Sing}(\mathcal{V}_f) = \varnothing$, then $\mathcal{V}_f$ is ρ-ribbon.* □

THEOREM 2.41. *Every covering space of an open Stein manifold is an open Stein manifold. A finite-sheeted branched covering space of a Stein disk branched along a non-singular holomorphic curve is a Stein surface with boundary.* □

3. Braids and braided surfaces

Much of the following material is treated (usually more generally and often from a different perspective) by Birman (1975) and Birman and Brendle (this Handbook), references to which should be assumed throughout.

3.1. *Braid groups*

DEFINITION 3.42. For any $n \in \mathbb{N}_{>0}$ and $X \in \binom{\mathbb{C}}{n}$, let $B_X := \pi_1\left(\binom{\mathbb{C}}{n}; X\right)$, and call B_X an *n-string braid group.*[9] The *standard* n-string braid group is $B_n := B_{\mathbf{n}}$. By convention, the (unique) 0-string braid group is $\mathrm{B}_0 := \{o\}$. Write o_X for the identity of B_X, and let $o^{(n)} := o_{\mathbf{n}}$.

Of course, since $\binom{\mathbb{C}}{n}$ is connected, every n-string braid group is isomorphic to B_n, but it is very convenient to allow more general basepoints. With the conventions in 3.42, B_0 is isomorphic but not identical to $B_1 = \{o^{(1)}\}$; this is consistent with the obvious fact that the groups $B_n, n \in \mathbb{N}_{>0}$, being fundamental groups of pairwise distinct spaces, are pairwise disjoint.

THEOREM 3.43.

$$
\begin{aligned}
B_n = \mathrm{gp}(\sigma_s, s \in \{1, \ldots, n-1\} \; : \; & [\sigma_s, \sigma_t](|s-t| > 1), \\
& \lll \sigma_s, \sigma_t \ggg (|s-t| = 1))
\end{aligned} \tag{3.11}
$$

□

It is usual to call (3.11) the *standard presentation* of B_n, the generators σ_s of (3.11) the *standard generators* of B_n, and the relators of (3.11) the *standard relators* of B_n. (It is also usual, and perhaps regrettable, to conflate $\sigma_s \in B_n$ with $\sigma_s \in B_{n'}$ for all $n, n' > s$. A more precise notation was proposed by Rudolph, 1985b, see **3.52**.) The detailed proof of 3.43 given by Fox and Neuwirth (1962) is, more or less exactly, an application of the usual algorithm (as in Magnus, Karrass, and Solitar, 1976), which produces a presentation of the fundamental group of a 2-dimensional cell complex with one 0-cell (the basepoint) having a generator for each 1-cell and a relator for each 2-cell, to the 2-skeleton of the cellulation of $\binom{\mathbb{C}}{n}$ that is dual to $\binom{\mathbb{C}}{n}/\equiv_{\mathbb{R}}$.

3.2. *Geometric braids and closed braids*

Let $I \subset \mathbb{R}$ be a closed interval. p: $I \to \binom{\mathbb{C}}{n}$ be a closed path. The multigraph $\mathrm{gr}(p) \subset I \times \mathbb{C}$ of p is called an n-string *geometric braid* for the (algebraic) braid in $B_{p(\partial I)}$ represented by p.

[9] As Magnus (1974, 1976) points out, in effect $\pi_1\binom{\mathbb{C}}{n}$ was investigated, and recognized as a "braid group", by Hurwitz (1891). Apparently, this insight had been long forgotten when Fox and Neuwirth (1962, p. 119) described as "previously unnoted" their "remark that B_n may be considered as the fundamental group of the space … of configurations of n undifferentiated points in the plane." It is interesting to speculate as to possible reasons for this instance of what Epple calls "elimination of contexts" (see especially Epple, 1995, p. 386, n. 32).

Let $q\colon \mathbb{O}' \to \binom{\mathbb{C}}{n}$ be a loop with domain the horizontal unknot $\mathbb{O}' \subset S^3$. The multigraph $\mathrm{gr}(q) \subset S^1 \times \mathbb{C}$ of q is a simple collection of closed curves in the open solid torus $\mathbb{O}' \times \mathbb{C}$. Using the vertical unbook $\mathbf{o}$, it is easy to create a (nearly standard) identification of $\mathbb{O}' \times \mathbb{C}$ with $S^3 \backslash \mathbb{O}$, and thus a (nearly standard) embedding of $\mathrm{gr}(q)$ into $S^3 \backslash \mathbb{O}$. This embedding has the property that $\arg(\mathbf{o})|\mathrm{gr}(q)\colon \mathrm{gr}(q) \to S^1$ is a covering map of degree n, and every simple collection of closed curves $L \subset \mathbb{O}' \times \mathbb{C}$ such that $\arg(\mathbf{o})|L\colon L \to S^1$ is a covering map of degree n arises in this way from some q. Call any such L an n-string *closed* **o**-*braid* (Rudolph, 1988). In general, if **a** is an *unbook* (that is, an open book on S^3 with unknotted binding A), then **a** is equivalent to **o**, and the class of *closed* **a**-*braids* (sometimes called, slightly abusively, simply "closed braids with axis A") is defined by any such equivalence. Given a basepoint $* \in \mathbb{O}'$, q naturally represents an element of $\beta_q \in B_{q(*)}$. The notation $\hat{\beta}(\beta_q)$ is often used for the closed **o**-braid $\mathrm{gr}(q)$.

3.3. *Bands and espaliers*

For $n = 2$, (3.11) says that B_2 is infinite cyclic. More specifically and directly, 2.34(2) shows that, for any edge $\mathbf{e} \subset \mathbb{C}$, the 2-string braid group $B_{\partial \mathbf{e}}$ is infinite cyclic with a preferred generator, say $\sigma_{\mathbf{e}}$; in fact $\sigma_{\mathbf{e}}$ depends only on $\partial \mathbf{e}$. In particular, any two 2-string braid groups are canonically isomorphic. For $n > 2$, typically no isomorphism between distinct n-string braid groups has much claim to be called canonical; however, the following is an immediate consequence of 2.32.

PROPOSITION 3.44. *If* $X \in \binom{\mathbb{R}}{n}$, *then* $B_X \cong B_n$, *and any path in* $\binom{\mathbb{R}}{n}$ *from* X *to* **n** *induces this canonical isomorphism.* □

For $n \geq 2$, if $X \cap \mathbf{e} = \partial \mathbf{e}$, then there is a *natural injection* $\mathfrak{i}_{\mathbf{e};X}\colon B_{\partial \mathbf{e}} \to B_X$, and $\mathfrak{i}_{\mathbf{e};X}$ depends only on the isotopy class of $\mathbf{e}$ (rel. $\partial \mathbf{e}$) in $\mathbb{C} \backslash (X \backslash \mathbf{e})$.

DEFINITIONS 3.45. A *positive X-band* is any $\sigma_{\mathbf{e};X} := \mathfrak{i}_{\mathbf{e};X}\ (\sigma_{\mathbf{e}}) \in B_X$. (When X is understood, or irrelevant, $\sigma_{\mathbf{e};X}$ may be abusively abbreviated to $\sigma_{\mathbf{e}}$.) A *negative X-band* is the inverse of a positive X-band. An *X-band* is a positive or negative X-band. Let $|\sigma_{\mathbf{e};X}^{\pm 1}| := \sigma_{\mathbf{e};X}$ denote the *absolute value* of the band $\sigma_{\mathbf{e};X}^{\pm 1}$, $\varepsilon(\sigma_{\mathbf{e};X}^{\pm 1}) := \pm$ the *sign* (positive or negative) of $\sigma_{\mathbf{e};X}^{\pm 1}$, and $\mathrm{e}(\sigma_{\mathbf{e};X}^{\pm 1})$ the *edge-class* of $\sigma_{\mathbf{e};X}^{\pm 1}$, that is, the isotopy class (rel. $\partial \mathbf{e}$) of $\mathbf{e}$ in $\mathbb{C} \backslash (X \backslash \mathbf{e})$. An *X-bandword* of *length k* is a k-tuple $\mathbf{b} =: (b(1),\ldots,b(k))$ such that each $b(i)$ is an X-band.[10] An X-bandword **b** is *quasipositive* in case each $b(i)$ is a positive X-band. The *braid* of **b** is $\beta(\mathrm{b}) := b(1)\cdots b(k) \in B_X$. Every braid in B_X is the braid of some X-bandword. A braid in B_X is *quasipositive* in case it is the braid of a quasipositive X-bandword.

[10] In particular, **n**-bandwords are just what (since Rudolph 1983a) I have long called "band representations" – a coinage which I would like to suppress, due to its misleading suggestion of a relation to group representation theory.

Orevkov (2004) gives an algorithm to determine whether or not a given braid in B_3 is quasipositive. Bentalha (2004) has announced a generalization of this to all B_n. At this writing it is not certain that Bentalha (2004) is correct.

PROPOSITION 3.46. *Any two positive X-bands are conjugate in* B_X. □

An n-string braid group is the knotgroup of $\Delta_n(\mathbb{C})$ in $\binom{\mathbb{C}}{n}$ (see Rudolph, 1983b); a positive band is a meridian.

PROPOSITION 3.47. *Let* $X \in \binom{\mathbb{C}}{n}$, $n \geq 2$. *Let* $\mathbf{e}$ $\mathbf{f} \subset \mathbb{C}$ *be two edges with* $X \cap \mathbf{e} = \partial\mathbf{e}$, $X \cap \mathbf{f} = \partial\mathbf{f}$. *If* $\mathbf{e} \cap \mathbf{f} = \varnothing$ *(resp.,* $\mathbf{e} \cap \mathbf{f} = \{x\} \subset \partial\mathbf{e} \cap \partial\mathbf{f}$*), then* $[\sigma_{\mathbf{e};X}, \sigma_{\mathbf{f};X}] = o_X$ *(resp.,* $\lll \sigma_{\mathbf{e};X}, \sigma_{\mathbf{f};X} \ggg = o_X$*).*

PROOF. This is readily proved directly. Alternatively, note that: (1) the general case is isotopic to a special case in which $X = \mathbf{n}$, $\mathbf{e} = [1, 2]$, and $\mathbf{f} = [s, s + 1]$ with $s \in \{2,\ldots,n\}$; (2) in such a special case, $B_X = B_n$, $\sigma_{\mathbf{e};X} = \sigma_1$ and $\sigma_{\mathbf{f};X} = \sigma_s$ are two standard generators of (3.11), and the claimed commutator and yangbaxter relators are two standard relators of (3.11). □

PROPOSITION 3.48. *Let* $\mathcal{T} \subset \mathbb{C}$ *be a planar tree. The positive* $\mathsf{V}(\mathcal{T})$*-bands* $\sigma_{\mathbf{e};\mathsf{V}(\mathcal{T})}$, $\mathbf{e} \in \mathsf{E}(\mathcal{T})$, *generate* $B_{\mathsf{V}(\mathcal{T})}$; *no proper subset of them does so. If* $\mathbf{e} \cap \mathbf{f} = \varnothing$ *(resp.,* $\mathbf{e} \cap \mathbf{f} = \{z\}$, $z \in \mathsf{V}(\mathcal{T})$*), then* $[\sigma_{\mathbf{e};\,\mathsf{V}(\mathcal{T})}, \sigma_{\mathbf{f};\mathsf{V}(\mathcal{T})}] = o_{\mathsf{V}(\mathcal{T})}$ *(resp.,* $\lll \sigma_{\mathbf{e};\mathsf{V}(\mathcal{T})}, \sigma_{\mathbf{f};\mathsf{V}(\mathcal{T})} \ggg = o_{\mathsf{V}(\mathcal{T})}$*).* □

Call the $\mathsf{V}(\mathcal{T})$-bands $\sigma_{\mathbf{e};\mathsf{V}(\mathcal{T})}$, $\mathbf{e} \in \mathsf{E}(\mathcal{T})$, the *$\mathcal{T}$-generators* of $B_{\mathsf{V}(\mathcal{T})}$. Note that 3.48 asserts that the braid group $B_{\mathsf{V}(\mathcal{T})}$ is a quotient of

$$\begin{aligned}\operatorname{gp}(\sigma_{\mathbf{e}}, \mathbf{e} \in \mathsf{E}(\mathcal{T}) : [\sigma_{\mathbf{e}}, \sigma_{\mathbf{f}}]&(\operatorname{card}(\mathbf{e} \cap \mathbf{f}) = 0),\\ &\lll \sigma_{\mathbf{e}}, \sigma_{\mathbf{f}} \ggg (\operatorname{card}(\mathbf{e} \cap \mathbf{f}) = 1))\end{aligned} \tag{3.12}$$

but not that these groups are identical. In fact, they are easily seen to be so if and only if $\mathcal{T}$ has no intrinsic vertices.

An *espalier* is a planar tree $\mathcal{T}$ such that each $\mathbf{e} \in \mathsf{E}(\mathcal{T})$ is a proper edge in $\mathbb{C}_-$ and $\operatorname{Re} |(\mathbf{e}\setminus\operatorname{Col}(\partial\mathbf{e} \subset \mathbf{e}))$ is injective. Rudolph (2001a) gives proofs of the following facts about espaliers.

PROPOSITION 3.49. (1) *Every planar tree is isotopic, in* $\mathbb{C}$, *to an espalier.* (2) *The embedding of an espalier* $\mathcal{T}$ *in* $\mathbb{C}_-$ *is determined, up to isotopy of* $(\mathcal{T}, \mathsf{V}(\mathcal{T}))$ *in* $(\mathbb{C}_-, \mathbb{R})$, *by the combinatorial structure of its cellulation together with the order induced on* $\mathsf{V}(\mathcal{T})$ *by its embedding in* $\mathbb{R} = \partial\mathbb{C}_-$. *In particular, given* $X = \{x_1 < \cdots < x_n\} \subset \mathbb{R}$ *and* $n - 1$ *pairs* $\{x_{i(p)} < x_{j(p)}\} \subset X$, *the following are equivalent.* (a) *There is*

an espalier $\mathcal{T}$ *with* $\mathsf{V}(\mathcal{T}) = \{x_1, \ldots, x_n\}$, $\mathsf{E}(\mathcal{T}) = \{\mathbf{e}_1, \ldots, \mathbf{e}_{n-1}\}$, *and* $\partial_{\mathbf{e}_p} = \{x_{i(p)}, x_{j(q)}\}$.
(b) *For* $1 \leq p < q \leq n-1$ *the pairs* $\{x_{i(p)} < x_{j(p)}\}$ *and* $\{x_{i(q)} < x_{j(q)}\}$ *are not linked (i.e., they are either in touch or unlinked).* □

DEFINITION 3.50. Let $X \in \binom{\mathbb{R}}{n}$. An X-band σ is *embedded* provided that some $\mathbf{e} \in \mathbf{e}(\sigma)$ is a proper edge in $\mathbb{C}_-$; when it is given that $\sigma = \sigma^{\pm 1}_{\mathbf{e};X}$ is embedded, it is assumed without further comment that $\mathbf{e} \in \mathbf{e}(\sigma)$ is such an edge. Clearly, there are exactly $n(n-1)/2$ embedded positive X-bands, one – which may be denoted $\sigma_{x_i, x_j; X}$ – for each $\{x_i < x_j\} \subset X$. The embedded positive X-bands generate B_X. (A presentation (2.1) of B_n with the positive $\mathbf{n}$-bands as generators is given by Birman et al., 1998, and many interesting conclusions drawn therefrom.) An X-bandword $\mathbf{b}$ is *embedded* in case each $b(i)$ is embedded. An X-bandword $\mathbf{b}$ is called *positive* in case it is both quasipositive and embedded.

Let $X_0 \subset X \in \binom{\mathbb{C}}{n}$. Unless $n_0 := \text{card}(X_0)$ equals n, typically no non-trivial homomorphism $B_{X_0} \to B_X$ has much claim to be called canonical. For $X \subset \mathbb{R}$, however, the situation is much better (cf. 3.44). In fact, given a positive embedded X_0-band $\sigma_{x_i, x_j; X_0} \in B_{X_0}$, let $\mathfrak{i}_{X_0;X}(\sigma_{x_i, x_j; X_0}) := \sigma_{x_i, x_j; X} \in B_X$.

PROPOSITION 3.51. *There is a unique homomorphism* $\mathfrak{i}_{X_0;X} : B_{X_0} \to B_X$ *extending* $\mathfrak{i}_{X_0;X}$ *as defined on the embedded* X_0*-bands, and it is injective.* □

Call $\mathfrak{i}_{X_0;X}$ the *canonical injection* of B_{X_0} into B_X. (The collision of notation with $\mathfrak{i}_{\mathbf{e};X}$ is unproblematic; if $X \subset \mathbb{R}$ and $\mathbf{e}$ is embedded with $\partial\mathbf{e} = \{x_i, x_j\}$, then $\mathfrak{i}_{x_i, x_j; X} = \mathfrak{i}_{\partial\mathbf{e}; X}$.) 3.51 excuses the conflation, under the single name σ_{x_i, x_j}, of all the positive embedded bands $\sigma_{x_i, x_j; X}$ for $\xi_i < \xi_j \subset X \subset \mathbb{R}, X$ finite.

EXAMPLE 3.52. For $n \geq m$, the canonical injection $\mathfrak{i}_{\mathbf{m};\mathbf{n}}$ is implicit in the identification of B_m as a subgroup of B_n, discussed following 3.43. Rudolph (1985b) proposed the notation $\beta^{(n-m)}$ for $\mathfrak{i}_{\mathbf{m};\mathbf{n}}(\beta)$, and the convention (extending the notations o and $o^{(n)}$ introduced in 3.42) that for each $m \in \mathbb{N}$, σ_{m-1} denote only an element of B_m, the other standard generators of B_m being $\sigma_1^{(m-2)}, \ldots, \sigma_{m-2}^{(1)}$. This notation and convention have been widely unadopted.

DEFINITION 3.53. Let $\mathcal{T}$ be an espalier. A $\mathcal{T}$*-bandword* is a $\mathsf{V}(\mathcal{T})$-bandword $\mathbf{b}$ such that every $|b(i)|$ is a $\mathcal{T}$-generator (so, in particular, a $\mathcal{T}$-bandword is an embedded $\mathsf{V}(\mathcal{T})$-bandword). A positive $\mathcal{T}$-bandword $\mathbf{b}$ is called *strictly* $\mathcal{T}$*-positive* in case every $\mathcal{T}$-generator appears among the bands $b(s)$.

For $X = \{x_1 < \cdots < x_n\} \in \mathcal{MP}^{[n]}(\mathbb{R})$, let $\mathcal{I}_X$ (resp., $\mathcal{Y}_X$) denote any espalier $\mathcal{T}$ with $\mathsf{V}(\mathcal{T}) = X$ and $\{\partial\mathbf{e}\colon\ \mathbf{e} \in \mathsf{E}(\mathcal{T})\} = \{\{x_p, x_{p+1}\}\colon\ 1 \leq p < n\}$ (resp., $\{\{x_1, x_p\}\colon 1 < p \leq n\}$). Among the combinatorial types of trees $\mathcal{T}$ with $\mathsf{V}(\mathcal{T}) = X$, $\mathcal{I}_X$ and $\mathcal{Y}_X$ represent two extreme types, which may be called *linear* (minimal number of endpoints) and *star-like* (maximal number of endpoints), respectively; further, among the linear (resp., star-like) espaliers, $\mathcal{I}_X$ (resp., $\mathcal{Y}_X$) is again extreme, in a sense that is obvious and easily formalized.

EXAMPLE 3.54. The $\mathcal{J}_n$-generators of B_n are the standard generators of (3.11), an $\mathcal{J}_n$-bandword is just a "braid word" in the usual sense (Birman, 1975, p. 70 ff.), and the braid of an $\mathcal{J}_n$-positive (or, for some authors, strictly $\mathcal{J}_n$-positive) $\mathcal{J}_n$-bandword is a "positive braid" in the usual sense (Birman, 1975; Rudolph, 1982b; Franks and Williams, 1987, etc.).

QUESTION 3.55. As noted after 3.43, the standard generators and standard relators of (3.11) correspond naturally to the codimension-1 and codimension-2 cells of $\binom{\mathbb{C}}{n}/\equiv_{\mathbb{R}}$. As noted in 3.54, the standard generators also correspond naturally to the edges of $\mathcal{J}_n$. Clearly, given any linear planar tree $\mathcal{T}$ with $\mathrm{card}(\mathsf{V}(\mathcal{T})) = n$, from an isotopy of $\mathcal{T}$ to $\mathcal{J}_n$ may be contrived a cellulation $\binom{\mathbb{C}}{n}/\equiv_{\mathcal{T}}$ with a unique codimension-0 cell, such that (1) the codimension-1 cells of $\binom{\mathbb{C}}{n}/\equiv_{\mathcal{T}}$ correspond to the $\mathcal{T}$-generators of $B_{\mathsf{V}(\mathcal{T})}$, and (2) the codimension-2 cells of $\binom{\mathbb{C}}{n}/\equiv_{\mathcal{T}}$ correspond to the relators of (3.11). Now suppose that $\mathcal{T}$ is a planar tree for which there exists a (natural or contrived) cellulation $\binom{\mathbb{C}}{n}/\equiv_{\mathcal{T}}$, with a unique codimension-0 cell, that has property (1) but, rather than property (2), satisfies both (3) some of the codimension-2 cells of $\binom{\mathbb{C}}{n}/\equiv_{\mathcal{T}}$ correspond to the relators of (3.12), and (4) the remaining codimension-2 cells of $\binom{\mathbb{C}}{n}/\equiv_{\mathcal{T}}$ correspond to some family of extra relators exactly sufficient to convert (3.12) into a presentation of $B_{\mathsf{V}(\mathcal{T})}$. Does this imply that, in fact, $\mathcal{T}$ is linear? Specifically, does there exist a cellulation $\binom{\mathbb{C}}{4}/\equiv_{\mathcal{Y}_4}$ with properties (1), (3), and (4)?

The methods of Birman et al. (1998) might help answer 3.55 (although the presentation they give is not obviously associated to a cellulation).

3.4. *Embedded bandwords and braided Seifert surfaces*

Early versions of the construction of braided Seifert surfaces described in this section appeared in Rudolph (1983b,c).

CONSTRUCTION 3.56. Let $X \in \binom{\mathbb{R}}{n}$, $T \in \binom{\mathbb{R}}{n}$, say $X =: \{x_1 < \cdots < x_n\}$, $T =: \{t_1 < \cdots < t_k\}$. Let $0 < \varepsilon < \min\{|t - t'|/2\colon t, t' \in T, t \neq t'\}$; in case $k > 0$, let $I = [\min T - \varepsilon, \max T + \varepsilon]$. Let $\mathbf{b}$ be an embedded X-bandword of length k. To implement this construction of braided Seifert surfaces, choose a proper arc $\mathbf{e}_s \subset \mathbb{C}_-$ in the edge-class $\mathbf{e}(b(s))$ for each $s \in \mathbf{k}$, and embeddings $\eta_x\colon h^{(0)} \hookrightarrow \mathbb{C}_+ \times \mathbb{R}$ $(x \in X)$

and η_t: $h^{(1)} \hookrightarrow \mathbb{C}_- \times]t - \varepsilon, t + \varepsilon[$ $(t \in T)$, subject to the following conditions. For each $x \in X$,

(1) η_x is proper along a boundary arc of $h^{(0)}$,

(2) $\eta_x(h^{(0)}) \subset \{x + iy\colon y \in \mathbb{R}_{\geq 0}\} \times \mathbb{R}$,

(3) and $I \subset \partial\eta_x(h^{(0)})$.

For each $t \in T$, say $t = t_p$, $p \in \mathbf{k}$,

(4) η_{t_p} is proper along the attaching arcs of $h^{(1)}$,

(5) $\mathrm{pr}_1 \circ \eta_{t_p} | \kappa(h^{(1)}) : \kappa(h^{(1)}) \to \mathbb{C}_-$ is a diffeomorphism onto $\mathbf{e}_p$,

(6) $(\mathrm{Re} \circ \mathrm{pr}_1, \mathrm{pr}_2) \circ \eta_t$: $h^{(1)} \to \mathbb{R} \times \mathbb{R}$ is a bowtie, and

(7) the sign of the crossing of $(\mathrm{Re} \circ \mathrm{pr}_1, \mathrm{pr}_2) \circ \eta_t |\partial h^{(1)}$ is equal to $\varepsilon(b(t))$.

It follows that

$$\Sigma(\mathbf{b}) := \bigcup_{x \in X} \eta_x(h^{(0)}) \cup \bigcup_{t \in T} \eta_t(h^{(1)}) \subset \mathbb{C} \times \mathbb{R} \tag{3.13}$$

is a (0, 1)-handle decomposition (2.3) of a surface. Any Seifert surface $S[\mathbf{b}] := \delta(\Sigma(\mathbf{b})) \subset \mathbb{R}^3 \subset S^3$, where δ: $\mathbb{C} \times \mathbb{R} \to \mathbb{R}^3$: $(z, t) \mapsto (\mathrm{Re}\, z, \mathrm{Im}\, z, t)$ and $\Sigma(\mathbf{b})$ is constructed as in 3.56, is called a *braided Seifert surface of* the embedded X-bandword $\mathbf{b}$. It is easy to see that $\gamma(\mathbf{b}) := \Sigma(\mathbf{b}) \cap (\mathbb{C}_- \times I)$ is a geometric braid for $\beta(\mathbf{b})$, and that the link $\partial S[\mathbf{b}]$ is isotopic in $S^3 \supset \mathbb{R}^3$ to $\hat{\beta}(\mathbf{b})$.

The *braid diagram of* $\mathbf{b}$ is the pair $\mathbf{BD}(\mathbf{b}) =: (\mathbf{P}(\mathbf{BD}(\mathbf{b})), \mathbf{I}(\mathbf{BD}(\mathbf{b})))$, where $\mathbf{P}(\mathbf{BD}(\mathbf{b})) := (\mathrm{Re}, \mathrm{pr}_2)(\gamma(\mathbf{b})) \subset \mathbb{R} \times I$ and $\mathbf{I}(\mathbf{BD}(\mathbf{b}))$ is the information about the signs of the crossings of $\mathbf{P}(\mathbf{BD}(\mathbf{b}))$ (indicated graphically in the style of 1.11(2); see Figure 15). A *braid diagram for* $\beta \in B_X$ is a braid diagram of any bandword $\mathbf{b}$ with $\beta(\mathbf{b}) = \beta$. A *standard braid diagram for* $\beta \in B_n$ is a braid diagram of any $\mathcal{I}_{\mathbf{n}}$-bandword $\mathbf{b}$ with $\beta(\mathbf{b}) = \beta$.

Of course $\mathbf{b}$ determines $S[\mathbf{b}]$ and $\hat{\beta}(\mathbf{b})$ up to isotopy; and clearly, up to isotopy (even isotopy through braided Seifert surfaces), none of X, T, or the collection of specific edge-class representatives $\mathbf{e}_s$ is necessary *per se* to the construction of $S[\mathbf{b}]$ – all that is needed is a modicum of combinatorial information extracted from $\mathbf{b}$. That information can be encoded as the k-tuple of triples $((i_{\mathbf{b}}(1), j_{\mathbf{b}}(1), \varepsilon_{\mathbf{b}}(1)),\ldots,(i_{\mathbf{b}}(k), j_{\mathbf{b}}(k), \varepsilon_{\mathbf{b}}(k)))$ with $\{x_{i_{\mathbf{b}}(s)} < x_{j_{\mathbf{b}}(s)}\} := \partial\, \mathbf{e}_s \subset X$ and $\varepsilon_{\mathbf{b}}(s) := \varepsilon(b(s))$. Another convenient encoding is graphical, using *charged fence diagrams* (see Rudolph, 1992a, 1998). Figure 16 pictures a (very simple) charged fence diagram.[11]

[11] In Rudolph (1992a), I described fence diagrams as "my synthesis of some diagrams that H. Morton used to describe certain Hopf-plumbed fiber surfaces in 1982… and 'square bridge projections' as described by H. Lyon" (Lyon, 1980). By Rudolph (2001a), I had recalled that I first saw fences (and braid groups!) "c. 1959, in one of Martin Gardner's 'Mathematical Games' columns" (reprinted in Gardner, 1966, as Chapter 2, "Group Theory and Braids"). Since 2001, I have become aware of "amida-diagrams" (introduced by Yamamoto, 1978, precisely to construct certain special Seifert surfaces) and "wiring diagrams" (cf. Cordovil and Fachada, 1995, and other literature from the theory of line arrangements), both closely related to fences, as are (via their essential identity with square bridge projections) the "barber-pole projections" attributed in Rudolph (1992a) to Thurston (unpublished), Erlandsson (1981), and Kuhn (1984). Is this another instance of Epple's "elimination of contexts", or a mere multiplication of contexts?

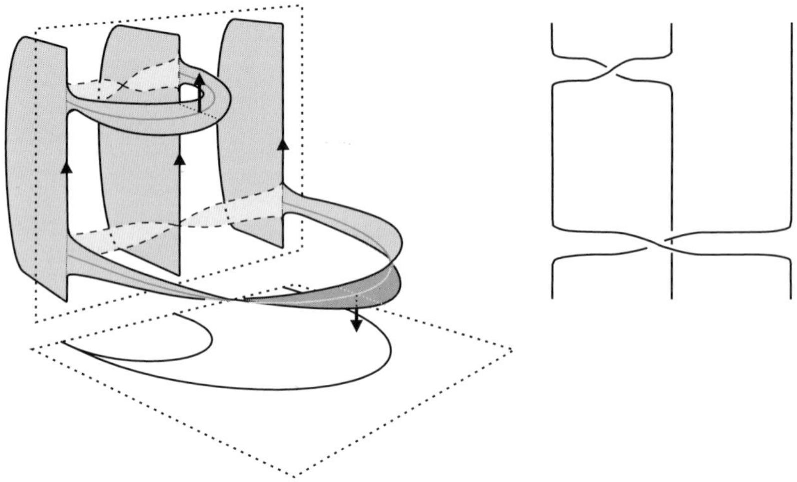

Fig. 15. Left: a surface $\Sigma(\mathrm{b})$; $\varepsilon(b(1)) = -$, $\varepsilon(b(1)) = +$. Right: a braid diagram for $\beta(\mathrm{b})$.

There is a more or less canonical way to turn an n-string braid diagram $\mathbf{BD(b)}$ into a standard link diagram $\mathbf{D}(\hat{\beta}(\mathbf{b}))$: identify $\mathbb{C}$ with $\mathbb{R}^2 \supset \mathbf{P}(\mathbf{BD(b)})$ by (Re, Im); in $\mathbb{C}$, attach n arcs to $\mathbf{P}(\mathbf{BD(b)}) \subset \mathbb{C}$, preserving orientation, so as to create a normal collection $\mathbf{P}(\mathbf{D}(\hat{\beta}(\mathbf{b})))$ of closed curves with no new crossings; and let $\mathbf{I}(\mathbf{D}(\hat{\beta}(\mathbf{b})))$ be $\mathbf{I}(\mathbf{BD(b)})$.

PROPOSITION 3.57. (1) *Let* $\mathbf{b}$ *be an* $\mathcal{I}_{\mathbf{n}}$*-bandword. Let* $L = \hat{\beta}(\mathbf{b})$. *The arcs* $\mathbf{P}(\mathbf{D}(L)) \setminus \operatorname{Int} \mathbf{P}(\mathbf{BD(b)})$ *can be chosen so that* $\mathsf{O}_{\mathbf{D}(L)} = \mathsf{O}^{+}_{\mathbf{D}(L)}$; *if they are, then* $\mathbf{D}(L)$ *is nested and the diagrammatic Seifert surface* $S(\mathbf{D}(L))$ *is isotopic to* $S[\mathbf{b}]$. (2) *Conversely, if* $\mathbf{D}(L)$ *is nested and* $\mathsf{O}_{\mathbf{D}(L)} = \mathsf{O}^{+}_{\mathbf{D}(L)}$, *then (up to isotopy of* $\mathbf{P}(\mathbf{D}(L))$*) there exists an* $\mathcal{I}_{\mathbf{n}}$*-bandword* $\mathbf{b}$, $n := \operatorname{card}(\mathcal{O}_{\mathbf{D}(L)})$, *such that* $L = \hat{\beta}(\mathbf{b})$ *and* $S(\mathbf{D}(L))$ *is isotopic to* $S[\mathbf{b}]$. (3) *If* $\mathbf{b}$ *is an embedded* $\mathbf{n}$*-bandword but not an* $\mathcal{J}_{\mathbf{n}}$*-bandword, and* $L = \hat{\beta}(\mathbf{b})$, *then* $\mathbf{D}(L)$ *is nested but* $\chi(S(\mathbf{D}(L))) < \chi(S[\mathbf{b}])$, *so* $S(\mathbf{D}(L))$ *and* $S[\mathbf{b}]$ *are not diffeomorphic, let alone isotopic.* □

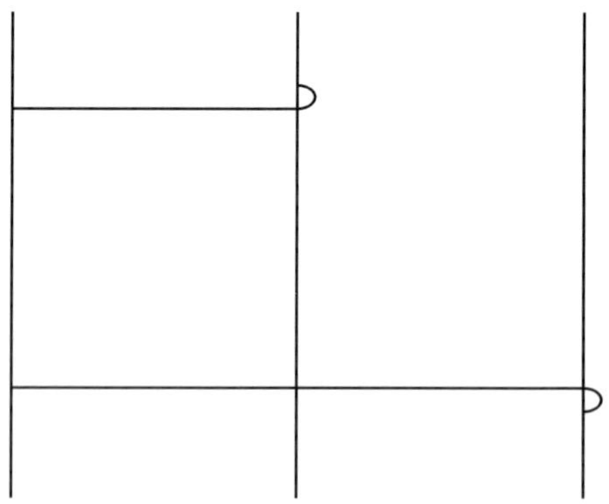

Fig. 16. A charged fence diagram for the surface in Figure 15.

3.57 shows that 2.25 is equivalent to the following

THEOREM 3.58. *For all n, for all $\beta \in B_n$,*

$$\mathrm{ord}_v P_{\hat{\beta}} \geq e(\beta) - n + 1 \tag{3.14}$$

where $e\colon \mathrm{B}_n \to \mathbb{Z}$ is abelianization (exponent sum). □

Proofs of the following theorem appear in Rudolph (1983b,c); it also follows from results of Bennequin (1983) on "Markov surfaces" (see Rudolph, 1985b).

THEOREM 3.59. *If $S \subset \mathbb{R}^3 \subset S^3$ is a Seifert surface, then there exist X and an embedded X-bandword* **b** *such that S is isotopic to $S[\mathbf{b}]$.* □

QUESTION 3.60. The *braid index* $\mathrm{brin}(L)$ of a link L is the minimum $n \in \mathbb{N}_{>0}$ such that L is isotopic to $\hat{\beta}(\mathbf{b})$ for some n-bandword **b** (which can, since its length is not an issue, be taken to be embedded); see Birman and Brendle (this Handbook). Let the *braided Seifert surface index* of a Seifert surface S be the minimum $n \in \mathbb{N}_{>0}$ such that S is isotopic to $S[\mathbf{b}]$ for some embedded **n**-bandword **b**; by 3.59, this invariant of S is an integer. How can it be calculated or estimated? This question is the analogue for Seifert surfaces of "Open Problem 1" of Birman and Brendle (this Handbook), which asks how to calculate $\mathrm{brin}(L)$. Clearly, the braided Seifert surface index of S is at least as great as $\mathrm{brin}(\partial S)$. Examples of Hirasawa and Stoimenow (2003) show that this inequality can be strict.

DEFINITION 3.61. A Seifert surface S is *quasipositive* in case S is isotopic to a braided Seifert surface $S[\mathbf{b}]$ for some positive X-bandword **b**. A link is *strongly quasipositive* in case it has a quasipositive Seifert surface.

3.5. *Plumbing and braided Seifert surfaces*

Let $X \in \binom{\mathbb{R}}{n}$. For $x \in X$, write $X_{\leq x} := X \cap \mathbb{R}_{\leq x}$, $X_{\geq x} := X \cap \mathbb{R}_{\geq x}$.

DEFINITIONS 3.62. Let $x \in X$. Let **b** be an embedded X-bandword of length k. Let $T_{\mathbf{b};\leq x} := \{s \in \mathbf{k}\colon\ b(s) \in \mathrm{i}_{X_{\leq x},X}\}$; and $T_{\mathbf{b};\geq x} := \{s \in \mathbf{k}\colon\ b(s) \in \mathrm{i}_{X_{\geq x},X}\}$; let $k_{\leq} := \mathrm{card}(T_{\mathbf{b};\leq x})$, $k_{\geq} := \mathrm{card}(T_{\mathbf{b};\geq x})$, so in every case $k_{\leq} + k_{\geq} \leq k$. In case $k_{\leq} + k_{\geq} = k$, say that **b** is *deplumbed by* $x \in X$. Writing $T_{\mathbf{b};\leq x} =: \{u_1 < \cdots < u_{k_{\leq}}\}$, $T_{\mathbf{b};\geq x} =: \{v_1 < \cdots < v_{k_{\geq}}\}$, let $\mathbf{b}_{\leq x} := (b_{\leq x}(1),\ldots,b_{\leq x}(k_{\leq}))$ (resp., $\mathbf{b}_{\geq x} := (b_{\geq x}(1),\ldots,b_{\geq x}(k_{\geq}))$) be the embedded $X_{\leq x}$-bandword (resp., $X_{\geq x}$-bandword) for which $\mathrm{i}_{X_{\leq x}},x)b_{\leq x}(s) = b(u_s)$ (resp., $\mathrm{i}_{X_{\geq x}},x)b_{\geq x}(t) = b(v_t)$). By construction 3.56, if **b** is deplumbed by $x \in X$, then $\Sigma(\mathbf{b}) = \Sigma(\mathbf{b}_{\leq x}) \cup \Sigma(\mathbf{b}_{\geq x})$ and $\Sigma(\mathbf{b}_{\leq x}) \cap \Sigma(\mathbf{b}_{\geq x}) = \Sigma(\mathbf{b}) \cap \mathbb{C} \times \{x\}$ is a disk (in fact, a 0-handle of the constructed (0, 1)-handle decompositions of $\Sigma(\mathbf{b}_{\leq x}) \cup \Sigma(\mathbf{b}_{\geq x})$);

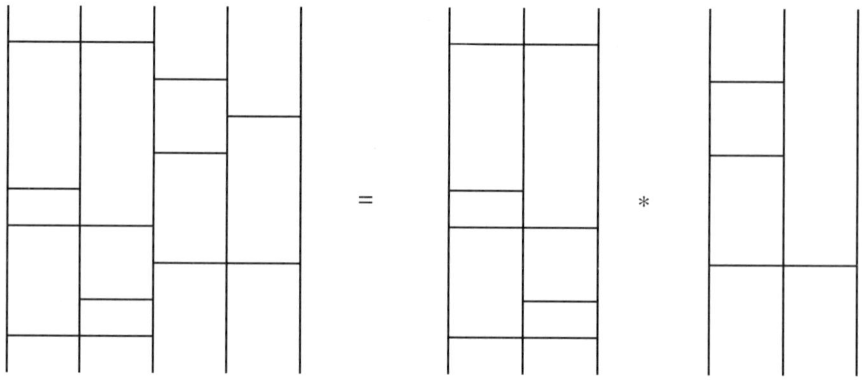

Fig. 17. An example of deplumbing.

the situation is illustrated using fence diagrams in Figure 17. The inverse of this operation, which combines two braided Seifert surfaces into one, is *braided Stallings plumbing*.

Although it appears to be a rather special case of *Stallings plumbing*, or *Murasugi sum*, a geometric operation on Seifert surfaces that has been extensively studied (by Murasugi, 1963; Conway, 1970; Siebenmann, 1975; Stallings, 1978; Gabai, 1983a,b, 1986, and many others; notations, definitions, and some history are given by Rudolph, 1998), braided Stallings plumbing is in fact perfectly general, according to Rudolph (1998).

THEOREM 3.63. *Up to isotopy, every Murasugi sum* $S_0 * S_1$ *of Seifert surfaces* S_0, S_1 *is a braided Stallings plumbing of braided Seifert surfaces.* □

3.6. *Labyrinths, braided surfaces in bidisks, and braided ribbons*

Let X be a topological space. Given f: $X \to MP_n$, let f_Δ: $X \times \mathbb{C} \to \mathbb{C}: (x, w) \mapsto f(x)(w)$, so that $\mathrm{gr}(R \circ f) = f_\Delta^{-1}(0)$. Generally, it can be a subtle matter to determine whether a map F: $X \times \mathbb{C} \to \mathbb{C}$ (with fibers of cardinality bounded by $n \in \mathbb{N}$) is f_Δ for some (continuous) f: $X \to MP_n$, but in case F is known to be such, write $f = F_\nabla$.

DEFINITION 3.64. A map f: $X \to MP_n$ is *amazing* provided that the partition of M into connected components of inverse images f^{-1} (C) of the various cells C of $MP_n/\equiv_\mathbb{R}$ has some finite refinement that is a stratification of X. A *labyrinth* for f is any such stratification.

EXAMPLES 3.65. Amazing maps have proved useful in several applications.

3.65.1. If $F(z, w) := f_0(z)w^n + f_1(z)w^{n-1} + \cdots + f_n(z) \in \mathbb{C}[z, w]$ is a polynomial in Weierstrass form, then F_∇: $\mathbb{C} \to MP_\mathbb{C}$ is (in effect) F considered as an element of $\mathbb{C}[z][w] \cong \mathbb{C}[z, w]$, and is amazing. Classically, in this situation w is said to be the

"n-valued algebraic function of z without poles" determined by the algebraic curve $\mathcal{V}_F$ (Bliss, 1933; Hansen, 1988), and a more or less canonical set of "branch cuts" for w can be extracted systematically from the labyrinth of F_∇. Rudolph (1983a) determined the topology of the labyrinth of F_∇ in the case $F_\varepsilon(z, w) = (w-1)(w-2)\cdots(w-n+1)(w-z)+\varepsilon$, first for $\varepsilon = 0$, then for small $\varepsilon \neq 0$, and used it to give the first proof of 3.77.

3.65.2. The term labyrinth was introduced by Dung and Hà (1995), in their study of line arrangements in $\mathbb{C}^2$. In a linear coordinate system chosen so that no line in the arrangement is vertical, such an arrangement is $\mathcal{V}_F$, where $F(z, w) \in \mathbb{C}[z, w]$ is a product of (unrepeated) factors $w + az + b$ (as in 3.65.1, where F_0 corresponds to an arrangement of $n-1$ horizontal lines crossed by a single diagonal line). Again, F_∇ exists and is amazing. (See Section 7.4 for further references on arrangements.)

3.65.3. Orevkov (1988) made excellent use of the observation that, if $V \subset \mathbb{C}^2$ is both a $\mathbb{C}$-algebraic curve and a nodal surface, then, after an arbitrarily small linear change of coordinates in $\mathbb{C}^2$, $V = \mathcal{V}_F$ where $F \in \mathbb{C}[z,w]$ is in Weierstrass form and the labyrinth of the amazing map F_∇ has a very special form. (See 7.2 for further discussion of this application.)

3.65.4. Orevkov (1996) developed the theory of labyrinths in careful detail, and applied it to the Jacobian Conjecture. (See 7.3.)

Amazing maps, and the facts about the low-codimension cells of $MP_n/\equiv_{\mathbb{R}}$ laid out in 1.35, together allow two closely related constructions – of braided surfaces in bidisks, and braided ribbons in D^4 – to be described and carried out more precisely than in the original sources (Rudolph, 1983b, 1985b).

PROPOSITION 3.66. *Let M be a surface. If $f: M \to MP_n$ is transverse to all the cells of $MP_n/\equiv_{\mathbb{R}}$, then f is amazing. More specifically, f has a labyrinth $M/\equiv$ with the following properties.*

(1) *The union of the vertices and edges of $M/\equiv$ is a graph $\Lambda(f)$ with no vertex of degree $d \notin \{1,4,6\}$.*
(2) *The association by f to each edge* **e** *of $\Lambda(f)$ of a codimension-*1 *cell of $MP_n/\equiv_{\mathbb{R}}$ endows* **e** *with* (a) *a natural transverse orientation (and therefore a natural orientation, so that* **e** *is naturally either a simple closed curve or an arc) and* (b) *a clew $\sigma(\mathbf{e})$ in the set $\{\sigma_1, \ldots, \sigma_{n-1}\}$ of standard generators of the presentation* (3.11) *of B_n.*
(3) *The association by f to each vertex x of $\Lambda(f)$ of a codimension-2 cell of $MP_n/\equiv_{\mathbb{R}}$ endows x with either* (a) *a standard relator of* (3.11), *in case* $\mathrm{val}_{\Lambda(f)}(x) \in \{4, 6\}$, *or* (b) *a trivial relator $\sigma_s^{\pm 1}\sigma_s^{\mp 1}$ in case* $\mathrm{val}_{\Lambda(f)}(x) = 1$.
(4) *The clews* (2b) *are consistent with the relators* (3) *as shown in Figure 18. (In each local picture, the mirror image of the illustrated situation is also allowed, as is simultaneous reversal of all edge orientations.)* □

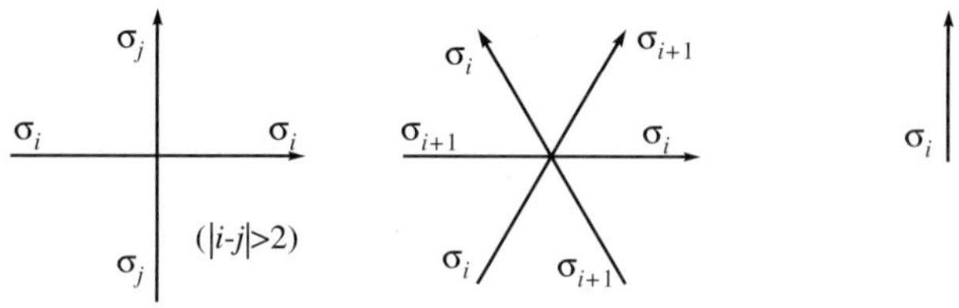

Fig. 18. Clews in the labyrinth of a $MP_n/\equiv_{\mathbb{R}}$-transverse map on a surface.

CONSTRUCTION 3.67. A *braided surface* of *degree n* in $D^2 \times \mathbb{C}$ is $S_f := f_\Delta^{-1}(0)$, where $f\colon D^2 \to MP_n$ is transverse to all the cells of $MP_n/\equiv_{\mathbb{R}}$ as in 2.66.

PROPOSITION 3.68. *A braided surface S_f in $D^2 \times \mathbb{C}$ is a surface, and* $\mathrm{pr}_1|S\colon S \to D^2$ is a simple branched covering map branched over $\mathsf{V}(\Lambda(f))$. □

In these terms, various facts about the constructions in (Rudolph, 1983b, 1985b) can be phrased as follows. Since any braided surface S_f is compact, it is actually contained in $D^2 \times rD^2$ for some r; clearly only braided surfaces in $D^2 \times D^2$ need to be considered. The obvious identification of $S^1 \times D^2 \subset \partial(D^2 \times D^2)$ with $\mathrm{Nb}(\mathbb{O}' \subset S^3)$ identifies the boundary of a braided surface $S_f \in D^2 \times D^2$ with a 1-submanifold of $\mathrm{Nb}(\mathbb{O}' \subset S^3)$, which is a closed n-string $\mathbf{o}$-braid isotopic to some $\hat{\beta}(\mathbf{b})$ of length $\mathrm{card}(\mathsf{V}(\Lambda(f)))$. The bandword $\mathbf{b}$ can be read off from the clews of $\Lambda(f)$. In particular, $\beta(\mathbf{b})$ is the ordered product of the clews on the (oriented) edges of $\Lambda(f)$ that intersect S^1. Every bandword determines a braided surface which is unique up to isotopy (through braided surfaces). Rudolph (1983a) characterizes quasipositive braided surfaces as follows.

THEOREM 3.69. *A braided surface S_f, where f is holomorphic, determines a quasipositive bandword. Any braided surface that determines a quasipositive bandword is isotopic (through braided surfaces) to S_f with f a polynomial.* □

There is a straightforward way to "round the corners" of $D^2 \times D^2$ by a *smoothing* $D^2 \times D^2 \to D^4$ – that is, a homeomorphism which is a diffeomorphism off $S^1 \times S^1$ – so that a braided surface in $D^2 \times D^2$ is carried to a ribbon surface in D^4. The converse is also true (Rudolph, 1983b, 1985b).

THEOREM 3.70. *Up to isotopy, every ribbon surface in D^4 is the image of a braided surface in $D^2 \times D^2$ by a smoothing $D^2 \times D^2 \to D^4$.* □

THEOREM 3.71. *Let* $\mathbf{b}$ *be an embedded bandword. The ribbon surface S in D^4 with $\partial S = \partial S[\mathbf{b}] \subset S^3$, produced by pushing* $\mathrm{Int}\, S[\mathbf{b}]$ *into* $\mathrm{Int}\, D^4$*, is isotopic to the ribbon surface which is the image of the braided surface in $D^2 \times D^2$ determined by the bandword* $\mathbf{b}$. □

DEFINITIONS 3.72. A *node* (resp., *cusp*) in an n-string braid group B_X is the square (resp., cube) of an X-band. A *nodeword* (resp., *cuspword*) in B_X is a k-tuple $\mathbf{b} =: (b(1),\ldots,b(k))$ such that each $b(i)$ is a band or a node (resp., a band, a node, or a cusp) in B_X.

The proof of 2.70 can be generalized (cf. Rudolph, 1983b) to show that every "nodal ribbon surface" in D^4 can be realized, up to smoothing and isotopy, by a "nodal braided surface" in $D^2 \times D^2$. Orevkov (1998) proved an immense generalization of 3.69; the following special case of Orevkov's result is easy to state and more than adequate for this review.

THEOREM 3.73. *If* $\mathbf{b}$ *is a quasipositive cuspword in* B_n, *then there is a holomorphic cusp curve in* $D^2 \times D^2$ *that corresponds to* $\mathbf{b}$ *via an appropriate holomorphic map* $f\colon D^2 \to MP_n$. □

The map f in 3.73 is amazing; it is transverse to all the cells of $MP_n/\equiv_{\mathbb{R}}$ if and only if the cuspword $\mathbf{b}$ is a bandword.

4. Transverse $\mathbb{C}$-links

DEFINITION 4.74. A link $L \subset S^3$ is a *transverse* $\mathbb{C}$*-link* provided that (S^3, L) is diffeomorphic to $(\Sigma, \mathcal{V}_f \cap \Sigma)$, where: (1) $D \subset \mathbb{C}^2$ is a Stein disk bounded by a strictly pseudoconvex 3-sphere Σ, and (2) $f \in \mathcal{O}(D)$ is a holomorphic function with $\mathrm{Sing}(\mathcal{V}_f) \cap \Sigma = \varnothing$, such that (3) the complex manifold $\mathrm{Reg}(\mathcal{V}_f)$ intersects Σ transversely and (4) $\mathcal{V}_f \cap \Sigma$ is non-empty (and therefore a smooth, naturally oriented closed 1-submanifold of Σ).

Various modifications may be made to the specifications in 4.74 without changing the class of links so specified.

PROPOSITION 4.75. (1) *The requirement that* Σ *be strictly pseudoconvex can be somewhat relaxed.* (2) *The requirement that* Σ *be strictly pseudoconvex can be considerably strengthened:* Σ *can be required to be convex, or even to be a round sphere.* (3) $\mathcal{V}_f$ *can be required to be non-singular, to be algebraic, or to be both at once.*

PROOF (Sketches). (1) Here, no attempt will be made to state the most general theorem (see 4.76.2). A sufficient example for present purposes is the following. Let $\gamma \subset \mathbb{C}$ be an arbitrary simple closed curve and $G \subset \mathbb{C}$ the 2-disk that it bounds, so that for any $r > 0$ the *bicylinder* $D := G \times rD^2$ is diffeomorphic (indeed, biholomorphic) to the bidisk $D_{1,1}$. Although D is not a Stein domain, Int D is an open Stein manifold, and $(D, \partial D)$ can be arbitrarily closely approximated by Stein domains. The result follows from 2.39 and transversality. (2) As pointed out by Boileau and Orevkov (2001), theorems of Eliashberg (1990) show that for this purpose all Stein disks are equally good. (3) A nearby level set of f will have no singularities and give the same link type; by 2.39, f is arbitrarily closely approximated by its sufficiently high-degree Taylor polynomials, and again we get the same link type. □

QUESTIONS 4.76. 4.74 and 4.75 immediately suggest a number of questions.

4.76.1. Is there an algorithm for determining whether or not a given link is a transverse $\mathbb{C}$-link? (It follows from 4.77 that the set of isotopy classes of transverse $\mathbb{C}$-links is recursive: there exists an algorithm that produces a link in every such isotopy class. Thus the question becomes: Is the set of isotopy classes of transverse $\mathbb{C}$-links recursively enumerable?)

4.76.2. How much can the requirement that Σ be strictly pseudoconvex be relaxed (as in 4.75(1))? See Rudolph (1985a) for a cautionary result.

4.76.3. Let L be a transverse $\mathbb{C}$-link. By 4.75(3), up to isotopy $L = S^3 \cap \mathcal{V}_f$ for some $f \in \mathbb{C}[z, w]$. What is the minimum degree of such a polynomial f? In light of 4.75(2), the same question can be asked with the roundness of S^3 weakened to convexity or strict pseudoconvexity of a 3-sphere Σ. Can either of these weakenings strictly decrease the minimum? Calculations of, or even good upper bounds for, this (or these) invariant (s) of L would have applications to finding embeddings of certain Stein domains (namely, cyclic branched covers of D^4 branched over holomorphic curves) into algebraic surfaces of unexpectedly low degree in $\mathbb{C}^3$ (Boileau and Rudolph, 1995).

4.1. *Transverse $\mathbb{C}$-links are the same as quasipositive links*

THEOREM 4.77. *Every quasipositive link is a transverse $\mathbb{C}$-link.* □

Proofs of 4.77 (with some variation in the details) are presented in (Rudolph, 1983a, 1984, 1985b). A remarkable theorem of Boileau and Orevkov (2001) asserts the converse.

THEOREM 4.78. *Every transverse $\mathbb{C}$-link is a quasipositive link.* □

COROLLARY 4.79. *L is a transverse $\mathbb{C}$-link if and only if L is quasipositive.* □

COROLLARY 4.80. *Every isotopy class of transverse $\mathbb{C}$-links is represented by a transverse intersection $\mathcal{V}_f \cap \{(z, w) \in \mathbb{C}^2 : \|(z, w)\| = 1\}$, where $f \in \mathbb{C}[z, w]$ and* $\operatorname{Sing}(\mathcal{V}_f) = \varnothing$. □

As Boileau and Orevkov note, their proof (the only one known to date) is completely non-constructive, relying strongly as it does on the theory of pseudo-holomorphic curves (Gromov, 1985).

QUESTION 4.81. Is there a constructive proof of 4.78? Biding such a proof, can an upper bound on $\{n$: L is isotopic to a quasipositive n-string braid$\}$ be deduced from other invariants of a transverse $\mathbb{C}$-link L?

4.2. *Slice genus and unknotting number of transverse $\mathbb{C}$-links*

THEOREM 4.82. *If $f \in \mathcal{O}(D^4)$ is such that* Sing $(\mathcal{V}_f) = \varnothing$, *and L is the transverse $\mathbb{C}$-link $\mathcal{V}_f \cap S^3$, then $X_s(L) = \chi(\mathcal{V}_f \cap D^4)$.* □

COROLLARY 4.83. *If* **b** *is a quasipositive n-string bandword of length k, then $X_s(\hat{\beta}(\mathbf{b})) = n - k$.* □

It had been known at least as early as 1982 (Boileau and Weber, 1983; Rudolph, 1983d) that these results would follow from a local version of the Thom Conjecture (see Section A.6): either (in the case of 3.82) that L be a torus link $O\{n, n\}$ for some $n \in \mathbb{N}_{>0}$, or (in the case of 4.83) that **b** be the important $\mathcal{J}_\mathbf{n}$-bandword[12]

$$(\overbrace{\sigma_1, \sigma_2, \ldots, \sigma_{n-1}, \ldots, \sigma_1, \sigma_2, \ldots, \sigma_{n-1}}^{n \text{ repetitions}}) =: \nabla_n \tag{4.15}$$

with closed braid $\hat{\beta}(\nabla_n) = O\{n,n\}$. The first proof of the local Thom Conjecture was given by Kronheimer and Mrowka (1993), using gauge theory for embedded surfaces, a 4-dimensional technique. The full Thom Conjecture (and more) was proved using Seiberg–Witten theory – also a 4-dimensional technique – by Kronheimer and Mrowka (1994); Morgan et al. (1996). Recently Rasmussen (2004) announced a purely 3-dimensional combinatorial proof based on Khovanov homology.

COROLLARY 4.84. *If L is a transverse $\mathbb{C}$-link then $X_r(L) = X_s(L)$.* □

It was also known (Boileau and Weber, 1983; Rudolph, 1983d) that the truth of the local Thorn Conjecture implies an affirmative answer to "Milnor's Question" (see Section A.5) on unknotting numbers of links of singularities (Section 4.1) or – more generally (see Rudolph, 1983b) – closed positive braids in the sense of 3.54.

COROLLARY 4.85. *If L is the link of a singularity (e.g., a torus link $O\{m,n\}$ with $0 \leq m < n$), then $\ddot{u}(L) =$* node$(L) =$ (card$(\pi_0(L)) - X(L))/2$. □

[12] There is a long and worthy tradition of designating (the braid of) this $\mathcal{I}_\mathbf{n}$-bandword as Δ^2 (Birman, 1975; Birman and Brendle, this Handbook, etc.), a usage apparently introduced by Fadell (1962) in homage to P. A. M. Dirac (who wrote that he "first thought of the string problem about 1929" in a letter quoted by Gardner, 1966, addendum to Chapter 2; see also Newman, 1942). Nonetheless, in the context of the knot theory of complex plane curves, where discriminants (which have their own, algebro-geometric, traditional claim on the notation Δ) abound, I believe that the makeshift of using ∇ to stand in for Δ^2 is preferable to the further overloading of the symbol Δ.

4.3. *Strongly quasipositive links*

4.82–4.85 immediately imply the following.

THEOREM 4.86. *If L is a strongly quasipositive link with quasipositive Seifert surface S, then $\chi(S) = X(L) = X_r(L) = X_s(L)$. If K is a strongly quasipositive knot with quasipositive Seifert surface S, then $g(S) = g(K) = g_r(K) = g_s(K) \leq \text{node}(K) \leq \ddot{u}(K)$.* □

To fully exploit 4.86 it is necessary to have a good supply of transverse $\mathbb{C}$-links or, equivalently, quasipositive (closed) braids. Clearly, every strongly quasipositive link (3.61) is quasipositive. The converse is false; for instance, the closed braid of ${}^{\sigma_2^3}\sigma_1\sigma_1 \in B_3$ is a quasipositive knot, but not strongly quasipositive (by 4.83). Nonetheless, the class of strongly quasipositive links is very varied.

4.3.1. *S-equivalence and strong quasipositivity*

PROPOSITION 4.87. *No link invariant calculable from a Seifert matrix (for instance, the Alexander polynomial of a knot, the signatures of a knot or link, etc.) can tell whether or not a link is quasipositive.*

PROOF. Let $[(L_i, L_j)_S]$ be a Seifert matrix for a link L with Seifert surface S. By 3.59, we may assume that $S = S[\mathbf{b}]$, for some embedded X-bandword $\mathbf{b}$ of length k. Let there be $m \leq k$ negative bands in $\mathbf{b}$. If $m = 0$, we are done. Otherwise, let $b(s)$ be negative, say $b(s) = \sigma_{x_p,x_q}^{-1}$, with $\{x_p < x_q\} \subset X$. Let $X' := X \cup \{x'_1, x'_2, x'_3, x'_4\}$, where $x_p < x'_1 < x'_2 < x'_3 < x'_4 < x_q$. Let $b'(t) = \mathrm{i}_{X,X'}b(t)$ for $1 \leq t < s$, $b'(s) = \sigma_{x'_3,x_q}$, $b'(s+1) = \sigma_{x'_2,x'_4}$, $b'(s+2) = \sigma_{x'_1,x'_3}$, $b'(s+3) = \sigma_{x_p,x_2}$, $b'(s+4) = \sigma_{x'_1,x'_4}$, and $b'(t) = \mathrm{i}_{X,X'}b(t+4)$ for $s+4 < t \leq k+4$ (see Figure 19 for a rendition of this operation in terms of fence diagrams). Manifestly, $S[\mathbf{b}]$ is diffeomorphic to $S[\mathbf{b}']$ by a diffeomorphism that is the identity of a single 1-handle, and an easy calculation shows that they have identical Seifert matrices. There are only $m - 1$ negative bands in $\mathbf{b}'$, so this result of Rudolph (1983c) is true by induction on m. □

COROLLARY 4.88. (1) *Every S-equivalence class of knots contains strongly quasipositive knots.* (2) *Every knot can be converted to a strongly quasipositive knot by a sequence of "doubled-delta moves".*

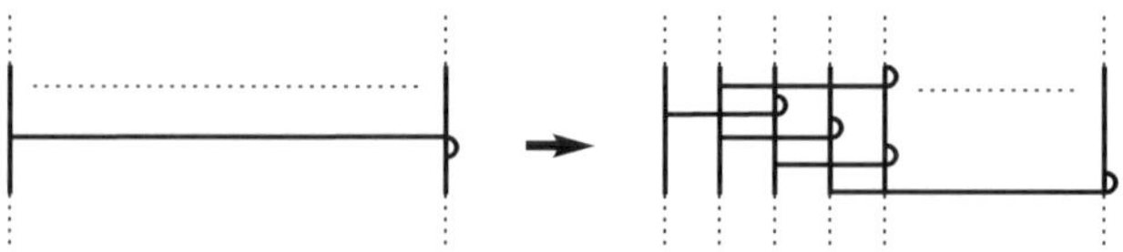

Fig. 19. "Trefoil insertion" on a fence diagram preserves the Seifert form while eliminating a negative band.

PROOF. Knots K_0 and K_1 are *S-equivalent* (in the first instance, Trotter, 1962, "*h*-equivalent") just in case K_i has a Seifert surface S_i such that some homology bases of S_0 and S_1 produce identical Seifert matrices; so (1) is immediate from (the proof of) 3.87. Naik and Stanford (2003) introduced the "doubled-delta move", an operation on standard link diagrams illustrated in Figure 20, and proved that the equivalence relation on knots generated by applying such moves is precisely *S*-equivalence; (2) follows directly. □

Baader (2004) has proved an analogue of 4.87: for any knot K and $n \in \mathbb{N}$, there is a quasipositive knot Q whose Vassiliev invariants of order less than or equal to n coincide with those of K.

QUESTION 4.89. Conway (1970) introduced an operation on knots and links called *mutation*. The result L' of applying mutation to a link L is called a *mutant* of L. Although mutant links can be non-isotopic, they are indistinguishable by a wide variety of link invariants. Mutation was generalized by Anstee et al. (1989). The result L' of applying the operation defined by Anstee et al. to L is called a *rotant* of L. The doubled-delta move is a special case of a rotant move. Obviously, 3.88(2) can be rephrased as "the doubled-delta move need not preserve strong quasipositivity". In what circumstances is a rotant of a strongly quasipositive link strongly quasipositive? In particular, is a mutant of a strongly quasipositive link necessarily strongly quasipositive?

4.3.2. *Characterization of strongly quasipositive links.* There is a simple characterization of quasipositive Seifert surfaces, and thus – in some sense – of strongly quasipositive links.

THEOREM 4.90. *A Seifert surface S is quasipositive if and only if, for some $n \in \mathbb{N}_{>0}$, S is ambient isotopic to a full subsurface of the fiber surface of the torus link $O\{n,n\}$.* □

QUESTIONS 4.91. Does there exist an algorithm to determine whether or not a given Seifert surface is quasipositive? Does there exist an algorithm to determine whether a given link is strongly quasipositive?

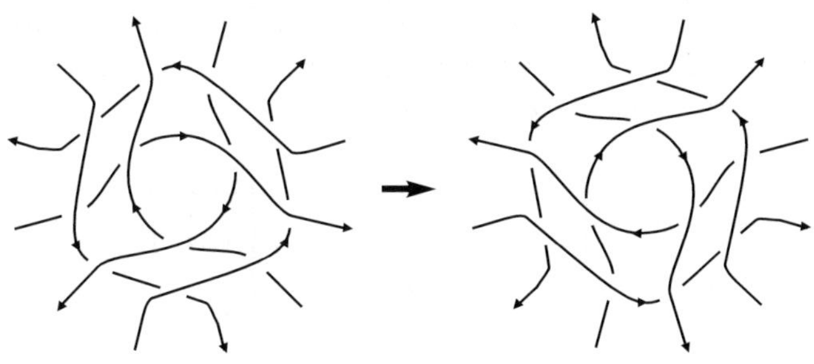

Fig. 20. The "doubled-delta" move.

4.3.3. *Quasipositive annuli*

LEMMA 4.92. *Let (L, g) be a framed link such that the annular Seifert surface $A(L, g)$ is quasipositive. If the framing f of L is less twisted than g, then $A(L,f)$ is quasipositive.*

PROOF. This is the "Twist Insertion Lemma" of Rudolph (1983a) (with the sign conventions in force here, it should be called the "Twist Reduction Lemma"). A proof using fence diagrams is shown in Figure 21. □

DEFINITION 4.93. The *modulus of quasipositivity* of a knot K is

$$q(K) := \sup\{t \in \mathbb{Z} : A(K, t) \text{ is a quasipositive Seifert surface}\}. \tag{4.16}$$

Rudolph (1990) applied 2.58 (via 1.26) and 3.90 to deduce the following.

PROPOSITION 4.94. *If $A(L,f)$ is quasipositive, then $\mathrm{ord}_v\{L, f\} \geq 0$.* □

COROLLARY 4.95. *If K is a knot, then $q(K) \leq \frac{1}{2}\mathrm{ord}_v\ \{K\} \leq -1 - \deg_a F_K^*(a, x)$.* □

COROLLARY 4.96. *If K is a knot, then $q(K) \in \mathbb{Z}$.* □

Since $q(K) = \mathrm{TB}(K)$ (Rudolph, 1995), 4.96 also follows from Bennequin (1983).

COROLLARY 4.97. $q(O) = -1$. □

PROPOSITION 4.98. *For $K \neq O$, $q(K) = \sup\{f \in \mathbb{Z}$: the framed knot (K,f) embeds on the fiber surface of $O\{n, n\}$ for some $n \in \mathbb{N}_{>0}\}$.*

PROOF. If a framed knot (K,f) embedded on a Seifert surface S and K is not full on S, then $(K,f) = (O, 0)$; so 4.98 follows immediately from the characterization theorem 4.90. □

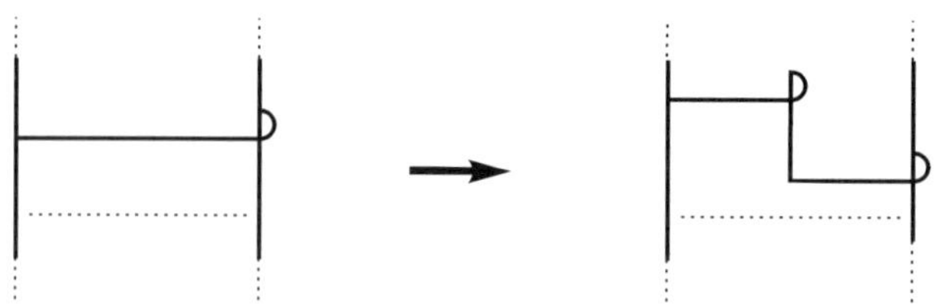

Fig. 21. "Twist reduction" on a fence diagram.

COROLLARY 4.99. *A quasipositive Seifert surface is incompressible.* □

Of course 4.99 also follows from 4.85. Yet another proof can be extracted from Bennequin (1983).

4.3.4. *Quasipositive plumbing.* A somewhat lengthy but straightforward combinatorial proof of the following theorem appears in Rudolph (1998).

THEOREM 4.100. *A Murasugi sum is quasipositive if and only if its summands are quasipositive.* □

COROLLARY 4.101. *Iterated Murasugi sums of quasipositive annuli are quasipositive Seifert surfaces.* □

EXAMPLES 4.102. 4.101 covers a surprising amount of ground.

4.102.1. A *positive Hopf-plumbed fiber surface* is an iterated Murasugi sum of positive Hopf annuli $A(O,-1)$ starting from the trivial Murasugi sum D^2. A fiber surface S is *stably positive Hopf-plumbed* in case some positive Hopf-plumbed fiber surface F can also be constructed as an iterated Murasugi sum of positive Hopf annuli $A(O,-1)$ starting from S. A remarkable theorem of Giroux (2002) states that a fibered link L has a stably positive Hopf-plumbed fiber surface if and only if a certain contact structure on S^3 (constructed from the fibration in a way he describes) is the standard contact structure. In combination with 3.90 and 3.100, Giroux's theorem implies that a fibered link is strongly quasipositive if and only if it is stably positive Hopf-plumbed.

4.102.2. A *basket* is a Seifert surface produced by repeatedly plumbing unknotted annuli $A(O,k_i)$ to a single fixed $D^2 \subset S^3$. (See Rudolph, 2001a, for details and examples.) A fundamental theorem of Gabai (1983b, 1985) implies that a basket is a fiber surface if and only if it is Hopf-plumbed (allowing both positive and negative Hopf annuli as plumbands). In particular, according to Rudolph (2001a), a fiber surface is a quasipositive basket if and only if it is isotopic to a braided Seifert surface $S[\mathbf{b}]$ where $\mathbf{b}$ is a strictly $\mathcal{T}$-positive $\mathcal{T}$-bandword for some espalier $\mathcal{T}$.

4.102.3. General *arborescent links* are constructed as boundaries of unoriented (possibly non-orientable) 2-manifolds formed by *unoriented plumbing*, in which not just unknotted annuli $A(O,k_i)$ but also unknotted Möbius bands are used as building blocks, while their mode of assembly is suitably restricted (it is coded by a planar tree with integer weights; see Conway, 1970; Siebenmann, 1975; Gabai, 1986). So-called *special arborescent links* (Sakuma, 1994) are those obtained by disallowing Möbius plumbands. As noted by Rudolph (2001a), the surfaces defining special arborescent links are baskets; in particular, when all weights of an arborescent presentation of an arborescent link L are strictly negative even integers, then L is strongly quasipositive.

4.102.4. The *positively clasped, k-twisted Whitehead double* $\mathcal{D}(K; k; +)$ of a knot K is defined to be the boundary of the non-trivial plumbing (that is, not a boundary-connected sum) of $A(K, k)$ and $A(O, -1)$ (Figure 22 illustrates this operation with a fence diagram). By 4.101 and 4.90, $\mathcal{D}(K; k; +)$ is strongly quasipositive if and only if $k \leq \mathrm{TB}(K)$ (see Rudolph, 1993).

4.102.5. If a set of Seifert surfaces contains $A(O, -1)$ and is closed under isotopy, Murasugi sum, and the operation of passing from a surface to a full subsurface, then it contains all quasipositive Seifert surfaces; and the set of all quasipositive Seifert surfaces is the smallest set with these properties.

4.3.5. *Positive links are strongly quasipositive.* A standard link diagram $\mathbf{D}(L)$ is called *positive* in case every crossing $\mathsf{x} \in \mathsf{X}_{\mathbf{D}(L)}$ is positive; a link L is positive if L has a positive standard link diagram. Nakamura (1998, 2000) and Rudolph (1999) independently proved the following.

THEOREM 4.103. *If the standard link diagram $\mathbf{D}(L)$ is positive, then the diagrammatic Seifert surface $S(\mathbf{D}(L))$ is quasipositive.* □

COROLLARY 4.104. *A positive link is strongly quasipositive.* □

QUESTIONS 4.105. Two questions posed by Rudolph (1999) remain open and are relevant here.

4.105.1. Can positive links be characterized as strongly quasipositive links that satisfy some extra geometric conditions?

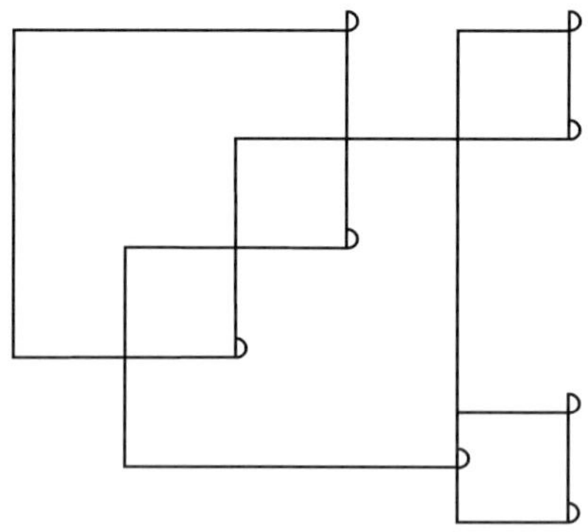

Fig. 22. A fence diagram for the plumbed Seifert surface of $\mathcal{D}(O\{2,3\}; 0; +)$.

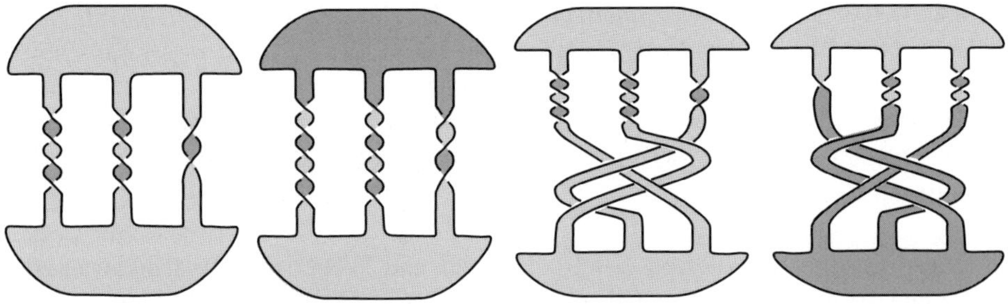

Fig. 23. The quasipositive pretzel surfaces $\mathfrak{P}(4, 4, -2)$ and $\mathfrak{P}(5, 5, -3)$; the braidzel surfaces $\mathfrak{P}(\sigma_1, \sigma_2^{-1}, \sigma_1, \sigma_2^{-1}; 4, 4, -2)$ and $\mathfrak{P}(\sigma_2^{-1}, \sigma_1, \sigma_2^{-1}, \sigma_1; 1, -3, -3)$.

4.105.2. Does $K \geq O\{2, 3\}$ imply that K is strongly quasipositive? (Here $K_1 \geq K_2$ means that K_1 is concordant to K_2 inside a 4-manifold with positive intersection form; this partial order was defined and studied by Cochran and Gompf (1988), who showed that if K is a positive knot, then $K \geq O\{2, 3\}$.)

4.3.6. *Quasipositive pretzels.* *Braidzels* are generalizations of the well-known (oriented) pretzel surfaces, with braiding data supplementing the twisting data that specifies an ordinary pretzel surface (a braidzel with trivial braiding); two typical pretzel surfaces and two typical braidzel surfaces are pictured in Figure 23. Braidzels were defined by Rudolph (2001b) and have been further studied by Nakamura (2004), who showed that every link has a Seifert surface which is a braidzel. Using braidzels, Rudolph (2001b) proved the following.

PROPOSITION 4.106. *The oriented pretzel surface $\mathfrak{P}(t_1, \ldots, t_k)$ is quasipositive if and only if the even integer $t_i + t_j$ is less than 0 for $1 \leq i < j \leq k$.* □

The question 4.89 was somewhat motivated by 4.106, since up to repeated mutation the pretzel link $\partial\mathfrak{P}(t_1, \ldots, t_k)$ depends only on $\{t_1, \ldots, t_k\}$.

QUESTION 4.107. What are necessary and sufficient conditions on the defining data of a braidzel that it be quasipositive?

4.3.7. *Links of divides.* Identify $\mathbb{C}^2$ with the tangent bundle of $\mathbb{C}$. A'Campo (1998) calls a normal collection P of edges in D^2 a *divide*, and constructs the *link of the divide P* as $L(P) := \{(z, w) \in S^3 : z \in P, w \in T_z(P)\}$. This construction has been extended – first by allowing non-proper edges ("free divides"; Gibson and Ishikawa, 2002a), more generally by allowing immersed unoriented circle components (Kawamura, 2002), and more generally yet (and most recently) by allowing immersions of graphs that are not 1-manifolds ("graph divides"; Kawamura, 2004).

THEOREM 4.108. *If P is a divide, then:* (1) *L(P) is fibered, and every link L of a singularity is of the form L(P) for some divide P* (A'Campo, 1998); (2) *L(P) is strongly quasipositive* (Hirasawa, 2000); (3) *L(P) is stably positive Hopf-plumbed* (Hirasawa, 2002); *and indeed* (4) *L(P) is positive Hopf-plumbed* (Ishikawa, 2002). *If P is a free divide, possibly with circle components, then* (5) *L(P) is strongly quasipositive* (Kawamura, 2002). *If P is a graph divide, then* (6) *L(P) is quasipositive* (Kawamura, 2004). □

See also Couture and Perron (2000); Hongler and Weber (2000); Ishikawa (2001); Gibson and Ishikawa (2002b); Gibson (2002); Goda, Hirasawa, and Yamada (2002); Chmutov (2003).

4.4. *Non-strongly quasipositive links*

Other than the tautologous construction by forming the closed braid of a quasipositive braid, little is yet known about systematic constructions of non-strongly quasipositive transverse $\mathbb{C}$-links.

EXAMPLE 4.109. There exist fibered links which are quasipositive but not strongly quasipositive (compare with 4.102.1). Examples include all the links $H_{n+m,m}$ obtained from $O\{n+2m, n+2m\}$ $(m, n \geq 1)$ by reversing the orientation of (any) m components. Figure 24 pictures a ribbon surface $R = f(S^1 \times [0,1] \coprod D^2)$ for $H_{1,1}$ and a fence diagram for the fiber surface $\mathrm{sm}(R)$ of $H_{1,1}$.

EXAMPLE 4.110. There exist arborescent links which are quasipositive but are not special arborescent and not strongly quasipositive (compare with 4.102.3). An example is the quasipositive arborescent ribbon knot pictured, along with its defining weighted tree, in Figure 25.

QUESTION 4.111. Every pretzel link is arborescent. (Caution: this does not mean that every pretzel link bounds an oriented pretzel surface, and in fact there is a natural sense in which those which do are a small minority. Nor is a typical oriented pretzel surface itself an arborescent Seifert surface.) What are necessary and sufficient conditions on an arborescent link that it be quasipositive, or strongly quasipositive?

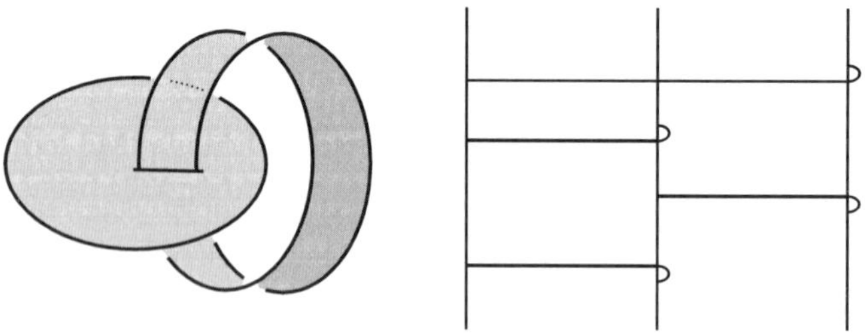

Fig. 24. A quasipositive fibered link which is not strongly quasipositive.

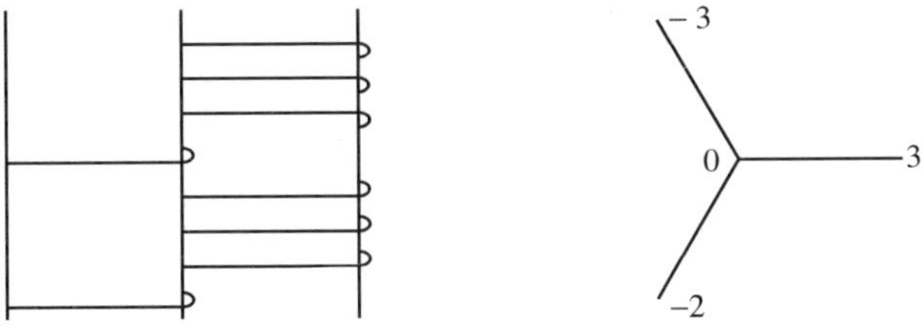

Fig. 25. Two representations of a quasipositive Lover's Knot.

Besides 4.106, some partial but systematic progress towards answering this question has been made by Tanaka (1998) (for the class of rational links).

5. Complex plane curves in the small and in the large

As noted in 4.80, every transverse $\mathbb{C}$-link L is "algebraic" in the sense that, up to isotopy, L is the transverse intersection of $rS^3 = \{(z, w) \in \mathbb{C}^2: \|(z, w\}\| = r\}$ and a complex algebraic curve $\mathcal{V}_f$; clearly the same class of links arises whether $r > 0$ is fixed once for all, or allowed to vary throughout $\mathbb{R}_{>0}$, so long as f is allowed to vary throughout $\mathbb{C}[z, w]$. By contrast, interesting additional restrictions on the isotopy type of L arise in case, for each fixed $\mathcal{V}_f$, r is constrained to be either "very small" or "very large". More precisely, the situation is as follows (essentially this is proved by Milnor, 1968, in general dimensions). Let $F \in \mathbb{C}[z, w]$ be non-constant and without repeated factors. The *argument* of F is $F/|F|$: $\mathbb{C}^2 \setminus \mathcal{V}_F \to S^1$. The *Milnor map of F* is $\varphi_F := \arg(F)|(S^{2n-1} \setminus \mathbb{D}(F))$; for $r > 0$, the *Milnor map of F at radius r*, denoted by $\varphi_{F,r}$, is the Milnor map of $(z, w) \mapsto F(z/r, w/r)$. Let $m(F) := \inf\{\|(z, w)\|^2: F(z, w) = 0\} = \sup\{r: rS^3 \cap \mathcal{V}_F = \varnothing\}$.

PROPOSITION 5.112. *There is a finite set $\mathcal{X}(F) \subset \mathbb{R}_{>m(F)}$ of radii r with the following properties.* (1) *If $r \in \mathbb{R}_{>m(F)} \setminus \mathcal{X}(F)$, then $\mathcal{V}_F$ intersects rS^3 transversally, so that $L(F, r) := (1/r)(\mathcal{V}_F \cap rS^3)$ is a link in S^3.* (2) *If r and r′ are in the same component of $\mathbb{R}_{>m(F)} \setminus \mathcal{X}(F)$ then $L(F, r)$ and $L(F, r')$ are isotopic.* (3) *If $m(F) < r \notin \mathcal{X}(F)$, then there is a trivialization n: $L(F,r) \times \mathbb{C} \to \nu(L(F,r))$, as in 1.18, that lies in the homotopy class corresponding to any Seifert surface for $L(F, r)$ and such that $\varphi_{F,r}$ is adapted to $\varphi_{F,r}$ with $d(K) = 1$ for all $K \in \pi_0(L(F, r))$.* □

DEFINITIONS 5.113. In case $F(0, 0) = 0$, for any $r \in \,]0, m(F)[$ the transverse $\mathbb{C}$-link $L(F, r)$ is called the *link of the singularity* of F (or of $\mathcal{V}_F$) at (0, 0). In any case, for any $r > \max \mathcal{X}(F)$ the transverse $\mathbb{C}$-link $L(F, r)$ is called the *link at infinity* of F (or of $\mathcal{V}_F$).

5.1. *Links of singularities as transverse $\mathbb{C}$-links*

It can happen that $\varphi_{F,r}$ has degenerate critical points, and so is not a Morse map, or that $\varphi_{F,r}$ is a Morse map but not a fibration (the latter in fact being common: as noted in the first paragraph of this section, every transverse $\mathbb{C}$-link can be realized as $L(F, r)$ for some F and $r > 0$, while according to the previous section many, many transverse $\mathbb{C}$-links are not fibered links). However, Milnor (1968) proved the following (and its analogue in higher dimensions).

THEOREM 5.114. *If $m(F) < r < \min \mathcal{X}(F)$, then $L(F, r)$ is a fibered link and $\varphi_{F,r}$ is a fibration.* □

In other words, if L is a link of a singularity, then L is fibered.

Let L be the link of a singularity. There is a huge literature devoted to studying the algebraic and geometric topology L, and especially of its fibration (which, at the geometric level, completely determines L and all its invariants – though not necessarily in a perspicuous fashion). Some starting points for the dedicated reader are Eisenbud and Neumann (1985); Durfee (1999); Neumann (2001); Lê Dũng Tráng (2003). Here it may simply be noted that many – though certainly not all – of the interesting topological features of the fibration of L can profitably be explored using one or another of the representations of L alluded to in earlier sections. On one hand, L is the link of a divide (A'Campo, 1998), from which one representation of L as a positive Hopf-plumbed link can be derived and exploited (A'Campo, 1998). On the other hand, it is easy to use labyrinths to see that L is a positive closed braid (and not overwhelmingly difficult actually to derive a specific positive braidword for L), whence by Rudolph (2001a) one may derive (and, at least potentially, exploit) an apparently different – though surely closely related – representation of L as a positive Hopf-plumbed link.

5.2. *Links at infinity as transverse $\mathbb{C}$-links*

Let L be a link at infinity. In general, L need not be fibered (for instance, every unlink $O^{(n)}$ is a link at infinity, but only $O = O^{(1)}$ is fibered), but as Neumann (1989) puts it, L is "nearly" fibered. The literature on links at infinity is smaller than that on links of singularities, but still too large and multifaceted to summarize here. The reader is referred to Boileau and Fourrier (1998) for a review up to 1998. Some more recent articles on various aspects of the subject are Bodin (1999); Némethi (1999); Neumann (1999); Bartolo and Cassou-Noguès (2000); Neumann and Norbury (2000); Păunescu and Zaharia (2000); Gwoździewicz and Płoski (2001, 2002); Rudolph (2002); Neumann and Norbury (2003); Cimasoni (2004).

6. Totally tangential $\mathbb{C}$-links

DEFINITION 6.115. A link $L \subset S^3$ is a *totally tangential $\mathbb{C}$-link* provided that $L = \mathcal{V}_g \cap S^3$, where $g \in \mathcal{O}(\mathbb{C}^2)$ is such that $\mathcal{V}_g \cap \operatorname{Int} D^4 = \varnothing$ and $\mathcal{V}_g \cap S^3$ is a non-empty non-degenerate critical manifold of index 1 of $\rho|\operatorname{Reg}(\mathcal{V}_g)$, with $\rho(z, w) := \|(z, w)\|^2$.

This section follows Rudolph (1992b, 1995, 1997).

LEMMA 6.116. *L is a totally tangential $\mathbb{C}$-link if and only if (1) L is Legendrian with respect to the standard contact structure on S^3, and (2) L is an $\mathbb{R}$-analytic submanifold of S^3.* □

By 6.116, a totally tangential $\mathbb{C}$-link has a natural framing $f = f_L^{\mathcal{L}}$.

PROPOSITION 6.117. *Let $\rho : U \to \mathbb{R}$ be an exhausting strictly plurisubharmonic function on an open set $U \subset C^2$ such that $D = \rho^{-1}(\mathbb{R}_{\leq 1})$ is a Stein disk, so $\Sigma := \partial D$ is a strictly pseudoconvex 3-sphere with a natural contact structure. If $g \in \mathcal{W}(D) \subset \mathcal{O}(U)$ is such that $\mathcal{V}_g \cap \Sigma$ is a non-empty non-degenerate critical manifold M of index 1 of $\rho|\mathrm{Reg}(\mathcal{V}_g)$, then M is a Legendrian 1-submanifold of the contact manifold Σ. Further, there exists a totally tangential $\mathbb{C}$-link $L \subset S^3$ such that (Σ, M) is diffeomorphic to (S^3, L) by a diffeomorphism carrying the natural normal line field of M in Σ to the natural normal line field of L in S^3.* □

QUESTION 6.118. Does every totally tangential $\mathbb{C}$-link arise (up to isotopy) as in 6.115 with g required to be a polynomial? (Rudolph, 1995 shows that g may be required to be entire.)

THEOREM 6.119. *If $L \subset S^3$ is Legendrian, then L is isotopic through Legendrian links to a totally tangential $\mathbb{C}$-link L' with* $\mathrm{tb}(L') = \mathrm{tb}(L)$.

PROOF. The proof for knots (Rudolph, 1995) extends directly to links. □

COROLLARY 6.120. *If $L \subset S^3$ is Legendrian, then there is a polynomial map p: $\mathbb{C}^2 \to \mathbb{C}$ such that* (1) $\mathrm{Sing}(\mathcal{V}_p) = \varnothing$, (2) *$\mathcal{V}_p$ intersects S^3 transversely,* (3) *$\mathcal{V}_p \cap D^4$ is an annular surface, and* (4) $\mathcal{V}_p \cap S^3 = \partial A(L, \mathrm{tb}\,(L))$. □

Coupled with 4.82–4.85, the connection 6.120 between totally tangential $\mathbb{C}$-links and transverse $\mathbb{C}$-links leads to several interesting conclusions.

THEOREM 6.121. *If* $\mathrm{TB}(K) \geq 0$, *then K is not slice.* □

A *front* (see Eliashberg, 1993; Etnyre, this Handbook) is a piecewise-smoothly immersed curve $\gamma \subset \mathbb{R}^2$ of the sort illustrated in Figure 26: γ has finitely many cuspidal corners and finitely many normal crossings; the corners are exactly the local extrema of $\mathrm{pr}_2 | \gamma$; at a corner, the common tangent line of the two branches is parallel to $\{0\} \times \mathbb{R}$; and elsewhere the tangent lines are parallel neither to $\{0\} \times \mathbb{R}$ nor to $\mathbb{R} \times \{0\}$. A *front diagram* is a pair $\mathbf{FD}(L) =: (\mathbf{P}(\mathbf{FD}(L)), \mathbf{I}(\mathbf{FD}(L)))$, where $\mathbf{P}(\mathbf{FD}(L))$ is a front, and $\mathbf{I}(\mathbf{D}(L))$ is the

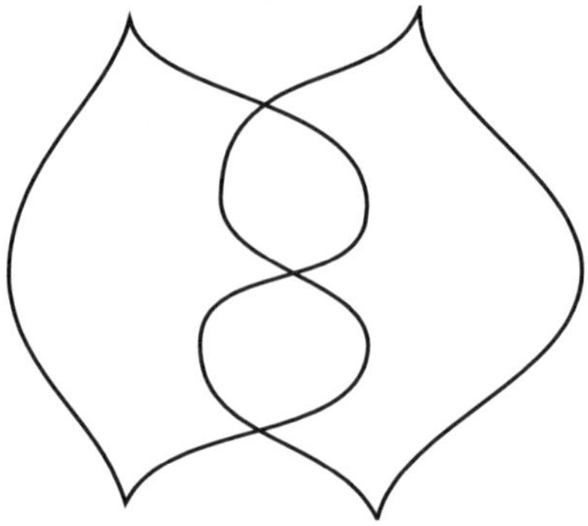

Fig. 26. A front.

information that L is a closed Legendrian 1-submanifold of $\mathbb{R}^3$ equipped with a certain standard contact structure (consistent with its embedding as the complement of a point in S^3 equipped with its standard contact structure).

LEMMA 6.122. (1) *Given any front* $\mathbf{P}(\mathbf{FD}(L))$, *the information* $\mathbf{I}(\mathbf{D}(L))$ *is true, and* $\mathbf{FD}(L)$ *determines* $|L|$ *up to a vertical translation in* $\mathbb{R}^3$. (2) *A standard link diagram* $\mathbf{D}(L)$ *for* L *can be derived from* $\mathbf{P}(\mathbf{FD}(L))$ *by* (a) *letting* $\mathbf{P}(\mathbf{D}(L))$ *be* $\mathbf{P}(\mathbf{FD}(L))$ *with its cusps rounded off, and* (b) *letting* $\mathbf{I}(\mathbf{D}(L))$ *declare that at each crossing of* $\mathbf{P}(\mathbf{D}(L))$, *the over-arc is the branch of which the tangent line has larger positive slope.* □

PROPOSITION 6.123. *For any knot* K, *if* $r \leq \mathrm{TB}(K)$ *then* $\mathrm{TB}(D(K, r, +)) \geq 1$. □

THEOREM 6.124. *For any knot* $K \subset S^3$, $g_s(K) \geq (\mathrm{TB}(K) + 1)/2$. □

7. Relations to other research areas

This section is a bibliography of papers that relate the knot theory of complex plane curves to other research areas; in some of the subsections, one (or a few) paper(s) are highlighted. The choice of highlights, and indeed the selections themselves, reflect both my bias and my ignorance; the latter, at least, is corrigible, and I welcome pointers to sources I have neglected. The division into subsections is necessarily somewhat arbitrary.

7.1. *Low-dimensional real algebraic geometry; Hilbert's 16th problem*

An interesting series of recent papers (Orevkov, 1999; Orevkov and Polotovskiĭ, 1999; Orevkov, 2000, 2001a,b; Orevkov and Shustin, 2002) has applied quasipositivity to Hilbert's 16th problem.

7.2. *The Zariski Conjecture; knotgroups of complex plane curves*

THEOREM 7.125. *Let $f \in \mathbb{C}[z,w]$. If* (1) *$\mathcal{V}_f \subset \mathbb{C}^2$ is a node curve, and* (2) *the link at infinity of $\mathcal{V}_f$ is a positive* (*and not merely quasipositive*) *closed braid, then the knotgroup $\pi_1(\mathbb{C}^2 \backslash \mathcal{V}_f)$ is abelian.* □

7.125 is the version of "Zariski's Conjecture" proved by Orevkov (1988). The original version of the conjecture (in which the link at infinity is required to be, not just any positive closed braid, but actually $\hat{\beta}(\nabla_{\deg f})$) had been proved by Fulton (1980); Deligne (1981); Nori (1983) and Harris (1986). The particular beauty of Orevkov's proof consists, first, in his use of a labyrinth for f (assumed to be in Weierstrass form) to construct a presentation of $\pi_1(\mathbb{C}^2 \backslash \mathcal{V}_f)$ with very many generators ($\deg_w(f)$ generators for each 0-dimensional stratum of the labyrinth) and very many relations, which are however all of very simple forms; and, second, in his ingenious application of the clews of the labyrinth, coupled with the positivity of the closed braid at infinity, to inductively reduce that presentation to a smaller presentation that visibly presents an abelian group.

The background and motivation for the conjecture, and related material on knotgroups of complex plane curves, may be found in Enriques (1923); Lefschetz (1924); Brauner (1928); Zariski (1929); van Kampen (1933); Burau (1934); Zariski (1936, 1971); Reeve (1954/1955); Cheniot (1973); Lê Dũng Tráng (1974); Oka (1975/76, 1974); Oka and Sakamoto (1978); Chang (1979); Randell (1980); Kaliman (1992); Moishezon and Teicher (1996); Kulikov (1997) and Garber (2003). The ideas of Orevkov (1998) are extended by Orevkov (1990a), Dethloff et al. (1998) and Neumann and Norbury (2003).

7.3. *Keller's Jacobian Problem; embeddings and injections of $\mathbb{C}$ in $\mathbb{C}^2$*

For somewhat unclear reasons, Keller's Jacobian Problem inspired Segre (1956/1957) to state – though not to prove correctly – a theorem classifying embeddings of $\mathbb{C}$ in $\mathbb{C}^2$, which has now been proved and generalized to injections.

THEOREM 7.126. *Let $p, q \in \mathbb{C}[t]$ be such that $F : \mathbb{C} \to \mathbb{C}^2 : t \mapsto (p(t), q(t))$ is a* (*set-theoretical*) *injection.* (1) *If F is a smooth embedding, then one of* $\deg(p)$, $\deg(q)$ *divides the other; as a consequence, there is a polynomial change of coordinates Q: $\mathbb{C}^2 \to \mathbb{C}^2$ such that $Q(F(\mathbb{C}))$ is a straight line.* (2) *If F is not smooth, then either one of* $\deg(p)$, $\deg(q)$ *divides the other or* $\deg(p)$ *and* $\deg(q)$ *are relatively prime; as a consequence, there is a polynomial change of coordinates $Q : \mathbb{C}^2 \to \mathbb{C}^2$ such that $Q(F(\mathbb{C}))$ is one of the curves $\{(z,w) \in \mathbb{C}^2 : z^m + w^n = 0\}$ for some relatively prime positive integers m and n.* □

7.126(1) was first proved by Abhyankar and Moh (1975) (algebraically) and by Suzuki (1974) (analytically); a topological (knot-theoretical) proof was given by Rudolph (1982a). 7.126(2) was first proved by Zaĭdenberg and Lin (1983) (analytically); a topological (knot-theoretical) proof was given by Neumann and Rudolph (1987).

Many other proofs have appeared since. Those proofs, related theorems, and some of the more knot-theoretical work on the original Jacobian Problem may be found in Abhyankar and Singh (1978); Oka and Sakamoto (1978); Orevkov (1990b, 1991); Chang (1991); Kang (1991); Chang and Wang (1993); Kaliman (1993); Abhyankar (1994); Artal–Bartolo (1995); Gurjar and Miyanishi (1996); Orevkov (1996); A'Campo and Oka (1996); van den Essen (1997) and Żołądek (2003).

7.4. *Chisini's statement; braid monodromy*

It was known early on, by algebraic geometers (see Zariski 1929) that, for proposed applications to the study of algebraic surfaces, the most important knotgroups of complex plane curves were those of cusp curves. Starting in 1933 and continuing through the 1950s, Oscar Chisini and his school published a series of papers around that subject (Dedò, 1950; Chisini, 1952; Tibiletti, 1952; Chisini, 1954; Tibiletti, 1955a,b,c,d and Chisini 1955). The tools they brought to bear were what would later be called "braid monodromy" and "the arithmetic of braids" (Moishezon 1981, 1983, 1985; Moishezon and Teicher 1988, 1991; Moishezon, 1994 and Moishezon and Teicher, 1994a,b,1996).

THEOREM 7.127. *Let* **b** *be a quasipositive cuspword in* B_n *such that* $\beta(\mathbf{b}) = \nabla_n$. (1) *For any* n *and any such* **b**, *there exists a holomorphic cusp curve in* $D^2 \times D^2$ *that corresponds to* **b** *via an appropriate holomorphic map* $f : D^2 \to MP_n$. (2) *There exist* n *and* **b** *such that no cusp curve in* $\mathbb{C}^2$ *corresponds to* **b** *via a polynomial map* $f: D^2 \to MP_n$. □

7.127(1) is of course a special case of 3.73, as proved (in much greater generality) by Orevkov (1998). 7.127(2) was proved by Moishezon (1994), in contradiction to "a statement of Chisini" (1954).

There is an enormous, interesting, and constantly growing literature on braid monodromy and its many offsprings, including applications to line arrangements, the Alexander invariants of complex plane curves, and so on; see Libgober (1986); Loeser and Vaquié (1990); Dung (1994); Dũng (1994); Dung and Hà (1995); Cohen and Suciu (1997); Dung (1999); Artal et al. (2001); Kharlamov and Kulikov (2001); Dung (2002); Kulikov and Kharlamov (2003) and Yetter (2003).

7.5. *Stein surfaces*

By 2.41, 3.69, 3.70, and 3.71, a branched covering of the bidisk branched over a quasipositive braided surface is diffeomorphic to a Stein surface. Loi and Piergallini (2001) proved the following converse.

THEOREM 7.128. *If* X *is a Stein surface, then* X *is diffeomorphic to a branched covering of* $D^2 \times D^2$ *branched over a quasipositive braided surface.* □

8. The future of the knot theory of complex plane curves

This section suggests some possible new directions for the knot theory of complex plane curves.

8.1. *Transverse $\mathbb{C}$-links and their Milnor maps*

As observed before 5.114, the Milnor map $\varphi_{F,r}$ at radius $r > m(F)$ of a non-constant reduced polynomial $F \in \mathbb{C}[z, w]$ can have degenerate critical points. However, this can happen for only finitely many $r \in \mathbb{R}_{>m(F)}$. Every other $\varphi_{F,r}$ is a Morse map, and typically *not* a fibration (even if $\mathcal{V}_F \cap rS^3$ is a fibered link; see Hirasawa and Rudolph, 2003). These Morse maps deserve further study.

MOTTO 8.129. *A Milnor map is interesting even if it is not a fibration.*

QUESTION 8.130. Here are a couple of specific questions about Milnor maps that are Morse. Is such a map necessarily topless? Under what conditions is such a map minimal (that is, possessed of exactly the minimum number of critical points necessary for any Morse map on its domain; see Pajitnov et al., 2001; Hirasawa and Rudolph, 2003, and references cited therein)?

Rational functions and non-reduced polynomials also have Milnor maps. Cimasoni's study (Cimasoni, 2004) of Alexander invariants of links of infinity of (reduced) polynomials leads him to investigate, in passing, the Milnor maps at infinity of non-reduced polynomials, and Pichon (2003) has begun the study of the Milnor map of rational functions at their points of indeterminacy, but a general study of these maps remains to be undertaken.

8.2. *Transverse $\mathbb{C}$-links as links at infinity in the complex hyperbolic plane*

To date, with the exception of Rudolph (2002), all investigations of links at infinity have implicitly or explicitly concerned themselves with behavior at "affine infinity", that is, in affine space $\mathbb{C}^2$. Is that an oversight?

MOTTO 8.131. *A transverse $\mathbb{C}$-link is a complex hyperbolic link at infinity.*

The meaning of 8.131 is this. As in the case of the real hyperbolic plane $\mathbb{H}^2(\mathbb{R})$, there is a Klein model of the complex hyperbolic plane $\mathbb{H}^2(\mathbb{C})$ which displays $\mathbb{H}^2(\mathbb{C})$ as holomorphically equivalent to Int $D^4 \subset \mathbb{C}^2$. From a purely topological – even complex-analytic – viewpoint, therefore, a link at infinity in the complex plane is nothing more (nor less) than a transverse $\mathbb{C}$-link in S^3. But, like $\mathbb{H}^2(\mathbb{R})$, $\mathbb{H}^2(\mathbb{C})$ has a geometry which makes it very different from the affine plane over the same field. Rudolph (2002) exploited (in a very minor way) the linear structure in $\mathbb{H}^2(\mathbb{C})$. What about more subtle geometric features, like any of several different variants on curvature?

QUESTION 8.132. One way to interpret the main theorem of Osserman (1981) is that, if $f \in \mathcal{O}(\mathbb{C}^2)$ is entire, then $\mathcal{V}_f$ has a well-defined link at infinity (i.e., is an algebraic curve) if and only if $\mathcal{V}_f$ has finite total absolute curvature. Given $f \in \mathcal{O}(\mathbb{H}^2(\mathbb{C}))$, possibly subject to further conditions (e.g., finite topological type), are there reasonable geometric conditions – for instance, integral curvature conditions – on $\mathcal{V}_f$ that are equivalent to the existence of a well-defined transverse $\mathbb{C}$-link "at infinity" for $\mathcal{V}_f$?

8.3. *Spaces of $\mathbb{C}$-links*

The sets of functions defined in 2.10 can be used to describe sets of singular transverse and totally tangential $\mathbb{C}$-links, as follows. In fact, it is obvious that the quotient set $(\mathcal{O}(D^4)\setminus\{0\})/\mathcal{U}(D^4)$ is naturally identified with the set of "singular transverse $\mathbb{C}$-links" (with positive multiplicities), and likewise that the quotient set $\mathcal{W}(D^4)/\mathcal{U}(D^4)$ is naturally identified with the set of "singular totally tangential $\mathbb{C}$-links". Moreover, both quotient sets are (more or less naturally) partitioned by the severity of the "singularity" of the $\mathbb{C}$-link. It is perhaps less obvious what topologies are most appropriate to turn these partitions (or suitable refinements thereof) into good stratifications of some infinite-dimensional manifolds, in which (for instance) the codimension-0 strata represent restricted isotopy types of ordinary non-singular transverse, or totally tangential, $\mathbb{C}$-links. Nonetheless, I proclaim the following.

MOTTO 8.133. *Naturally stratified spaces of singular $\mathbb{C}$-links exist and should be studied à la Vassiliev.*

A significant distinction between "singular links" in the present style, and in the style of Vassiliev, is that because these singular links are the 0-loci of (restrictions of holomorphic) functions $S^3 \to \mathbb{C}$ rather than the images of functions $S^1 \times \mathbf{k} \to S^3$, at a generic bifurcation (i.e., a codimension-1 stratum in one of the proposed spaces of singular links) the number of components changes (by one).

8.4. *Other questions*

Here are two last open questions ranging somewhat further afield than the earlier ones.

QUESTIONS 8.134. It is a long-standing conjecture (closely related to the Whitehead Conjecture; see Howie (1985)) that, if $S \subset D^4$ is a ribbon 2-disk, then $D^4 \setminus S$ is aspherical. A simple homological argument shows that this conjecture is equivalent to the conjecture that $H^2(\widetilde{D^4 \setminus S}; \mathbb{Z}) = 0$, where for any M, $\tilde{M}$ here denotes the universal covering space of M. Now, as Rudolph (1983) points out, if $S \subset D^4$ is any ribbon 2-disk, then there is a piece of complex plane curve $\Gamma \subset D^4$ (diffeomorphic to D^2) such that $D^4 \setminus S$ is homotopy-equivalent to $D^4 \setminus \Gamma$. If a piece of complex plane curve $\Gamma \subset D^4$ is diffeomorphic to D^2, is $D^4 \setminus \Gamma$ aspherical? Of course, $D^4 \setminus \Gamma$ is homotopy equivalent to Ext $(\Gamma \subset D^4)$. For an appropriate choice of Nb($\Gamma \subset D^4$), Ext ($\Gamma \subset D^4$) is a Stein domain. By 2.41 and 2.38, to answer this question affirmatively it would suffice to show that every complex curve in the universal cover X of Int (D^4 Γ) is $\mathcal{V}_f$ for some $f \in \mathcal{O}(X)$.

QUESTION 8.135. Call $K \subset S^3$ a *counter-Smith knot* in case K is a non-trivial knot and, for some $p > 1$, the p-sheeted cyclic branched cover of S^3 branched along K is again S^3. Assuming the Poincaré Conjecture, the Smith Conjecture is equivalent to the assertion that there are no counter-Smith knots. A 3-dimensional proof of the Smith Conjecture has of course been given by Thurston *et al.*; it is beautiful but lengthy, and depends on separate analyses of a number of different cases. In 1983 I noted that a simple appeal to Donaldson's results from 4-dimensional gauge theory shows that, if the knot K is a non-slice transverse $\mathbb{C}$-link, then K is not counter-Smith. Can techniques that have been developed in the meantime be used to give a 4-dimensional proof of the Smith Conjecture (modulo the Poincaré Conjecture) that avoids a case-by-case analysis?

Acknowledgements

It is a pleasure to be able to thank Claude Weber for his support and encouragement since my first trip to Switzerland in 1982. My continuing attempts to understand the knot theory of complex plane curves have been greatly furthered over the years by much work and talk with Michel Boileau, Walter Neumann, and Stepan Orevkov, and more recently Mikami Hirasawa. I have also had many useful conversations with Norbert A'Campo, Thomas Fiedler, Francoise Michel, and Andrei Pajitnov. A number of graduate students (several of them advised by mathematicians already mentioned), having written theses contributing to the knot theory of complex plane curves, have been kind enough to explain their work to me; these include Laurence Fourrier, Laure Eichenberger, Arnaud Bodin, and others. Many other workers in this and related fields have helped me keep reasonably in touch with developments; I thank them all.

Completion of this survey was partially supported by the Fonds National Suisse at the University of Geneva, and by a National Science Foundation Interdisciplinary Grant in the Mathematical Sciences, DMS-0308894.

Appendix A. And now a few words from our inspirations

A.1. *Hilbert's 16th problem (1891, 1900)*

> 16. Problem der Topologie algebraischer Curven und Flächen.
> Die Maximalzahl der geschlossenen und getrennt liegenden Züge, welche eine ebene algebraische Curve n ter Ordnung haben kann, ist von Harnack...bestimmt worden; es ensteht die weitere Frage nach der gegenseitigen Lage der Curvenzüge in der Ebene. ... *Eine gründliche Untersuchung der gegenseitigen Lage bei der Maximalzahl von getrennten Zügen scheint mir ebenso sehr von Interesse zu sein ....*
>
> (Hilbert, 1900, p. 283).

16. Problem of the topology of algebraic curves and surfaces.
The maximum number of closed and separate branches which a plane algebraic curve of the nth order can have has been determined by Harnack. ... There arises the further question as to the relative position of the branches in the plane. ... *A thorough investigation of the relative position of the separate branches when their number is the maximum seems to me to be of very great interest ....*

(Hilbert, 1902, p. 23).

A.2. *Zariski's Conjecture (1929)*

Theorem 7. The fundamental group of an irreducible curve f of order n, possessing ordinary double points only, is cyclic of order n. ... To prove the 7, we prove first the following lemma:

Lemma. The *fundamental group of n lines in arbitrary position is abelian.*

... Now let f be an irreducible curve of order n with ordinary double points only. The continuous system $\{f\}$... contains in particular curves which degenerate into n arbitrary lines.[*]...

---

[*] This follows from a noted principle, announced by F. Enriques in 1904, "Sulla proprietá caratteristica delle superficie algebriche irregolari," *Rendiconto delle sessioni della R. Accademia delle Science dell'Instituto di Bologna*, nuova serie, Vol. 9 (1904–1905), pp. 5–13, and completed later by F. Severi, *Vorlesungen über algebraische Geometrie*: Anhang F, No. 7. Leipzig, Teubner, 1921. ....

(Zariski, 1929, pp. 316–318)

A.3. *Keller's Jacobian Problem (1939)*

Es sind nur die Faelle 6., 7. noch unentschieden, also die Frage, ob Polynome mit der Funktionaldeterminante 1 sich stets durch Polynome umkehren lassen. ... Mir scheint die Frage eine Untersuchung sehr zu lohnen, sie scheint jedoch bereits im ebenen Fall sehr schwierig zu sein.

(Keller, 1939, p. 301).

[It is only cases 6., 7. that remain undecided, that is, the question of whether a polynomial mapping with Jacobian determinant 1 always has a polynomial inverse. ... To me it seems that this is a very worthwhile question, but it seems to be very difficult already in the case of the plane.]

A.4. *Chisini's Statement (1954)*

É naturale che la esperienza abbia indotto i ricercatori ad ammettere (si pure nella sola fase di ricerca) che le trecce dedotte dalla treccia canonica con modifica dell'ordine dei tratti e con fusione di questi (curve che la treccia stessa mostra se irridicibili o non) corrispondano a curve algebriche esistenti, cioè siano trecce davvero algebriche. La dimostrazione del fatto indicato è lo scopo di questo lavoro. ...

(Chisini, 1954, p. 144).

[Naturally, experience has led researchers to hypothesize (if only at this stage of research) that the braids derived from the canonical braid by modification of the order of the strings and by the fusion thereof (curves which the braid itself shows to be irreducible or not) correspond to existing algebraic curves, that is, they are truly algebraic braids. The proof of the indicated fact is the purpose of this work... .]

A.5. *Milnor's Question (1968)*

> Now let us look at the algebraic geometry of our singular point. To any singular point z of a curve $V \subset \mathbb{C}^2$ there is associated an integer $\delta_z > 0$ which intuitively measures the number of double points of V concentrated at z. ...
>
> **Remark 10.9.** It would be nice to have a better topological interpretation of the integer δ. It can be shown that the link $K = V \cap S_\epsilon$ bounds a collection of r smooth 2-cells in the disk D_ϵ having no singularities other than δ ordinary double points. *Question*: Is δ perhaps equal to the "Überschneidungszahl" of K: the smallest number of times which K must be allowed to cross itself during a smooth deformation so as to transform K into a collection of r unlinked and unknotted circles? ...
>
> (Milnor, 1968, pp. 85,92)

A.6. *The Thom Conjecture (c. 1977)*

> **Problem 4.36**. (A) *Conjecture*: The minimal genus of a smooth imbedded surface in CP^2 representing $n \in H_2(CP^2; Z) = Z$ is $(n - 1)(n - 2)/2$.
>
> (Kirby, 1978, p. 309)

According to Kirby (1997), "... when this problem was written in 1977, Thom's name was not generally associated with the conjecture; however, the association developed later although it is not clear whether Thom ever made the conjecture." Indeed, Sullivan (1973) states the conjecture with no name attached. On the other hand, Massey (1973) calls it both "the Thom–Atiyah conjecture" and "the Atiyah–Thom conjecture", a joint attribution which seems to be otherwise unattested in the literature. Both Thom (1982) and Massey (1992) have disclaimed being the first to state this conjecture.

References

Items tagged with Ⓐ (resp., Ⓖ) are freely and permanently available in electronic form at http://arxiv.org/math (resp., http://gdz.sub.uni-goettingen.de/); in the online version of this survey, each such tag is hyperlinked to the relevant electronic resource.

Abhyankar, S.S., *Some remarks on the Jacobian question*, Proc. Ind. Acad. Sci. Math. Sci. **104** (3) (1994), 515–542, with notes by Marius van der Put and William Heinzer, updated by Avinash Sathaye.

Abhyankar, S.S. and Moh, T.T., *Embeddings of the line in the plane*, J. Reine Angew. Math. **276** (1975), 148–166.

Abhyankar, S.S. and Singh, B., *Embeddings of certain curves in the affine plane*, Amer. J. Math. **100** (1) (1978), 99–175.

Abraham, R. and Robbin, J., *Transversal Mappings and Flows*, W.A. Benjamin, New York (1967).

Ⓐ A'Campo, N., *Generic immersions of curves, knots, monodromy and gordian number*, Inst. Hautes Etudes Sci. Publ. Math. **88** (1998), 151–169.

A'Campo, N. and Oka, M., *Geometry of plane curves via Tschirnhausen resolution tower*, Osaka J. Math. **33** (4) (1996), 1003–1033.

Ⓐ Adams, C., *Hyperbolic Knots*, Handbook of Knot Theory, W.W. Menasco and M.B. Thistlethwaite, eds, Elsevier, Amsterdam (2005), 1–18.

Alexander, J.W., *Note on Riemann spaces*, Bull. Amer. Math. Soc. **26** (1920), 370–373.

Ⓐ Anstee, R.P., Przytycki, J.H. and Rolfsen, D., *Knot polynomials and generalized mutation*, Topol. Appl. **32** (3) (1989), 237–249.

Arnol'd, V.I., *Braids of algebraic functions and cohomologies of swallowtails*, Uspehi Mat. Nauk **23** (4 (142)) (1968), 247–248.

Ⓐ Artal, E., Carmona, J. and Cogolludo, J.I., *Effective invariants of braid monodromy and topology of plane curves*, (2001).

Artal-Bartolo, E., *Une démonstration géométrique du théorème d'Abhyankar-Moh*, J. Reine Angew. Math. **464** (1995), 97–108.

Baader, S., *A note on Vassiliev invariants of quasipositive knots*, preprint (December 2004).

Bartolo, E.A. and Cassou-Noguès, P., *Polynôme d'Alexander à l'infini d'un polynôme à deux variables*, Rev. Mat. Complut. **13** (2) (2000), 267–285.

Bennequin, D., *Entrelacements et équations de Pfaff*, Third Schnep-fenried Geometry Conference, Vol. 1, Society of Mathematics, France, Paris (1983), (Schnepfenried, 1982), 87–161.

Bennequin, D., L'instanton gordien (d'après P.B. Kronheimer et T.S. Mrowka). Astérisque 216 (Exp. No. 770, 5), 233–277, Séminaire Bourbaki, Vol. 1992/93, (1993).

Ⓐ Bentalha, A., *Solution of the Quasipositivity Problem in the Braid Group*, (2004), preprint.

Ⓐ Birman, J., Ko, K.H. and Lee, S.J., *A new approach to the word and conjugacy problems in the braid groups*, Adv. Math. **139** (2) (1998), 322–353.

Birman, J.S., *Braids, Links, and Mapping Class Groups*, Annals of Mathematics Studies, Vol. 82, Princeton University Press, Princeton, NJ (1975).

Ⓐ Birman, J.S. and Brendle, T., *Braids: A Survey*, Handbook of Knot Theory, W.W. Menasco and M.B. Thistlethwaite, eds, Elsevier, Amsterdam (2005), 19–103.

Bliss, G.A., *Algebraic Functions*, AMS Colloquium Publications, Vol. XVII, American Mathematical Society, New York (1933), published with the cooperation of the National Research Council.

Bodin, A., *Milnor fibration and fibred links at infinity*, Int. Math. Res. Notices **11** (1999), 615–621.

Boileau, M. and Fourrier, L., *Knot theory and plane algebraic curves*, Chaos Solitons Fractals **9** (4–5) (1998), 779–792.

Boileau, M. and Orevkov, S.Y., *Quasipositivité d'une courbe analytique dans une boule pseudo-convexe*, C.R. Acad. Sci. Paris **332** (2001), 825–830.

Boileau, M. and Rudolph, L., *$\mathbb{C}^3$-Algebraic Stein Fillings via Branched Covers and Plumbing,* (1995), preprint.

Boileau, M. and Weber, C., *Le problème de J. Milnor sur le nombre gordien des nceuds algébriques*, Knots, Braids and Singularities (Plans-sur-Bex, 1982), Enseignement Math., Geneva (1983), 49–98.

Brauner, K., *Zur Geometric der Funktionen zweier komplexer Veränderlicher. II: Das Verhalten der Funktionen in der Umgebung ihrer Verzweigungsstellen. III: Klassifikation der Singularitäten algebroider Kur-ven. IV: Die Verzweigungsgruppen*, Abhandlungen Hamburg **6** (1928), 1–55.

Burau, W., *Kennzeichnung der Schlauchverkettungen*, Abhandlungen Hamburg **10** (1934), 285–297.

Cartan, H., *Quotient d'un espace analytique par un groupe d'automorphismes*, Algebraic Geometry and Topology: A Symposium in Honor of S. Lefschetz, No. 12 in Princeton Mathematical Series, R.H. Fox, D.C. Spencer and A.W. Tucker, eds, Princeton University Press, Princeton, NJ (1957), 90–102.

Chang, H.C., *On two classical facts by Zariski and van Kampen*, Chinese J. Math. **7** (2) (1979), 153–161.

Chang, H.-C., *On equisingularity, analytical irreducibility and embedding line theorem*, Chinese J. Math. **19** (4) (1991), 379–389.

Chang, H.C. and Wang, L.C., *An intersection-theoretical proof of the embedding line theorem*, J. Algebra **161** (2) (1993), 467–479.

Cheniot, D., *Une démonstration du théorème de Zariski sur les sections hyperplanes d'une hypersurface projective et du théorème de Van Kampen sur le groupe fondamental du complémentaire d'une courbe projective plane*, Compositio Math. **27** (1973), 141–158.

Chisini, O., *Una suggestiva rappresentazione reale per le curve algebriche piane*, 1st. Lombardo, Rend., II. Ser. **66** (1933), 1141–1155.

Chisini, O., *Sulla costruzione a priori delle trecce caratteristiche*, Ann. Mat. Pura Appl. **33** (4) (1952), 353–366.

Chisini, O., *Il teorema di esistenza delle trecce algebriche. I, II*, Atti Accad. Naz. Lincei. Rend. Cl. Sci. Fis. Mat. Nat. **8** (17) (1954), 143–149, see also 307–311.

Chisini, O., *Il teorema d'esistenza delle trecce algebriche. III*, Atti Accad. Naz. Lincei. Rend. Cl. Sci. Fis. Mat. Nat. **18** (8) (1955), 8–13.

Chmutov, S., *Diagrams of divide links*, Proc. Amer. Math. Soc. **131** (5) (2003), 1623–1627, electronic.

Ⓐ Cimasoni, D., *The Alexander module of links at infinity*, Int. Math. Res. Not. **20** (2004), 1023–1036.

Cochran, T.D. and Gompf, R.E., *Applications of Donaldson's theorems to classical knot concordance, homology 3-spheres and property P*, Topology **27** (1988), 495–512.

Ⓐ Cohen, D.C. and Suciu, A.I., *The braid monodromy of plane algebraic curves and hyperplane arrangements*, Comment. Math. Helv. **72** (2) (1997), 285–315.

Conway, J.H., *An Enumeration of Knots and Links, and Some of their Algebraic Properties*, Computational Problems in Abstract Algebra (Proceedings of Conference, Oxford, 1967), Pergamon, Oxford (1970), 329–358.

Cordovil, R. and Fachada, J.L., *Braid monodromy groups of wiring diagrams*, Boll. Un. Mat. Ital. B (7) **9** (2) (1995), 399–416.

Couture, O. and Perron, B., *Representative braids for links associated to plane immersed curves*, J. Knot Theory Ramificat. **9** (1) (2000), 1–30.

Crick, F.H.C., *Linking numbers and nucleosomes*, Proc. Nat. Acad. Sci. USA **73** (8) (1976), 2639–2643.

Crowell, R.H. and Fox, R.H., *Introduction to Knot Theory*, Graduate Texts in Mathematics, No. 57, Springer, New York (1977), reprint of the 1963 original.

Dedò, M., *Algebra delle trecce caratteristiche: relazioni fondamentali e loro applicazione*, Ist. Lombardo Sci. Lett. Rend. Cl. Sci. Mat. Nat. (3) **14** (83) (1950), 227–258.

Deligne, P., *Le groupe fondamental du complément d'une courbe plane n'ayant que des points doubles ordinaires est abélien*, Bourbaki Seminar, Vol. 1979/80, d'après W. Fulton, ed., Springer, Berlin (1981), 1–10.

Descartes, R., *Géométrie*, Dover Publications, New York (1637), facsimile of original edition, with a translation by David Eugene Smith and Marcia L. Latham (1954).

Ⓐ Dethloff, G., Orevkov, S. and Zaïdenberg, M., *Plane curves with a big fundamental group of the complement*, Voronezh Winter Mathematical Schools, Amer. Math. Soc., Providence, RI (1998), 63–84.

Dung, N.V., *The fundamental group of the complement of complexified real arrangements*, Ann. Sci. Math. Québec **18** (2) (1994), 157–167.

Dũng, N.V., *On the fundamental group of the complement of arrangements*, Kodai Math. J. **17** (3) (1994), 428–431, workshop on Geometry and Topology (Hanoi, 1993).

Dung, N.V., *Braid monodromy of complex line arrangements*, Kodai Math. J. **22** (1) (1999), 46–55.

Dung, N.V., *A model for homotopy type of the complement*, Acta Math. Vietnam **27** (3) (2002), 289–295, dedicated to the memory of Le Van Thiem (Hanoi, 1998).

Dung, N.V. and Hà, H.V., *The fundamental group of complex hyperplanes arrangements*, Acta Math. Vietnam. **20** (1) (1995), 31–41.

Ⓖ Durfee, A. and Kauffman, L., *Periodicity of branched cyclic covers*, Math. Ann. **218** (2) (1975), 157–174.

Ⓐ Durfee, A.H., *Singularities*, History of Topology, I.M. James, ed., North-Holland, Amsterdam (1999), 417–434.

Eisenbud, D. and Neumann, W., *Three-Dimensional Link Theory and Invariants of Plane Curve Singularities*, Annals of Mathematics Studies, Vol. 110, Princeton University Press, Princeton, NJ (1985).

Eliashberg, Y., *Filling by Holomorphic Discs and its Applications, Geometry of Low-Dimensional Manifolds, 2 (Durham, 1989)*, London Mathematical Society Lecture Note Series, Vol. 151, Cambridge University Press, Cambridge (1990), 45–67.

Eliashberg, Y., *Legendrian and Transversal Knots in Tight Contact 3-Manifolds*, Topological Methods in Modern Mathematics (Stony Brook, NY, 1991), Publish or Perish, Houston, TX (1993), 171–193.

Enriques, F., *Sulla Costruzione Delle Funzioni Algebriche Possedenti Una Data Curva Di Diramazione*, Ann. Mat. **1** (1923), 185–198.

Epple, M., *Branch points of algebraic functions and the beginnings of modern knot theory*, Historia Math. **22** (4) (1995), 371–401.

Erlandsson, T., *Geometry of contact transformations in dimension three*, PhD thesis, Uppsala (1981).

Ⓐ Etnyre, J., *Legendrian and Transversal Knots*, Handbook of Knot Theory, W.W. Menasco and M.B. Thistlethwaite, eds, Elsevier, Amsterdam (2005), 105–185.

Fadell, E., *Homotopy groups of configuration spaces and the string problem of Dirac*, Duke Math. J. **29** (1962), 231–242.

Ⓖ Fox, R. and Neuwirth, L., *The braid groups*, Math. Scand. **10** (1962), 119–126.

Fox, R.H., *Covering spaces with singularities*, Algebraic Geometry and Topology: A Symposium in Honor of S. Lefschetz, No.12 in Princeton Mathematical Series, R.H. Fox, D.C. Spencer and A.W. Tucker, eds, Princeton University Press, Princeton, NJ (1957), 243–257.

Fox, R.H., *A quick trip through knot theory*, Topology of 3-Manifolds and Related Topics (Proc. Univ. Georgia Institute, 1961), Prentice-Hall, Englewood Cliffs, NJ (1962), 120–167.

Fox, R.H., review of Murasugi (1960), Math. Rev. **25**.

Ⓖ Frankl, P. and Pontrjagin, L., *Ein Knotensatz mit Anwendung auf die Dimensionstheorie*, Mathe. Annalen **102** (1930), 785–789.

Franks, J. and Williams, R.F., *Braids arid the Jones-Conway polynomial*, Trans. Amer. Math. Soc. **303** (1987), 97–108.

Freyd, P., Yetter, D., Hoste, J., Lickorish, W.B.R., Millett, K. and Ocneanu, A., *A new polynomial invariant of knots and links*, Bull. Amer. Math. Soc. (N.S.) **12** (2) (1985), 239–246.

Fuchs, D.B., *Cohomology of the braid group mod 2*, Funkcional Anal. i Priložen. (Functional Analysis Appl.) **4** (1970), 62–73, see also pp. 143–151.

Fulton, W., *On the fundamental group of the complement of a node curve*, Ann. Math. (2) **111** (2) (1980), 407–409.

Gabai, D., *Foliations and the topology of 3-manifolds*, J. Diff. Geom. **18** (3) (1983a), 445–503.

Gabai, D., *The Murasugi Sum is a Natural Geometric Operation*, Low-Dimensional Topology (San Francisco, California, 1981), American Mathematical Society, Providence, RI (1983b), 131–143.

Gabai, D., *The Murasugi sum is a natural geometric operation. II*, Combinatorial Methods in Topology and Algebraic Geometry (Rochester, NY, 1982), American Mathematical Society, Providence, RI (1985), 93–100.

Gabai, D., *Genera of the arborescent links*, Mem. Amer. Math. Soc. **59** (339) (1986), i–viii, see also 1–98.

Garber, D., *Plane curves and their fundamental groups: generalizations of Uludag's construction*, Algebr. Geom. Topol. **3** (2003), 593–622.

Gardner, M., *Martin Gardner's New Mathematical Diversions from Scientific American*, Simon and Schuster, New York (1966).

Gibson, W., *Oriented divides and plane curve braids*, J. Knot Theory Ramificat. **11** (6) (2002), 973–1016, knots 2000 Korea, Vol. 3 (Yongpyong).

Gibson, W. and Ishikawa, M., *Links and Gordian numbers associated with generic immersions of intervals*, Topol. Appl. **123** (3) (2002a), 609–636.

Gibson, W. and Ishikawa, M., *Links of oriented divides and fibrations in link exteriors*, Osaka J. Math. **39** (3) (2002b), 681–703.

Giroux, E., *Géométrie de contact: de la dimension trois vers les dimensions supérieures*, Proceedings of the International Congress of Mathematicians, Vol. II (Beijing, 2002), Higher Education Press, Beijing (2002), 405–414.

Goda, H., Hirasawa, M. and Yamada, Y., *Lissajous curves as A'Campo divides, torus knots and their fiber surfaces*, Tokyo J. Math. **25** (2) (2002), 485–491.

Ⓐ Gompf, R.E., *Handlebody construction of Stein surfaces*, Ann. Math. (2) **148** (2) (1998), 619–693.

Ⓖ Gromov, M., *Pseudoholomorphic curves in symplectic manifolds*, Invent. Math. **82** (2) (1985), 307–347.

Grundberg, J., Hansson, T.H., Karlhede, A. and Lindström, U., *Spin, statistics and linked loops*, Phys. Lett. B **218** (3) (1989), 321–325.

Gunning, R.C. and Rossi, H., *Analytic Functions of Several Complex Variables*, Prentice-Hall, Englewood Cliffs, NJ (1965).

Gurjar, R.V. and Miyanishi, M., *On contractible curves in the complex affine plane*, Tôhoku Math. J. (2) **48** (3) (1996), 459–469.

Gwoździewicz, J. and Ploski, A., *Formulae for the singularities at infinity of plane algebraic curves*, Univ. Iagel. Acta Math. **39** (39) (2001), 109–133, effective methods in algebraic and analytic geometry, 2000 (Kraków).

Gwoździewicz, J. and Płoski, A., *On the singularities at infinity of plane algebraic curves*, Rocky Mountain J. Math. **32** (1) (2002), 139–148.

Hansen, V.L., *Polynomial Covering Maps*, Braids (Santa Cruz, CA, 1986), Contemporary Mathematics, Vol. 78, American Mathematical Society, Providence, RI (1988), 229–243.

Ⓖ Harris, J., *On the Severi problem*, Invent. Math. **84** (3) (1986), 445–461.

Hass, J., *The geometry of the slice-ribbon problem*, Math. Proc. Cambridge Philos. Soc. **94** (1) (1983), 101–108.

Heegard, P., *Forstudier til en topologisk Teori for de algebraiske Fladers Sammenhaeng*, PhD thesis, København (1898).

Ⓖ Hilbert, D., *Mathematische Probleme. Nachrichten von der Gesellschaft der Wissenschaften zu Göttingen*, Mathematisch-Physikalische Klasse **3** (1900), 253–297.

Hilbert, D., *Mathematical problems*, Bull. Amer. Math. Soc. **8** (4) (1902), 437–479.

Hirasawa, M., *Links of divides are strongly quasipositive*, Paper presented at the Research-in-Pairs-Workshop on 3-Manifolds and Singularities, Oberwolfach, November 21, (2000).

Hirasawa, M., *Visualization of A'Campo's fibered links and unknotting operation*, Topol. Appl. **121** (2002), 287–304, Proceedings of the First Joint Japan–Mexico Meeting in Topology (Morelia, 1999).

Ⓐ Hirasawa, M. and Rudolph, L., *Constructions of Morse Maps for Knots and Links, and Upper Bounds on the Morse–Novikov Number*, (2003).

Hirasawa, M. and Stoimenow, A., *Examples of knots without minimal string Bennequin surfaces*, Asian J. Math. **7** (2003), 435–446.

Hongler, C.V.Q. and Weber, C., *The link of an extrovert divide*, Ann. Fac. Sci. Toulouse Math. (6) **9** (1) (2000), 133–145.

Hoste, J., *The Enumeration and Classification of Knots and Links*, Handbook of Knot Theory, W.W. Menasco and M.B. Thistlethwaite, eds, Elsevier, Amsterdam (2005), 209–232.

Howie, J., *On the asphericity of ribbon disc complements*, Trans. Amer. Math. Soc. **289** (1) (1985), 281–302.

Ⓖ Hurwitz, A., *Über Riemannsche Flächen mit Gegebenen Verzweignungs–Punkten*, Math. Annalen **39** (1891), 1–61.

Ishikawa, M., *The $\mathbb{Z}_2$-Coefficient Kauffman State Model on Divides*, (2001), preprint.

Ishikawa, M., *Plumbing constructions of connected divides and the Milnor fibers of plane curve singularities*, Indag. Math. (N.S.) **13** (4) (2002), 499–514.

Kaliman, S., *Fundamental group of the complement of affine plane curves*, Topology' 90 (Columbus, OH, 1990), Ohio State Univ. Math. Res. Inst. Publ., Vol. 1, de Gruyter, Berlin (1992), 207–214.

Kaliman, S., *On the Jacobian conjecture*, Proc. Amer. Math. Soc. **117** (1) (1993), 45–51.

Kang, M.C., *On Abhyankar-Moh's epimorphism theorem*, Amer. J. Math. **113** (3) (1991), 399–421.

Ⓐ Kauffman, L.H., *Knot Diagrammatics*, Handbook of Knot Theory, W.W. Menasco and M.B. Thistlethwaite, eds, Elsevier, Amsterdam (2005), 233–318.

Kauffman, L.H., *On Knots*, Annals of Mathematics Studies, Vol. 115, Princeton University Press, Princeton, NJ (1987).

Kauffman, L.H. and Neumann, W.D., *Products of knots, branched fibrations and sums of singularities*, Topology **16** (4) (1977), 369–393.

Kawamura, T., *Quasipositivity of links of divides and free divides*, Topol. Appl. **125** (1) (2002), 111–123.

Kawamura, T., *Links associated with generic immersions of graphs*, Algebr. Geom. Topol. **4** (1) (2004), 571–594.

Keller, O.-H., *Ganze Cremona-Transformationen*, Monatsh. Math. Phys. **47** (1939), 299–306.

Kharlamov, V. and Kulikov, V., *Diffeomorphisms, isotopies, and braid monodromy factorizations of plane cuspidal curves*, C.R. Acad. Sci. Paris Sér. I Math. **333** (9) (2001), 855–859.

Kirby, R., *Problems in low dimensional manifold theory*, Algebraic and Geometric Topology (Proceedings of the Symposium on Pure Mathematics, Stanford University, Stanford, California, 1976), Part 2, American Mathematical Society, Providence, RI (1978), 273–312.

Kirby, R., *Problems in low-dimensional topology*, Geometric Topology (Athens, GA, 1993), AMS/IP Study on Advance Mathematics, Vol. 2, American Mathematical Society, Providence, RI (1997), 35–473.

Kronheimer, P.B. and Mrowka, T.S., *Gauge theory for embedded surfaces. I*, Topology **32** (4) (1993), 773–826.

Kronheimer, P.B. and Mrowka, T.S., *The genus of embedded surfaces in the projective plane*, Math. Res. Lett. **1** (6) (1994), 797–808.

Kuhn, N., *A conjectural inequality on the slice genus of links,* PhD thesis, Princeton University, (1984).

Kulikov, V.S., *On fundamental groups of complements to toric curves*, Izv. Ross. Akad. Nauk Ser. Mat. **61** (1) (1997), 89–112.

Kulikov, V.S. and Kharlamov, V.M., *On braid monodromy factorizations*, Izv. Ross. Akad. Nauk Ser. Mat. **67** (3) (2003), 79–118.

Lê Dũng Tráng, *Three Lectures on Local Monodromy*, Matematisk Institut, Aarhus Universitet, Aarhus, (1974), lectures 1974, Lecture Notes Series, No. 43.

Lê Dũng Tráng, *Plane Curve Singularities and Carousels*, Proceedings of the International Conference in Honor of Frédéric Pham (Nice, 2002), Vol. 53, (2003), 1117–1139.

Lefschetz, S., *L'analysis Situs et la Géométrie Algébrique*, Gauthier-Villars, Paris (1924).

Libgober, A., *On the homotopy type of the complement to plane algebraic curves*, J. Reine Angew. Math. **367** (1986), 103–114.

Lickorish, W.B.R., *A relationship between link polynomials*, Math. Proc. Cambridge Philos. Soc. **100** (1) (1986), 109–112.

Litherland, R.A., *Signatures of Iterated Torus Knots*, Topology of Low-Dimensional Manifolds (Proceedings of the Second Sussex Conference, Chelwood Gate, 1977), Springer, Berlin (1979), 71–84.

Ⓐ Livingston, C., *A Survey of Classical Knot Concordance*, Handbook of Knot Theory, W.W. Menasco and M.B. Thistlethwaite, eds, Elsevier, Amsterdam (2005), 319–347.

Loeser, F. and Vaquié, M., *Le polynôme d'Alexander d'une courbe plane projective*, Topology **29** (2) (1990), 163–173.

Ⓐ Loi, A. and Piergallini, R., *Compact Stein surfaces with boundary as branched covers of B^4*, Invent. Math. **143** (2001), 325–348.

Lyon, H.C., *Torus knots in the complements of links and surfaces*, Mich. Math. J. **27** (1) (1980), 39–46.

Magnus, W., *Braid groups: a survey*, Proceedings of the Second International Conference on the Theory of Groups (Australian National University, Canberra, 1973), Lecture Notes in Mathematics, Vol. 372, Springer, Berlin (1974), 463–487.

Magnus, W., *Review of Birman (1975)*, Bull. Amer. Math. Soc. **82** (1976), 42–45.

Magnus, W., Karrass, A. and Solitar, D., *Combinatorial Group Theory: Presentations of Groups in Terms of Generators and Relations*, Dover Publications, New York (1976), revised edition.

Massey, W.S., *Surfaces in 4-dimensional manifolds*, Symposia Mathematica, Vol. XI (Convegno di Geometria, INDAM, Rome, 1972), Academic Press, London (1973), 257–270, text of a lecture presented on May 22, 1972.

Massey, W.S., Personal communication, (1992).

Milnor, J., *Singular points of complex hypersurfaces*, Annals of Mathematics Studies, Vol. 61, Princeton University Press, Princeton, NJ (1968).

Moishezon, B., *Stable Branch Curves and Braid Monodromies*, Algebraic Geometry (Chicago, Ill., 1980), Springer, Berlin (1981), 107–192.

Moishezon, B., *Algebraic surfaces and the arithmetic of braids. I*, Arithmetic and Geometry, Vol. II, Birkhäuser, Boston, MA (1983), 199–269.

Moishezon, B., *Algebraic surfaces and the arithmetic of braids. II*, Combinatorial Methods in Topology and Algebraic Geometry (Rochester, NY, 1982), American Mathematical Society, Providence, RI (1985), 311–344.

Moishezon, B., *The arithmetic of braids and a statement of Chisini*, Geometric Topology (Haifa, 1992), Contemporary Mathematics, Vol. 164, American Mathematical Society, Providence, RI (1994), 151–175.

Moishezon, B. and Teicher, M., *Braid group technique in complex geometry. I. Line arrangements in CP^2*, Braids, (Santa Cruz, CA, 1986), Contemporary Mathematics, Vol. 78, American Mathematical Society, Providence, RI (1988), 425–555.

Moishezon, B. and Teicher, M., *Braid group technique in complex geometry. II. From arrangements of lines and conics to cuspidal curves*, Algebraic Geometry (Chicago, IL, 1989), Lecture Notes in Mathematics, Vol. 1479, Springer, Berlin (1991), 131–180.

Moishezon, B. and Teicher, M., *Braid group techniques in complex geometry. III. Projective degeneration of V_3*, Classification of Algebraic Varieties (L'Aquila, 1992), Contemporary Mathematics, Vol. 162, American Mathematical Society, Providence, RI (1994a), 313–332.

Moishezon, B. and Teicher, M., *Braid group techniques in complex geometry. IV, Braid monodromy of the branch curve S_3 of $V_3 \to CP^2$ and Application to $\pi_1(CP^2 - S_3, *)$*, Classification of Algebraic Varieties (L'Aquila, 1992), Contemporary Mathematics, Vol. 162, American Mathematical Society, Providence, RI (1994b), 333–358.

Moishezon, B. and Teicher, M., *Braid group technique in complex geometry. V. The fundamental group of a complement of a branch curve of a Veronese generic projection*, Comm. Anal. Geom. **4** (1–2) (1996), 1–120.

Morgan, J.W., Szabó, Z. and Taubes, C.H., *A product formula for the Seiberg–Witten invariants and the generalized Thom conjecture*, J. Diff. Geom. **44** (1996), 706–788.

Morton, H.R., *Seifert circles and knot polynomials*, Math. Proc. Cambridge Philos. Soc. **99** (1) (1986), 107–109.

Ⓖ Murakami, H., *Some metrics on classical knots*, Math. Ann. **270** (1) (1985), 35–45.

Murasugi, K., *On alternating knots*, Osaka Math. J. **12** (1960), 277–303.

Murasugi, K., *On a certain subgroup of the group of an alternating link*, Amer. J. Math. **85** (1963), 544–550.

Ⓐ Naik, S. and Stanford, T., *A move on diagrams that generates S-equivalence of knots*, J. Knot Theory Ramificat. **12** (5) (2003), 717–724.

Nakamura, T., *Positive knots*, Master's thesis, Keio University, February 1998.

Nakamura, T., *Four-genus and unknotting number of positive knots and links*, Osaka J. Math. **37** (2) (2000), 441–451.

Nakamura, T., *Notes on Braidzel Surfaces for Links*, (2004), preprint.

Napolitano, F., *Topology of complements of strata of the discriminant of polynomials*, C.R. Acad. Sci. Paris Sér. I Math. **327** (7) (1998), 665–670.

Napolitano, F., *Topologie du compléments du discriminant des espaces d'applications*, PhD thesis, Universié Paris IX Dauphine Ceremade (2000).

Narasimhan, R., *Imbedding of holomorphically complete complex spaces*, Amer. J. Math. **82** (1960), 917–934.

Némethi, A., *On the Seifert form at infinity associated with polynomial maps*, J. Math. Soc. Jpn **51** (1) (1999), 63–70.

Ⓖ Neumann, W. and Rudolph, L., *Unfoldings in knot theory*, Math. Ann. **278** (1–4) (1987), 409–439; Corrigendum, Math. Ann. **282** (1988), 349–351.

Ⓖ Neumann, W.D., *Complex algebraic plane curves via their links at infinity*, Invent. Math. **98** (3) (1989), 445–489.

Neumann, W.D., *Irregular links at infinity of complex affine plane curves*, Quart. J. Math. Oxford Ser. (2) **50** (199) (1999), 301–320.

Neumann, W.D., *Topology of Hypersurface Singularities*, (2001), http://math.columbia.edu/~neumann/preprints/kaehlerl.ps (to appear in a collection of works of Kaehler).

Ⓐ Neumann, W.D. and Norbury, P., *Unfolding polynomial maps at infinity*, Math. Ann. **318** (1) (2000), 149–180.

Ⓐ Neumann, W.D. and Norbury, P., *The Orevkov invariant of an affine plane curve*, Trans. Amer. Math. Soc. **355** (2) (2003), 519–538, electronic.

Newman, M.H.A., *On a string problem of Dirac*, J. Lond. Math. Soc. **17** (1942), 173–177.

Nori, M.V., *Zariski's conjecture and related problems*, Ann. Sci. École Norm. Sup. (4) **16** (2) (1983), 305–344.

Ⓖ Oka, M., *The monodromy of a curve with ordinary double points*, Invent. Math. **27** (1974), 157–164.

Oka, M., *On the fundamental group of the complement of a reducible curve in P^2*, J. Lond. Math. Soc. (2) **12** (2) (1975/1976), 239–252.

Oka, M. and Sakamoto, K., *Product theorem of the fundamental group of a reducible curve*, J. Math. Soc. Jpn **30** (4) (1978), 599–602.

Orevkov, S.Y., *The fundamental group of the complement of a plane algebraic curve*, Mat. Sb. (N.S.) **137** (179)(2) (1988), 260–270, see also p. 272.

Orevkov, S.Y., *The commutant of the fundamental group of the complement of a plane algebraic curve*, Uspekhi Mat. Nauk **45** (1(271)) (1990a), 183–184.

Orevkov, S.Y., *An example in connection with the Jacobian conjecture*, Mat. Zametki **47** (1) (1990b), 127–136, see also p. 173.

Orevkov, S.Y., *Letter to the editors: "An example in connection with the Jacobian conjecture"*, Mat. Zametki **49** (6) (1991), 148.

Orevkov, S.Y., *Rudolph diagrams and analytic realization of the Vitushkin covering*, Mat. Zametki **60** (2) (1996), 206–224, see also p. 319.

Orevkov, S.Y., *Realizability of a braid monodromy by an algebraic function in a disk*, C.R. Acad. Sci. Paris Sér. I Math. **326** (7) (1998), 867–871.

Orevkov, S.Y., *Link theory and oval arrangements of real algebraic curves*, Topology **38** (4) (1999), 779–810.

Orevkov, S.Y., *Markov moves for quasipositive braids*, C.R. Acad. Sci. Paris Sér. I Math. **331** (7) (2000), 557–562.

Orevkov, S.Y., *Complex Orientations of M-Curves of Degree 7*, Topology, Ergodic Theory, Real Algebraic Geometry, American Mathematical Society Translations Series 2, Vol. 202, American Mathematical Society, Providence, RI (2001a), 215–227.

Orevkov, S.Y., *Quasipositivity test via unitary representations of braid groups and its applications to real algebraic curves*, J. Knot Theory Ramificat. **10** (7) (2001b), 1005–1023.

Orevkov, S.Y., *Quasipositivity problem for 3-braids*, Turk. J. Math. **28** (1) (2004), 89–93.

Orevkov, S.Y. and Polotovskiĭ, G.M., *Projective M-cubics and M-quartics in general position with a maximally intersecting pair of ovals*, Algebra I Analiz **11** (5) (1999), 166–184.

Orevkov, S.Y. and Shustin, E.I., *Flexible, algebraically unrealizable curves: rehabilitation of Hilbert–Rohn–Gudkov approach*, J. Reine Angew. Math. **551** (2002), 145–172.

Osserman, R., *The total curvature of algebraic surfaces*, Contributions to analysis and geometry (Baltimore, MD, 1980), Johns Hopkins University Press, Baltimore, MD (1981), 249–257.

Pajitnov, A., Weber, C. and Rudolph, L., *The Morse–Novikov number for knots and links*, Algebra i Analiz **13** (3) (2001), 105–118.

Păunescu, L. and Zaharia, A., *Remarks on the Milnor fibration at infinity*, Manuscripta Math. **103** (3) (2000), 351–361.

Pichon, A., *Germes Analytiques Réels et Décompositions En Livre Ouvert de la Sphère de Dimension Trois*, talk at Université de Genéve, Séminaire singulier, atelier des 26 et 27 fèvrier, (2003).

Przytycki, J.H. and Traczyk, P., *Invariants of links of Conway type*, Kobe J. Math. **4** (2) (1988), 115–139.

Randell, R., *On the fundamental group of the complement of a singular plane curve*, Quart. J. Math. Oxford Ser. (2) **31** (121) (1980), 71–79.

Ⓐ Rasmussen, J.A., *Khovanov Homology and the Slice Genus*, (2004).

Reeve, J.E., *A summary of results in the topological classification of plane algebroid singularities*, Univ. e Politec. Torino. Rend. Sem. Mat. **14** (1954/1955), 159–187.

Reidemeister, K., *Knoten und Gruppen*, Abhandlungen Hamburg **5** (1926), 7–23.

Reshetikhin, N.Y. and Turaev, V.G., *Ribbon graphs and their invariants derived from quantum groups*, Comm. Math. Phys. **127** (1) (1990), 1–26.

Rudolph, L., *Embeddings of the line in the plane*, J. Reine Angew. Math. **337** (1982a), 113–118.

Rudolph, L., *Nontrivial positive braids have positive signature*, Topology **21** (3) (1982b), 325–327.

Rudolph, L., *Algebraic functions and closed braids*, Topology **22** (2) (1983a), 191–202.

Ⓖ Rudolph, L., *Braided surfaces and Seifert ribbons for closed braids*, Comment. Math. Helv. **58** (1) (1983b), 1–37.

Rudolph, L., *Constructions of Quasipositive Knots and Links. I.*, Knots, Braids and Singularities (Plans-sur-Bex, 1982), University of Genéve, Geneva (1983c), 233–245.

Ⓐ Rudolph, L., *Some knot theory of complex plane curves*, Enseignement Mathématique **29** (1983d), 185–208.

Ⓖ Rudolph, L., *Some topologically locally-flat surfaces in the complex projective plane*, Comment. Math. Helv. **59** (4) (1984), 592–599.

Rudolph, L., *Plane curves in fancy balls*, L'Enseignement Mathématique **31** (1985a), 81–84.

Rudolph, L., *Special positions for surfaces bounded by closed braids*, Rev. Mat. Iberoamericana **1** (3) (1985b), 93–133.

Rudolph, L., *Mutually Braided Open Books and New Invariants of Fibered Links*, Braids (Santa Cruz, CA, 1986), American Mathematical Society, Providence, RI (1988), 657–673.

Rudolph, L., *A congruence between link polynomials*, Math. Proc. Cambridge Philos. Soc. **107** (2) (1990), 319–327.

Ⓐ Rudolph, L., *Quasipositive annuli (Constructions of quasipositive knots and links, IV)*, J. Knot Theory Ramificat. **1** (4) (1992a), 451–466.

Rudolph, L., *Totally tangential links of intersection of complex plane curves with round spheres*, Topology '90 (Columbus, OH, 1990), de Gruyter, Berlin (1992b), 343–349.

Ⓐ Rudolph, L., *Quasipositivity as an obstruction to sliceness*, Bull. Amer. Math. Soc. (N.S.) **29** (1) (1993), 51–59.

Ⓖ Rudolph, L., *An obstruction to sliceness via contact geometry and "classical" gauge theory*, Invent. Math. **119** (1) (1995), 155–163.

Rudolph, L., *The slice genus and the Thurston–Bennequin invariant of a knot*, Proc. Amer. Math. Soc. **125** (10) (1997), 3049–3050, http://www.ams.org/proc/1997-125-10/S0002-9939-97-04258-5/.

Rudolph, L., *Quasipositive plumbing (Constructions of quasipositive knots and links. V)*, Proc. Amer. Math. Soc. **126** (1) (1998), 257–267, http://www.ams.org/proc/1998-126-01/S0002-9939-98-04407-4/.

Ⓐ Rudolph, L., *Positive Links are Strongly Quasipositive*, Geometry and Topology Monographs 2, Proceedings of the Kirbyfest, J. Hass and M. Scharlemann, eds, (1999), 555–562.

Ⓐ Rudolph, L., *Hopf plumbing, baskets, arborescent Seifert surfaces, espaliers, and homogeneous braids*, Topol. Appl. **116** (2001a), 255–277.

Ⓐ Rudolph, L., *Quasipositive pretzels*, Topol. Appl. **115** (2001b), 115–123.

Ⓐ Rudolph, L., *Some fibered and non-fibered links at infinity of hyperbolic complex line arrangements*, Topol. Appl. **118** (2002), 187–197.

Sakuma, M., *Minimal genus Seifert surfaces for special arborescent links*, Osaka J. Math. **31** (4) (1994), 861–905.

Ⓐ Scharlemann, M., *Thin Position in the Theory of Classical Knots*, Handbook of Knot Theory, W.W. Menasco and M.B. Thistlethwaite, eds, Elsevier, Amsterdam (2005), 429–459.

Segre, B., *Corrispondenze di Möbius e trasformazioni cremoniane intere*, Atti Accad. Sci. Torino. Cl. Sci. Fis. Mat. Nat. **91** (1956/1957), 3–19.

Ⓖ Seifert, H., *Über das Geschlecht von Knoten*, Math. Annalen **110** (1934), 571–592.

Seifert, H., *Die Verschlingungsinvarianten der zyklischen Knotenüberlagerungen*, Abh. Math. Sem. Univ. Hamburg **11** (1935), 84–101.

Shibuya, T., *Some relations among various numerical invariants for links*, Osaka J. Math. **11** (1974), 313–322.

Siebenmann, L., Exercices sur les nœuds rationels, mimeographed notes, Orsay, (1975).

Stallings, J.R., *Constructions of Fibred Knots and Links*, Algebraic and Geometric Topology (Proceedings of the Symposium Pure Mathematics, Stanford University Stanford, California 1976), Part 2, American Mathematical Society, Providence, RI (1978), 55–60.

Sullivan, D., Problems. Mimeographed notes (M.I.T.), final and complete list of all the problems presented and discussed in the summer problem session at M.I.T, June 1973.

Suzuki, M., *Propriétés Topologiques Des Polynômes de Deux Variables Complexes, et Automorphismes Algébriques de L'espace* $\mathbb{C}^2$, J. Math. Soc. Jpn **26** (1974), 241–257.

Tanaka, T., *Unknotting numbers of quasipositive knots*, Topol. Appl. **88** (3) (1998), 239–246.

Thom, R. Personal communication (November 19), (1982).

Tibiletti, C., *Complementi all'algebra delle trecce caratteristiche e loro applicazione*, Ist. Lombardo Sci. Lett. Rend. Cl. Sci. Mat. Nat. (3) **16** (85) (1952), 249–259.

Tibiletti, C.M., *La irregolaritá di un piano multiplo dedotta dalla treccia diramante*, Atti Accad. Naz. Lincei. Rend. Cl. Sci. Fis. Mat. Nat. **18** (8) (1955a), 480–486.

Tibiletti, C.M., *Trecce algebriche di curve di diramazione: costuzioni ed applicazioni*, Rend. Sem. Mat. Univ. Padova **24** (1955b), 183–214.

Tibiletti, C.M., *Trecce relative a forme canoniche del gruppo di monodromia*, Ist. Lombardo Sci. Lett. Rend. Cl. Sci. Mat. Nat. (3) **19** (88) (1955c), 25–40.

Tibiletti, C.M., *La irregolarità di un piano multiplo dedotta dalla treccia diramante*, Atti Accad. Naz. Lincei. Rend. Cl. Sci. Fis. Mat. Nat. **18** (8) (1955d), 480–486.

Tietze, H., *Über die topologischen Invarianten mehrdimensionaler Mannigfaltigkeiten*, Monatshefte für Math. und Physik **19** (1908), 1–118.

Tristram, A.G., *Some cobordism invariants for links*, Proc. Cambridge Philos. Soc. **66** (1969), 251–264.

Trotter, H.F., *Homology of group systems with applications to knot theory*, Ann. Math. (2) **76** (1962), 464–498.

Vaĭnšteĭn, F.V., *The cohomology of braid groups*, Funktsional. Anal. i Prilozhen. (Functional Analysis Application) **12** (1978), 72–73, see also 135–137.

van den Essen, A., *Polynomial automorphisms and the Jacobian conjecture*, Algèbre non commutative, groupes quantiques et invariants (Reims, 1995), Sémin. Congr. Soc. Math. Vol. 2, France, Paris (1997), 55–81, http://smf.emath.fr/Publications/SeminairesCongres/1997/1.html.

van Kampen, E.R., *On the fundamental group of an algebraic curve*, Am. J. Math **55** (1933), 255–260.

Viéte, F., De æquationum Recognitione et Emendatione Tractatus Duo, Kent State University Press, Kent, OH, included in The Analytic Art: Nine studies in algebra, geometry and trigonometry from the *Opus restitutae mathematicae analyseos, seu algebrâ novâ* (1646), translated from the Latin and with an introduction by T. Richard Witmer (1983), (1593).

Ⓐ Weeks, J., *Computation of Hyperbolic Structures in Knot Theory*, Handbook of Knot Theory, W.W. Menasco and M.B. Thistlethwaite, eds, Elsevier, Amsterdam (2005), 461–480.

Ⓖ Wendt, H., *Die gordische Auflösung von Knoten*, Math. Z. **42** (1937), 680–696.

Whitney, H., *Elementary structure of real algebraic varieties*, Ann. of Math. (2) **66** (1957), 545–556.

Whitney, H., *Complex Analytic Varieties*, Addison-Wesley, Reading, MA (1972).

Wirtinger, W., *Über die Verzweigungen bei Funktionen von zwei Veränderlichen*, Jahresbericht d. Deutschen Mathematiker Vereinigung **14** (1905), 517.

Yamamoto, M., *Amida-diagrams and Seifert matrices of positive iterated torus knots*, Proc. Jpn Acad. Ser. A Math. Sci. **54** (8) (1978), 256–261.

Ⓐ Yetter, D.N., *Quandles and monodromy*, J. Knot Theory Ramificat. **12** (2003), 523–541.

Zariski, O., *On the problem of existence of algebraic functions of two variables possessing a given branch curve*, Am. J. Math **51** (1929), 305–328.

Zariski, O., *On the Poincaré.group, of rational plane curves*, Amer. J. Math. **58** (1936), 607–619.

Zariski, O., *Algebraic Surfaces*, Springer, New York (1971), (supplemented edition) with appendices by S.S. Abhyankar, J. Lipman, and D. Mumford, Ergebnisse der Mathematik und ihrer Grenzgebiete, Band 61.

Zaĭdenberg, M.G. and Lin, V.Y., *An irreducible, simply connected algebraic curve in C^2 is equivalent to a quasihomogeneous curve*, Dokl. Akad. Nauk SSSR **271** (5) (1983), 1048–1052.

Żołądek, H., *A new topological proof of the Abhyankar–Moh theorem*, Math. Zeitschrift **244** (2003), 689–695.

Index of Definitions and Notations

Pages references are to first, or defining, instances of terms and symbols.

CHAPTER 9

Thin Position in the Theory of Classical Knots

Martin Scharlemann*

Department of Mathematics, South Hall, University of California, Santa Barbara, CA 93106, USA
E-mail: mgscharl@math.ucsb.edu

Contents

Abstract

This is a survey of the role that the notion of "thin position" has played in the theory of classical knots and, more generally, to the understanding of knotted graphs in 3-space. The extension of thin position to graphs, beyond being of interest in its own right, also is shown to have applications in knot theory.

* Research supported in part by an NSF grant.

HANDBOOK OF KNOT THEORY
Edited by William Menasco and Morwen Thistlethwaite

For our purposes, "knot theory" will have the narrowest interpretation: the study of isotopy classes of locally flat embeddings of the circle S^1 in S^3. The distinctions between smooth, PL, and locally flat topological embeddings are usually unimportant in these dimensions so, for convenience, we will use language that is typically associated with the smooth category. For example, a knot K is a smooth submanifold of S^3 diffeomorphic to S^1. The focus of this article will be on a particular technique for analyzing knots, called "thin position". Much like the study of crossing diagrams, it is a technique that exploits very heavily the fact that the ambient manifold is S^3 and not another manifold, not even a possibly alternate homotopy sphere. The roots of the technique might be traced back to Alexander's proof [1] of the Schönfliess theorem, in which he imagines a horizontal plane sweeping across a sphere embedded in 3-space and examines the intersection set during the sweep-out. The modern use began with a stunning application by David Gabai, in his proof that knots in S^3 satisfy Property R [2].

This is mostly an expository survey; one bit of new mathematics is an updated application of thin position to the proof that Heegaard splittings of S^3 are standard (cf. Corollary 5.13).

1. From crossing number to bridge number

One way, perhaps historically the first way, of thinking about knots in S^3 is this: choose a point p in S^3. A knot in S^3 is generically disjoint from p, as is an isotopy between knots. Thus knot theory in S^3 (narrowly defined, as above) is equivalent to knot theory in $\mathbb{R}^3 \cong S^3 - \{p\}$. Once we think of a knot as lying in $\mathbb{R}^3$, it is natural to imagine projecting it to $\mathbb{R}^2$; this is what we do when we draw the knot. Such a generic projection will look like an immersed closed curve in the plane having only double points. If we keep track, at the double points, of which strand in the original knot passes over the other we have the classical description of a knot via its crossing diagram. Crossing diagrams are one of the oldest and most naive ways of trying to classify knots, but their importance has been re-emphasized by the modern discovery of new knot invariants, invariants that are most easily described via these projections. Unlike more sophisticated invariants (coming, e.g, from the algebraic topology of the knot complement) it is clear that knot projections make immediate use of the fact that the ambient space is the 3-sphere, and not some other manifold.

The first knot invariant that is suggested by knot projection is the crossing number of the knot. For a generic projection of the knot to the plane, the crossing number of the projection is just the number of double points of the projection. An isotopy of the knot may reduce this number; the crossing number of the knot is defined as the minimum number that can be achieved via an isotopy of the knot in 3-space. The crossing number is the most natural invariant for cataloguing knots via their projections (it is the basis of the standard knot tables) but otherwise it is not a particularly good invariant. For a given knot, while it is easy to find an upper bound for the crossing number (just use any projection), there is no natural way to find the projection that minimizes the crossing number. In particular, the behavior of the invariant under, e.g., knot sum is not well understood.

There is another invariant, much like the crossing number and only slightly more difficult to describe, that is in fact remarkably well behaved under knot sum. Suppose one starts with a knot projection, and a point on the knot, and starts moving along the knot, recording at each crossing whether one is on the upper strand (an overcrossing, say marked with a $+$) or on the lower strand (an undercrossing, say marked with a $-$). Continue in this way around the entire knot and examine the result. It is a sequence of signs, e.g., $+, +, -, +, -, -$. The number of signs recorded is twice the number of crossings of the projection since in a trip around the knot, one passes through each crossing twice. For the same reason there will be as many $+$ as $-$ signs. Now, instead of considering the number of crossings of the knot (i.e., focusing on the number of $+$ and $-$ signs) consider the number of times the sign recorded *changes*. If the last sign differs from the first, record that also as a change, as if one were viewing the pattern of $+$ and $-$ on a circle. Thus for the sequence $+, +, -, +, -, -$ the number of changes is four. Of course the number of changes is necessarily even. Half that number (i.e., the number of strings of consecutive $+$s, say) is called the *bridge number* of the knot projection. Just as with the crossing number, the bridge number of the projection may change as the knot is isotoped; the minimum that can be achieved by an isotopy of the knot is called the bridge number of the knot. For reasons that will be apparent, the bridge number of the unknot, which has no crossings, is conventionally set to be 1. This invariant was introduced by Schubert [18].

The terminology "bridge number" is meant to evoke the following picture: if a knot projection has bridge number n, that means there are $2n$ sign changes from $+$ to $-$ or vice versa; equivalently, each $+$ sign lies in one of n strings of consecutive $+$ signs. Each string of consecutive $+$ signs corresponds to a strand of the knot which can be thought of as lying just above the plane. Between these strands are strands with only $-$ crossings; these can be thought of as lying on the plane. Thus the bridge number $\beta(K)$ of the knot K is the minimum number of bridges that one would need to erect on the plane so that the entire knot K could be put on the plane and on the bridges, with K crossing over each bridge exactly once. Schubert [18] proved that this invariant is essentially additive. To be precise, $\beta - 1$ is additive, i.e.,

$$\beta(K_1 \# K_2) - 1 = (\beta(K_1) - 1) + (\beta(K_2) - 1)$$

or

$$\beta(K_1 \# K_2) = \beta(K_1) + \beta(K_2) - 1$$

A modern proof is given in [19].

Schubert's remarkable result suggests that there should be a more natural way of viewing bridge number. Consider the description of the knot just given: lying mostly on the plane, but with certain sections of it elevated above the plane on bridges. The original perspective on the knot is the bird's-eye view in which we look down on the knot from above. Imagine instead the perspective of someone standing on the plane, looking sideways at the knot and thinking about the height of the knot above the plane. In other words, instead of the projection $\mathbb{R}^3 \to \mathbb{R}^2$ that described the original knot projection, consider instead the projection $h: \mathbb{R}^3 \to \mathbb{R}$ defined by height above the plane. Then $h|K$ will achieve a single maximum on each bridge, and each pair of successive maxima is separated by a minimum corresponding to a strand of K that lies on the plane. In other words, if K is put on n bridges, then $h|K$ has n maxima. This number of maxima can be preserved even when K is moved slightly to make it generic with respect to the height function by, e.g., putting a bit of a dip into the level strands lying on the plane. In other words, we have the following:

PROPOSITION 1.1. *Suppose $K \subset \mathbb{R}^3$ has a projection with bridge number n. Then K may be isotoped so that the standard height function $h: \mathbb{R}^3 \to \mathbb{R}$ restricts to a Morse function on K with n maxima and n minima.*

Conversely, we have

PROPOSITION 1.2. *Suppose $K \subset \mathbb{R}^3$, and the standard height function $h: \mathbb{R}^3 \to \mathbb{R}$ restricts to a Morse function on K with n maxima and n minima. Then $\beta(K) \leq n$.*

PROOF. Each point on K at which $h|K$ has a local maximum can be pushed even higher by an isotopy of K, say along a path that rises from the maximum and, by general position, misses the rest of K as it rises. Similarly, each point at which $h|K$ has a local minimum can be pushed lower. So with no loss, K may be isotoped so that all the maxima occur near the same height (say 1) and similarly all the minima occur near the same height (say -1). Consider how the plane $P = h^{-1}(0)$ divides K : P cuts K into $2n$ strands, n above the plane, each containing a single maximum, and n below the plane, each having a single minimum.

The plane $P_{\varepsilon-1}$ slightly above height -1 cuts off n tiny strands of K, one for each minimum, that lie below the plane. These are clearly isotopic rel their endpoints to n disjoint arcs in $P_{\varepsilon-1}$. See Figure 1. Between the heights $\varepsilon - 1$ and 0 there are no critical points of h on K. It follows that in fact all n strands of K lying below P can be simultaneously isotoped rel their ends to lie on P. Similarly, all strands of K lying above P can be isotoped rel their end points so that they consist of level arcs in a plane P_ε just above P, together with vertical arcs at their end points between P and P_ε. When viewed from above as projected onto P, K then has a projection with bridge number no higher than n. □

Combining the two propositions above gives a more natural definition of bridge number of a knot K : take the number of maxima that the standard height function h has on the knot (in general position with respect to h) and minimize that number via an isotopy of K. The result is the bridge number $\beta(K)$ of K. If there is a horizontal plane P with the property that all maxima of $h|K$ lie above P and all minima lie below it, then K is said to be in *bridge position* with respect to h; the plane P is called a *dividing plane* for K. Any knot K can be isotoped so that it is in bridge position, with $\beta(K)$ maxima and $\beta(K)$ minima.

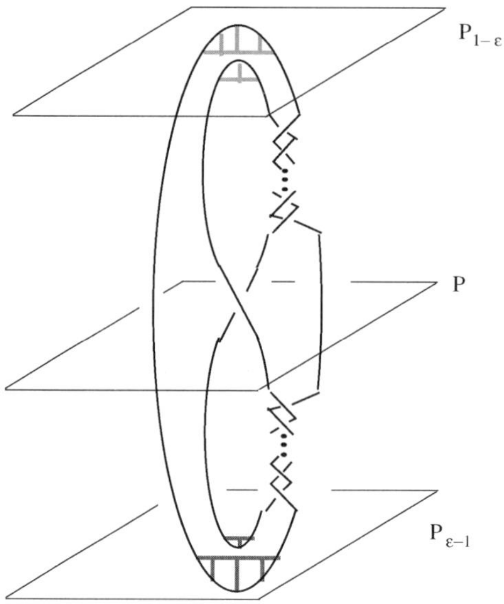

Fig. 1. Level planes for a 2-bridge knot.

2. From bridge number to width

Consider the critical values of $K \subset \mathbb{R}^3$ in general position with respect to $h : \mathbb{R}^3 \to \mathbb{R}$, the standard height function. As noted above, it is always possible to isotope K so as to raise the height of a maximum or to lower the height of a minimum, without affecting the height of any other critical point. Indeed there is such an isotopy whose support on K is limited to a neighborhood of the critical point whose height is changed. Similarly, there is no difficulty lowering the height of a maximum through an interval that contains no other critical values, since by standard Morse theory, the preimage of an interval without critical values is a simple product. Combining these two observations, it is easy to isotope K to interchange the heights of critical points whose critical values are adjacent, so long as both critical points are maxima, or both are minima.

The only difficulty in rearranging the heights of critical points is the interchange of two adjacent critical values in which the higher value is a maximum and the lower value is a minimum. Informally, this can be described as moving a maximum down past a minimum. Such a move may or may not be possible, depending on the structure of the knot (cf. Figure 2). It is reasonable then to think of this move (pushing a maximum down past a

Fig. 2. Getting stuck.

minimum) as simplifying the picture of the knot, when it can be done; the point of thin position is to formally capture this idea in a useful way.

Let $K \subset \mathbb{R}^3$ be a knot in general position with respect to the standard height function $h: \mathbb{R}^3 \to \mathbb{R}$. That is, $h|K$ is a Morse function for which no two critical points have the same critical value. For each $t \in R$ let P_t denote the plane $h^{-1}(t)$. If t is a regular value for $h|K$ then K crosses P_t transversally, necessarily in an even number $w(t) \in \mathbb{N}$ of points. The number $w(t)$ changes only at critical values, where it increases by two at each minimum and decreases by two at each maximum. So if $c_0 < c_1 < \cdots < c_n$ are the critical values of $h|K$ and values $r_1, \ldots, r_n$ are chosen so that $c_{i-1} < r_i < c_i$, $i = 1, \ldots, n$, then the function $w(t)$ is determined by the sequence $w(r_i)$, $i = 1, \ldots, n$. That sequence is unchanged by pushing a maximum down past another maximum (or a minimum past another minimum) but is affected by pushing a maximum down past a minimum. In the last case, if the critical values are c_{i-1}, c_i, then the reordering changes $w(r_i)$ to $w(r_i) - 4$ and has no other effect. See Figure 3. More dramatically, if K can be isotoped so that the maximum and its adjacent minimum cancel, then both c_{i-1} and c_i disappear and so both $w(r_i), w(r_{i+1})$ disappear from the sequence.

These changes are so straightforward, they suggest the following definition:

DEFINITION 2.1. Suppose, as above, $K \subset \mathbb{R}^3$ is in general position with respect to the standard height function h, $c_0 < c_1 < \cdots < c_n$ are the critical values of $h|K$ and regular values $r_i \in R$ are chosen so that $c_{i-1} < r_i < c_i$, $i = 1, \ldots, n$. The *width of K with respect to h*, denoted by $w(K, h)$, is $\sum_i w(r_i)$. The *width of K*, denoted by $w(K)$, is the minimum of $w(K', h)$ over all knots K' isotopic to K. We say that K is in *thin position* if $w(K, h) = w(K)$.

There is a small technical advantage in modifying this definition slightly so that it can be applied directly and a bit more usefully in the compact situation $K \subset S^3$. Define the standard height function $h: S^3 \to \mathbb{R}$ as the composition $S^3 \subset \mathbb{R}^4 \to \mathbb{R}$ of the inclusion and the standard projection onto the last factor of $\mathbb{R}^4$. Then h has two critical points on S^3 (called the north and south poles). A knot K in general position with respect to h will be disjoint from these poles, and the above definition can be restated for this height function. The most significant difference with this definition is that (for $-1 < t < 1$), $h^{-1}(t)$ is now a 2-sphere P_t instead of a plane; if $K \subset S^3$ is in bridge position, then what divides the maxima from the minima is a *dividing sphere*

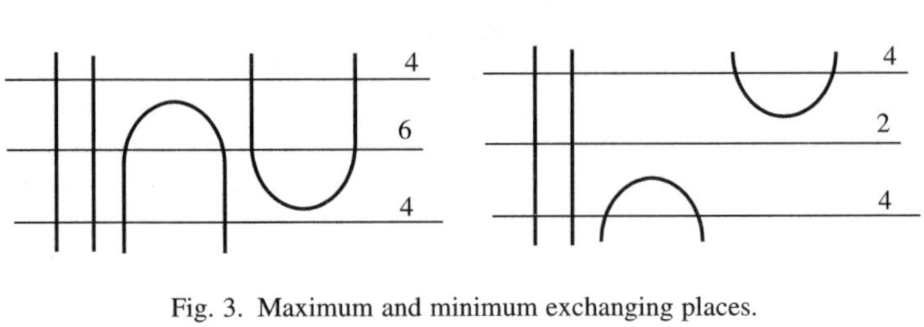

Fig. 3. Maximum and minimum exchanging places.

instead of a dividing plane. Whenever t is a regular value for $h|K$, we continue to denote $|P_t \cap K|$ by $w(t)$.

Note that if $K \subset \mathbb{R}^3$ or $K \subset S^3$ has been isotoped to be in thin position with respect to the standard height function h, it is impossible to have a limited isotopy that simply pushes a maximum below a minimum or cancels a maximum with a minimum, since either move decreases the width by at least 4.

3. Application: thinning the unknot

To show the power of this idea, we begin with a simple exercise that illustrates how thin position interacts with geometric properties of a knot, in particular with an essential surface in the knot complement.

It will be useful to have the following notation and definitions:

NOTATION. For M a manifold and $X \subset M$ a polyhedron, let $\eta(X)$ denote a closed regular neighborhood, whereas (abusing notation slightly) $M - \eta(X)$ will mean the closed complement of $\eta(X)$ in M.

DEFINITION 3.1. Let $K \subset S^3$ be a knot, $P \subset S^3$ be a sphere that is level with respect to the standard height function and is transverse to K. Let B_{u} and B_{l} denote the balls which are the closures of the region above P and below P, respectively. An *upper disk* (resp. *lower disk*) with respect to P is a disk $D \subset S^3 - \eta(K)$ transverse to P such that $\partial D = \alpha \cup \beta$, where β is an arc imbedded on $\partial\eta(K)$, parallel to a subarc of K, α is an arc properly imbedded in $P - \eta(K)$, $\partial\alpha = \partial\beta$ and a small product neighborhood of α in D lies in B_{u} (resp. B_{l}), i.e., it lies *above* (resp. *below*) P.

Note that int(D) may intersect P in simple closed curves or indeed in other arcs. An innermost simple closed curve cuts off a disk that lies either entirely above or below P. Such a disk is called, respectively, an *upper cap* or *lower cap*.

Natural upper disks with interiors disjoint from P arise, e.g., in the case where the arcs $K \cap B_{\mathrm{u}}$ each have exactly one maximum, so in particular the collection of arcs is the untangle. To see the upper disks, consider what happens as a descending level sphere P_t sweeps across a maximum. Join the descending arcs of K from that maximum by an arc in P_t and continue to carry that arc down all the way to P. The result is a disjoint collection of upper disks, one for each component of $K \cap B_{\mathrm{u}}$; each intersects a level sphere in at most one arc. These are called a family of *descending disks* for $K \cap B_{\mathrm{u}}$. Choices are involved in this construction: each time a new maximum is encountered as P_t sweeps down, a choice is made about how existing arcs from earlier descending disks lie in relation to this maximum. In fact, if one allows isotopies that raise and lower maxima (but never introduce minima) one has the general observation, whose proof is mostly left as an exercise.

LEMMA 3.2. *Suppose P is a level sphere for $K \subset S^3$ and each component of $K \cap B_{\mathrm{u}}$ has a single maximum. Suppose Δ is a collection of disjoint upper disks contained* *entirely in* B_u. *Then* Δ *can be isotoped rel the arcs* $\Delta \cap P$ *so that it becomes part of a complete collection of descending disks. Moreover, such an isotopy can be found so that during the isotopy no new critical points of* $h|K$ *are introduced.*

A proof hint is this: start with any complete collection of descending disks Δ' and alter Δ and Δ' to reduce $|\Delta' \cap \Delta|$, i.e., the number of components in $\Delta' \cap \Delta$.

PROPOSITION 3.3. *Suppose* $K \subset S^3$ *is the unknot, in bridge position with respect to the standard height function* $h : S^3 \to \mathbb{R}$. *There is a dividing sphere* P *for* K *so that a maximum and a minimum can simultaneously be isotoped to lie on* P. *During the generic isotopy, the width remains unchanged.*

PROOF. Since K is the unknot, it bounds a disk D. By a small isotopy of D we can arrange that D is in general position with respect to h and near each critical point of $h|K$, $h|D$ has a half-center singularity. (See Figure 9, applied there in another context.) That is, near each maximum of K, D is incident to K from below and near each minimum, D is incident to K from above. In particular, for a level sphere P just below a maximum (resp. just above a minimum) one of the components of $D - P$ is an upper disk (resp. lower disk) contained entirely above (resp. below) P.

Let $t_l < t_u$ be the heights of, respectively, the highest minimum and the lowest maximum of $h|K$ and for each $t_l < t < t_u$, let P_t, as above, denote the level sphere $h^{-1}(t)$. We have just seen that for t slightly less than t_u, $D - P_t$ contains an upper disk among its components and similarly, for t slightly greater than t_l, $D - P_t$ contains a lower disk among its components.

CLAIM. *There is a value of* t, $t_l < t < t_u$ *for which* P_t *admits disjoint upper and lower disks* D_u, D_l *so that no component of* $int(D_u) \cap P$ *or* $int(D_l) \cap P$ *is an arc.*

PROOF OF CLAIM. For any generic value of t, $t_l < t < t_u$ (i.e., a value of t for which P_t is transverse to D), consider an outermost arc of $P_t \cap D$ in D. The disk it cuts off from D is either an upper disk or a lower disk; moreover the subarc of K incident to the disk lies either entirely above or entirely below P_t. Hence for each generic value there is either an upper or a lower disk as desired.

Now imagine t ascending from t_l up to t_u. Since near t_l an outermost arc of P_t in D cuts off a lower disk and near t_u one cuts off an upper disk, and at any generic t in between, one or the other is cut off, there are two possibilities. One is that there is a generic value for which outermost arcs of $P \cap D$ in D cut off both an upper and a (disjoint) lower disk from D; then we are done with proving the claim. The second possibility is that there is a critical value t_0, $t_l < t_0 < t_u$ for $h|D$, whose critical point is necessarily an interior point of D, so that for small ε, outermost arcs of $P_{t_0+\varepsilon}$ and $P_{t_0-\varepsilon}$ in D cut off, respectively, an upper and a lower disk. In this case, thicken D slightly and let $D_\pm$ be the boundary disks of the thickened region. In other words, $D_\pm$ are two copies of D, very near to D but on opposite sides. Then P_{t_0} is transverse to both D_+ and D_- and outermost arcs of P_{t_0} in D_+ and D_- will cut off (disjoint) upper and lower disks, completing the proof of the claim.

In the special case in which both upper and lower disks have interiors entirely disjoint from P, Lemma 3.2 applies, and K may be isotoped rel $K \cap P$, never changing the width, so that afterwards the upper disk is a descending disk and, dually, the lower disk is an ascending disk. These disks then define the isotopy of a maximum and minimum to disjoint arcs in P, as required to complete the proof of the proposition.

If the interiors of the upper disk D_u or the lower disk D_l are not disjoint from P, the argument is only moderately more complicated. In that case, each component of intersection is a closed curve, and such an innermost closed curve on D_u or D_l cuts off a disk that is an upper or lower cap. Suppose, e.g., that there is an upper cap C. A standard innermost disk, outermost arc argument will alter a complete collection of descending disks for $K \cap B_u$ to a collection of upper disks disjoint from C. Via Lemma 3.2 there is an isotopy of $K \cap B_u$ rel P which does not increase width so that afterwards, these disks are descending disks, possibly now again intersecting C, but only in their interiors. Then alter C, via an innermost disk argument, isotoping C so that afterwards the set of descending disks is disjoint from C. This establishes that, after an isotopy of K and C with support away from P and never increasing the width, there is a complete collection of descending disks for $K \cap B_u$ that is disjoint from C.

Suppose then that there are disjoint upper and lower caps C_u and C_l. Modify $K \cap B_u$, $K \cap B_l$ away from P so that all descending disks (resp. all ascending disks) are disjoint from C_u (resp. C_l). ∂C_u and ∂C_l bound disjoint disks E_u, E_l in the sphere P. Pick a component of $K \cap B_u$ incident to E_u and a component of $K \cap B_l$ incident to E_l. A descending disk for the former will intersect P inside of P_u and an ascending disk for the latter will intersect P inside of P_l. In particular, they will be disjoint, and so they can be used to isotope arcs to P as required. A similar argument applies if there is an upper cap and a lower disk whose interior is disjoint from P, or symmetrically.

The only remaining case to consider is when there are no upper caps, say, but the interior of the upper disk D_u intersects P, so there are lower caps. We will show that in this case, one of the cases we have already considered also applies.

Let Δ be a complete collection of descending disks for $K \cap B_u$. We argue by induction on $|D_u \cap \Delta|$ that there are both an upper disk (or an upper cap) and a lower cap so that their boundaries are disjoint. If $|D_u \cap \Delta| = 0$ then each component of $D_u \cap B_u$ lies in the ball $B_u - \Delta$. After compressing in $B_u - \Delta$ each component becomes a disk. Since a neighborhood of ∂D_u lies in B_u, it follows that after the compressions, ∂D_u bounds a disk in $B_u - \Delta$, hence in $B_u - K$, as required.

So suppose $D_u \cap \Delta \neq \emptyset$. A simple innermost disk argument could eliminate a closed curve of intersection, so we can take all components of intersection to be arcs. Surprisingly, we may also assume that the lower cap is a slight push-off of a disk component of $D_u \cap B_l$. Indeed, consider an innermost disk of $D_u - P$. If it lies in B_u then it is an upper cap disjoint from the lower cap and we are done. If it lies in B_l then we may as well take a slight push-off as our lower cap.

This surprising fact means that an outermost arc of $D_u \cap \Delta$ in Δ can be used to ∂-compress $D_u \cap B_u$ to an arc that is disjoint from the lower cap. This boundary compression defines an isotopy on the interior of D_u that reduces $|D_u \cap \Delta|$ without disturbing the disjoint lower cap. After the isotopy, the result follows by induction. □

Note that if the ends of the maximum and minimum arcs given by Proposition 3.3 both coincide then they constitute all of K and K already was in thin position. Otherwise, the minimum and maximum can be pushed on past each other, or just cancelled if the arcs have a single end in common, reducing the width. Thus we have:

COROLLARY 3.4. *If the unknot is in bridge position, then either it is in thin position (and so has just a single minimum and maximum) or it may be made thinner via an isotopy that does not raise the width.*

The corollary begs the question: is the hypothesis that the unknot is in bridge position really needed? In other words

QUESTION 3.5. Suppose $K \subset S^3$ is the unknot. Is there an isotopy of K to thin position (i.e., a single minimum and maximum) via an isotopy during which the width is never increasing?

One suspects there are counterexamples, though it would be difficult to prove for such a counterexample that no such isotopy exists.

4. Thick and thin regions

If $K \subset S^3$ is in bridge position, then $w(t)$, $-1 < t < 1$ is constant on intervals that contain no critical values of $h|K$. $w(t)$ always increases by 2 at each critical value $h|K$ that lies below the height of a dividing sphere and then decreases by 2 at each critical value that lies above the height of a dividing sphere. If K is not in bridge position, $w(t)$ will still increase or decrease by 2 at each critical value of $h|K$, but $w(t)$ will have one or more local minima as well as more than one local maximum. For example, if t_0 is a regular value of $h|K$ and the critical values above and below t_0 correspond, respectively, to a minimum and maximum of $h|K$, then $w(t)$ will be greater if t is either increased or decreased past the adjacent critical values of $h|K$. In other words, $w(t_0)$ is a local minimum of $w(t)$; the interval of regular values for h on which it lies is called a thin region (and the corresponding heights the thin levels). The level sphere P_{t_0} is called a thin sphere.

Symmetrically, if t_0 is a regular value of $h|K$ and the critical values above and below correspond, respectively, to maxima and minima of $h|K$, then the interval of regular values for h on which it lies is called a thick region (and the corresponding heights the thick levels). The level sphere P_{t_0} is then called a thick sphere.

To repeat in the notation of Definition 2.1, call the level r_i a *thin level* of K with respect to h if the critical point c_i is a local maximum for h and c_{i+1} is a local minimum for h. Dually r_i is a *thicklevel* of K with respect to h if c_i is a local minimum and c_{i+1} is a local maximum. Many r_i may be neither thin nor thick. Since the lowest critical point of $h|K$ is a minimum and the highest is a maximum, there is one more thick level than thin level.

There is an alternative way, using thin and thick levels, to calculate the width $w(K, h)$ of a knot in S^3. Choose values $r_{i_1}, \ldots, r_{i_k}$ in Definition 2.1 to be those of the thick levels of K and $r_{j_1}, \ldots, r_{j_{k-1}}$ to be those of the thin levels, so $r_{i_l} < r_{j_l} < r_{i_{l+1}}$, $1 \leq l \leq k-1$.

LEMMA 4.1. *Let $a_l = w(r_{i_l})$ and $b_l = w(r_{j_l})$. Then*

$$2w(K) = \sum_{l=1}^{k} a_l^2 - \sum_{l=1}^{k-1} b_l^2$$

PROOF. See the last section of [14] for McCrory's simple proof; indeed a contemplative look at the last figure there should suffice. □

If K is in thin position, thin and thick levels of the height function have important geometric properties. For example, Thompson [20] showed that if thin position for K is not bridge position, so K has thin levels, then there is an essential meridinal planar surface for K. One way of finding such a surface was recently identified by Ying-Qing Wu [22]:

THEOREM 4.2. *Suppose $K \subset S^3$ is in thin position but not in bridge position, so there are thin levels. Suppose $P_{r_{j_l}}$ is the thinnest thin sphere (that is, among all values at thin levels r_{j_i}, $w(r_{j_l})$ is the lowest). Then the planar surface $P_- = P_{r_{j_l}} - \eta(K)$ is essential in $S^3 - \eta(K)$. In other words, P_- is incompressible in $S^3 - \eta(K)$ and is not a boundary parallel annulus.*

A sample application for this result comes from the work of Gordon and Reid [5]. A knot $K \subset S^3$ is said to have *tunnel number one* if there is a properly imbedded arc $\tau \subset S^3 - \eta(K)$ so that $S^3 - (\eta(K) \cup \eta(\tau))$ is a genus two handlebody. Gordon and Reid showed that a tunnel number one knot has no incompressible planar surfaces in its complement. Combining the two, we have Thompson's result [20]:

THEOREM 4.3. *Suppose a tunnel number one knot $K \subset S^3$ is in thin position. Then it is also in minimal bridge position.*

A second feature of thin and thick spheres is that they intersect essential surfaces in the knot complement in a controlled way. For example, suppose F is a Seifert surface for K, i.e., an orientable surface in S^3 whose boundary is K. Suppose F is in general position with respect to the height function $h: S^3 \to \mathbb{R}$. In other words, all the critical points of $h|F$ and $h|\partial F = K$ are non-degenerate and no two occur at the same height. First consider thin levels:

THEOREM 4.4. *Suppose $K \subset S^3$ is in thin position but not in bridge position, so there are thin levels. Suppose F is a Seifert surface for K, in general position with respect to h, and P_r is a thin sphere. Then every arc component of $F \cap P_r$ is essential in F.*

PROOF. The argument is most easily described in the compact manifold $S^3 - \eta(K)$ so let P^- be the planar surface $P_r - \eta(K)$. Since F is a Seifert surface we can assume that

$F \cap \partial\eta(K) = \partial F$ is a longitude, and so ∂F intersects each component of ∂P^- exactly once. In particular, every arc component of $F \cap P^-$ is essential in P^-; indeed such components pair up the components of ∂P^-. Suppose some arc component is inessential (i.e., ∂-parallel) in F; let α be an outermost such component, i.e., a component cutting off a subdisk E of F which contains no other arc component of $F \cap P^-$, though it may contain circle components.

Aside: An experienced 3-manifold topologist would expect first to eliminate these circle components, but in fact we do not know that we can, for although Theorem 4.2 tells us that the thinnest level sphere gives rise to an incompressible surface, we do not know this to be true for an arbitrary thin level sphere and so we cannot automatically eliminate a circle in $F \cap P_-$ just because it bounds a disk in $F - P_-$.

Now ∂E consists of two arc components: $\alpha \subset P_-$ and a subarc β of $\partial F = K$. Hence the disk E can be used to isotope β to α, though note that this isotopy may move β through P_- since the interior of E may have circles of intersection with P_-. Nonetheless, we do know that β lies entirely on one side of P_-, say above P_-. So the effect of moving β to α is at least the effect of moving a maximum (namely a maximum of β) past a minimum (namely the minimum that is the lowest critical value of $h|K$ above height r). In fact, much more may be accomplished, e.g., the elimination of other critical points from $h|\beta$, but the net effect is to lower the width of K. Since we have assumed that K begins in thin position, this is impossible, proving the theorem. □

Note that the fact that the isotopy of β may pass through P_- is an example of why this argument cannot be directly applied to Question 3.5: we can always thin the unknot via an arc by arc isotopy as just described, but we have little control over the width during each of these isotopies.

It is a bit more surprising that there is a version of Theorem 4.4 that also applies at a thick level:

THEOREM 4.5. *Suppose $K \subset S^3$ is in thin position and F is a Seifert surface for K in general position with respect to h. Suppose P_r is a thick sphere for K. Let $c_- < r < c_+$ be critical values of $h|K$ that are adjacent to r, so in particular c_+ is the height of a maximum and c_- is the height of a minimum. Then either K is the unknot or there is a value r', $c_- < r' < c_+$ so that every arc component of $F \cap P_{r'}$ is essential in F.*

PROOF. Consider a level sphere P_+ just below height c_+, so in particular there are no critical values for F between the level of P_+ and c_+. If all arc components of $F \cap P_+$ are essential in F, we are done, so suppose $E \subset F$ is a disk cut off by an outermost inessential arc α, with $\partial E = \alpha \cup \beta$, $\beta \subset K$. Let $\gamma \subset (K - P_+)$ be the interval containing the maximum at height c_+. Suppose first that β lies below P_+, so in particular E is a lower disk. If the ends of β coincide with the ends of γ then E describes an isotopy of β up to P_+; after the isotopy K has a single maximum and minimum and so is the unknot. If a single end of β coincides with a single end of γ then E can be used to isotope β up past c_+, canceling the maximum in γ, as well as one or more critical points in β. This would reduce the width of K, which is impossible. Finally, if the ends of β and γ are disjoint, then E can be used to move β above c_+ (since there are no critical values of $h|F$ between the level of P_+ and c_+) thereby moving a minimum past a maximum, and possibly canceling other critical points on β. Again this would contradict the assumption that K is in thin position. We therefore conclude that in fact β lies above P_+ (indeed perhaps $\beta = \gamma$) so E is an upper disk.

Similarly, a level sphere P_- just above the level of c_- either cuts off a lower disk or we are done. If there is a generic height between $P_\pm$ for which all arcs of intersection with F with the corresponding level sphere are essential in F we are done. On the other hand, if at every generic level there is at least one inessential arc of intersection then there is always a disk cut off from F that is an upper or a lower disk. Then, as in the Proof of Proposition 3.3 (perhaps after thickening F as there we thickened D), there is a level sphere P that cuts off simultaneously an upper disk E_u and a disjoint lower disk E_l, via arcs $\alpha_u, \alpha_l \subset (P \cap F)$. If both ends of α_u and α_l coincide, then K is the unknot. Otherwise, E_u and E_l can be used to isotope arcs of K to P, lowering the width as described in the Proof of Proposition 3.3. Since K was assumed to be thin, this is impossible. □

The original application by Gabai that prompted the definition of thin position is in a similar setting [2]. Gabai's application was in the proof of the Poenaru conjecture: for $K \subset S^3$ there is an essential (e.g., non-separating) planar surface $(Q, \partial Q) \subset (S^3 - \eta(K), \partial\eta(K))$ whose boundary components are all longitudes (i.e., of slope 0) on $\partial\eta(K)$ only if K is the unknot. For technical reasons, Gabai wished to exhibit a planar surface $(P, \partial P) \subset (S^3 - \eta(K), \partial\eta(K))$ transverse to Q for which ∂P is a collection of meridians of $\eta(K)$ and every arc of $P \cap Q$ is essential in both P and Q. Gabai applied essentially the argument above, substituting Q for F. The upshot is a pair of planar surfaces, $P, Q \subset S^3 - \eta(K)$ so that in each surface, viewing the boundary components as large vertices and the intersection arcs as edges connecting the vertices, we have what appears to be a planar graph, with all the rich structure that this implies. (It must be said, though, that this structure is only the starting point of Gabai's deep and complex argument using sutured manifold theory.) This application is an echo, in some sense, of Laudenbach's [10] seminal introduction of graph theory into such arguments. Laudenbach proved the Poenaru conjecture, even for knots in a mere homotopy 3-sphere, in the simplest interesting case: when Q has just three boundary components.

There are two important directions in which Gabai's application generalizes. To formulate the first extension, note first that the Poenaru conjecture implies in particular that if 0-framed surgery on a knot K yields a manifold containing a non-separating sphere, then K itself has genus 0. Gabai generalizes this to show that if 0-framed surgery on K yields a manifold containing a non-separating genus g surface, then the genus of K is no larger than g. The relevant extension of the thin position part of his argument is to replace the punctured sphere Q with an essential punctured genus g surface; P remains a planar surface intersecting Q in arcs that are essential in both surfaces.

A second extension is used in the celebrated proof by Gordon and Luecke [4], that a knot is determined by its complement. In their setting, surgery on K with slope ± 1 hypothetically yields S^3 again, with the core of the solid torus representing a new knot $K' \subset S^3$. They seek to simultaneously find meridinal planar surfaces P and Q in $S^3 - \eta(K)$

and $S^3 - \eta(K')$, respectively, so that P and Q are transverse and all arcs in $P \cap Q$ are essential in both surfaces. Viewed in $S^3 - \eta(K)$, each component of ∂Q has slope the same as the surgery slope. In order to find the pair of surfaces, they must simultaneously use thin position on height functions for $K \subset S^3$ and for $K' \subset S^3$.

It is important to repeat that in both the Gabai argument and the Gordon–Luecke argument, the use of thin position is only one of the many parts of the full proof, and far from the deepest.

To conclude this section, consider again the property of bridge number that first attracted Schubert: its good behavior under summation of knots. If K_1, K_2 are knots in minimal bridge position, there is a natural way to get a bridge positioning of the sum $K_1\#K_2$. Namely, arrange (as one can) that in the bridge positioning of the K_i, the right-most vertical strand of K_1 has no crossings and the left-most vertical strand of K_2 has no crossings. Put K_1 to the left of K_2 and do the connected sum along these vertical strands. The result is a bridge positioning of $K_1\#K_2$ with $\beta(K_1) + \beta(K_2) - 1$ bridges; Schubert's theorem says that this is a minimal bridge positioning. See Figure 4.

There is a similar construction which one might easily conjecture would do the same for width. Put K_1 and K_2 in thin position and position them so that K_1 lies entirely above K_2. Then do the connected sum of the knots via a monotonic band from the lowest minimum of K_1 to the highest maximum of K_2. See Figure 5. Of course we do not immediately know that the result is a minimal width presentation for $K_1\#K_2$, but this simple picture does show:

LEMMA 4.6.

$$w(K_1\#K_2) \leq w(K_1) + w(K_2) - 2$$

With the precedent of Schubert's theorem before us, it is natural to ask

QUESTION 4.7. Is the inequality in Lemma 4.6 ever strict, or is it always true that

$$w(K_1\#K_2) = w(K_1) + w(K_2) - 2?$$

In other words, does the positioning shown in Figure 5 always minimize width?

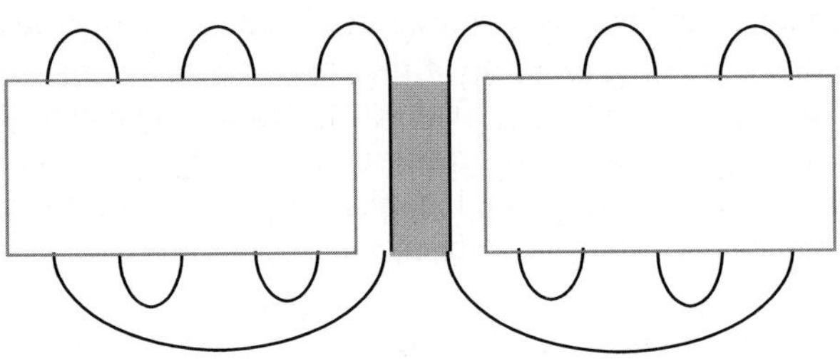

Fig. 4. Adding knots and bridge numbers.

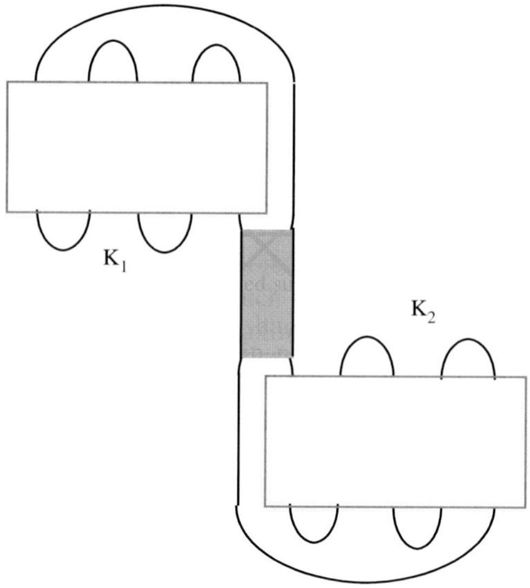

Fig. 5. Adding knots and knot width.

Rieck and Sedgwick [13] show that the answer is yes when neither K_1 nor K_2 has in its exterior an essential meridional planar surface; Wu notes that their result now follows easily from Theorem 4.2. See also [6]. Without the assumption on essential planar surfaces in the knot exteriors, only a little is currently known, e.g.,

$$w(K_1 \# K_2) \geq \max\{w(K_1), w(K_2)\},$$

cf. [14], though there is strong evidence that the answer is no, cf. [17].

5. From knots to graphs

5.1. *Normal form for trivalent graphs*

Thin position techniques outlined above for knots $K \subset S^3$ do not particularly make use of the fact that K is connected (though some of the applications do) and so thin position can be applied to links as well as knots in S^3. On the other hand, some thought is needed if thin position is to be applied to imbedded graphs in S^3. We will restrict our discussion to trivalent graphs; presumably more general graphs can be treated similarly but so far there seems to be no notable application to higher valence graphs.

DEFINITION 5.1. Let Γ be a finite trivalent graph in $S^3 - \{\text{poles}\}$ and let $h : S^3 \to \mathbb{R}$ be the standard height function. Γ is in *normal form* with respect to h if

(1) the critical points of $h|$edges are non-degenerate and each lies in the interior of an edge,

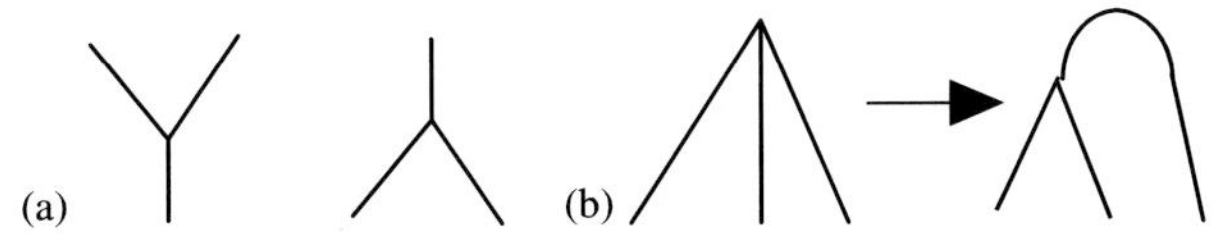

Fig. 6. Vertices in normal form.

(2) the critical points of $h|$edges and the vertices of Γ all occur at different heights and
(3) at each (trivalent) vertex v of Γ either two ends of incident edges lie above v (we say v is a *Y-vertex*) or two ends of incident edges lie below v (we say v is a *λ-vertex*). (See Figure 6a.)

Any $\Gamma \subset S^3$ can be perturbed by a small isotopy to be normal; for example, note that if three edges are incident to the same vertex from below, then a small isotopy moves the end of one edge so that it is incident from above and has a maximum near the vertex. In other words, such a vertex is replaced by an interior maximum adjacent to a λ-vertex. (See Figure 6b.)

Suppose Γ is in normal form with respect to h.

DEFINITION 5.2. The *maxima* of Γ consist of all local maxima of $h|$edges and all λ-vertices. The minima of Γ consist of all local minima of $h|$edges and all Y-vertices.

A maximum (resp. minimum) that is not a λ-vertex (resp. Y-vertex) will be called a *regular* maximum (resp. minimum). The set of all maxima and minima (hence including all the vertices of Γ) is called the set of *criticalpoints* of Γ. The heights of the critical points are called the *critical values* or *critical heights*.

Γ is in *bridge position* if there is a level sphere, called a *dividing sphere*, that lies above all minima of Γ and below all maxima.

There are two sorts of complications introduced when valence 3 vertices are allowed: there is some subtlety in finding an appropriate calculation of width; and surfaces that are properly imbedded in the graph complement may behave in a less orderly fashion. We treat each of these in turn.

5.2. *Width for graphs*

A naive way to define width is to proceed just as in the case of knots or links: For each generic $-1 < t < 1$ let $w(t) = |\Gamma \cap P_t|$, a function that increases by two at a regular minimum of h, and by one at a Y-vertex. Similarly $w(t)$ decreases by two at a regular maximum of h, and by one at a λ-vertex. Now pick generic heights $r_1, \ldots, r_n$, each between a distinct adjacent pair of critical heights, and calculate the sum $\sum_i w(r_i)$. A complication that this definition introduces is this: there are now two types of maxima in Γ, regular maxima and λ-vertices. If one is pushed down past the other, the width changes. In particular, if a maximum of unknown type is pushed down past other maxima and other

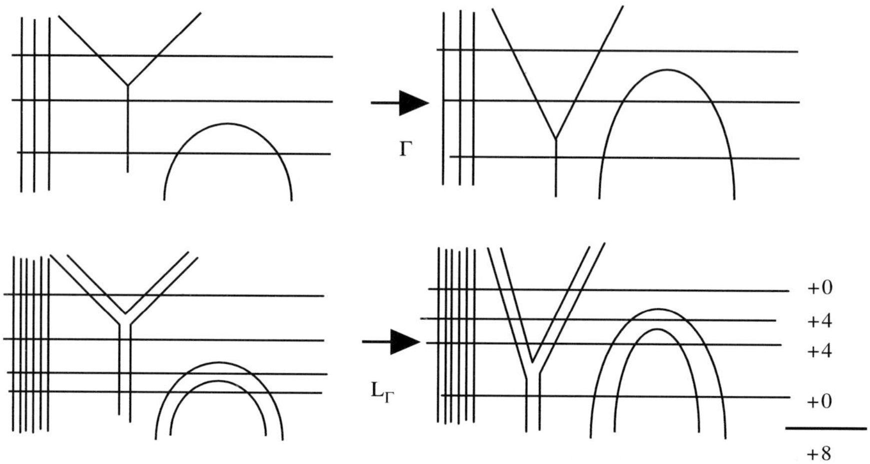

Fig. 7. Vertex and maximum exchanging places.

minima, we do not a priori know that the width is decreased. (Symmetric statements are true, of course, for minima.) Such a definition then would make arguments using upper and lower disks, arguments that worked so successfully in the case of knots, pretty useless for graphs.

A fix for this is to alter the definition slightly, taking into account what the nature of the adjacent critical height is. A motivating thought is this: if we widen the graph to look like a ribbon whose core is the graph, then the boundary of the ribbon is just a standard link L_Γ, with no vertices. Exchanging the heights of two critical points of Γ will rearrange heights of several critical points on L_Γ but in a predictable way. Moving maxima past maxima in Γ will rearrange only maxima of L_Γ (and so have no effect on height); moving a maximum past a minimum in Γ will similarly move maxima past minima in L_Γ. See Figure 7. It is relatively easy to express the width of L_Γ via the number of intersections of Γ with level planes, and we will use that as the definition of the width of Γ. The upshot is a somewhat more complicated definition, but one that automatically has the property we seek: it is indifferent to the exchange in heights of adjacent maxima or adjacent minima, but will go down if a maximum is moved below a minimum.

DEFINITION 5.3. Let $c_0 < \cdots < c_n$ be the successive critical heights of Γ. Let r_i, $1 \leq i \leq n$ be generic levels chosen so that $c_{i-1} < r_i < c_i$ and let P_i denote the critical point at height c_i. For each i, $1 \leq i \leq n$ define ρ_i by:

$$\rho_i = \begin{cases} 2 & \text{when } p_{i-1}, p_i \text{ are both vertices} \\ 3 & \text{when exactly one of } p_{i-1}, p_i \text{ is a vertex} \\ 4 & \text{when } p_{i-1}, p_i \text{ are both regular critical points} \end{cases}$$

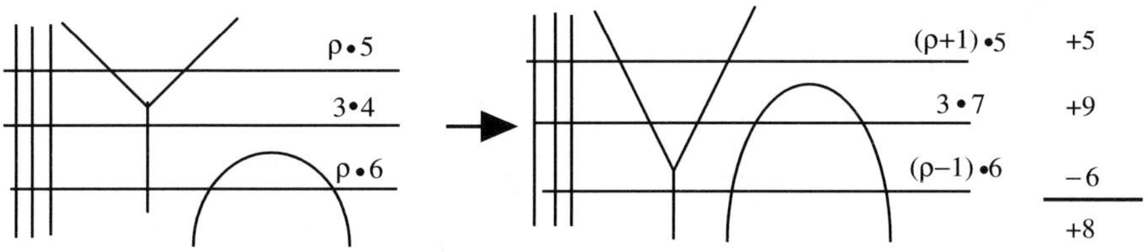

Fig. 8. An alternative calculation.

Define the width of Γ with respect h to be

$$W(\Gamma, h) = \sum_i \rho_i w_{r_i}$$

This is essentially the width introduced in Section 3 of [3], but made symmetric with respect to reflection through a horizontal plane.

DEFINITION 5.4. A graph $\Gamma \subset S^3$ is in thin position (with respect to h) if $W(\Gamma, h)$ cannot be lowered by an isotopy of Γ. In that case, $W(\Gamma, h)$ is denoted $W(\Gamma)$.

REMARK. In practice, the chief property of the width $W(\Gamma, h)$ that we will need is this: the width is decreased if a maximum is pushed below a minimum, but the width is unaffected by pushing one maximum above or below another maximum, or one minimum above or below another minimum. See Figures 7 and 8.

5.3. *Surfaces in graph complements*

Surfaces that lie in a knot complement have the pleasant feature that boundary components are either horizontal (if the boundary component is a meridian) or the height function on the boundary circles roughly follow that of the knot, having maxima where the knot has a maximum and similarly with minima. It is easy to locally isotope the surface so that all singularities of h on its boundary are local half-centers.

In contrast, surfaces in a graph complement may have minima at λ-vertex maxima and maxima at Y-vertex minima. See Figure 9.

A graph $\Gamma \subset S^3$ in normal form with respect to h can be thickened slightly to give a solid handlebody $\eta(\Gamma) \subset S^3$ with the predictable height structure. (e.g., very near any regular maximum of Γ there are two non-degenerate critical points of $h|\partial(\eta(\Gamma))$, one a saddle just below and one a maximum just above.) We will be concerned with simple closed curves on $(\partial\eta(\Gamma)$ and with properly imbedded surfaces in $S^3 - \eta(\Gamma)$.

DEFINITION 5.5. Suppose Γ is a graph, in normal form with respect to h, and $c \subset \partial\eta(\Gamma)$ is a simple closed curve. Then c is in normal form on $\partial\eta(\Gamma)$ if either it is a horizontal meridian circle or each critical point of h on c is non-degenerate, and occurs near an associated critical point of Γ in $\partial\eta(\Gamma)$. Furthermore, the number of critical points of c has been minimized via isotopy of c in $\partial\eta(\Gamma)$.

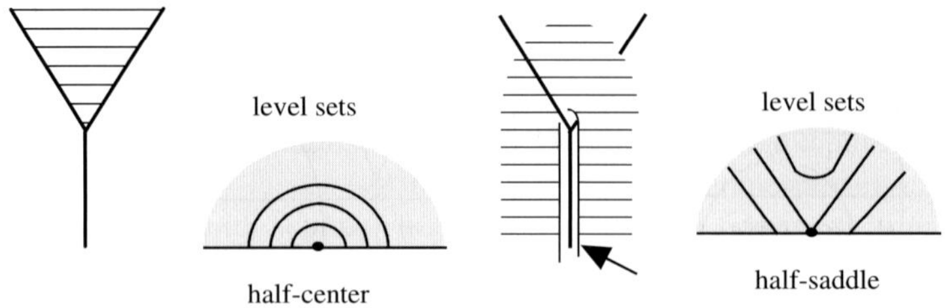

Fig. 9. Normal form critical points at the boundary.

DEFINITION 5.6. A properly imbedded surface

$$(F, \partial F) \subset (S^3 - \eta(\Gamma), \partial\eta(\Gamma))$$

is in *normal form* if

(1) each critical point of h on F is non-degenerate,
(2) ∂F is in normal form with respect to h,
(3) no critical point of h on $\mathrm{int}(F)$ occurs near a critical height of h on Γ,
(4) no two critical points of h on $\mathrm{int}(F)$ or ∂F occur at the same height,
(5) the minima (resp. maxima) of $h|\partial F$ at the minima (resp. maxima) of Γ are also local extrema of h on F, i.e., "half-center" singularities,
(6) the maxima of $h|\partial F$ at Y-vertices and the minima of $h|\partial F$ at λ-vertices are, on the contrary, "half-saddle" singularities of h on F.

Standard Morse theory ensures that, for Γ in normal form, any properly imbedded surface $(F, \partial F)$ can be properly isotoped to be in normal form.

5.4. *Upper and lower triples*

The definition of upper and lower disks naturally extends to the context of graphs:

DEFINITION 5.7. Given Γ in normal form and P a level sphere for h at a generic height, let B_u and B_l denote the balls which are the closures of the region above P and below P, respectively. An *upper disk* (resp. *lower disk*) for P is a disk $D \subset S^3 - \eta(\Gamma)$ transverse to P such that $\partial D = \alpha \cup \beta$, where β is a normal arc imbedded on $\partial\eta(\Gamma)$, α is an arc properly imbedded in $P - \eta(\Gamma)$, $\partial\alpha = \partial\beta$, and a small product neighborhood of α in D lies in B_u (resp. B_l) i.e., it lies *above* (resp. *below*) P.

Note that $D \cap P$ consists of simple closed curves and arcs with ends in β. A natural occurrence of upper (or, symmetrically, lower) disks is this: according to Definition 5.6, a maximum of ∂F near a maximum of Γ is a half-center singularity on ∂F. In particular, a sphere P just below this maximum will cut off an upper disk from F.

As was noted above, curves in $\partial\eta(\Gamma)$ can be quite complicated, so a somewhat more elaborate notion than upper or lower disk will be needed.

DEFINITION 5.8. Suppose, as above, Γ is in normal form and P is a level sphere for h at a generic height. An *upper triple* (resp. lower triple) (v, α, E) for P is an upper (resp. lower) disk E with these properties:

(1) the arc $\alpha \subset \partial E$ of Definition 5.7 has its ends at different points of $\Gamma \cap P$ (i.e., α is not a loop),

(2) v is one of the points of $\Gamma \cap P$ at an end of α and

(3) although there may be arc components of $\text{int}(E) \cap P$, none of them is incident to v.

For example, in the old setting of, say, F a Seifert surface for a knot K, any arc α of $F \cap P$ that is inessential in F cuts off either an upper or a lower disk E. If v is either point of $K \cap P$ at the ends of α, then (v, α, E) is an upper or lower triple.

5.5. *An application*

As an illustration of how thin position can be used for graphs – in particular, why it is useful to have a definition that is indifferent to pushing maxima past maxima – we will offer an updated proof of the key theorem in [15], which leads to the proof that any Heegaard splitting of S_3 is standard. (The roots of this proof go back to Otal [12].) For the proof in [15] we did not have in hand the efficient Definition 5.3 of width; instead we used a rather clumsy alternative, examining the entire function $w(t)$ and minimizing its maximum, together with the number of times the function achieves that maximum.

The setting we consider is this: $\Gamma \subset S^3$ is a finite graph in normal form whose complement is ∂-reducible, so there is a disk $(D, \partial D) \subset (S^3 - \eta(\Gamma), \partial\eta(\Gamma))$ in which ∂D is essential in $\partial\eta(\Gamma)$. The edges of the graph Γ are allowed to slide over each other.

LEMMA 5.9. *Suppose P is a generic level sphere for $\Gamma \subset S^3$, $S^3 - \eta(\Gamma)$ is ∂-reducible, and the ∂-reducing disk D has been chosen to minimize $|D \cap P|$ and is properly isotoped to be in normal form. Then either there is an edge of Γ that is disjoint from ∂D or there is a point $v \in P \cap \Gamma$ with the following property: Suppose α is an arc of $P \cap D$ that is outermost among the set of arcs of $P \cap D$ that are incident to v. Let E be the disk it cuts off from D. Then (v, α, E) is either an upper or a lower triple.*

PROOF. Since D was chosen to minimize $|D \cap P|$ it follows that every component of $D \cap P$ is essential in the planar surface $P - \eta(\Gamma)$. If any intersection point of Γ with P is incident to no arc component of $D \cap P$ then the edge of Γ containing that point is disjoint from ∂D and we are done. So we may as well assume that each point of $\Gamma \cap P$ is incident to some arc component of $D \cap P$; it follows that some point $v \in \Gamma \cap P$ is incident to no loops at all. Among all arcs of $D \cap P$ that are incident to v, let α be the arc that is outermost on D. Then by construction, α cuts off an upper disk E (say) from D in which no other arc of intersection is incident to v and α is not a loop in P. □

LEMMA 5.10. *Suppose the edges of Γ have been slid and isotoped so as to minimize $W(\Gamma)$ (cf. Definition 5.3) and suppose $S^3 - \eta(\Gamma)$ is ∂-reducible. Then either the edges of Γ can be slid until there is a ∂-reducing disk whose boundary is disjoint from an edge, or Γ is in bridge position.*

PROOF. Suppose on the contrary that Γ is not in bridge position. Let P be a thin level sphere, i.e., a level sphere intersecting Γ so that the adjacent critical heights above and below P are a minimum (possibly a Y-vertex) and a maximum (possibly a λ-vertex), respectively. Choose a ∂-reducing disk D so as to minimize $|D \cap P|$. If any edge of Γ is not incident to ∂D then we are done. If every edge is incident to ∂D then in particular $D \cap P$ is incident to every point in $\Gamma \cap P$. In that case, let (v, α, E) be the upper (say) triple given by Lemma 5.9. Then E may be used to slide one end ε of the edge of Γ on which v lies down to $\alpha \subset P$. (The details of this move, involving possible "broken edge slides" are a bit more complicated than it might first appear, cf. Proposition 2.2 of [15].)

Unfortunately this move, so similar to the one used in Theorem 4.4, does not in this case necessarily thin Γ. To see why, suppose that, before the slide, the end ε simply ascends from v into a Y-vertex, from below. Then the slide we have just described will move the end to α (creating a λ-vertex just below P and lowering $|P \cap \Gamma|$ by 1); but also the Y-vertex merely becomes a regular minimum. Both the new λ-vertex and the transformation of the Y-vertex into a regular minimum will actually raise the width, by a total of $4|P \cap \Gamma| - 2$. See Figure 10. This is a technical setback, but not a devastating one, as we now briefly outline.

Note that the slide we have just described lowers $w_P = |P \cap \Gamma|$, so we cannot repeat the process indefinitely. The argument stops either because there is a ∂-reducing disk disjoint from an edge (and we are done) or when P is no longer a thin sphere. In the latter case, either all the minima above P or all the maxima below P have been removed by the sequence of edge slides. With no loss, assume that the process stops because all the minima above P have been removed. (We do not assume that an upper disk arises at each stage, but of course an upper disk does happen to be needed at the last stage, since a lower disk would give rise to a minimum just above P.) We will show that by the time the process stops, the width has been reduced.

We will assume that there are no further thin levels above P and leave it to the reader to adjust the following argument for the general case (by counting for a and b only minima just above P and then subtracting a further such term for each thin sphere above P). Let a be the number of regular minima and b be the number of Y-vertices lying above P. Define

$$W_P(\Gamma) = W(\Gamma) - 4(2a + b)w_P.$$

CLAIM. *Every move in the process (whether on an upper or a lower disk) decreases $W_P(\Gamma)$.*

PROOF OF CLAIM. Each of the moves is either a simple isotopy of an arc of Γ to an arc lying just above or below P, or it is a slide of an end ε of an edge of Γ to an arc just above

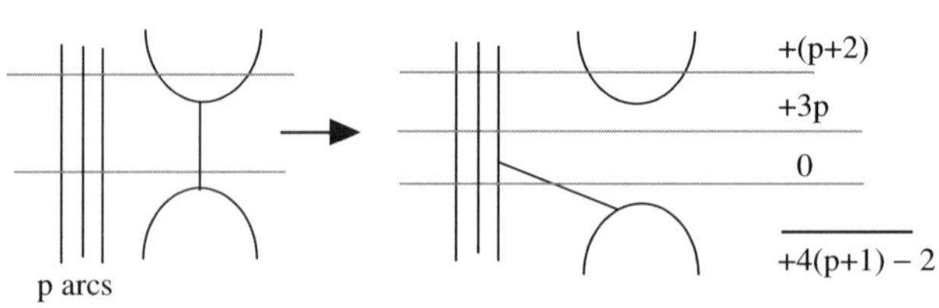

Fig. 10. A useful slide that, sadly, raises width.

or below P. In the former case, no vertex moves and the proof is almost immediate. The move reduces W_Γ and we only have to check that the reduction is greater than the increase of $-4(2a+b)w_P$ which may result from the elimination of regular minima of Γ that lie above P. These regular minima necessarily lie on the arc $\beta \subset \Gamma$ being isotoped. For every minimum of β above P there will be a maximum (and in toto one more maximum than minimum on each component of $\beta - P$ that lies above P). Eliminating both a minimum and a maximum above P will reduce $W(\Gamma)$ by a total of at least $8w_P + 16$. (Since width is not altered by rearranging orders of maxima or of minima, we may assume for the purposes of calculation that the canceling critical points are, respectively, the lowest maximum and the highest minimum, i.e., at adjacent heights.) At the same time, $-4(2a+b)w_P$ will go up by only $8w_p$. Thus, in any case, the isotopy of β will reduce $W_P(\Gamma)$.

Next suppose the move is an edge slide. If the end ε of an edge that is slid descends into a Y-vertex or ascends into a λ-vertex, then the edge slide does not create more critical levels and again the argument is fairly straightforward: $W(\Gamma)$ is always reduced, and if ε ascends into a λ-vertex or if it descends into a Y-vertex lying below P then the only minima above P that disappear are internal minima for which the above argument applies. If the end of ε descends into a Y-vertex above P, that minimum is eliminated (raising $-4(2a+b)w_P$ by $4w_P$) but it can be viewed as being cancelled with an adjacent maximum, which reduces $W(\Gamma)$ by at least $4w_P + 6$.

So, not surprisingly, to prove the claim we are reduced to the case in which an extra critical level may be created, because the end ε either descends into a λ-vertex or ascends into a Y-vertex. (In these cases the slide does not eliminate the critical point at its end, and creates a new one near P.) Let us call the terminating vertex v. There are four cases: v may lie above or below P and the disk defining the move may be an upper or lower one. But, for example, if v lies above ε and the disk is a lower one, we may imagine the slide as the composition of one based on an upper disk taking v down below P, followed by one based on a lower disk bringing v back up to P. In other words, it suffices to consider the two cases where v lies below (resp. above) P and the disk determining the slide is a lower disk (resp. upper disk).

Suppose first that v is below P and the move is via a lower disk, so the vertex is moved to a Y-vertex just above P. We have already argued that eliminating internal critical points on ε can only improve the situation, so we may as well assume that ε either descends from p straight down into a λ-vertex or ε has a single internal minimum, adjacent to its ascent from below into a Y-vertex. In the latter case the internal minimum of ε is also eliminated so again no new critical level is really created: in fact the slide is equivalent to a move that just brings the Y-vertex up to P. Since raising a Y-vertex cannot raise the width, and a new Y-vertex above P reduces $-4(2a+b)$, the overall effect is to reduce $W_P(\Gamma)$. In the case where the end ε descends straight into v, a λ-vertex, the slide that moves v up to P can be viewed as the composition of two moves: first move v up to a new Y-vertex just above the lowest thin sphere P' above v. We have already seen (cf. Figure 10) that this raises the width by $4w_{P'} - 2$. Next raise v up above P. This lowers the width every time v passes a maximum: by 4 for every λ-vertex passed and by 8 for every regular maximum passed (cf. Figure 8). But then the total amount of the reduction, determined by the number and type of maxima between P' and P is at least $4(w_{P'} - w_P)$ so in the end the move raises the width by at most $4w_P - 2$. When combined with the effect of raising b by 1, the result is that $W_P(\Gamma)$ goes down by at least 2 (indeed exactly 2 only if $P' = P$).

Finally, suppose v is above P and the move is via an upper disk. Again we may as well assume, because eliminating internal critical points only improve the situation, that ε either simply ascends from P into a Y-vertex or its terminating end descends into a λ-vertex from a single adjacent internal maximum. We have seen that in the former case the width increases by exactly $4w_P - 2$. On the other hand, the Y-vertex at v becomes a regular vertex, reducing $-4(2a + b)$ by 4. The net effect is to reduce $W_P(\Gamma)$ by 2. In the latter case, again the slide effectively just moves the λ-vertex below P, reducing the width by moving a λ-vertex down past minima, but having no effect on a or b. This finally proves the claim in all cases.

Now let Γ' be the graph when the process stops, with no further minima above P. We have just seen $W_P(\Gamma') < W_P(\Gamma) < W(\Gamma)$. But since there are no minima of Γ' above P, $W_P(\Gamma') = W(\Gamma')$. Hence $W_P(\Gamma') < W(\Gamma)$, a contradiction to the original assumption that Γ was in thin position. □

LEMMA 5.11. *Suppose Γ is in bridge position and $S^3 - \eta(\Gamma)$ is ∂-reducible. Then the edges of Γ can be slid rel a dividing sphere until either*

- *there is a ∂-reducing disk whose boundary is disjoint from an edge or*
- *for some dividing sphere P there are both upper and lower triples (v_u, α_u, E_u), (v_l, α_u, E_l) so that the disks E_u, E_l are disjoint in $S^3 - \eta(\Gamma)$.*

In the latter case, the triples may further be chosen so that either no arc of $(E_u - \alpha_u) \cap P$ is incident to v_l or, vice versa, no arc of $(E_l - \alpha_l) \cap P$ is incident to v_u.

PROOF. Let P be a dividing sphere and choose D among all ∂-reducing disks for $S^3 - \Gamma$ so that $|\partial D \cap P|$ is minimal. In particular, this guarantees that no arc component of $D \cap P$ is a trivial loop in the planar surface $P - \Gamma$.

CLAIM. *Γ can be slid and isotoped rel P so that it is still in bridge position and at the lowest maximum (resp. the highest minimum) of Γ, ∂D also has a maximum (resp. minimum). So for D in normal form, both the lowest maximum and highest minimum of Γ are incident to half-center singularities on ∂D.*

PROOF OF CLAIM. Choose any component Γ_0 of $\Gamma - P$, say one lying above P. Since Γ is in bridge position, Γ_0 is necessarily a tree. In particular, $\eta(\Gamma_0) - P$ is a planar surface. If ∂D is not incident to an edge of Γ_0 we are done, so assume it is incident to every edge; it follows (by examining an innermost loop, if any, on the planar surface $\eta(\Gamma_0) - P$) that there is a component of $\partial D - P$ that runs from the end ε_1 of one edge of Γ_0 to the end of another ε_2. Imagine collapsing the edges of Γ_0 that are not incident to P to a single vertex (so Γ_0 is simply the cone on its ends) then sliding to recreate a trivalent graph with only maxima, in which a single pair of edges (forming a maximum that we may isotope to be the lowest maximum) contains the entire subarc of ∂D that connects ε_1 to ε_2. See Figure 11. The new graph is again in bridge position and is homeomorphic to the original, so the width has not been altered. This establishes the claim.

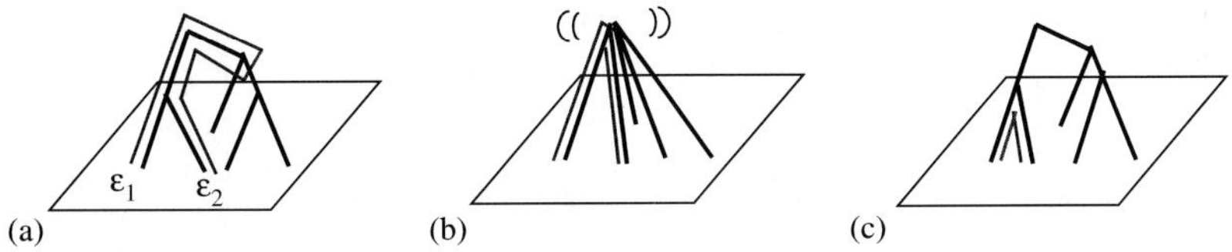

Fig. 11. A slide simplifying the disk boundary.

Following the Claim, note that a level sphere just below the lowest maximum will cut off an upper disk entirely contained above the sphere; moreover the arc of intersection with P is not a loop. Hence it is an upper triple. Similarly, a level sphere just above the highest minimum will cut off a lower triple. According to Lemma 5.9 every generic level sphere in between cuts off either an upper or a lower triple. So, as usual, there is a level sphere that cuts off both an upper and a lower triple, (v_u, α_u, E_u) and (v_l, α_l, E_l). Moreover at least one of the two, say (v_u, α_u, E_u), is obtained via Lemma 5.9. If, among the arcs incident to v_u, there is also an arc cutting off a lower triple, use this triple for $(v_l = v_u, \alpha_l, E_l)$. Then automatically no arc of $(E_l - \alpha_l)$ is incident to v_u, establishing the last property required. If, on the other hand, no arc incident to v_u cuts off a lower triple, then every arc outermost among the arcs incident to v_u must cut off an upper triple. In this case, to establish the last property of the lemma, suppose on the contrary that some arc of $(E_l - \alpha_l)$ is incident to v_u. Then an outermost such arc will cut off a (possibly different) upper triple (v'_u, α'_u, E'_u) with the property that no arc of $(E'_u - \alpha'_u) \cap P$ is incident to v_l. □

LEMMA 5.12. *Suppose Γ is not the unknot, and the edges of Γ have been slid and isotoped so as to minimize $W(\Gamma)$. Suppose further that $S^3 - \eta(\Gamma)$ is ∂-reducible. Then the edges of Γ can be further slid until there is a ∂-reducing disk whose boundary is disjoint from an edge.*

PROOF. Suppose not. Following Lemma 5.10, we may assume Γ is in bridge position. Then consider the upper and lower triples (v_u, α_u, E_u), (v_l, α_l, E_l) given by Lemma 5.11 with respect to a dividing sphere P. In particular, we assume with no loss that no arc of $(E_u - \alpha_u) \cap P$ is incident to v_l. Then E_u may be used to slide an end of the edge on which v_u lies down to α_u without affecting the end of the edge on which v_l lies, so afterwards the latter end can also be brought to P. But sliding one end down and the other end up will typically reduce the width, which is impossible. An alternate possibility is that the two moves actually level an entire edge, but again this would allow the graph to be thinned. See Figure 12a and b. The final possibility is that the two slides together simultaneously level two edges (when α_l and α_u have the same pair of end vertices), moving a cycle $\gamma \subset \Gamma$ in Γ onto P (see Figure 12c).

If either of the disk components of $P - \gamma$ is disjoint from Γ, then either $\Gamma = \gamma$ (and we are done) or that disk component is a ∂-reducing disk as required. But even if a disk component P_0 of $P - \gamma$ intersects Γ, we can just apply to P_0 the process we earlier applied to all of P to find a series of edge slides that will either finally exhibit an edge disjoint from

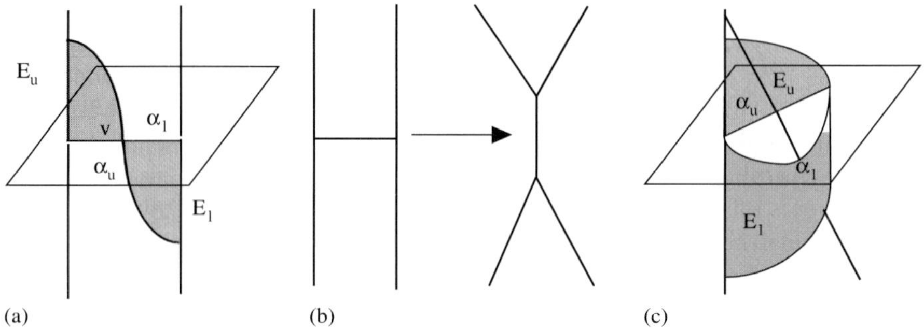

Fig. 12. A level cycle in the graph appears.

a ∂-reducing disk D (via a point in $\Gamma \cap P_0$ incident to no arc of $P_0 \cap D$) or will iteratively reduce the number of points in $\Gamma \cap P_0$ until $\Gamma \cap P_0 = \emptyset$ so, as above, P_0 is a ∂-reducing disk disjoint from an edge. □

COROLLARY 5.13. *Any Heegaard splitting of S^3 is standard.*

PROOF. Given a Heegaard splitting of S^3, let Γ be a spine of one of the handlebodies. Apply the above argument not to just a single ∂-reducing disk for the complement (i.e., a single meridian disk for the complementary handlebody) but to a complete collection of such disks. The argument is essentially the same and terminates either with an edge e disjoint from a complete collection or with Γ simply the unknot, i.e., the standard genus one splitting of S^3. In the former case, a meridian circle of the edge e is disjoint from a complete collection of meridians for the handlebody $S^3 - \eta(\Gamma)$, so it also bounds a disk in $S^3 - \eta(\Gamma)$. Thus there is a sphere in S^3 intersecting Γ in a single point in e. This is a reducing sphere for the Heegaard splitting which divides the splitting into two separate splittings of the 3-sphere. The conclusion follows by induction on the genus of the Heegaard splitting. □

6. From graphs back to knots

One might hope that thin position would be helpful in understanding the tunnel structure for knots in S^3. We noted above that for a tunnel number one knot, thin position is bridge position. That is, if K has tunnel number one and is in thin position, then there are no thin level spheres for K. It is natural to ask about the behavior of the tunnel arc with respect to the standard height function, once the knot is in thin position. The union of the knot K and the tunnel arc τ is of course a graph in S^3; moreover, one way of viewing the definition of unknotting tunnel is that the graph $\Gamma = K \cup \tau$ is the spine of a genus two Heegaard splitting of S^3. The reason that this does not just fall into the program leading to Corollary 5.13 is that in the knot tunnel case, whereas we are allowed to slide the ends of the tunnel over the knot, and over the other end of the tunnel, we can never regard a subarc of K as an edge of Γ that can be slid. Nonetheless, the answer is simple and direct: the tunnel may be made level with respect to the standard height function and this is the thinnest positioning of τ possible.

THEOREM 6.1. *Suppose that K is a knot with unknotting tunnel τ and K is in thin position with respect to the standard height function h. Then τ can be slid and isotoped without moving K until τ is level – either a level arc or a level "eyeglass" (the wedge of an arc and a circle). Moreover, after τ is perturbed slightly (to put $K \cup \tau$ in normal position) the graph $K \cup \tau$ cannot be made thinner by sliding τ.*

For a proof see [3]. In fact a similar theorem is true for arbitrary genus 2 spines of S^3 even when we do not allow edges to slide, see [16].

Theorem 6.1 raises the natural question whether a similar theorem is true for more than a single tunnel. In general, for K a knot in S^3, a collection $\tau_1, \ldots, \tau_n$ of disjoint properly embedded arcs in $S^3 - K$ is a system of unknotting tunnels if the graph $\Gamma = K \cup (\tau_1 \cup \cdots \cup \tau_n)$ is a Heegaard spine (i.e., the complement of $\eta(\Gamma)$ is a handlebody).

QUESTION 6.2. Suppose $\tau_1, \ldots, \tau_n$ is a system of unknotting tunnels for a knot $K \subset S^3$, in thin position with respect to the standard height function h. Suppose $\tau_1, \ldots, \tau_n$ are slid and isotoped to minimize the width of $\Gamma = K \cup (\tau_1 \cup \cdots \cup \tau_n)$. Is each of the tunnels a perturbed level arc?

Of course many versions of this question are possible, e.g., extending it to links or to arbitrary graphs in S^3. Even the case of a pair of tunnels seems difficult; it seems the first order of business would need to be a generalization of Morimoto's theorem [11] (so essential for the proof of theorem in [20]) to handlebodies of higher genus.

7. Graphs in other 3-manifolds

All our discussion so far revolves around objects (knots, links, graphs) in the 3-sphere, on which we have the standard height function. In fact, one can imagine using thin position in many other contexts. For example, if K is a knot in an arbitrary closed 3-manifold M, and $H_1 \cup_P H_2$ is a Heegaard splitting for M, one can describe the Heegaard splitting as a product structure on the complement of spines Σ_1, Σ_2 for the respective handlebodies H_1, H_2. That is, $M - (\Sigma_1 \cup \Sigma_2) \cong (P \times (-1, 1))$. Just as in the applications above, one can define the width of K with respect to this structure, and try to minimize the width. In effect, we are retrospectively viewing the whole discussion above as the special case in which M is S^3 and the splitting is of genus 0. Of course, many of the arguments above rely heavily on the fact that P is a sphere, so generalizing in this direction has not been particularly fruitful.

But there is a remarkable application of thin position that occurs as a crucial step in Thompson's recognition algorithm for the 3-sphere [21]. In this section we will briefly outline how it arises, and note some related applications to other decision problems in 3-manifold topology.

Suppose M is a closed 3-manifold with a given triangulation $\mathcal{T}$. Let Γ be the 1-skeleton of the triangulation. Recall that a compact surface $F \subset M$ is *normal* with respect to the triangulation if

- F is in general position with respect to $\mathcal{T}$ (so in particular F intersects Γ is a finite number of points, each on an edge of Γ),
- for each 2-simplex Δ_2 in $\mathcal{T}$, each component of $\Delta_2 \cap F$ is an arc with its ends on different faces of Δ_2,
- for each 3-simplex Δ_3 in $\mathcal{T}$, each component of $\Delta_3 \cap F$ is either a triangle (i.e., parallel to a face of Δ_3) or a square (i.e., it is incident to each face in a single arc). See Figure 13.

It is easy to show that any incompressible surface in M can be isotoped so that it is normal. The converse is not true, most obviously because the link in M of any vertex of $\mathcal{T}$ is a normal sphere. On the other hand, if F is a normal surface, then the complement $F - \Gamma$ is incompressible in $M - \Gamma$. That implication is essentially reversible: if F is a surface so that $F - \Gamma$ is incompressible in $M - \Gamma$ then F may be isotoped rel $F \cap \Gamma$ so that either F is normal or it is a sphere that bounds a ball intersecting Γ in a single unknotted arc.

There is an algorithm to find a maximal collection Σ of disjoint non-parallel normal surfaces in M; the roots of this algorithm go back to the early work of Kneser, establishing that there are at most a finite number of connected summands in M [8]. At the very least, Σ contains a linking sphere of each vertex, but typically there are many more. For example, if an edge of $\mathcal{T}$ is incident to two distinct vertices, tube together their linking vertices by a tube along the edge. Unless M is a specific 2-vertex triangulation of S^3 (cf. [7]) such a sphere is normal. In the end, it is possible to show that each component of $M - \Sigma$ is one of three types:

- a ball containing a single vertex, and bounded by a vertex-linking sphere,
- a punctured 3-ball with more than one boundary component,
- a single further component M_0, for which $|\partial M_0| = 1$.

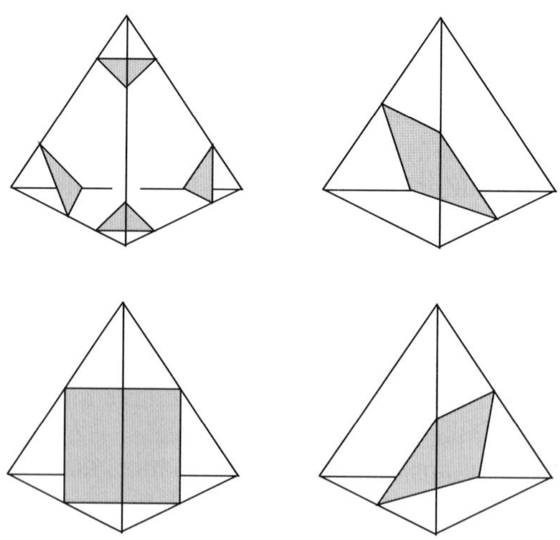

Fig. 13. A normal surface.

Then M is the 3-sphere if and only if M_0 is a 3-ball, and M_0 is algorithmically recognizable among the components of $M - \Sigma$ by the fact that it is the only component that has a single boundary component and contains no vertex. So in order to determine if M is a 3-sphere, it suffices to find an algorithm to decide if M_0 is a 3-ball.

Inside of M_0 is a proper collection of arcs, $K = \Gamma \cap M_0$ and, because $\Sigma - \Gamma$ is incompressible in the complement of Γ, we have that the planar surface $\partial M_0 - K$ is incompressible in $M_0 - K$. Suppose M_0 is a 3-ball and imagine putting K in thin position with respect to the radial height function on the ball. We know immediately that K is also in bridge position; that is, all the maxima lie above all the minima. For if not, consider the thin spheres in M_0. We have noted above (essentially Theorem 4.2) that some thin sphere P has the property that $P - K$ is incompressible in $M_0 - K$; hence P would be a normal sphere in M_0 not parallel in $M_0 - K$ to ∂M_0. But this would contradict the completeness of Σ.

This connection between thin spheres in $M_0 - K$ and normal spheres in M prompts this question: what would a thick sphere tell us? (Note that there has to be a thick sphere, since if there are only minima in K then $\partial M_0 - K$ would be compressible.) If P is the (unique) thick sphere, then a maximum of K can be pushed down to P and a minimum pushed up, but not simultaneously. Translating back into how P would appear in the triangulation, it turns out that it looks just like a normal sphere, except in a single 3-simplex Δ_3, where a single component is not a triangle or a square, but an octagon (cf. Figure 14). Such a surface is called an *almost normal surface*. Observe that arcs in the 1-skeleton of the 3-simplex can be pushed to the octagon from either side, but their images there necessarily intersect. Roughly the same algorithm that detects normal spheres can be used to detect almost normal spheres. The upshot is this:

FACT 1. *If M is the 3-sphere then there is an almost normal sphere in the component M_0 of $M - \Sigma$.*

To complete the argument that this is an algorithm, one needs to know that if there is an almost normal sphere P in the component M_0 of $M - \Sigma$ then M_0 is not the 3-ball and so M is not S^3. Observe first of all that since a subarc of Γ is parallel to an arc in P, it follows that $P - K$ is compressible in $M_0 - K$ on the side containing the arc: basically one constructs a compressing disk by doubling the disk defining the parallelism. This argument applies on both sides of P, so P is compressible in $M_0 - K$ in both directions (but compressing disks

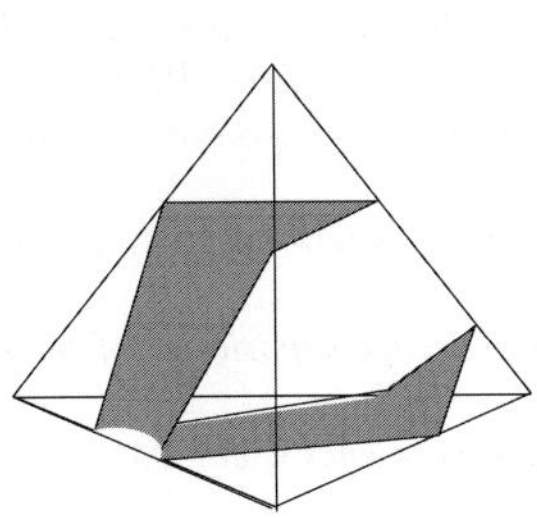

Fig. 14. An almost normal surface.

on opposite sides necessarily intersect). Thicken P to a collar $P \times I$, then maximally compress $(P \times \partial I) - K$ in $M_0 - K$; the result must be an incompressible planar surface (possibly with many components). Each component then just comes from a trivial sphere cutting off a ball intersecting K in an unknotted arc, or it becomes a normal 2-sphere, hence a sphere parallel in $M_0 - K$ to $\partial M_0 - K$. Filling in all the 3-balls, then, creates exactly a copy of M_0. In other words, M_0 can be obtained from $P \times I$ by attaching only 2 and 3-handles to $P \times \partial I$. In particular, if M_0 is a homology ball, it is a real ball. We conclude:

FACT 2. *If M is a homology sphere and there is an almost normal sphere in the component M_0 of $M - \Sigma$, then M is the 3-sphere.*

Since the homology of M is easily calculable, the combination of Facts 1 and 2 gives the Thompson algorithm for recognizing the 3-sphere.

Without wandering too far afield, note that the algorithm above is particularly straightforward if there are few normal spheres. A triangulation (broadly defined) of a closed 3-manifold is *0-efficient* if the only normal spheres are vertex linking. Clearly such a manifold must be irreducible. It is a theorem of Jaco and Rubinstein [7] that, with just a few specific exceptions, any triangulation of a closed, orientable, irreducible 3-manifold can be modified to be 0-efficient and such a triangulation has only a single vertex. More generally, if M is reducible, there is an algorithm to decompose M into a connected sum of 3-manifolds, each of which either has a 0-efficient triangulation or is visibly homeomorphic to S^3, $S^2 \times S^1$, RP^3 or $L(3, 1)$. Thus to get an algorithm that precisely describes the connected sum decomposition of M, one need only apply the Thompson algorithm above in the case in which the triangulation is 0-efficient.

There are other clever applications of thin position in settings that go well beyond the scope of this article. A favorite is [9], where Lackenby shows that the natural combinatorial ideal triangulation of a punctured torus bundle (with pseudo-Anosov monodromy) coincides with the natural hyperbolic ideal triangulation.

References

[1] J.W. Alexander, *On the subdivision of 3-space by a polyhedron*, Proc. Natl Acad. Sci. **10** (1924), 6–8.

[2] D. Gabai, *Foliations and the topology of 3-manifolds. III*, J. Diff. Geom. **26** (1987), 479–536.

[3] H. Goda, M. Scharlemann and A. Thompson, *Levelling an unknotting tunnel*, Geom. Topol. **4** (2000), 243–275.

[4] C.McA. Gordon and J. Luecke, *Knots are determined by their complements*, J. Am. Math. Soc. **2** (1989), 371–415.

[5] C.McA. Gordon and A.W. Reid, *Tangle decompositions of tunnel number one knots and links*, J. Knot Theory Ramificat. **4** (1995), 389–409.

[6] J. Hendricks, *Mp-small summands increase knot width*, Algebr. Geom. Topol. **4** (2004), 1041–1044.

[7] W. Jaco and H. Rubinstein, *0-efficient triangulations of 3-manifolds*, J. Diff. Geom. **65** (2003), 61–168.

[8] H. Kneser, *Geschlossene Flächen in dreidimensionalen Mannigfaltigkeiten*, Jahresber. Dtsch. Math. Verein. **38** (1929), 248–260.

[9] M. Lackenby, *The canonical decomposition of once-punctured torus bundles*, Comment. Math. Helv. **78** (2003), 363–384.
[10] F. Laudenbach, *Une remarque sur certains nœuds de $S^1 \times S^2$*, Compositio Math. 38 (1979), 77–82.
[11] K. Morimoto, Planar surfaces in a handlebody and a theorem of Gordon–Reid, *Proceedings of Knots '96*, S. Suzuki, ed., World Science Publishing, Singapore (1997), 127–146.
[12] J.-P. Otal, *Sur les scindements de Heegaard de la sphere S^3*, Topology **30** (1991), 249–258.
[13] Y. Rieck and E. Sedgwick, *Thin position for a connected sum of small knots*, Algebr. Geom. Topol. **2** (2002), 297–309.
[14] M. Scharlemann and J. Schultens, *3-manifolds with planar presentations and the width of satellite knots*, Trans. Am. Math. Soc. Math., GT/0304271 (in press).
[15] M. Scharlemann and A. Thompson, *Thin position and Heegaard splittings of the 3-sphere*, J. Diff. Geom. **39** (1994), 343–357.
[16] M. Scharlemann and A. Thompson, *Thinning genus two Heegaard spines in the 3-sphere*, J. Knot Theory Ramificat. **12** (2003), 683–708.
[17] M. Scharlemann and A. Thompson, On the additivity of knot width, *Proceedings of the Casson Fest, Geometry and Topology Monographs*, Vol. 7, (2004), 135–144.
[18] H. Schubert, *Über eine numerische knoteninvariante*, Math. Z. **61** (1954), 245–288.
[19] J. Schultens, *Additivity of bridge numbers of knots*, Math. Proc. Camb. Philos. Soc. **135** (2003), 539–544.
[20] A. Thompson, *Thin position and bridge number for knots in the 3-sphere*, Topology **36** (1997), 505–507.
[21] A. Thompson, *Thin position and the recognition problem for S^3, Math. Res. Lett.* **1** (1994), 613–630.
[22] Y.-Q. Wu, *Thin position and essential planar surfaces*, Proc. Am. Math. Soc. **132** (2004), 3417–3421.

CHAPTER 10

Computation of Hyperbolic Structures in Knot Theory

Jeff Weeks

www.geometrygames.org/contact.html

Contents

Abstract

Geometrical structures — typically hyperbolic — provide deep insight into the topology of knot and link complements. This chapter shows how to construct a hyperbolic structure on the complement of a topologically defined knot or link, and how to perform hyperbolic Dehn filling.

The exposition begins one dimension down, with 0-dimensional links on a 2-sphere, then applies the insights gained there to illustrate Thurston's now classic work on hyperbolic links and Dehn filling. To put these ideas into practice, the chapter explains SnapPea's algorithms for triangulating a link complement efficiently and for converging quickly to the hyperbolic structure while avoiding singularities in the parameter space.

HANDBOOK OF KNOT THEORY
Edited by William Menasco and Morwen Thistlethwaite

1. Introduction

Knot and link complements enjoy a geometry of crystalline beauty, rigid enough that simple cut-and-paste techniques meet geometrical as well as topological needs, yet surprisingly complex in their inexhaustible variety. The cut-and-paste approach makes computer exploration of their geometry easy: link complements become finite unions of tetrahedra, handled in a purely combinatorial way, with no need for the messy machinery of differential geometry.

The present article begins with the geometry of 2-dimensional link complements in Section 2 to provide an overview of all the main ideas. Section 3 explains an efficient algorithm for triangulating 3-dimensional link complements, Section 4 shows how to compute the hyperbolic structure, and finally Section 5 shows how the hyperbolic structure deforms to yield hyperbolic structures on closed manifolds obtained by Dehn filling.

Readers may consult other chapters in this volume for richer discussions of hyperbolic knots and links [1] and Dehn fillings [2]. Hyperbolic structures have found applications in knot tabulation [3] and more generally provide a fast and effective way to test hyperbolic knots and links for equivalence [4] and to compute their symmetry groups [5,6] and other invariants [7]. The computer program SnapPea [8] implements these applications based on the foundation described in the present article. The SnapPea source code contains detailed explanations of all algorithms used.

Even though the exposition in the present article is original, most of the mathematics was born in the work of Bill Thurston. Thurston's informal 1979 notes (see also [9,10]) contained the theory of hyperbolic knots and links, hyperbolic Dehn filling, and in particular his simple and elegant system (explained here in Sections 4 and 5) for finding hyperbolic structures in terms of complex edge angles.

2. Two-dimensional preview

Before tackling 3-dimensional knots and links, let us review the 2-dimensional ones. Their topology and geometry is far simpler, yet the insights they provide will serve us well in the 3-dimensional case.

Just as a 3-dimensional link is a collection of 1-spheres (circles) in a 3-sphere, a 2-dimensional link is a collection of 0-spheres (pairs of points) in a 2-sphere. The classification of 2-dimensional links is easy: two links are equivalent if and only if they have the same number of components (Figure 1).

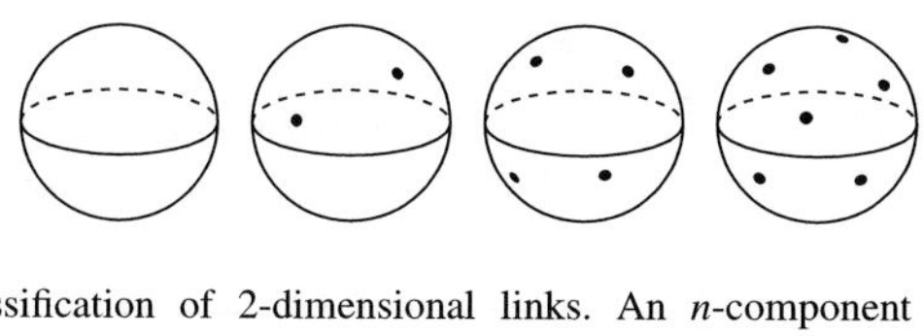

Fig. 1. Topological classification of 2-dimensional links. An n-component link consists of n 0-spheres on a 2-sphere.

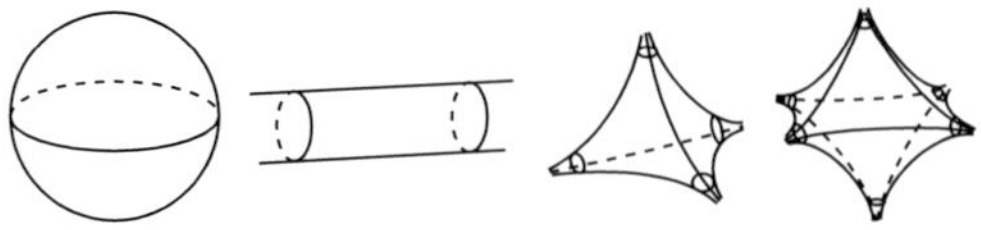

Fig. 2. Geometric structures on 2-dimensional link complements. The 0-component link complement has spherical geometry and the 1-component link complement admits Euclidean geometry. All the rest admit hyperbolic geometry.

Each link complement admits a constant curvature geometry. The 0-component link complement, which is simply an unpunctured 2-sphere, already has spherical geometry. At first glance the remaining n-component link complements, which are punctured 2-spheres, also seem to have spherical geometry, but these geometric structures are disallowed because they are incomplete. Many formal definitions of completeness appear in the literature, but intuitively a surface is *incomplete* if a traveler starting at some point on the surface can reach an "edge" (either a ragged edge or a boundary) within a finite distance. Conversely a surface is *complete* if a traveler starting at any point on the surface can travel any finite distance in any direction without hitting an edge. In the case of a punctured 2-sphere, a traveler easily reaches a puncture, so the surface is incomplete.

To construct a complete geometric structure on the 1-component link complement, pull the link itself (the pair of points) to infinity, dragging the link complement along with it (Figure 2, second frame). The link complement becomes an infinite cylinder, which has locally Euclidean geometry (constant zero curvature) and is complete.

Following the same technique, take the 2-component link and pull the link itself (two pairs of points) to infinity. Intuitively the stretched out link complement looks hyperbolic (Figure 2, third frame). To make this rigorous, cut the link complement into four ideal triangles (Figure 3, left) and then *define* the hyperbolic structure on the link complement to be the union of four hyperbolic ideal triangles with edges identified ("glued") in the appropriate way (Figure 3, right).

More generally, for every $n > 1$ the n-component link complement splits into a union of ideal triangles which then defines a hyperbolic structure on the link complement. The hyperbolic structure is never unique, but a simple Euler number argument shows that

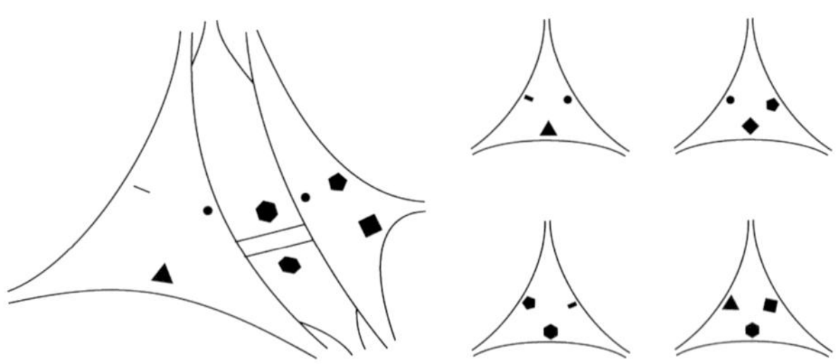

Fig. 3. The complement of a 2-component link splits into four ideal triangles. The small symbols show how to re-glue the edges in pairs to recover the original manifold.

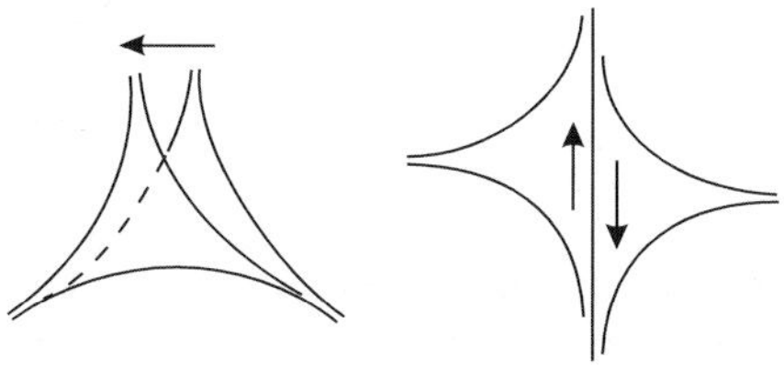

Fig. 4. Ideal triangles are rigid; any two are congruent. *Proof*: Given any two ideal triangles, move one so that one of its edges coincides with the corresponding edge of the other, and then slide it along until the third vertex coincides as well (left panel). The gluing between ideal triangles is flexible – two neighboring triangles may slide along each other like the two sides of the San Andreas fault (right panel).

the number of ideal triangles must be exactly $4(n - 1)$. An ideal triangle has area π, so the total area of the n-component link complement is $4\pi(n - 1)$ no matter what hyperbolic structure is chosen.

Assuming we identify the ideal triangles' edges midpoint-to-midpoint (the midpoint of an edge being the edge's intersection with an orthogonal line of mirror symmetry), the hyperbolic structure is complete, and is therefore called a *complete hyperbolic structure*. The parts stretching off to infinity are called *cusps*.

On the one hand, ideal triangles are rigid (Figure 4, left). On the other hand, the gluing between two ideal triangles is flexible: each edge is infinitely long so the neighboring triangles may slide past one another (Figure 4, right). These 2-dimensional facts are exactly the opposite of the situation in 3 dimensions, where ideal tetrahedra are flexible but the gluings between them are rigid. Nevertheless, deforming the hyperbolic structure produces analogous results in both 2 and 3 dimensions, and so examining the 2-dimensional case in this section will provide insight into the analogous 3-dimensional results in Section 5.

To deform a hyperbolic structure, start with one of the cusps. Slice the cusp open (Figure 5). Gluing opposite edges of the sliced-open cusp straight across (Figure 5, top) would restore the cusp to its original condition, but gluing opposite edges with a shift (Figure 5, bottom) yields a different result. If you physically construct a

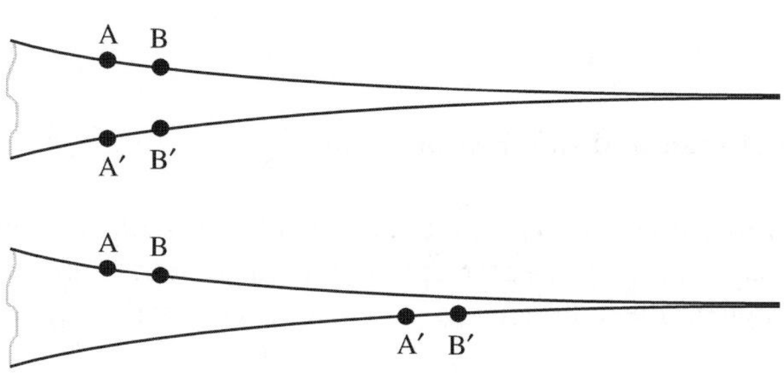

Fig. 5. A sliced open cusp. Glue the edges straight across (top) and the cusp is complete. Glue the edges with a shift (bottom) and the result is incomplete.

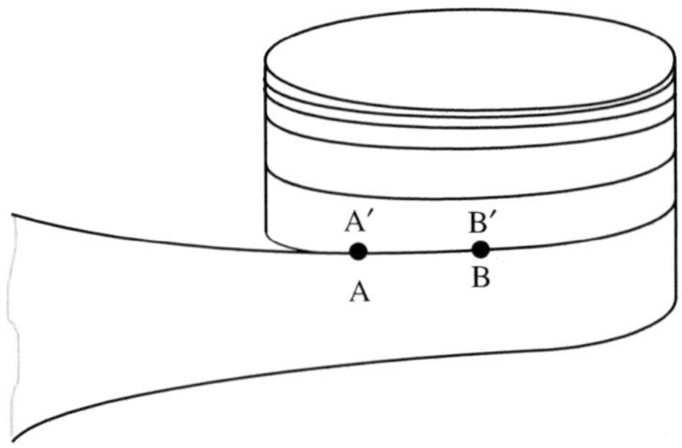

Fig. 6. Physical realization of a cut-open cusp whose edges are identified with a shift (Figure 5 bottom). The strip wraps around like a cylinder, approaching a limiting circle.

paper model[1] of the sliced-open cusp and wrap it around so that each point on one edge glues to a point on the opposite edge that's shifted by, say, 4 cm, you obtain a cylinder (Figure 6). Your paper strip, if infinitely long, would wrap around the cylinder infinitely many times, approaching but never reaching a limiting circle.

Globally, for each 0-sphere in an n-component link, pick a geodesic edge connecting the two cusps belonging to that 0-sphere. Cut, shift and reglue along each such edge (Figure 7, left). Locally the result at each cusp looks like Figure 6. Globally the result looks like Figure 7 right. That is, the result is a surface of genus n with n circles missing (namely the limiting circles of the former cusps). With foresight we may arrange for the geodesic edges (along which we cut, shifted and reglued) to be edges of the triangulation. That is, our original cusped surface was the union of $4(n-1)$ hyperbolic ideal triangles, and we cut, shifted and reglued along edges of that ideal triangulation. Thus the result (Figure 7, right) is also the union of $4(n-1)$ hyperbolic ideal triangles and therefore enjoys a hyperbolic structure, albeit an incomplete one. The missing circles are geodesics. Adding those n missing geodesic circles yields a complete hyperbolic structure on the closed surface of genus n.

In 2 dimensions most link complements admit a complete hyperbolic structure. Deforming a complete hyperbolic structure yields a hyperbolic structure (with missing geodesics) for a closed surface. Both these results generalize readily to three dimensions.

3. Triangulation of knot and link complements

Just as every 2-dimensional link complement splits into ideal triangles (Figure 3), every 3-dimensional link complement splits into ideal tetrahedra. For now we will work with "topological ideal tetrahedra" – that is, we will visualize them as ideal

[1]The paper model is of course only an approximation. The real cusp is intrinsically hyperbolic while your paper model is intrinsically flat. Nevertheless the paper model suffices for the demonstration at hand. Remember that the width of your paper strip must shrink exponentially fast as you move along it lengthwise.

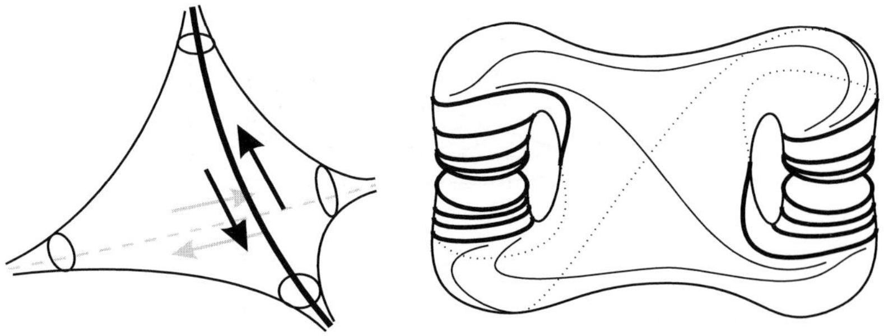

Fig. 7. For each 0-sphere (pair of points) in an n-component 2-dimensional link, shear along an edge connecting the two cusps belonging to that 0-sphere. Locally the result is as in Figure 6. Globally the result is a closed hyperbolic surface of genus n with n geodesic circles missing.

tetrahedra but will not worry about the exact hyperbolic geometry they carry. The latter will be the subject of Section 4.

The goal in triangulating a link complement is to produce a triangulation that quickly simplifies down to as few tetrahedra as possible. Minimizing the number of tetrahedra saves space and computational time, but more importantly it vastly improves the chances that all tetrahedra will be "positively oriented", a condition needed to rigorously guarantee that the subsequent hyperbolic structure is correct. The triangulation algorithm used in the computer program SnapPea has proven more effective than several alternatives the author tested, so that is the algorithm presented here.

Technical Note (which the reader may ignore): SnapPea's algorithm requires a connected link projection. If a given link projection consists of more than one component, SnapPea does Type II Reidemeister moves to make the projection connected. In the same spirit, to each obviously unknotted component SnapPea does a Type I Reidemeister move to add a nugatory crossing. Link projections requiring these moves never have hyperbolic complements; the moves are needed only for non-hyperbolic knots and links.

The link complement will have one cusp for each component of the link. For visual convenience we chop off the cusps. That is, rather than triangulating the complement of the link using ideal tetrahedra, we will triangulate the complement of a tubular neighborhood of the link using truncated ideal tetrahedra. Once such a truncated ideal triangulation is found, the extension to true ideal tetrahedra is easy and obvious.

Imagine the tubular neighborhood of the link lying near the equatorial 2-sphere of the 3-sphere S^3. To keep the construction simple and natural, the triangulation must adhere closely to the link projection itself. To accomplish this, let us triangulate it in $S^2 \times I$ rather than in S^3 (Figure 8). The two missing solid balls – one lying to either side of $S^2 \times I$ in S^3 – will be added back later.

To triangulate the link complement in $S^2 \times I$, cut straight down through it just as you would cut cookie dough with a cookie cutter. Figure 9 shows the pattern of the cuts: cut along the centerline of each link component and also along the dual graph. The resulting pieces are all identical. Figure 10 (first panel) shows one piece. It is not yet a truncated

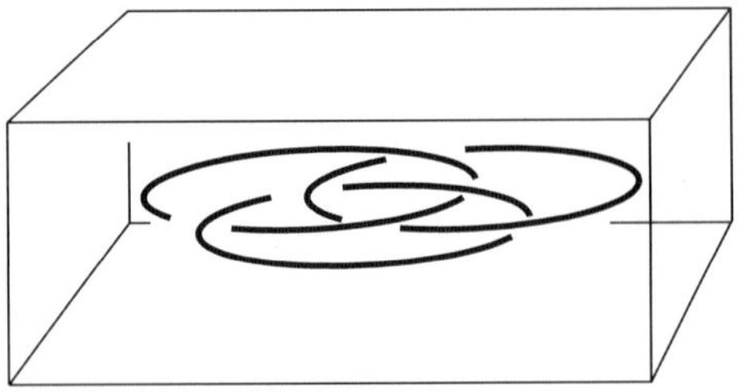

Fig. 8. Imagine the link projection lying near the equatorial 2-sphere of S^3. For convenience we will triangulate the link complement in a regular neighborhood of the equatorial 2-sphere (topologically $S^2 \times I$) and worry later about the neglected solid balls lying above and below it.

tetrahedron. It has four truncated vertices, two of which border the tubular neighborhood of the knot and two of which border the upper and lower surfaces of $S^2 \times I$, along with six ordinary faces, four of which are combinatorial hexagons and two of which are combinatorial squares.

Now deform each combinatorial square to become a tall narrow rectangle (Figure 10, second panel) that finally collapses to a vertical line segment (third panel). The resulting solid (last panel) has four truncated vertices (all are triangles) and four ordinary faces (all are hexagons), and is in fact combinatorially a truncated tetrahedron!

Collapsing all square faces to vertical lines does not change the topology of the manifold. If some link component had only overcrossings or only undercrossings, then collapsing the squares would indeed change the manifold's topology because we would be collapsing an embedded cylinder, but the preceding Technical Note excludes this possibility. Collapsing an embedded square or a series of embedded squares is safe.

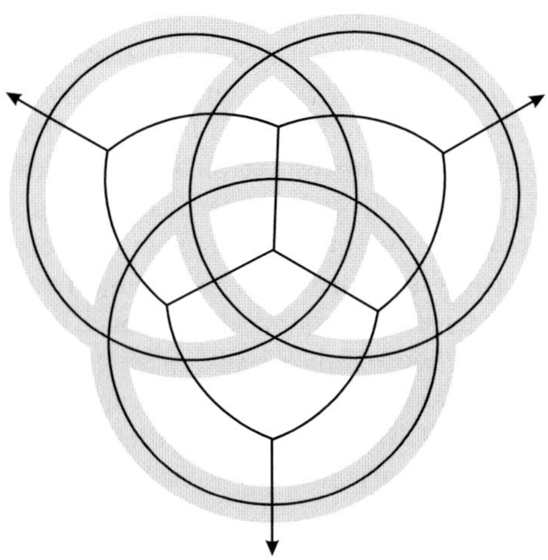

Fig. 9. View from above the complement of the link in $S^2 \times I$ (cf. Figure 8). Cut straight down through the link complement, with one set of cuts following the centerlines of the link projection itself while a second set of cuts follows the dual graph. The resulting pieces are all homeomorphic (up to reflection).

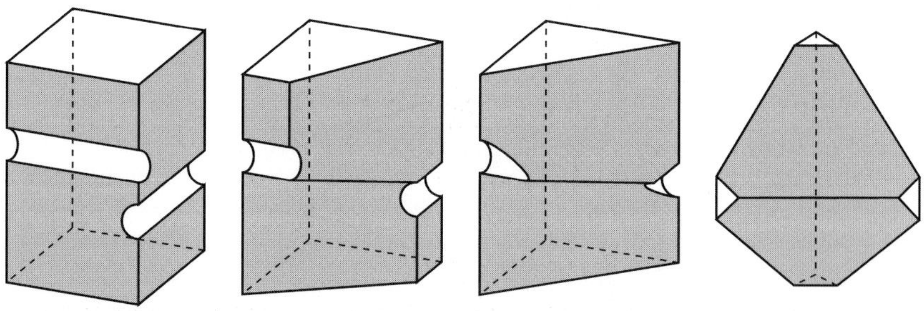

Fig. 10. The pieces in the decomposition of Figure 9 are all identical. Each has a square truncated vertex touching the upper surface of $S^2 \times I$, another square truncated vertex touching the lower surface of $S^2 \times I$, and two hemi-cylindrical truncated vertices running along the boundary of the (excised) tubular neighborhood of the link (far left panel). In addition, each piece has six regular faces: two squares and four hexagons. Deforming the two square faces to become narrow rectangles does not change the topology (center left panel). Collapsing those two narrow rectangles to become lines (center right panel) yields a polyhedron that is combinatorially a truncated tetrahedron (far right panel), with four truncated vertices (all triangles) and six faces (all hexagons).

A simple indexing system describes the triangulation in a format amenable to computer use. Label the four truncated vertices of each tetrahedron with the integers $\{0, 1, 2, 3\}$ and label each face with the index of its opposite vertex. To specify how a face of one tetrahedron glues to a face of another, simply specify the permutation of the vertex index set $\{0, 1, 2, 3\}$ induced by reflecting the vertices of the original tetrahedron across the face in question onto the vertices of the neighboring tetrahedron.

In general the vertex indices may be assigned arbitrarily, but in the case of a triangulated link complement we make the convention that vertex 0 is the truncated vertex at the south pole of S^3 (on the bottom of the polyhedron in Figure 10, left), vertex 1 is the truncated vertex at the north pole of S^3 (on the top of the polyhedron in Figure 10, left), and vertices 2 and 3 are the truncated vertices touching the link (the left and right hemi-cylindrical truncated vertices, respectively, in Figure 10, left). With this convention the symmetry of the decomposition guarantees that *all* gluing permutations are the same throughout the triangulation, namely $0123 \rightarrow 0132$.

We now have a nicely triangulated manifold, but we constructed it in $S^2 \times I$ rather than S^3, so in S^3 it is the complement of the link plus two solid balls. To remedy this problem, drill out a tube connecting the thickened link to the solid ball at the north pole (thus "cancelling" that solid ball) and a second tube connecting the thickened link to the solid ball at the south pole. The easiest way to drill out a tube is to splice into the triangulation a triangular pillow that already contains a pre-drilled tube (Figure 11). That is, let F be any 2-cell of the triangulation incident to both the northern spherical boundary of $S^2 \times I$ and one of the torus boundary components of the link complement. If we cut along F, insert an ordinary (un-drilled) triangular pillow, and reglue, then the topology of the manifold does not change. But if we instead insert a triangular pillow with a pre-drilled tube (Figure 11) then the boundary component at one end of the tube gets joined to the boundary component at the other end of the tube. In the present case this joins the spherical boundary component at the north pole (resp. south pole) to the torus boundary of the link, thereby neutralizing the former.

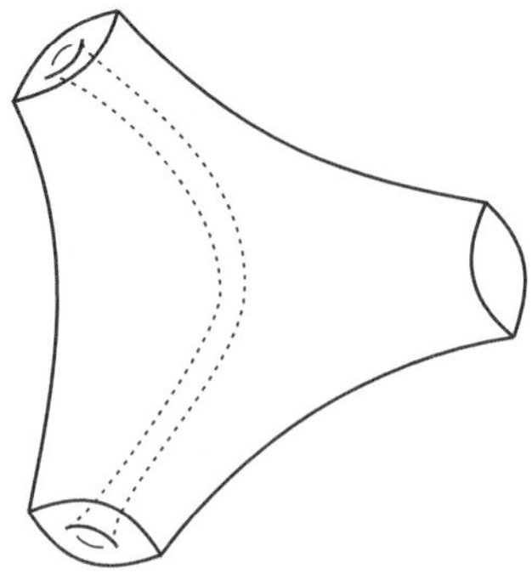

Fig. 11. A triangular pillow with a pre-drilled tube running through it provides a tunnel joining two of its truncated vertices.

A triangular pillow with a pre-drilled tube can be constructed from two truncated (or ideal) tetrahedra. Face 0 of the first tetrahedron glues to face 0 of the second tetrahedron via the gluing 0123 → 0213. Face 3 of the first tetrahedron glues to face 3 of the second tetrahedron via the gluing 0123 → 0213. Faces 1 and 2 of the second tetrahedron glue to each other via the gluing 0123 → 0213. Faces 1 and 2 of the first tetrahedron remain unglued and provide the two external boundary faces of the pillow.

Once the two triangular pillows are installed we have a valid triangulation of the link complement. The triangulation contains $4n + 4$ tetrahedra, but the number of tetrahedra decreases substantially when the triangulation is simplified. Two elementary operations serve to simplify the triangulation: the first operation replaces three tetrahedra surrounding a common edge with two tetrahedra sharing a common face, while the second operation cancels two "flattened" tetrahedra that share two adjacent faces. For an explanation of an effective high-level algorithm governing the application of the two elementary operations, please see the file `simplify_triangulation.c` in the SnapPea source code [8]. Note that the high-level algorithm sometimes requires the inverse of the first operation, which temporarily increases the number of tetrahedra – by replacing two tetrahedra sharing a face with three tetrahedra surrounding an edge – but ultimately leads to simplifications.

4. The complete hyperbolic structure

If the link complement admits a hyperbolic structure [1], then the "topological ideal tetrahedra" of the previous section may be replaced with honest hyperbolic ideal tetrahedra.

In principle this is easy: just imagine the topological ideal triangulation sitting in the hyperbolic manifold in some wiggly way, and pull all its 1-dimensional edges taut, so that the edges become geodesics running from infinity in one cusp, though the fat part of the manifold, and back to infinity either in the same cusp or in a different cusp.

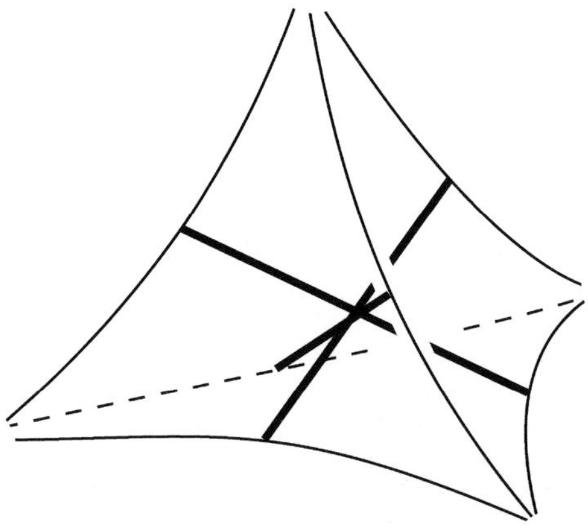

Fig. 12. Every ideal tetrahedron has three mutually perpendicular symmetry axes. To construct a symmetry axis, pick a pair of opposite edges and find the shortest geodesic γ connecting them. Because γ is the shortest path between those two edges, it must meet both the edges at right angles. A half-turn about γ therefore preserves each edge setwise while interchanging its endpoints-at-infinity. In other words, a half-turn about γ permutes the tetrahedron's ideal vertices and thus defines a symmetry of the tetrahedron. Elementary symmetry considerations then imply that the three symmetry axes – corresponding to the tetrahedron's three pairs of opposite edges – must meet orthogonally at the tetrahedron's center.

Each 2-dimensional face is now defined by its three geodesic edges, and each 3-dimensional ideal tetrahedron is defined by its faces.[2]

In practice we solve for the shapes of the honest hyperbolic ideal tetrahedra analytically. The shape of an ideal tetrahedron is determined by its dihedral angles. By symmetry (Figure 12) opposite dihedral angles are equal, so three of the angles determine the opposite three. Furthermore, if we examine a horospherical cross section of a cusp (Figure 13), we may replace the real dihedral angle with a complex dihedral angle (Figure 14). The three complex dihedral angles depend on each other: any one of them determines the other two (Figure 15). Thus a single complex dihedral angle completely parameterizes the shape of an ideal tetrahedron.

The real dihedral angles surrounding a single edge in a hyperbolic ideal triangulation sum to 2π. Analogously, the complex dihedral angles surrounding an edge multiply to 1. Figure 16 illustrates how the complex angles provide more information than the real angles do. In practice one replaces the product $\prod z_i = 1$ with its more powerful logarithmic form $\sum \log z_i = 2\pi i$ to insure that the arguments sum to 2π rather than, say, to 4π or 0. Geometrically this ensures that the sequence of tetrahedra wraps exactly once around the edge.

Our goal, in geometrical terms, is to take the topological ideal triangulation of a knot or link complement produced in Section 3 and realize each topological ideal tetrahedron as

[2]If the triangulation is inefficient, with more than a minimal number of tetrahedra, then there is some danger that one or more of the tetrahedra will become negatively oriented – in effect the triangulation folds over on itself and then double back at those places, leaving some points in the manifold covered three times, twice by ordinary positively oriented tetrahedra and once by a negatively oriented tetrahedron. But we need not worry about this problem here.

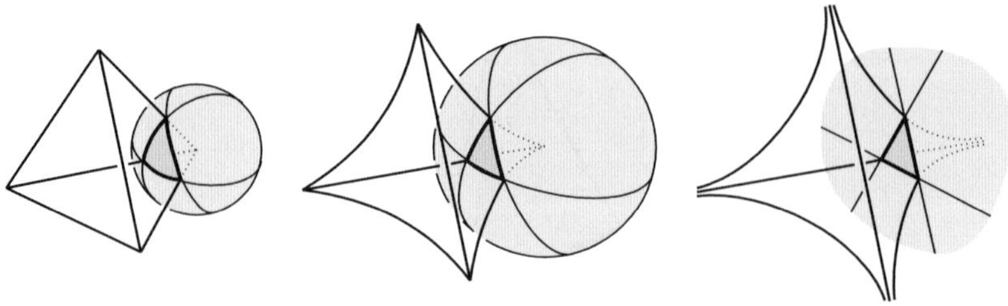

Fig. 13. In an ordinary tetrahedron (left) the cross section near a vertex – defined to be the intersection of the tetrahedron with a sphere centered at the vertex itself – is a spherical triangle. If we move the vertices outward (center), while keeping the cross section close to the tetrahedron's center, then the radius of the sphere must increase, making the cross section flatter. In the limit as the vertices go to infinity (right) the cross section becomes completely flat. Such a limiting sphere is called a *horosphere* and is in fact a 2-dimensional Euclidean plane sitting in 3-dimensional hyperbolic space. Even though it is intrinsically flat, the horosphere remains extrinsically convex to compensate for the ambient negative curvature of hyperbolic 3-space.

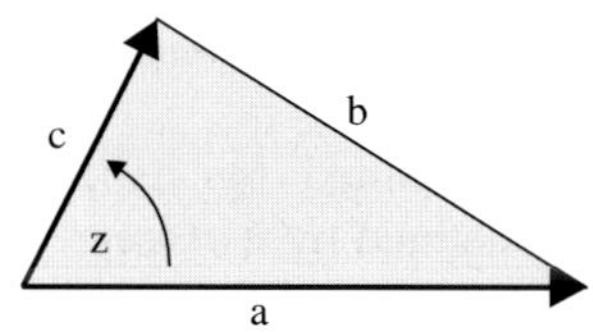

Fig. 14. Consider more carefully the horospherical cross section (defined in Figure 13) of the cusp of an ideal tetrahedron. The cross section is a Euclidean triangle whose three angles are simply the (real) dihedral angles of the tetrahedron. It turns out to be more convenient to replace each real dihedral angle θ with a complex dihedral angle z whose argument $\arg z$ is the real dihedral angle θ and whose modulus $|z|$ is the ratio of the lengths of the adjacent sides (in the example shown, $|z| = c/a$). In other words, z is the complex number that rotates one side of the triangle counterclockwise to an adjacent side, as viewed from the cusp.

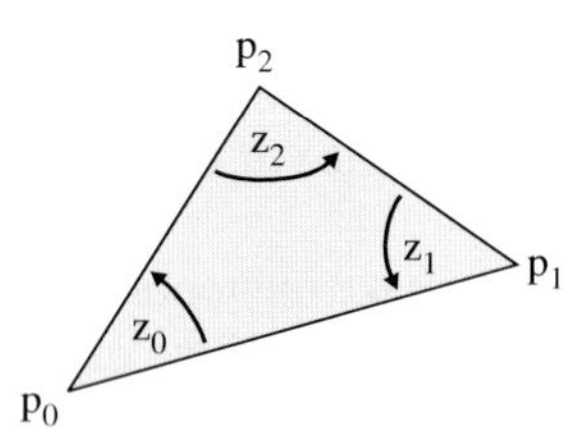

Fig. 15. Any one of the three complex dihedral angles z_0, z_1 and z_2 determines the other two. *Proof*: Position the cusp cross section arbitrarily in the complex plane and let its vertices be the complex numbers p_0, p_1 and p_2. The definition of the complex dihedral angle (recall Figure 14) implies $z_0 = (p_2 - p_0)/(p_1 - p_0)$, $z_1 = (p_0 - p_1)/(p_2 - p_1) = 1/(1 - z_0)$ and $z_2 = (p_1 - p_2)/(p_0 - p_2) = 1/(1 - z_1) = 1 - (1/z_0)$.

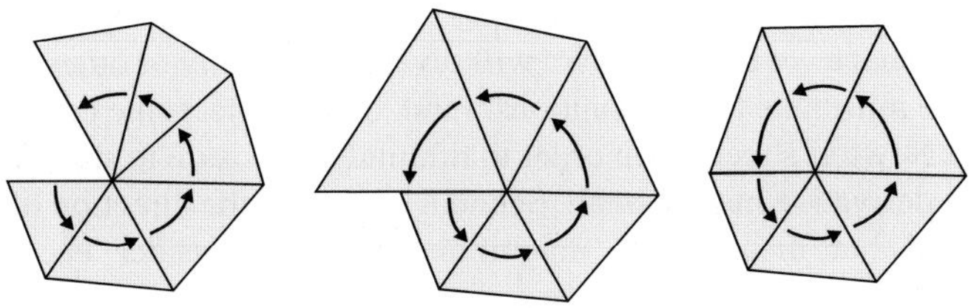

Fig. 16. Look at a cusp cross section to see how the dihedral angles meet near an edge. In a failed attempt at a hyperbolic ideal triangulation (left) the real dihedral angles do not sum to 2π. In a second failed attempt (center) the problem is more subtle: the real dihedral angles sum to 2π but the complex dihedral angles do not multiply to 1 (the sum of their arguments is correct but the product of their moduli is not). In a valid hyperbolic ideal triangulation (right) the product of the complex dihedral angles is 1.

an honest hyperbolic ideal tetrahedron in such a way that they fit together correctly around their common edges. Algebraically, this means we must find complex dihedral angles z for the tetrahedra such that the edge equations $\sum \log z_i = 2\pi i$ are satisfied. But we are not quite ready to solve those equations. Even though a simple Euler characteristic argument guarantees that the number of equations equals the number of variables, the equations are not independent. Rather the solution space contains one complex degree of freedom for each cusp (i.e., for each component of the original link). For now we will resolve this ambiguity by insisting that each cusp be complete, i.e., that its cross section be effectively a parallelogram (Figure 17, left) rather than some other quadrilateral (Figure 17, right). Section 5 will explore the more general case in detail.

In summary, to the set of edge equations we add a set of cusp equations ensuring that each cusp is complete. We then solve the equations using Newton's method and obtain the solution.

In practice the situation is somewhat delicate. When applied blindly, Newton's method usually fails. That is, if one begins with, say, regular ideal tetrahedra (all complex dihedral

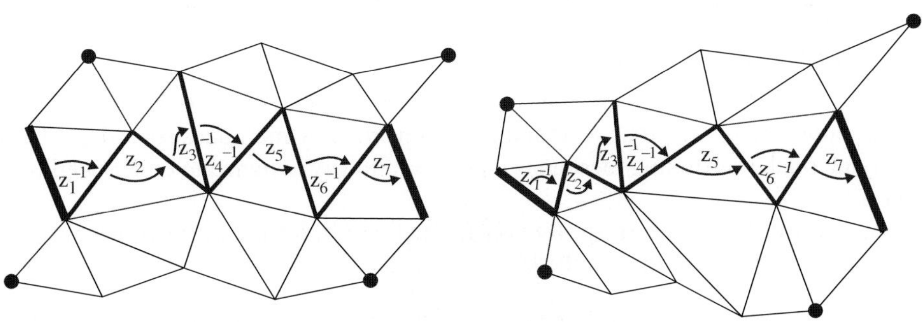

Fig. 17. The cross section of each cusp is a torus whose fundamental domain – viewed in the universal cover – is combinatorially a quadrilateral. In the special case that the quadrilateral is effectively a parallelogram (left) the covering transformations are pure translations. In the general case, by contrast, the quadrilateral is arbitrary (right) and the covering transformations are similarities, not isometries, of the Euclidean plane. Algebraically, the rotational–dilational part of a covering transformation may be computed as a product of complex dihedral angles; the product will be exactly 1 if and only if the covering transformation is a pure translation.

angles z equal to the sixth root of unity $\frac{1}{2} + \frac{\sqrt{3}}{2}i$) and applies the standard Newton's method, often the shapes of the tetrahedra will diverge to infinity or other nonsense values. To avoid this, one must take two precautions. First, one re-selects the coordinate system at each iteration of Newton's method in order to minimize exposure to singularities and keep the entries in the derivative matrix small. Second, one trusts the direction of the gradient in Newton's method but distrusts its magnitude. Let us consider each of these two precautions in detail.

Choice of coordinates. Complex dihedral angles of $z = 0, 1$, or ∞ correspond to degenerate tetrahedra. Near those values, bad things happen. The two main problems are that (1) some of the entries in the derivative matrix (used in Newton's method) approach infinity, and (2) incrementing the solution can move it too close to a singularity, resulting in wild swings in the real dihedral angles. Switching the coordinates from the complex dihedral angle z to its logarithm $\log z$ helps a bit. Rather than having two singularities (at $z = 0$ and $z = 1$) embedded in the parameter space, you have only one. The singularity that used to be at $z = 1$ is now at $\log z = 0$, but the singularity that used to be at $z = 0$ has been happily pushed out to infinity.

This strategy can be further improved by choosing the (logarithmic) coordinate system based on the current shape of the tetrahedron. The coordinate system is chosen so that the current shape of the tetrahedron stays as far away as possible from the one remaining singularity in the parameter space. Specifically, let the three complex dihedral angles be

$$z_0 = z, \quad z_1 = \frac{1}{1-z}, \quad z_2 = 1 - \frac{1}{z} \tag{4.1}$$

(note that those expressions are taken directly from the formulas in the caption of Figure 15) and divide the complex plane into three regions

Region A: $|z - 1| > 1$ and $\operatorname{Re} z < 1/2$

Region B: $|z| > 1$ and $\operatorname{Re} z > 1/2$

Region C: $|z - 1| < 1$ and $|z| < 1$

Viewed on the Riemann sphere, the singularities at 0, 1 and ∞ are equally spaced points on the equator, and the Regions A, B and C are separated by meridians spaced $2\pi/3$ apart. Points lying on the separating meridians may be arbitrarily assigned to either neighboring region.

When z lies in Region A (resp. Region B, Region C), let $\log z_0$ (resp. $\log z_1$, $\log z_2$) parameterize the shape of the tetrahedron.

PROPOSITION. If one chooses coordinates as in the preceding sentence, then the entries in the derivative matrix remain bounded.

PROOF. Each entry in the derivative matrix used in Newton's method is a fixed linear combination of the derivatives of $\log z_0$, $\log z_1$ and $\log z_2$ for several tetrahedra, so it suffices to show that each such derivative has modulus less than or equal to one.

First compute

$$\frac{d(\log z_0)}{dz} = \frac{1}{z}, \quad \frac{d(\log z_1)}{dz} = \frac{1}{1-z}, \quad \frac{d(\log z_2)}{dz} = \frac{1}{z(z-1)} \tag{4.2}$$

and then take ratios of the above to obtain

$$\begin{array}{lll} \dfrac{d(\log z_0)}{d(\log z_0)} = 1 & \dfrac{d(\log z_0)}{d(\log z_1)} = \dfrac{1-z}{z} & \dfrac{d(\log z_0)}{d(\log z_2)} = z-1 \\ \dfrac{d(\log z_1)}{d(\log z_0)} = \dfrac{z}{1-z} & \dfrac{d(\log z_1)}{d(\log z_1)} = 1 & \dfrac{d(\log z_1)}{d(\log z_2)} = -z \\ \dfrac{d(\log z_2)}{d(\log z_0)} = \dfrac{1}{z-1} & \dfrac{d(\log z_2)}{d(\log z_1)} = -\dfrac{1}{z} & \dfrac{d(\log z_2)}{d(\log z_2)} = 1. \end{array} \tag{4.3}$$

If z lies in Region A and we have chosen $\log z_0$ coordinates as required, then the derivatives in the first column of Eq. (4.3) have modulus less than or equal to 1. This is obvious for the first entry in the column. For the third entry it is an immediate consequence of the condition is $|z-1| > 1$. For the second entry, note that

$$|\mathrm{Im}\, z| = |\mathrm{Im}(1-z)|$$

and

$$|\mathrm{Re}\, z| < |\mathrm{Re}(1-z)| \quad \text{iff} \quad \mathrm{Re}\, z < 1/2$$

so $|z| < |1-z|$.

Similar arguments show that when z lies in Region B (resp. Region C) the derivatives in the second column (resp. third column) have modulus less than or equal to 1. □

Theoretical Note #1: The computed dihedral angles, given by the imaginary parts of $\log z_0$, $\log z_1$ and $\log z_2$, are not *a priori* limited to the range $(0, \pi)$. However, only when they fall in the range $(0, \pi)$ does the computed solution have a direct geometrical interpretation as a union of ideal tetrahedra comprising a cusped hyperbolic 3-manifold. When some or all of the angles fall outside the range $(0, \pi)$ the situation is more complicated: in most cases the hyperbolic structure still exists and has the computed volume but in rare cases spurious non-geometric solutions occur.

Theoretical Note #2: I briefly entertained the idea of finding a single coordinate system that avoids all three singularities. Unfortunately Picard's Little Theorem shows that this is not possible for an analytic function. It might be possible for a nonanalytic function – perhaps a simple function of z and $\bar{z}$ – but I have not pursued this idea and in any case such a function would not be conformal.

Avoiding singularities. When applying Newton's method to find a hyperbolic structure, one trusts the direction of the gradient but distrusts its magnitude. More precisely, one

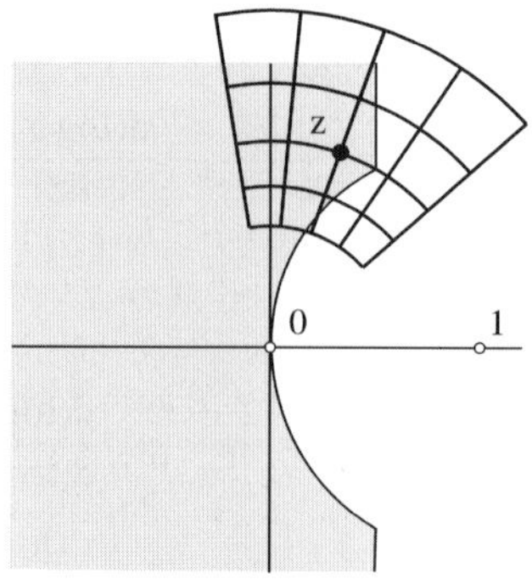

Fig. 18. Do not let Re(log z) or Im(log z) change by more than 1/2 during a single iteration of Newton's method. Our choice of coordinates (in Region A, B or C) guarantees that the parameter z lies in the shaded region relative to the chosen coordinates, away from the singularity at $z = 1$ or $\log z = 0$. More precisely, all points in the shaded region satisfy $|\mathrm{Im}(\log z)| \geq \pi/3 > 1$, so the restriction $|\Delta\mathrm{Im}(\log z)| \leq 1/2$ keeps z safely away from the singularity at $z = 1$. Similarly the restriction $|\Delta\mathrm{Re}(\log z)| \leq 1/2$ keeps the solution from approaching the singularities at $z = 0$ or $z = \infty$ too quickly.

insists that neither the real part nor the imaginary part of the parameter $\log z$ change by more than 1/2 for any tetrahedron. This restricts the change to a limited zone (Figure 18) and in particular keeps the parameter well away from the one singularity that remains in the parameter space (at $z = 1$ or $\log z = 0$). If Newton's method calls for a change exceeding those limits, then we rescale the proposed change (for all the tetrahedra, not just the offending one) so that the largest change in any Re(log z) or Im(log z) is 1/2.

5. Hyperbolic Dehn filling

Section 2 showed how to construct a hyperbolic structure on the complement of a k-component link of 0-spheres on a 2-sphere ($k \geq 2$) and then went on to show how deforming the hyperbolic structure (Figures 5 and 6) yields a hyperbolic structure on a closed surface of genus k with k closed geodesics missing (Figure 7). That was 2-dimensional hyperbolic Dehn filling. Three-dimensional hyperbolic Dehn filling is similar: deforming the complete hyperbolic structure on a k-component link complement (Section 4) will yield a hyperbolic structure on a closed manifold with k closed geodesics missing. In spite of the strong analogy between 2-dimensional and 3-dimensional hyperbolic Dehn filling, there are nevertheless a few differences. In the 2-dimensional case the ideal triangles were rigid while the gluings between them were flexible, whereas in the 3-dimensional case the shapes of the tetrahedra themselves are flexible while the gluings between them are rigid. More interestingly, in the 2-dimensional case different deformations all gave the same topological 2-manifold (namely the closed surface of genus k, once the k missing geodesics are filled in), whereas in the 3-dimensional case different deformations give topologically distinct 3-manifolds (filling in the missing geodesics realizes a Dehn filling on the link complement, with the Dehn filling coefficients depending on the deformation – more on this below).

The cut-open 2-dimensional cusps of Figure 5 become, in three dimensions, solid structures resembling the Eiffel tower (Figure 19). If the cusp cross section is effectively a parallelogram (Figure 17, left) the sides of the Eiffel tower match straight across (Figure 19, left) in the sense that people traveling across one face of the tower will re-enter the opposite face on the same level at which they left. This condition ensures that the hyperbolic structure is complete. If, on the other hand, the cusp cross section is not a parallelogram (Figure 17, right), then the sides of the Eiffel tower match with a vertical offset (Figure 19, right). For example, travelers leaving the tower at point B on one level will re-enter at point B' on a lower level. Note that the width w of the upper quadrilateral's long side equals the width w' of the lower quadrilateral's short side.

Henceforth we will restrict our attention to the case that the offset in one direction is a rational multiple of the offset in the transverse direction (Figure 20). Cutting along a consistent set of cross sections (like those illustrated in Figure 20) splits the cusp into an infinite set of bricks, all of equal height. Ignore for a moment the bricks' freshly cut top and bottom faces, and instead glue them together along their side faces. The result will be a solid cylinder (Figure 21) with infinitely many progressively narrower bricks spiraling in towards the center. The vertical geodesic at the exact center is missing, and indeed plays the role in three dimensions of the missing geodesics in Figure 7. Restoring the gluings on the bricks' top and bottom faces converts the cylinder into a solid torus, still with its central geodesic missing. Typically the cylinder's bottom glues to its top with some nonzero twist.

Filling in the missing geodesic at the center of the solid torus (Figure 21) realizes a Dehn filling on the link complement. To read off the Dehn filling coefficients, simply note how a meridian of the solid torus wraps around the cusp. In the example shown (see Figure 22 for a top view) any topological meridian will wrap 5 times around the

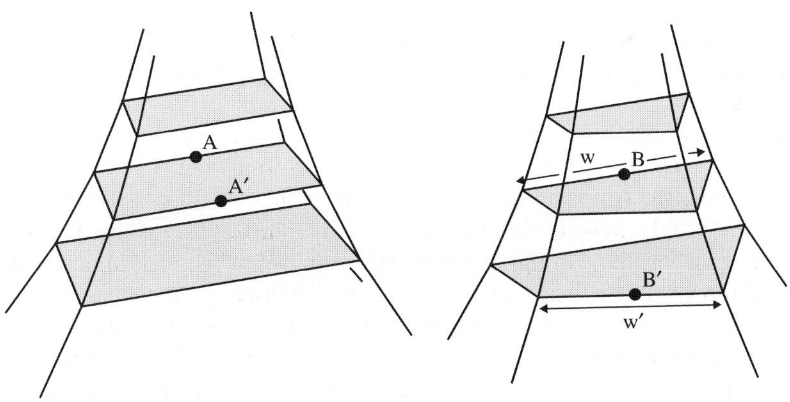

Fig. 19. A cut-open 3-dimensional cusp looks like an infinitely tall Eiffel tower. The cross sections are all quadrilaterals, which are similar to one another and whose size shrinks exponentially fast as you travel upward, in strong analogy to the cut-open 2-dimensional cusps of Fig. 5 whose cross sections are line segments shrinking exponentially fast. The cusp on the left has a parallelogram cross section (cf. Figure 17 left), while the cusp on right has a generic cross section (cf. Figure 17 right).

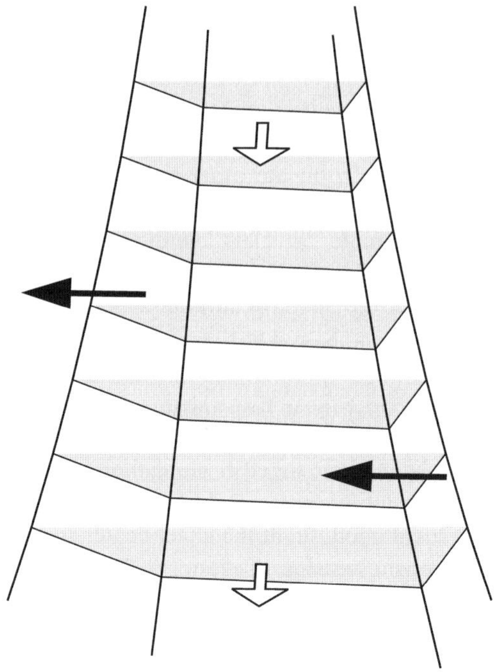

Fig. 20. In this example a traveller going to the left and returning from the right (black arrow) drops down two levels, while a traveller going towards the front and returning from the back (white arrow) goes up five levels. Because the offsets are rational multiples of each other (5/2 or 2/5) the cross-sectional quadrilaterals piece together to form a consistent surface. If the offsets were not rational multiples of each other, no such consistent set of equally spaced cross sections would be possible.

quadrilateral's "long direction" and 2 times around its "short direction". Typically the "long direction" is chosen to be a longitude of the original knot or link while the "short direction" is chosen to be a meridian of the original link, making this a (2, 5) Dehn surgery.

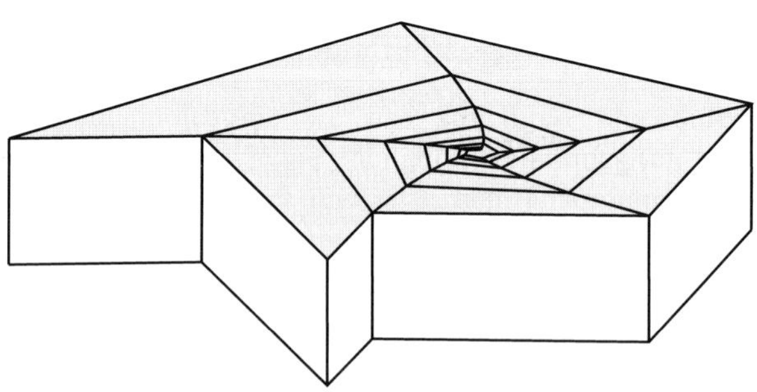

Fig. 21. Slice the cusp of Figure 20 into pieces along the illustrated cross sections and then re-glue the resulting pieces according to the identifications on their left, right, front and back sides. The result resembles a solid cylinder, with ever smaller pieces spiraling in towards the center.

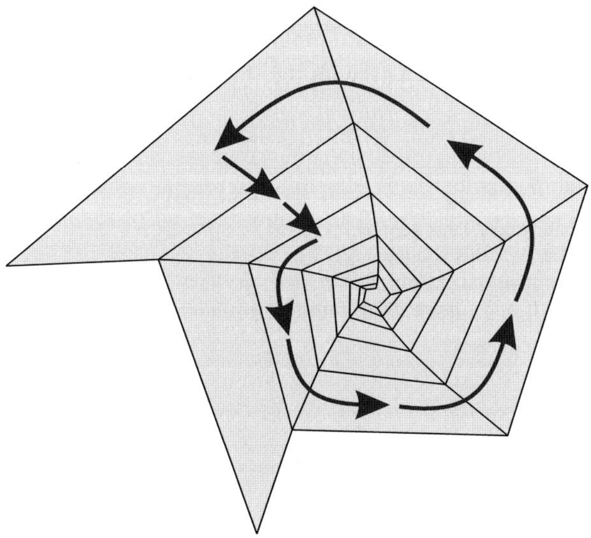

Fig. 22. View the solid cylinder of Figure 21 from above and note the path of a meridian encircling the missing central geodesic. That meridian defines the Dehn filling curve, so expressing it relative to the original (meridian, longitude) coordinates of the knot or link complement immediately gives the Dehn filling coefficients, in this case (2, 5).

In practice, of course, we do not randomly deform the hyperbolic structure and then wait to see what Dehn filling coefficients emerge. Instead, we choose the Dehn filling coefficients (p, q) in advance and ask what deformation of the hyperbolic structure will accommodate them. Recall from Figure 17 that a product of complex dihedral angles gives the rotational/dilational factor taking one side of the quadrilateral to the other. In Section 4 we insisted that that product be1; more precisely we replaced the naive product equation $\prod z_i = 1$ with the more powerful logarithmic equation $\sum \log z_i = 1$ to guard against stray multiples of $2\pi i$. Here we apply the same technique, but focusing on the arbitrary quadrilateral (Figure 17, right) instead of the parallelogram (Figure 17, left). We now get one expression $\sum \log z_i$ for the rotation/dilation in the meridional ("short") direction and a different expression $\sum \log z_i'$ for the rotation/dilation in the longitudinal ("long") direction. Tracing all the way around the loop in Figure 22 returns us to our starting point with a 2π rotation. The loop consists of p meridians and q longitudes, so the analytic condition is

$$p \sum \log z_i + q \sum \log z_i' = 2\pi i. \tag{5.1}$$

In Section 4 we had supplemented the edge equations with the cusp equations to solve for the hyperbolic structure on the cusped manifold. We now instead supplement the edge equations with the Dehn filling equations (5.1) to solve for the hyperbolic structure on the Dehn filled manifold.

Acknowledgement

I thank Adam Weeks Marano for help with the illustrations.

References

[1] C. Adams, "Hyperbolic knots", this volume.
[2] C. Gordon, "Dehn surgery", this volume.
[3] J. Hoste, "The enumeration and classification of knots and links", this volume.
[4] J. Weeks, *Convex hulls and isometries of cusped hyperbolic 3-manifolds*, Topology Appl. **52** (1993), 127–149.
[5] S. Henry and J. Weeks, *Symmetry groups of hyperbolic knots and links*, J. Knot Theory Ramifications **1** (1992), 185–201.
[6] C. Hodgson and J. Weeks, *Symmetries, isometries and length spectra of closed hyperbolic 3-manifolds*, Exp. Math. **3** (1994), 261–274.
[7] C. Adams, M. Hildebrand and J. Weeks, *Hyperbolic invariants of knots and links*, Trans. Am. Math. Soc. **326** (1991), 1–56.
[8] J. Weeks, "SnapPea: a computer program for creating and studying hyperbolic 3-manifolds", available for free download from www.geometrygames.org.
[9] W.P. Thurston, *Three dimensional manifolds, Kleinian groups and hyperbolic geometry*, Bull. (New Series) Am. Math. Soc. **6** (1982), 357–381.
[10] W.P. Thurston, *Three-dimensional Geometry and Topology*, Princeton Mathematical Series, Vol. 35, S. Levy, ed., Princeton University Press, Princeton, USA (1997).

Author Index

Subject Index